Biological Processes and Genetic Engineering

Biological Processes and Genetic Engineering

Edited by **David Rhodes**

SYRAWOOD
PUBLISHING HOUSE
New York

Published by Syrawood Publishing House,
750 Third Avenue, 9th Floor,
New York, NY 10017, USA
www.syrawoodpublishinghouse.com

Biological Processes and Genetic Engineering
Edited by David Rhodes

International Standard Book Number: 978-1-68286-278-0 (Hardback)

Printed in China.

Contents

Preface

There has been rapid progress in the fields of genetic engineering and biological processes, and their applications are finding their way across multiple industries. This book brings forth path-breaking studies conducted in these subjects. The chapters included herein discuss topics such as biodesign process, cellular design, bioproduction, etc. It will prove to be immensely beneficial to students and researchers pursuing genetic engineering, biology and allied sciences. This text aims to further the scope of research in these disciplines and contribute to their progress.

Significant researches are present in this book. Intensive efforts have been employed by authors to make this book an outstanding discourse. This book contains the enlightening chapters which have been written on the basis of significant researches done by the experts.

Finally, I would also like to thank all the members involved in this book for being a team and meeting all the deadlines for the submission of their respective works. I would also like to thank my friends and family for being supportive in my efforts.

Editor

2ab assembly: a methodology for automatable, high-throughput assembly of standard biological parts

Mariana Leguia[1,2,3], Jennifer AN Brophy[1], Douglas Densmore[1,4], Angel Asante[1] and J Christopher Anderson[1,2,3*]

Abstract

There is growing demand for robust DNA assembly strategies to quickly and accurately fabricate genetic circuits for synthetic biology. One application of this technology is reconstitution of multi-gene assemblies. Here, we integrate a new software tool chain with 2ab assembly and show that it is robust enough to generate 528 distinct composite parts with an error-free success rate of 96%. Finally, we discuss our findings in the context of its implications for biosafety and biosecurity.

Keywords: 2ab reaction, Automated assembly, DNA fabrication, Synthetic biology

Background

Synthetic biology uses a ground-up approach to genetic engineering with a vision to solve the technical and conceptual bottlenecks associated with the field. Standard biological parts have emerged as a mechanism through which researchers can emphasize the engineering component of genetic engineering. To viably implement such a vision, however, a new experimental approach that can systematically identify and optimize biological function is needed. Currently there are no established methodologies to encapsulate biological part function such that meaningful, quantitative predictions about genetic composition can be made a priori. Although in practice post-hoc analyses of projects provide insights into the individual properties of parts, most engineering strategies require either an ad-hoc search of intuitively-chosen constructs or library-based approaches. In the hands of experienced researchers these methods can be quite effective for systems containing up to 6 genes [1]. Nevertheless, beyond this scale these methods appear to be insufficient. We have implemented an alternate approach to experimental design that uncovers optimal solutions to design problems by linking automated DNA fabrication to effective construct screening. In the first step, we use cost effective, fast and scalable DNA fabrication methods to generate large sets of explicit DNA constructs automatically. In the second step, we functionally screen those DNAs via readily available assays that are also automatable, cost effective, fast and scalable. Additionally, we integrate these activities under a software tool chain specifically developed for these types of applications. By linking these functions together into a single process we not only uncover optimal solutions to a design problem, but also encapsulate important data about the parts from which they were composed. In turn, this leads to greater insight and faster optimization.

To illustrate how this alternate approach can be implemented in a synthetic biology lab we have focused on the production of multi-subunit protein complexes. The ability to produce functional protein from recombinant sources is essential in synthetic biology and for the scientific community at large. Nevertheless, producing these proteins quickly and successfully, particularly in the case of multi-subunit complexes, can be challenging. Given an overall lack of tools available to quickly and effectively identify appropriate conditions a priori, production efforts are often inefficient, resulting in poor or failed attempts. In these cases, original constructs need to be modified, which is often costly, labor-intensive and time-consuming, particularly if multiple rounds of modification are necessary. An attractive alternative to sequential, slow, expensive, multiple rounds of

* Correspondence: jcanderson@berkeley.edu
[1]Department of Bioengineering, University of California, Berkeley, CA 94720, USA
[2]QB3: California Institute for Quantitative Biological Research, University of California, Berkeley, CA 94720, USA
Full list of author information is available at the end of the article

modification is the generation, in parallel, of several permutations of a particular genetic architecture, followed by rapid functional screening of those permutations. Currently however, the assembly of long, error-free DNAs of the type required to build complex, tunable genetic circuits is the primary bottleneck in the field.

In recent years several DNA assembly strategies have been described, including Gibson, SLIC, Golden Gate, and various BioBrick approaches (for reviews see [2-5]). Each method has a range of advantages and disadvantages, and thus, different construction scenarios are best served by different technologies. Some of these methods can be used to assemble DNAs simultaneously. However, in the context of the design paradigm described here, a key parameter to consider is whether a particular strategy can be implemented in an automated platform that is amenable to scaling. Clearly, some chemistries will automate better than others. Thus, indentifying DNA fabrication protocols that are compatible with automation and scaling is an essential step towards easing the construction jam. Furthermore, quick functional screens of assembled permutations, via established assays that can also be automated and scaled, is essential for the isolation and control of most, if not all, of the variables affecting a particular project.

Here, we describe the construction of a set of 528 distinct composite devices that sample a large combinatorial space. Specifically, every device is a bi-cistronic operon constructed out of 8 standard basic parts. Operons encode two separate protein fusions containing an epitope tag either at the N- or C- terminus that is used to establish function. With support from software tools specifically developed for these types of DNA fabrication experiments [6,7], we have generated all constructs automatically, using a methodology called 2ab assembly. 2ab assembly is a single iterative process, based on the BglBricks standard [8], that is amenable to high-throughput automation and scaling [7]. Like other BioBricks™-based schemes, the BglBricks standard is based on unique restriction enzyme sites that flank the part (BglII at the 5′ end and BamHI at the 3′ end) and generate compatible cohesive ends that can be joined back together. When two parts are joined using simple restriction digestion and ligation reactions, the resulting composite part also contains unique BglII and BamHI sites flanking the 5′ and 3′ ends, respectively, as well as a small scar sequence encoding Gly-Ser separating input parts. As we show here however, 2ab assembly is not solely dependent on the BglBricks standard. In fact, 2ab assembly is flexible enough to accommodate alternate assembly standards as well. Using simple ELISA assays we have probed a variety of parameters that influence protein expression in functional form. Further, we surveyed and identified optimal architectures that result in the expression of functional protein complexes. We show that 2ab assembly is robust enough to generate hundreds of large error-free DNAs of the type required in a single experiment within the paradigm of synthetic biology with a 96% success rate.

Results

Selection and design of an appropriate DNA assembly set: bi-cistronic operons of Mediator complex for protein production

To test this new experimental approach we set out to generate a large set of DNA constructs that would allow us to measure the robustness of our assembly methodology. Further, we wanted to make use of assembled DNAs to address a challenging biological problem of significant current relevance. The expression and purification of multi-subunit protein complexes in soluble form is one such challenge, particularly because there are no tools that can be used at the outset of an experiment to select strategies that will yield successful results. Protein production is influenced by multiple variables, including expression host, construct composition and architecture, timing, rate, and temperature of expression, purification method, and so on. In the case of multi-subunit complexes, potential challenges can be exacerbated because individual subunits may require co-expression and/or independent regulation of expression. Thus, it may be necessary to consider separate control of additional variables, such as expression timing, rate and stoichiometry, which can further complicate a project.

As proof-of-concept for our 2ab assembly methodology we put together a set of 528 unique plasmids encoding permutations of bi-cistronic operons that could be used to express, identify and purify distinct epitope-tagged proteins known to form a stable complex in *E. coli*. The proteins selected were Med7 and Med21, two subunits of the *S. cerevisiae* Mediator complex. Mediator is a large protein machine (more than 1 MDa in size, containing 20+ independent subunits) that functions as a key regulator of eukaryotic transcription [9,10]. Despite its importance, however, the size and complexity of Mediator pose specific research challenges. Thus, tools that can help address such difficulties are useful to begin to address existing obstacles. Med7 and Med21 were specifically selected as a suitable pair for this assembly because they have been previously purified in *E. coli* and because *E. coli*-expressed Med7 and Med21 can form stable hetero-dimers [11]. Thus, such an assembly set would allow us to test the functionality of our constructs post-assembly in a variety of ways: for example, by measuring relative protein expression profiles of individual subunits; or by measuring the degree of protein-protein interaction between separate subunits.

We tagged each Mediator subunit with a small epitope that would enable a variety of downstream assays, including but not limited to, detection of individual subunit expression levels and protein-protein interactions. We used

12 different tags (6X-His, Avi, E, FLAG, HA, HSV, Myc, S, Strep, T7, V5 and VSV) (full details provided in Additional file 1) for which there are commercially-available antibodies (listed in Additional file 2). We varied the location of the tag relative to the protein ORF (either at the N- or the C-terminus) and we sampled all possible architectures: both subunits tagged at the N-terminus [NN] or C-terminus [CC]; first subunit tagged at the N-terminus and second subunit tagged at the C-terminus [NC]; and vice-versa [CN] (Figure 1). In each case, tags were separated from their corresponding ORFs by a TEV protease cleavage in order to facilitate tag removal if desired. Expression of every operon was placed under the control of an arabinose-inducible Pbad promoter and terminator. All bi-cistronic operons were composed of 8 basic parts each: a promoter to drive expression, followed by the first ORF (consisting of 3 basic parts: Med7 and one epitope tag, separated by a TEV cleavage site), followed by the second ORF (also consisting of 3 basic parts: Med21 and one epitope tag, separated by a TEV site), followed by a terminator. The Eugene programming language, which was specifically created for the specification of synthetic biological designs from a collection of individual standard parts [6], was used to generate our assembly set given a defined set of composition constraints. These constraints are declarative rules that either prevent or ensure the appearance of specific tag combinations in a construct. The general bi-cistronic operon architecture can then be permuted while adhering to these constraints. Given 12 tags and 4 possible tagging architectures (NN, NC, CN and CC), the number of bi-cistronic operons needed to sample the entire combinatorial space was 528.

All 528 unique bi-cistronic operons were assembled from a total of 31 basic parts (Additional file 1). Basic parts encoding tags and ORFs were gene synthesized in two versions in order to simplify construction: either

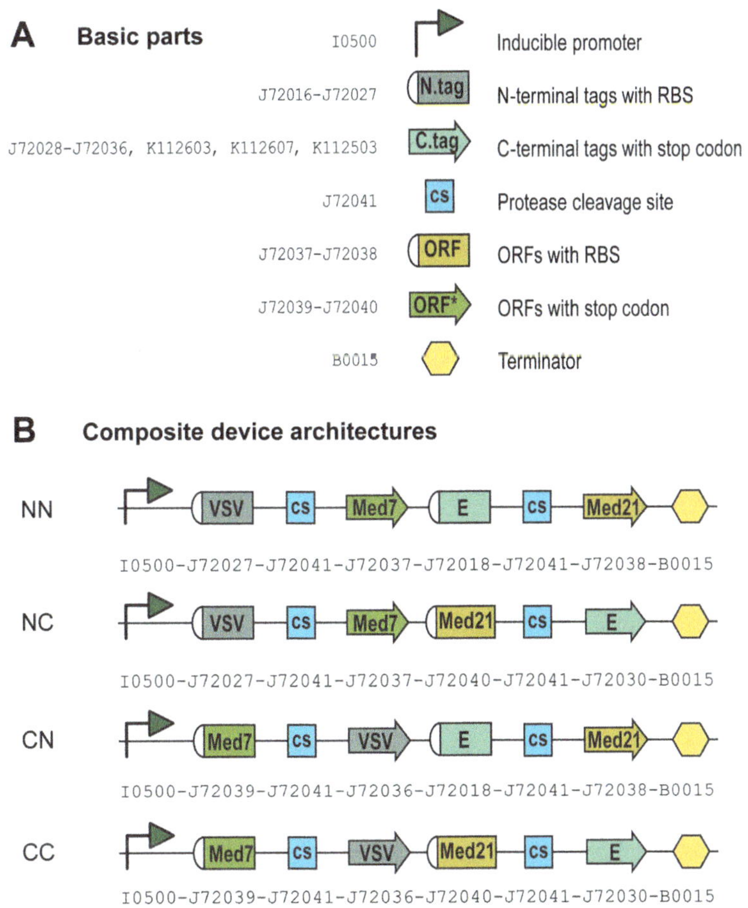

Figure 1 Assembly of bi-cistronic operons. (**A**) Basic parts used in the assembly. ORFs and tags were synthesized as either N- or C-terminal sets. N-terminal sets contain RBSs as a component of the part, while C-terminal sets contain stop codons at the end of the part. (**B**) Sample combinatorial architectures for a set of constructs containing two ORFs (Med7 in the first position and Med21 in the second position) tagged at the N- or C-terminus with different tags (VSV-tag for Med7 and E-tag for Med21). Every operon is composed of 8 basic parts. For this arrangement, 4 different construct architectures are possible: in the NN configuration both ORFs are tagged at the N-terminus; in the NC configuration the first ORFs is tagged at the N-terminus and the second ORF is tagged at the C-terminus; in the CN configuration the first ORFs is tagged at the C-terminus and the second ORF is tagged at the N-terminus; in the CC configuration both ORFs are tagged at the C-terminus.

with RBSs and start codons at the beginning, or with stop codons at the end. Tags in the first group, referred to as N-terminal tags, were designed to sit at the N-terminus of Mediator subunits. They included an RBS (determined using the RBS calculator algorithm [12] set at a target translation rate of 10,000), an ATG start codon for initiation of translation, and lacked an in-frame stop codon at the end. Tags in second group, referred to as C-terminal tags, were designed to sit at the C-terminus of Mediator subunits. They did not include an RBS or a start codon, but they did contain an in-frame stop codon at the end. Mediator subunits were similarly synthesized. N-terminal ORFs accepted N-terminal tags and did not include an RBS, but did contain an in-frame stop codon. C-terminal ORFs accepted C-terminal tags and included an RBS (determined as above), but lacked a stop codon at the end. Both Med7 and Med21 basic parts were made as truncations that express well in *E. coli* and retain their ability to interact with each other [11]. Med7 was truncated N- and C-terminally and included amino acids 102–205, while Med21 was truncated C-terminally only and included amino acids 1–132. The protease cleavage site part was gene synthesized without start or stop codons because it would separate tags from their corresponding Mediator subunits.

Assembly of all target plasmids was completed in 3 stages of 2ab assembly, according to an assembly tree generated by the AssemblyManager software tool, using a robotics platform containing a Biomek 3000 liquid handling robot [7,13]. AssemblyManager examined the set of devices to be constructed as a group and created an optimal assembly tree. Optimality was defined by a minimal depth tree, where the stages indicated sets of parallel 2ab assembly reactions. The number of steps (a single 2ab assembly reaction) in the tree was also minimized. Steps were further reduced by sharing intermediate reactions. The tree was then post-processed to assign antibiotic resistance markers to input nodes. In the event of a conflicting antibiotic assignment, the tree was modified to accommodate the correct resistance at the expense of the overall assembly cost. A set of heuristics were used to accomplish this process [13]. Specifically for this assembly, we used 31 basic parts to make 56 composite parts (made of 2 basic parts each) in stage 1. In stage 2, we used the parts made in stage 1 to assemble 48 composite parts (made of 4 basic parts each). In the final stage, we used the parts made in stage 2 to assemble 528 composite parts (made of 8 basic parts each). The target set of 528 plasmids contained a total of 3696 junctions between parts. However, most of these junctions were not unique, and therefore in practice, only 632 total junctions were needed to complete the set.

Definition of 2ab assembly

2ab assembly is a DNA fabrication methodology that uses 2ab reactions to build progressively more complex composite parts starting from basic standard biological parts. 2ab reactions proceed in iterative cycles of restriction digestion, ligation, transformation and selection of desired products. The reaction enables ligation by double antibiotic selection of 5′ and 3′ parts, designated as "lefty" and "righty" respectively, located on two different assembly vectors.

The core of 2ab assembly is a set of highly engineered plasmids containing two different antibiotic resistance genes (from a total of three: A = ampicillin, C = chloramphenicol, K = kanamycin) separated by restriction site. Given these three antibiotics, six different combinations of assembly vectors are possible: AC, CK, KA, AK, KC and CA. For a given vector pair to work properly during 2ab assembly, the choice of vector pair is pre-determined such that, once two plasmids recombine to form new architectures, desired child products can be selected away from undesired products, as well as from parents, using differential antibiotic selection.

Every assembly reaction is subject to the following three constraints: First, the antibiotic resistance markers on every assembly vector must be distinct from one another (vectors with two AA, CC, or KK genes are not permitted). Second, the resistance marker on the right position of the assembly parent donating the "lefty" part must be the same as the resistance marker on left position of the assembly parent donating the "righty" part. As illustrated in Figure 2, an AK-KC vector pair can be used to assemble parts 1 and 2 together, in that order, provided that part 1, the "lefty" or 5′ part, is on AK, while part 2, the "righty" or 3′ part, is on KC. Similarly, a CK-KA pair can be used to assemble parts 1 and 2 together, in that order, provided that part 1 is on CK and part 2 is on KA, and so on. The final constraint is that the antibiotic combination associated with the correctly assembled child must be distinguishable from the antibiotic combination associated with the child byproduct, as well as both of the parents. As illustrated in Figure 2, the AK-KC vector pair produces AC and KK combinations when recombined. Both of these are distinguishable from each other and from the parents. Although in theory 2ab assembly can be carried out using a variety of assembly standards, we have developed this methodology for use with the previously described BglBricks standard [8]. Thus, most of the parts in assembly pairs described here are BglBricks parts flanked by BglII and BamHI restriction sites on their 5′ and 3′ ends, respectively. Further, the two antibiotic markers in every assembly vector are separated by an XhoI site. Modifications to this set-up, where alternate assembly standards like the BBF RFC10 have been used to compare the robustness of BglBricks-based 2ab assembly, are specifically noted throughout.

Prior to the first round of assembly, "lefty" and "righty" parts are defined by specific methylation. For this, lefty

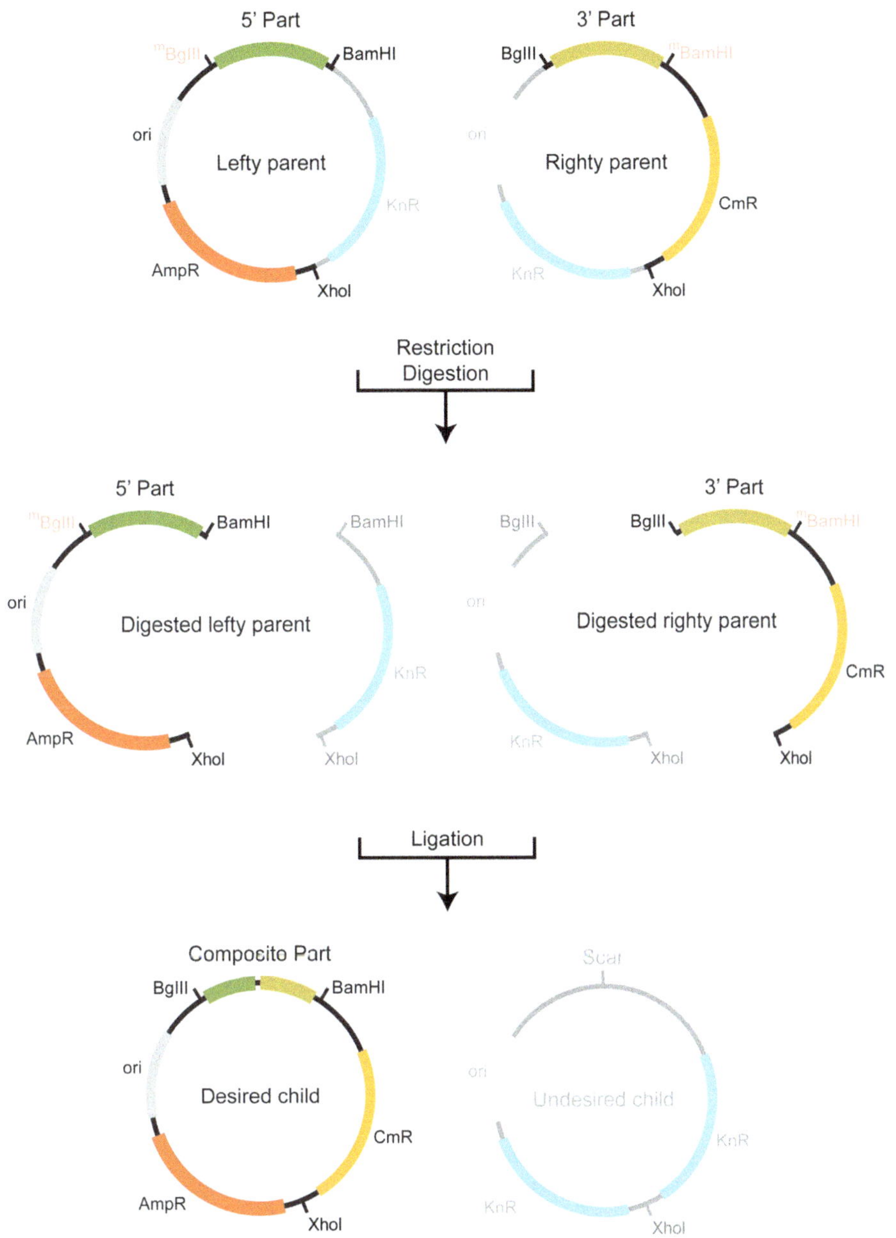

Figure 2 The 2ab reaction. These are carried out *in-vitro* and used to make junctions between parts located on different assembly vectors. The choice of vector pair for each junction is determined by an assembly tree generated by AssemblyManager. AssemblyManager takes into account antibiotic resistance markers on both lefty and righty input plasmids and generates a binary tree containing legal lefty and righty pairs. In practice, lefty and righty elements are made by harvesting plasmids from *E. coli* strains that specifically methylate either BglII (part J72015) or BamHI (part J72015) restriction sites (methylated sites are shown in red and grayed over to indicate protection from digestion) [7,8]. Isolated plasmids are then combined in one pot and digested with an enzyme cocktail containing BglII, BamHI and XhoI. Given the methylation state of each plasmid, lefties are cut with BamHI and XhoI, while righties are cut with BglII and XhoI. Following digestion, a ligation reaction recombines all fragments to generate a new composite part with a distinct permutation of antibiotic markers. The desired child product can then be selected away from undesired products, including parent plasmids, by growing it in the appropriate combination of antibiotics. Each new composite part can now be used iteratively in additional 2ab reactions to create progressively more complex parts.

and righty plasmids are transformed and harvested from strains of *E. coli* that specifically methylate BglII (part J72015) and BamHI (part J72015) restriction sites, respectively [7,8]. The methylation state of each plasmid, which is different in "lefty" and "right" parts, dictates differential patterns of restriction digestion. In "lefties," BglII restriction sites are methylated, and thus protected from digestion, while in "righties," BamHI restriction sites are methylated and protected from digestion.

In practice, 2ab reactions are carried out *in-vitro* in one pot. Following isolation from their corresponding methylation strains, "lefties" and "righties" are combined together and digested with an enzyme cocktail containing BglII, BamHI and XhoI. Given the methylation state of each, "lefty" is cut with BamHI and XhoI, while "righty" is cut with BglII and XhoI. This results in the generation of two linear fragments per plasmid. As illustrated in Figure 2, AK is digested such that the A resistance marker is now on the same fragment of DNA as the "lefty" or 5′ part, while the K resistance marker is on its own. Similarly, KC is digested such that the C resistance marker is now on the same fragment of DNA as the "righty" or 3′ part, while the K resistance marker is on its own. Following digestion, a ligation reaction recombines all the fragments to generate four distinct possible plasmid architectures. Two are the same as the parent plasmids, while the other two are recombined child products.

Child plasmids are generated when BamHI overhangs of the "lefty" vector and BglII overhangs of the "righty" vector are ligated together on one end, and the two XhoI overhangs are ligated together on the other end. The desired child product contains a new composite part and a new distinct permutation of antibiotic markers (AC in this example), while the child byproduct contains no part and two copies of the remaining antibiotic marker (KK in this example). Given that each possible plasmid architecture contains a different antibiotic combination, desired child products can be selected away from undesired products, as well as from parents, simply by growing them in the appropriate antibiotic combination. Each new composite part can now be used iteratively in additional 2ab reactions to create progressively more complex parts because all the elements necessary for 2ab assembly have been maintained: the new part is still flanked by BglII and BamHI sites on its 5′ and 3′ ends, respectively; and two antibiotic resistance genes are still present and separated by an XhoI site. In order for the next cycle of 2ab assembly to begin, new composite parts need to be specified as lefties or righties (and methylated accordingly), and compatible assembly vector partners selected. In practice, these choices have been determined by the assembly tree generated using AssemblyManager at the outset of the assembly.

Screening of 2ab assembly

To measure the robustness of 2ab assembly, every junction was assembled twice, in separate replicate experiments. For every junction made, at least three independent colonies containing putative composite parts were screened. All colonies were subjected to several screens: the first was growth phenotyping under various antibiotic combinations; the second was restriction mapping or colony PCR amplification; and the last was sequencing. These tests were performed in combination in order to ensure proper part assembly at every stage.

In the first screen, single colonies were streaked onto both double (AC, CK or KA, depending on the antibiotic combination marking the new 2ab junction generated) and triple (ACK) antibiotic combinations. Transformants containing correctly assembled parts were able to grow on the appropriate double antibiotic combination (indicating they contained growth markers associated with correctly assembled parts), but were not able to grow on the triple antibiotic combination (indicating possible co-transformation of one of the parent plasmids and/or inappropriate assembly). As shown in Table 1, the average co-transformation rate was about 5% (111 of 2208), indicating that the vast majority of transformants (2097 of 2208) contained growth markers associated with properly assembled parts. Co-transformation rates ranged from 2% for junctions between AK pairs, to 26% for junctions between KC pairs. We routinely observe a wide range of co-transformation rates. The variability is mostly linked to two factors: first, the identity of the antibiotic pair (CK and KC pairs show higher rates of co-transformation than AC, CA, AK and KA pairs when equivalent amounts of plasmid DNA are used); and second, the concentration of plasmid DNA used for transformation (higher amounts of plasmid DNA give rise to higher rates of co-transformation).

Transformants showing the correct growth phenotype were further screened either by restriction mapping with BglII, BamHI and XhoI, or by colony PCR amplification. Even though colony PCR amplification is preferable as a screening tool because it does not require extraction of plasmid DNA, which is costly and time-consuming, the sizes of many stage 1 and 2 products could not be clearly resolved from those of their parents with this technique, and thus, were restriction mapped. All products of stages 1 and 2 (104 total) were screened by restriction mapping, while all products of stage 3 (528 total) were screened by colony PCR.

A few clones from every stage of assembly, including those with correct, questionable and incorrect restriction maps or amplicons, were selected for sequencing to verify part identity. As shown in Table 1, about 96% of all colonies screened contained correctly assembly parts. The percentages of properly assembled junctions ranged from 93% for CA and KC pairs, to 100% for AK and KA pairs. The results obtained by sequencing verify that our restriction enzyme mapping and colony PCR screens were stringent enough to assess assembly success accurately. Of the clones showing correct restriction maps or PCR amplicons sequenced (103 total), 100% contained properly assembled parts. Of the clones with questionable maps or PCR amplicons sequenced (8 total), 50% contained properly assembled parts, indicating that we

Table 1 Summary of BglBricks-based 2ab assembly results

Assembly stage	1			2		3	Totals
junction type	CA	AK	KC	CK	KA	CA	
junctions made	28	14	14	24	24	528	632
colonies screened by growth phenotype	84	42	42	156	156	1728	2208
co-transformants	3	1	11	12	7	77	111
% co-transformation	3.57	2.38	26.19	7.69	4.49	4.46	5.03
colonies screened by mapping or PCR	28	14	14	48	48	622	774
colonies w/ correct map or PCR phenotype	26	13	13	47	48	535	682
colonies w/ questionable map or PCR phenotype	2	1	1	0	0	75	79
colonies w/ incorrect map or PCR phenotype	0	0	0	1	0	12	13
colonies sequenced	9	1	14	9	8	73	114
colonies w/ correct map or PCR sequenced	7	1	13	9	8	65	103
correct sequencing result	7	1	13	9	8	65	103
colonies w/ questionable map or PCR sequenced	2	1	1	n/a	n/a	4	8
correct sequencing result	0	1	0	n/a	n/a	3	4
colonies w/ incorrect map or PCR sequenced	n/a	n/a	n/a	0	n/a	4	4
correct sequencing result	n/a	n/a	n/a	n/a	n/a	0	0
% correct assembly	92.86	100	92.86	97.92	100	95.06	96.45*

*Average of % assembly rates for every junction at various assembly stages.

had been conservative in our scoring of proper assembly as assessed by restriction enzyme mapping and colony PCR. Of the clones with incorrect restriction maps or PCR amplicons sequenced (4 total), none contained properly assembled parts.

To further test the robustness of 2ab assembly, we compared the success rate obtained by BglBricks-based 2ab assembly to that obtained using a different assembly standard also in the context of 2ab assembly. For this, we chose the original BBF RFC10 standard described by Knight [14], which uses SpeI and XbaI restriction sites flanking the 5′ and 3′ ends of parts, respectively. Specifically, we carried out a small assembly using a split version of super folder-GFP [15], expressed under a constitutive promoter, that results in visibly green colonies in *E. coli* when properly assembled. The lefty input part consisted of the first 173 amino acids of sf-GFP driven by the tet promoter, while the righty input part consisted of the last 64 amino acids of sf-GFP. Input parts were subcloned into all six 2ab assembly vectors to test assembly of all junctions. Following assembly, lefty and righty input parts are separated by a small scar sequence (ACTAGA) encoding Thr-Arg, which does not interfere with the fluorophore. As shown in Table 2, the average co-transformation rate was about 14% (174 of 1199), indicating that the vast majority of transformants (1025 of 1199) contained growth markers associated with properly assembled parts. Co-transformation rates ranged from 12% for junctions between KA pairs, to 19% for junctions between CK pairs. Proper assembly was further assessed by counting green colonies. About 92% of colonies screened (3221 of 3486) were positive for GFP expression. The percentages of properly assembled junctions ranged from 91% for CK pairs, to 95% for AC pairs. To ensure that we were not under reporting the number of correct assemblies, we sequenced the assembly products extracted from 26 white colonies. Of these, none contained properly assembled parts, indicating that our screens were robust enough to assess assembly success accurately.

Characterization of errors

In the process of assessing the robustness of our 2ab assembly methodology we identified three main sources of error. The first was co-transformation of plasmids, which has been briefly mentioned above. Co-transformation results in colonies that are able to grow in 2ab combinations marking the newly assembled junction, but that may or may not contain properly assembled parts. Instead, these colonies contain antibiotic resistance markers associated with correctly assembled junctions in one or more plasmids that may or may not include the assembled plasmid of interest. Although there was a slight qualitative correlation between co-transformation and certain antibiotic combinations (CK and KC pairs showed higher rates of co-transformation than AC, CA, AK and KA pairs when equivalent amounts of plasmid DNA are used) there was no correlation to part size or part source. Instead, co-transformation was directly linked to the concentration of plasmid DNA used in transformation reactions. Higher amounts of plasmid DNA give rise to higher rates of co-transformation, thus, a simple way to reduce

Table 2 Summary of BBF RFC10-based 2ab assembly results

Assembly stage				1			Totals
junction type	CA	AK	KC	CK	KA	AC	
total junctions made	1	1	1	1	1	1	6
colonies screened by growth phenotype	200	198	201	200	199	201	1199
co-transformants	29	26	27	37	24	31	174
% co-transformation	14.50	13.13	13.43	18.5	12.06	15.42	14.51
total colonies counted	580	1011	274	366	688	567	3486
white colonies	45	83	24	34	48	31	265
green colonies	535	928	250	332	640	536	3221
total colonies sequenced	5	6	5	2	7	1	26
white colonies sequenced	5	6	5	2	7	1	26
correct sequencing result	0	0	0	0	0	0	0
green colonies sequenced	0	0	0	0	0	0	0
correct sequencing result	n/a	n/a	n/a	n/a	n/a	n/a	0
% correct assembly	92.24	91.79	91.24	90.71	93.02	94.53	92.40

them is to dilute the amount of DNA used for transformation. Although co-transformation rates were not high enough to present a problem in our assemblies, these colonies can be easily screened and eliminated by simultaneously re-streaking on double and triple antibiotic combinations. Colonies containing properly assembled parts will be able to grow on two antibiotics marking the newly assembled junction, but they will not be able to grow on all three. Any colony that grows on triple antibiotic combinations should be discarded.

A second, less common source of error (12 instances out of 774), results in colonies that exhibit the appropriate biotope, but that contain unrecognizable plasmids. In every one of these cases, the failure could be correlated back to poor quality parent input plasmid DNA and corrected by re-extracting the input plasmid.

The last source of error resulted in the production of junctions lacking whole input parts. In total, we identified 80 instances of BglBricks-based 2ab assembled junctions, and 9 instances of BBF RFC10-based 2ab assembled junctions, containing assembly scars but lacking either the entire lefty or righty input part. In the case of BglBricks-based assembly, empty junctions arise as a result of incomplete methylation of parent plasmid DNA. When input plasmids are incompletely methylated they are subject to part loss during digestion steps with BglII and BamHI, which effectively eliminates the input part from the reaction but maintains the vector backbone available for 2ab assembly. In the case of BBF RFC10-based assembly, empty junctions arise as a result of incomplete heat-inactivation of restriction enzymes following digestion steps, which has the same effect as incomplete methylation of parent input DNA. Given that partial methylation is a limitation of the methylation strains themselves (which have the potential to be significantly improved), rather than of 2ab assembly *per se*, our reported success rate of 96% is a conservative estimate of the actual rate, indicating that our technology is indeed robust enough to construct large error-free DNA sets with a very high success rate.

Screening the functionality of assembled bi-cistronic operons

Addressing the bottleneck around DNA fabrication is only the first step towards developing high-throughput solutions in synthetic biology. Thus, in addition to showing that our DNA assembly methodology is sufficiently robust to build large error-free DNAs with a ~96% success rate, we were interested in testing the functionality of assembled constructs. To do this in a cellular context, we induced expression and used indirect ELISA assays to quantify steady-state relative protein expression levels from every epitope-tagged Mediator subunit in all 528 bi-cistronic operons (entire data set is provided in Additional files 3 and 4). For a select subset of operons, we also used sandwich ELISA assays to probe protein-protein interactions and verify functionality of individually expressed proteins (data not shown). For the purposes of this assembly however, it was much more important to have a complete data set that would allow us to comparatively asses operon functionality (quantified by measuring relative protein expression levels via indirect ELISAs), than to asses individual subunit functionality (quantified by measuring protein-protein interactions via sandwich ELISAs). Given these priorities, and given the large number of operons in the assembly set (528 plasmids, assayed for two tags each, in triplicate), we used indirect ELISAs, which are simpler and faster to carry out, to generate a complete data set that allowed us to make side-by-side comparisons and predictions.

To assay the operons, they were first sorted by tag and grouped together such that those with tags in common were assayed in tandem in replicate experiments. Figure 3 shows indirect ELISA data for a representative set of constructs sharing VSV and E tags. Given that each operon contains two different tags, and given 4 possible tagging architectures (NN, NC, CN and CC), every pair of tags appears in 8 different constructs. In the first four constructs VSV-Tag is fused to the first ORF (Med7) and E-Tag to the second ORF (Med21). In the last four constructs, which are "reciprocal" architectures, VSV-Tag is fused to the second ORF (Med21) and E-Tag to the first (Med7). The results show that varying the location of the tag from the N- to the C-terminus of an ORF can have a dramatic effect on protein expression levels. In this set, Med7 expressed well when tagged at the C-terminus, but significantly less so when tagged at the N-terminus. In turn, Med 21 expressed at lower levels than Med7, either when tagged at the N- or C-terminus.

High-throughput approaches such as these enable rapid processing of large numbers of samples in parallel, which in turn allows the simultaneous visualization of multiple sets of data such that patterns can become readily apparent. From these patterns, many of the ways in which multiple variables influence experimental outcomes begin to emerge. To test the hypothesis that varying the location of the tags would have a more significant effect on protein expression levels than varying the identity of the tag itself, we plotted protein concentration values obtained via ELISAs into "heat maps" that included all possible tags and tagging combinations (Additional file 3). In the context of a heat map, the pattern of expression for Med7 becomes particularly obvious. Regardless of which tag was used, Med7 expression levels were significantly higher when the protein was tagged C-terminally. In some instances, Med7 expression levels in excess of 55 fold over background were observed. An exception to this pattern was FLAG, which resulted in expression levels close to background in all architectures tested. The pattern of expression for Med21 was a bit less obvious than that for Med 7 because overall Med21 expressed at lower levels. In cases where patterns were discernible, as in the case of tag combinations between Avi, HSV and T7, for example, it was apparent that Med21 expression tends to be higher when the protein is tagged C-terminally.

Given the observed patterns we hypothesized that the location of an ORF with respect to the driving promoter would also influence expression, and specifically, that expression from ORFs located proximally to the promoter would be higher than from ORFs located distally. To test this we assembled a small sub-set of 24 bi-cistronic operons where we switched the location of the two ORFs, placing Med21 in the first position, and Med7 in the second. We selected three tags (Avi, HSV and T7) that had shown discernible patterns in the first assembly and tested them as before. Figure 4 shows that, while the tag location patterns remained the same (both ORFs expressed better when tagged C- rather N-terminally), switching the location of the ORFs equalized protein expression significantly. In the first assembly, where Med 7 was in the first ORF position and Med21 in the second ORF position, the fold difference between the two highest expressing architectures in the sub-set (Med7 at 44.82 and Med21 at 12.06) was ~33 fold. In the second assembly, where Med 21 was in the first position and Med7 the second, the fold difference between the two highest expressing architectures in the sub-set (Med21 at 9.48 and Med7 at 8.35) decreased to ~1fold. Control of this variable alone can have a significant impact on experimental outcomes, which is particularly noteworthy given that producing protein in stoichiometric equivalences that enable formation of functional soluble complexes is often more important than producing large quantities of non-functional single subunits. Taken together, our results confirm that recombinant protein expression is influenced by multiple distinct variables that correlate with specific sequences of DNA at specific locations in a plasmid. By enabling the rapid and effective construction of plasmid variants, 2ab assembly enables access and further control over various experimental conditions which, in turn, facilitate tuning of gene expression circuits used in synthetic biology.

Discussion

We share the vision of synthetic biology as a field dedicated to solving the technical and conceptual bottlenecks of genetic engineering. Although we currently face several challenges, it is clear that many of these obstacles can be attributed either to an insufficient survey of the design space or to a lack of knowledge about the molecular and cellular functions of added components. Both of these problems can be addressed by robust, cost-effective, rapid, and outsourced DNA fabrication platforms that lower barriers to fabrication and streamline assembly. Basic standard biological parts, such as BioBricks™, as well as standard assembly strategies to join them together head-to-tail, have emerged as a way to deal with the potential cloning problems associated with the uniqueness of each DNA sequence. The rationale behind these approaches is that by standardizing parts and part junctions to conform to a particular set of rules, considerations of design and function can be separated away from those pertaining to assembly. Furthermore, because the products generated via these methods are themselves standardized parts, a single,

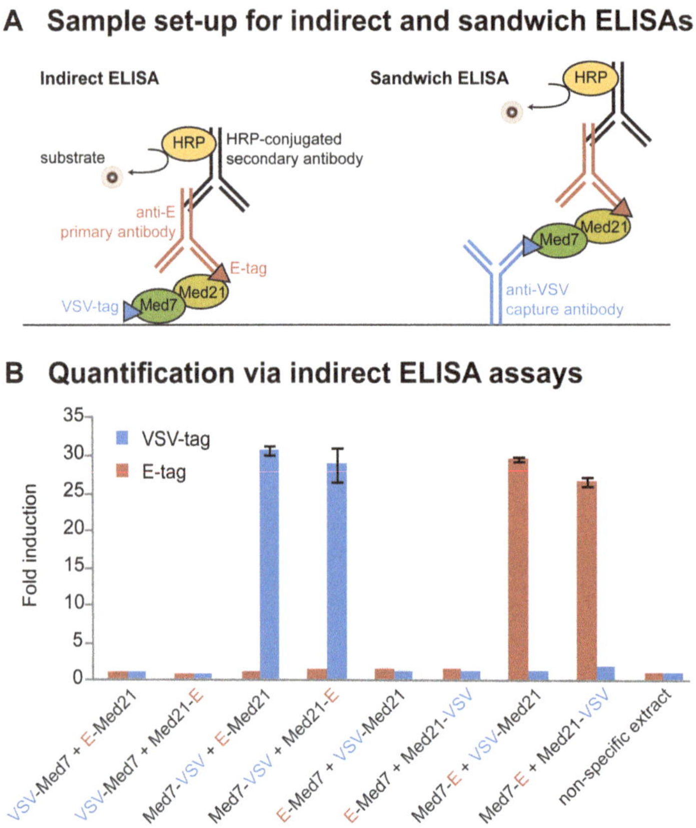

C Heat map representation of ELISA values

Quantification of first ORF: Med7

Med7		E		VSV		
		N	C	N	C	average
E	N			1.59	1.56	1.58
	C			29.53	26.55	28.04
VSV	N	1.11	1.00			1.06
	C	30.78	28.85			29.82

Quantification of second ORF: Med21

Med21		E		VSV	
		N	C	N	C
E	N			1.34	1.30
	C			1.17	1.66
VSV	N	1.06	0.93		
	C	1.60	1.60		
average		1.33	1.27	1.26	1.48

Standard deviations for Med7 values

Med7		E		VSV		
		N	C	N	C	average
E	N			0.27	0.08	0.18
	C			0.40	0.54	0.47
VSV	N	0.05	0.03			0.04
	C	0.58	2.29			1.44

Standard deviations for Med21 values

Med21		E		VSV	
		N	C	N	C
E	N			0.02	0.03
	C			0.03	0.08
VSV	N	0.07	0.03		
	C	0.03	0.03		
average		0.05	0.03	0.03	0.06

Figure 3 Quantification of protein expression from differentially tagged ORFs. (**A**) Sample set-up for indirect and sandwich ELISA assays used to quantify protein expression of individual subunits and protein-protein interactions, respectively. (**B**) Indirect ELISA data for a representative set of 8 constructs sharing VSV and E tags. For any two tags, 8 different architectures are possible. In the first four constructs VSV-Tag (in blue) is fused to the first ORF (Med7) and E-Tag (in red) to the second ORF (Med21). In the next four constructs, which are "reciprocal" architectures, VSV-Tag is fused to the second ORF (Med21) and E-Tag to the first (Med7). Constructs with tags in common are assayed in tandem, replicate experiments. Reported values are normalized for protein concentration and background. Control experiments include antibody drop-outs (not shown) and protein extracts lacking epitope tags. (**C**) ELISA values plotted into "heat maps" to better visualize patterns of expression. The top two maps show averaged values from replicate experiments, while the bottom two maps show the corresponding standard deviations. Maps on the left show Med7 expression patterns from the first ORF position (on the "y" axis) and should be read from left to right. Maps on the right show Med21expression patterns from the second ORF position (on the "x" axis) and should be read from top to bottom. Red and yellow colors indicate high and low relative expression levels, respectively.

Med7-Med21 architecture

Quantification of first ORF

Med7		Avi N	Avi C	HSV N	HSV C	T7 N	T7 C	average
Avi	N			1.05	0.91	4.07	3.40	2.36
Avi	C			34.12	28.50	32.95	10.19	26.44
HSV	N	1.16	1.40			1.61	1.72	1.47
HSV	C	31.64	25.85			44.82	44.14	36.61
T7	N	2.00	0.88	1.36	1.25			1.37
T7	C	31.00	7.61	22.69	13.02			18.58

Quantification of second ORF

Med21		Avi N	Avi C	HSV N	HSV C	T7 N	T7 C
Avi	N			1.20	9.05	1.05	1.09
Avi	C			0.98	8.16	0.84	1.09
HSV	N	1.12	1.10			1.32	1.32
HSV	C	1.18	1.70			1.40	2.04
T7	N	1.08	1.02	1.52	9.41		
T7	C	0.97	0.94	1.79	12.06		
average		1.09	1.19	1.37	9.67	1.15	1.38

Med21-Med7 architecture

Med21		Avi N	Avi C	HSV N	HSV C	T7 N	T7 C	average
Avi	N			1.24	2.13	0.85	0.90	1.28
Avi	C			6.66	6.44	1.27	0.85	3.81
HSV	N	2.38	3.65			0.90	0.84	1.94
HSV	C	3.90	4.10			4.13	0.93	3.26
T7	N	2.47	3.04	1.72	2.08			2.33
T7	C	4.21	3.75	9.48	8.65			6.52

Med7		Avi N	Avi C	HSV N	HSV C	T7 N	T7 C
Avi	N			2.80	4.10	2.65	4.15
Avi	C			3.40	3.67	3.40	3.81
HSV	N	2.16	8.35			1.35	7.88
HSV	C	1.97	8.09			2.01	7.61
T7	N	1.01	7.47	1.17	6.80		
T7	C	1.01	6.84	1.23	7.15		
average		1.54	7.69	2.15	5.43	2.35	5.86

Figure 4 Heat maps illustrating the effect of switching the location of specific ORFs relative to the driving promoter. Maps include all possible tagging combinations for 24 bi-cistronic operons sharing tags Avi, HSV and T7. The top two maps illustrate patterns of expression for a set of constructs where Med7 and Med 21 are located in the first and second ORF positions, respectively. The bottom two maps illustrates patterns of expression for the same set of epitope tags only that Med7 and Med 21 are now located in the second and first ORF positions, respectively. Maps on the left quantify expression levels for the first ORF (in the "y" axis) and should be read from left to right, whereas maps on the right quantify expression levels from the second ORF (in the "x" axis) and should be read from top to bottom. As before, red and yellow colors indicate high and low relative expression levels, respectively.

iterative, standard assembly reaction can be used to concatenate parts together to form progressively more complex genetic devices. Thus, the problem of composition can be pursued without consideration for how the DNAs are assembled together. Our results indicate that it is possible to robustly and effectively automate the fabrication of hundreds of genes using BioBricks™-based approaches. Several alternate chemistries that can also be used as standard fabrication methods, including SOEing, SLIC, CPEC, Golden Gate Shuffling, and the isothermal method reported by Gibson [16-21], to name a few, have been recently developed. We anticipate that many of these could be similarly automated, and thus, that specific types of experiments will be better served by different assembly methodologies.

Although we currently lack specific benchmarks to compare our results to those obtained using the alternate chemistries mentioned above, a success rate above 96% for BioBricks™-based approaches is very encouraging, particularly since this is a conservative estimate of efficiency that could be improved upon through minor modifications of the existing set-up. The majority of failures observed were related to colonies containing unrecognizable plasmids, which were always traced back to poor quality input mini-prep DNA. Thus, these errors could likely be mitigated using DNA extraction protocols that perform more consistently. A second source of failure was traced back to incomplete methylation of input plasmid DNA, which gave rise to junctions lacking either lefty or righty parts. These errors can be improved upon with better methylation strains. Several of these strains are currently under development in the lab. In a few instances we also observed that particular combinations of antibiotics performed better than others, which could likely be resolved using DIAL strains or through modifications to regulatory elements, including promoter and/or RBS mutations.

Automated assembly enables cumulative information gain in every design cycle by lowering existing barriers to construction and screening. Large numbers of data points reveal patterns that reflect specific underlying properties of the elements contained in a system. Patterns can reflect several distinct variables. In the examples presented here, two of the variables observed to have a significant effect on relative protein expression levels were the location of an epitope tag within an ORF, and the location of that ORF relative to its driving promoter (first or second position in a bi-cistronic operon). As demonstrated previously, other variables can also be used to manipulate expression systems, including RBS and/or promoter mutations [22]. Further, these variables can be used to test whether an engineered device is behaving in a manner that is quantitatively consistent with a particular model. Despite the fact that observed patterns can be useful tools for the interpretation of results, particularly because they enable informed decisions about how the system should be manipulated in subsequent cycles, these trends can rarely be reliably mapped onto specific biophysical parameters. Thus, one of the main challenges we still face is how to encapsulate the root of observed patterns. Ideally, combinatorial sets could be used to quantify parameters that reflect inherent properties of a particular part. As shown with the test sets described here, the patterns observed had, at the very least, qualitative utility for deciding whether expression levels of independent ORFs should be increased or decreased, and by how much, in order to obtain stoichiometric equivalences. Although it is tempting to try to directly encapsulate measurements like ELISA or fluorescence reads as relative measures of particular biophysical values, like transcription, translation or folding rates, we want to root our theory of encapsulation in real numbers that can be directly mapped to actual biophysical parameters reflecting molecular function, such as a molecule's K_m, K_{cat}, folding rate, etc. Ironically perhaps, the most central parameter that we currently manipulate in synthetic biology, expression level, cannot be easily characterized or encapsulated in terms of such parameters. To get there, one would be fully justified to express a sigma70 binding constant for a promoter, for example. Nevertheless, that alone would not encapsulate everything the promoter could do. In practice, transcription rates would depend on many factors simultaneously exerting an effect within the context of the cell, including load and stress, to name just a few. Although reading trends and patterns within a well-defined context is clearly useful, we still have a long way to go before we can encapsulate information that seeks to define narrow parameters, like whether a protein prefers to be tagged at the N- or C-terminus, or whether using a particular tag will work on a particular percentage of proteins or times. Given that undoubtedly these are statements about molecular function that are related to folding kinetics, it will be necessary to develop theory that will eventually enable us to properly measure these types of parameters, such that they can be reported in biophysically-justifiable terms.

Here we have successfully automated fabrication and characterization of a large set of DNAs based on standard biological parts. Although the process was neither easy nor straightforward, it is clear that it was doable. In fact, the theory of how to do this is simple, and implementation difficulties stem from the fact that it is challenging to automate this type of fabrication and characterization in a single, streamlined liquid handling platform, not from any intrinsic complications associated with the building of large sets of DNAs on automated platforms. To do this well in practice, it would need to be implemented in the context of a centralized DNA fabrication facility that has an assembly line of robots dedicated to specific tasks along the assembly process. Given advances in acoustic liquid handling and microfluidics, more streamlined hardware solutions may soon be available. In the mean time, we should state that the most striking observation coming out of these studies was the lack of appropriate software solutions available to integrate automation tools in a traditional academic laboratory. Electronic information held in our lab, like in many labs, is stored within an aggregation of tools that provide reasonable organization for small teams of researchers. These include sequence-management programs like ApE, Google docs for tables of sequences, parts, and samples, and wikis for experimental data. Though tools such as the Registry of Standard Biological Parts and ICE platforms provide excellent solutions for the dissemination of sequence and qualitative usage information, they do not support the full range of data types that must be considered, tracked, and persisted for managing a synthetic biology experiment. For large sets like ours, the data entry process alone becomes unrealistically burdensome when using only these tools, and the lack of cross-tool communication leads to an enormous amount of custom script writing simply to port information from existing tools into the software needed to run the automation hardware. Further, it was clear that user errors, custom notes, and non-standardized formats to freeform documents make conversion into a standardized form difficult and cumbersome. Despite rapid progress in the area of BioCAD tools [6,23-27], basic day-to-day usability and integration into wet-lab workflows remains one of the outstanding challenges to fully realizing the benefits of these tools.

BioCAD-driven wetlab automation has been identified in both academia and industry as the primary driver of new advances in genetic engineering. Considering the biosafety and biosecurity risks associated with DNA fabrication at this level of sophistication, it is important that we begin to draft road maps of the major events in the development of fabrication standards so that proper practices can be anticipated and interventions implemented *a priori*. This is particularly important given the high level of interest that such technologies generate in the DIY community. The question of whether individuals, either acting alone or in small groups, could make use of these advances to set-up clandestine operations is legitimate. In the absence of guarantees the answer seems to be that although not easy at all, it would be possible for teams of 2–3 people to generate a few hundred DNA constructs a month using a similar, minimal automation setup. Clearly, DNA fabrication is just the first hurdle to overcome in a series of required steps before a functional microbial chemical factory is possible. Thus, although initial barriers to entry may not be as high as desirable, increasing the levels of throughput required for success bring about a series of complications that are not easily overcome unless one gets to a much larger operational scale, such as those found in the context of successful and well-capitalized biotech companies. When acting alone, members of the DIY community could potentially work on applications using BioCAD-driven design processes. However, efforts to optimize processes using combinatorial approaches will likely require external support from an angel investor or well-capitalized fund in order to have a meaningful effect. As we demonstrate with our particular data set, full integration of automated tools into a wet-lab environment will require not only hardware, bioware, and software, but also a complete set of tools that are responsive to the practical needs and practices of wet-lab researchers and can manage the information already held in synthetic biology labs. Fortunately, there is great interest within the synthetic biology community to develop community-shared data standards and interoperating procedures [28] that should significantly facilitate progress in moving forward in an ethical and safe manner.

Conclusions

We have shown the use of an integrated tool chain from abstract design through to fabrication and analysis. The composite devices assembled in our set contained a total of 3696 junctions between parts (out of which 632 are unique) and were assembled in three steps in about 1 month at a cost (without labor) around $4000. From this we have learned that, as it currently stands, the methodology is robust enough to construct hundreds of large error-free DNAs with a 96% success rate. Though there remain many challenges to fully incorporating these automation workflows into the wet-lab work flow, it is clear that this is doable and provides rich information to accelerate the process of engineering specific devices and a critical tool for the development of a robust encapsulation theory.

Methods

Materials

Unless otherwise specified, all enzymes were purchased from NEB (Ipswich, MA). Oligos were purchased from IDT (San Diego, CA) and used unpurified. Antibodies were purchased from GenScript (Piscataway, NJ), Bethyl (Montgomery, TX) and Abcam (Cambridge, MA) and used at the dilutions specified below.

Construction of basic parts

The Registry of Standard Biological Parts (http://partsregistry.org), which is searchable by part number, provides details on the construction of all basic parts used in this study. Additional file 1 lists both part number for the Registry, and ID number for the non-profit plasmid repository Addgene (www.addgene.org). Sequences for the oligos needed to gene synthesize basic parts were generated using OligoDesigner (http://andersonlab.qb3.berkeley.edu/andersonSoftware.html). All parts were codon optimized for expression in *E. coli.* Undesirable restriction enzyme sites (BamHI, BglII, EcoRI and XhoI) were removed using GeneDesign (http://baderlab.bme.jhu.edu/gd/). Part identity was confirmed by sequencing, which was outsourced to Quintara (Berkeley, CA). For maintenance, parts were digested with EcoRI and BamHI, and directionally cloned into various 2ab assembly vectors prior to transformation into MC1061 or derivatives [7,8]. All part and vector sequences, as well as physical DNAs and relevant bacterial strains, are available through Addgene.

2ab assembly of composite parts

All junctions between parts were generated by 2ab assembly using a minimal hardware platform consisting of a liquid handling robot (Beckman Coulter Biomek 3000 Laboratory Automation Workstation), a plate-spinning bench-top centrifuge (Beckman Coulter Allegra™ 25R) and a 96-well plate thermocycler (MJ Research PTC-200 Peltier Thermal Cycler), with additional support from a software tool called AssemblyManager. Detailed methods for the implementation of 2ab assembly on the Biomek 3000 liquid handling robot have been previously described [7], and sample robot files required to carry out these protocols, as well as the AssemblyManager software tool, are available for download (http://andersonlab.qb3.berkeley.edu/andersonSoftware.html). Briefly, 2ab assembly proceeds in iterative cycles of 3 steps per cycle: in the first step plasmid DNA is digested with restriction enzymes to generate lefty and righty fragments; in the second step DNA fragments are ligated back together to generate new plasmid combinations; and in the last step plasmids are transformation into *E. coli* for selection and screening. For BglBricks-based 2ab assembly, digestion reactions were carried out in one-pot at 37°C for 1 hr, in 96-well PCR plates, in 16 μL volumes containing 10 μL of digestion cocktail (1 μL 10X NEB Buffer 2, 0.5 μL BamHI, 0.5 μL BglII, 0.5 μL XhoI, 7.5 μL water) and 3 μL each lefty and rightly input plasmids. Lefty and righty parts were generated by transforming and harvesting plasmids from BglII- and BamHI-methylating strains (parts J72007 and J72013), respectively [7,8]. These strains are *pir* +

MC1061 derivatives that replicate R6K origins found in all assembly vectors. Following digestion, restriction enzymes were heat-killed at 65°C for 20 min. Ligation reactions were carried out at room temperature for 30 min by adding 4 μL of ligation cocktail (0.4 μL NEB Buffer 2, 2 μL 10 mM ATP, 0.5 μL T4 DNA ligase, 1.1 μL water) into each heat-killed 16 μL digestion reaction. Transformation reactions were carried out by adding 30 μL of chemically competent cells, made using either the BglII- or the BamHI-methylating strain, into each 20 μL ligation. Following incubation on ice for 10 min, cells were heat shocked at 42°C for 3 min, and then rescued by adding 100 μL of 2YT and incubating at 37°C for 30 min. Transformations were then plated onto LB agar strips supplemented with appropriate antibiotic combinations (ampicillin at 100 ug/mL, chloramphenicol at 25 ug/mL and kanamycin at 25 ug/mL) using 24-well strip plates (Analytical Sales & Services, cat# 47025). BBF RFC10-based 2ab assembly was carried out similarly to BglBricks-based 2ab assembly, but with minor modifications due to a lack of strains capable of methylating SpeI or XbaI sites. Lefty and righty plasmids were digested separately, at 37°C for 1 hr, in 8 μL volumes containing 5 μL of SpeI and XhoI or XbaI and XhoI digestion cocktail (0.5 μL 10X NEB Buffer 2, 0.25 μL each enzyme, 4 μL water) and 3 μL each lefty or rightly input plasmids. Following digestion, restriction enzymes were heat-killed at 80°C for 20 min. Righty and lefty products were then combined in one pot prior to ligation and transformation, which were carried out exactly as described above.

Screening of assembled composite parts

Transformants containing putative composite parts were subjected to various screens at every stage of assembly: the first screen was growth under various antibiotic combinations; the second was restriction enzyme mapping or colony PCR amplification; and the last was sequencing. For growth screens, single colonies were spotted onto LB agar plates containing both double (AC, CK or KA) and triple (ACK) antibiotic combinations and grown overnight at 37°C prior to scoring. Transformants with the correct growth phenotype were then screened either by restriction enzyme mapping or by colony PCR. For restriction enzyme mapping, plasmid DNA was extracted using NucleoSpin plasmid columns (Machery-Nagel), either in single or Multi-8 format, according to the manufacturer's directions, and digested with BglII, BamHI and XhoI digestion cocktail as above. For colony PCR amplification, single colonies were grown overnight at 37°C in 96 well blocks (Analytical Sales & Services, cat# 27P687) containing 1 mL LB liquid medium supplemented with appropriate antibiotics. PCR assays were carried out in 20 μL volumes using 96-well PCR plates. The reaction mixture contained 0.5 units of Taq DNA polymerase, 1.5 mM MgCl, 200 uM dNTPs, 200uM each oligo (ca998: GTATCACGAGGCA GAATTTCAG and G00101: ATTACCGCCTTTGAGT GAGC) and approximately 0.13 μL of saturated bacterial culture, transferred into the 96-well plate containing the reaction mixture using a 96-pin tool (V&P Scientific, cat#246A). Sequencing was outsourced to Quintara (Berkeley, CA).

Cell lysates

Prior to ELISAs, cell lysates were prepared from induced *E. coli* strain pir-116 containing the composite part of interest. Briefly, frozen glycerol stocks were used to inoculate 1 mL 2YT liquid medium supplemented with antibiotics and grown overnight at 37°C in 96 well blocks. The next morning, saturated cultures were diluted 1:500 into fresh 1 mL aliquots of 2YT liquid medium, supplemented with antibiotics and 0.2% arabinose. These were grown overnight again at 37°C in 96 well blocks. Induced cultures were then pelleted by centrifugation for 5 min at 5400 g and resuspended in 100 μL of Bugbuster HT (Novagen) by vortexing. After a 20–30 min incubation period at room temperature, lysates were cleared by centrifugation for 5 min at 5400 g, and protein concentration determined by BCA assay (Pierce) according to the manufacturer's directions. Absorbance was measured at 562 nm using a Tecan Safire 2 plate reader. Lysates were diluted 1:10 in TBST (50 mM Tris pH 7.8, 150 mM NaCl, 0.05% Tween-20) prior to use.

ELISAs

All ELISA assays were carried out using commercially available affinity purified antibodies (Additional file 2). Indirect ELISAs were used to quantify protein expression, whereas sandwhich ELISAs were used to quantify protein-protein interactions. For indirect ELISAs, 40 μL of cleared cell lysate were absorbed onto a 96-well ELISA plate (NUNC) at room temperature for 1 hr. Following absorption, wells were washed 3X with 200 μL TBST at room temperature for 10 min. Wells were then blocked using 200 μL of StartingBlock™ Buffer (Thermo Scientific) according to the manufacturer's directions. Following blocking, wells were washed 3X as described and 100 μL of TBST containing primary antibodies against one of twelve epitope tags (diluted 1:3000) were applied at room temperature for 1 hr. Wells were washed 3X and 100 μL of TBST containing HRP-conjugated secondary antibodies against the constant region of the detecting primary antibody (diluted 1:5000) were applied at room temperature for 1 hr. Wells were washed 3X and 100 μL of 1-Step Ultra TMB-ELISA reagent (Thermo Scientific) were applied, followed by 100 μL of 2 M sulfuric acid to stop the reaction. Absorbance was measured at 450 nm using a Tecan Safire 2 plate reader. For sandwich ELISAs, 100 μL of carbonate buffer containing antibodies against one of twelve epitope tags (diluted 1:500) were absorbed onto a 96-well ELISA plate at 4°C overnight. Following absorption, wells were washed 3X with 200 μL TBST at room temperature for 10 min and then blocked using 200 μL of StartingBlock™ Buffer. Following blocking, wells were washed 3X and 40 μL of cleared cell lysate applied at room temperature for 1 hr. Wells were washed 3X and 100 μL of TBST containing detecting primary antibodies against the second tag in the complex (diluted 1:3000) were applied at room temperature for 1 hr. The remaining washes, addition of the secondary HRP-conjugated antibodies (diluted 1:5000), addition of the TMB-ELISA reagent and of the sulfuric acid, as well as the absorbance measurements, were carried out as described in the indirect ELISA protocol. All ELISAs were performed in triplicate. Values were normalized for protein concentration and background. Control experiments included antibody drop-outs and the use of protein extracts lacking the epitope tags being assayed.

Additional files

Additional file 1: Basic parts used for BglBricks-based 2ab assembly of bi-cistronic operons of Mediator complex.

Additional file 2: Antibodies used in ELISA assays.

Additional file 3: Heat maps of protein concentration values obtained via indirect ELISAs. Maps include all possible tags and tagging combinations in order to visualize expression patterns for each epitope-tagged ORF in all 528 bi-cistronic operons. The top map illustrates patterns of expression for Med7 located in the first ORF position and should be read from left to right. The bottom map illustrates patterns of expression for Med21 located in the second ORF position and should be read from top to bottom. Red color indicates high relative expression, whereas yellow color indicates low relative expression.The complete data set, including standard deviation values, are provided in Additional file 4.

Additional file 4: Complete data set of protein concentration values obtained via indirect ELISAs from all 528 bi-cistronic operons.

Abbreviations

2ab: 2 antibiotic; A: Ampicillin; C: Chloramphenicol; K: Kanamycin; ORF: Open reading frame; RBS: Ribosome binding site; BBF: BioBricks™ Foundation; RFC: Request for comments; DIY: Do-it-yourself.

Competing interests

The authors declare that they have no competing interests.

Authors' contributions

The experiments were designed by ML and JCA. Computational tools were developed by DD and JCA. DNA assemblies were performed by ML and JB. ELISAs were performed by ML and AA. The manuscript was drafted by ML and JCA. All authors read and approved the final manuscript.

Acknowledgements

We thank Thomas Earnest, David Bushnell, and Roger Kornberg for helpful discussions on the design of Mediator constructs; Tim Hsiau and Christopher Batten for computational help with assembly trees; Jin Huh for development of the BamHI- and BglII-methylation strains. This work was funded by the SynBERC NSF ERC.

Author details

[1]Department of Bioengineering, University of California, Berkeley, CA 94720, USA. [2]QB3: California Institute for Quantitative Biological Research, University of California, Berkeley, CA 94720, USA. [3]Physical Biosciences Division, Lawrence Berkeley National Laboratory, Berkeley, CA 94720, USA. [4]Current address: Department of Electrical and Computer Engineering, Boston University, Boston, MA 02215, USA.

References

1. Purnick PE, Weiss R: **The second wave of synthetic biology: from modules to systems.** *Nat Rev Mol Cell Biol* 2009, **10**:410–422.
2. Ellis T, Adie T, Baldwin GS: **DNA assembly for synthetic biology: from parts to pathways and beyond.** *Integr Biol (Camb)* 2011, **3**:109–118.
3. Czar MJ, Anderson JC, Bader JS, Peccoud J: **Gene synthesis demystified.** *Trends Biotechnol* 2009, **27**:63–72.
4. Tian J, Ma K, Saaem I: **Advancing high-throughput gene synthesis technology.** *Mol Biosyst* 2009, **5**:714–722.
5. Xiong AS, Peng RH, Zhuang J, Liu JG, Gao F, Chen JM, Cheng ZM, Yao QH: **Non-polymerase-cycling-assembly-based chemical gene synthesis: strategies, methods, and progress.** *Biotechnol Adv* 2008, **26**:121–134.
6. Bilitchenko L, Liu A, Cheung S, Weeding E, Xia B, Leguia M, Anderson JC, Densmore D: **Eugene - a domain specific language for specifying and constraining synthetic biological parts, devices, and systems.** *PLoS One* 2011, **6**:e18882.
7. Leguia M, Brophy J, Densmore D, Anderson JC: **Automated assembly of standard biological parts.** *Methods Enzymol* 2011, **498**:363–397.
8. Anderson JC, Dueber JE, Leguia M, Wu GC, Goler JA, Arkin AP, Keasling JD: **BglBricks: A flexible standard for biological part assembly.** *J Biol Eng* 2010, **4**:1.
9. Kim YJ, Bjorklund S, Li Y, Sayre MH, Kornberg RD: **A multiprotein mediator of transcriptional activation and its interaction with the C-terminal repeat domain of RNA polymerase II.** *Cell* 1994, **77**:599–608.
10. Sato S, Tomomori-Sato C, Parmely TJ, Florens L, Zybailov B, Swanson SK, Banks CA, Jin J, Cai Y, Washburn MP, *et al*: **A set of consensus mammalian mediator subunits identified by multidimensional protein identification technology.** *Mol Cell* 2004, **14**:685–691.
11. Baumli S, Hoeppner S, Cramer P: **A conserved mediator hinge revealed in the structure of the MED7.MED21 (Med7.Srb7) heterodimer.** *J Biol Chem* 2005, **280**:18171–18178.
12. Salis HM: **The ribosome binding site calculator.** *Methods Enzymol* 2011, **498**:19–42.
13. Densmore D, Hsiau TH, Kittleson JT, DeLoache W, Batten C, Anderson JC: **Algorithms for automated DNA assembly.** *Nucleic Acids Res* 2010, **38**:2607–2616.
14. Knight TF Jr: **Idempotent vector design for standard assembly of Biobricks.** 2003, In DSpace http://hdlhandlenet/17211/21168.
15. Pedelacq JD, Cabantous S, Tran T, Terwilliger TC, Waldo GS: **Engineering and characterization of a superfolder green fluorescent protein.** *Nat Biotechnol* 2006, **24**:79–88.
16. Engler C, Gruetzner R, Kandzia R, Marillonnet S: **Golden gate shuffling: a one-pot DNA shuffling method based on type IIs restriction enzymes.** *PLoS One* 2009, **4**:e5553.
17. Gibson DG: **Synthesis of DNA fragments in yeast by one-step assembly of overlapping oligonucleotides.** *Nucl Acids Res* 2009, **37**(20):6984–6990. doi:10.1093/nar/gkp687.
18. Horton RM: **PCR-mediated recombination and mutagenesis. SOEing together tailor-made genes.** *Mol Biotechnol* 1995, **3**:93–99.
19. Horton RM, Cai ZL, Ho SN, Pease LR: **Gene splicing by overlap extension: tailor-made genes using the polymerase chain reaction.** *Biotechniques* 1990, **8**:528–535.
20. Quan J, Tian J: **Circular polymerase extension cloning for high-throughput cloning of complex and combinatorial DNA libraries.** *Nat Protoc* 2011, **6**:242–251.
21. Quan J, Tian J: **Circular polymerase extension cloning of complex gene libraries and pathways.** *PLoS One* 2009, **4**:e6441.
22. Callura JM, Cantor CR, Collins JJ: **Genetic switchboard for synthetic biology applications.** *Proc Natl Acad Sci U S A* 2012, **109**:5850–5855.
23. Bilitchenko L, Liu A, Densmore D: **The Eugene language for synthetic biology.** *Methods Enzymol* 2011, **498**:153–172.
24. Czar MJ, Cai Y, Peccoud J: **Writing DNA with GenoCAD.** *Nucleic Acids Res* 2009, **37**:40–47.
25. Smith LP, Bergmann FT, Dhandran D, Sauro HM: **Antimony: a modular model definition language.** *Bioinformatics* 2009, **25**:2452–2454.

26. Umesh P, Naveen F, Rao CU, Nair AS: **Programming languages for synthethic biology.** *Syst Synth Biol* 2010, **4**:265–269.
27. Wilson ML, Hertzberg R, Adam L, Peccoud J: **A step-by-step introduction to rule-based design of synthetic genetic constructs using GenoCAD.** *Methods Enzymol* 2011, **498**:173–188.
28. Galdzicki M, Rodriguez C, Chandran D, Sauro HM, Gennari JH: **Standard biological parts knowledgebase.** *PLoS One* 2011, **6**:e17005.

2

Dual organism design cycle reveals small subunit substitutions that improve [NiFe] hydrogenase hydrogen evolution

Isaac T Yonemoto, Christopher W Matteri, Thao Amy Nguyen, Hamilton O Smith and Philip D Weyman*

Abstract

Background: Photosynthetic microorganisms that directly channel solar energy to the production of molecular hydrogen are a potential future biofuel system. Building such a system requires installation of a hydrogenase in the photosynthetic organism that is both tolerant to oxygen and capable of hydrogen production. Toward this end, we have identified the [NiFe] hydrogenase from the marine bacterium *Alteromonas macleodii* "Deep ecotype" that is able to be heterologously expressed in cyanobacteria and has tolerance to partial oxygen. The *A. macleodii* enzyme shares sequence similarity with the uptake hydrogenases that favor hydrogen uptake activity over hydrogen evolution. To improve hydrogen evolution from the *A. macleodii* hydrogenase, we examined the three Fe-S clusters found in the small subunit of many [NiFe] uptake hydrogenases that presumably act as a molecular wire to guide electrons to or from the active site of the enzyme. Studies by others altering the medial cluster of a *Desulfovibrio fructosovorans* hydrogenase from 3Fe-4S to 4Fe-4S resulted in two-fold improved hydrogen evolution activity.

Results: We adopted a strategy of screening for improved hydrogenase constructs using an *Escherichia coli* expression system before testing in slower growing cyanobacteria. From the *A. macleodii* enzyme, we created a mutation in the gene encoding the hydrogenase small subunit that in other systems is known to convert the 3Fe-4S medial cluster to 4Fe-4S. The medial cluster substitution did not improve the hydrogen evolution activity of our hydrogenase. However, modifying both the medial cluster and the ligation of the distal Fe-S cluster improved *in vitro* hydrogen evolution activity relative to the wild type hydrogenase by three- to four-fold. Other properties of the enzyme including thermostability and tolerance to partial oxygen did not appear to be affected by the substitutions.

Conclusions: Our results show that substitution of amino acids altering the ligation of Fe-S clusters in the *A. macleodii* [NiFe] uptake hydrogenase resulted in increased hydrogen evolution activity. This activity can be recapitulated in multiple host systems and with purified protein. These results validate the approach of using an *E. coli*-cyanobacteria shuttle system for enzyme expression and improvement.

Keywords: Hydrogenase, Cyanobacteria, Iron-sulfur Cluster, *Alteromonas Macleodii* "Deep Ecotype"

Background

Hydrogen produced photobiologically from sunlight and water is a potential future biofuel and is an alternative to traditional carbon-based fuels [1-4]. Enzymes that produce hydrogen include nitrogenases and hydrogenases, the latter comprising the unrelated families of [Fe-Fe] hydrogenases, Fe-only hydrogenases, and [NiFe] hydrogenase enzymes [5]. Cyanobacteria such as *Synechococcus elongatus* are ideal host strains for a biological system that converts solar energy to hydrogen because they are genetically tractable and easily cultured [6]. Because cyanobacteria also perform oxygenic photosynthesis, the oxygen produced will inactivate most hydrogenases [7,8]. Therefore, installing a hydrogenase that is less sensitive to oxygen is a primary design constraint of this system. Among the hydrogenase families, oxygen tolerance has been described only among the [NiFe] hydrogenases, although not all [NiFe] hydrogenases are oxygen tolerant [9-11]. There is some evidence the native cyanobacterial bidirectional hydrogenase may be

* Correspondence: pweyman@jcvi.org
J. Craig Venter Institute, 10355 Science Center Dr, San Diego, CA 92121, USA

capable of hydrogen production in the presence of oxygen [12], but the energetic consequences of using NADPH as an electron donor makes it less attractive.

We previously reported the discovery and characterization of a novel hydrogenase isolated from Sargasso Sea environmental DNA with high identity to the stable [NiFe] uptake hydrogenase from *Thiocapsa roseopersicina* [13]. A closely-related sequence was later identified in the *Alteromonas macleodii* 'Deep ecotype' genome (hereafter referred to as the *A. macleodii* hydrogenase), and was characterized as active in the presence of 2% oxygen [14]. The DNA encoding the enzyme was then cloned and used to express active enzyme in both *Escherichia coli* [15] and *S. elongatus* [16]. Expressing active *A. macleodii* hydrogenase in both *E. coli* and cyanobacteria requires the co-expression of at least nine additional maturation factors to properly assembly an active enzyme [17]. Due to the complex maturation requirements of [NiFe] hydrogenases, only a few other examples exist of heterologous expression in *E. coli* [18-20]. The dual expression system is ideal for rapidly generating variant hydrogenases using powerful *E. coli* genetic tools, quickly screening for desirable properties in *E. coli*, and confirming the persistence of the property in the final target cyanobacterial host.

Although some uptake [NiFe] hydrogenases are oxygen-tolerant, they are known to be generally directionally biased against hydrogen evolution [21,22]. Enzyme bias in uptake hydrogenases appears to have been altered in separate experiments investigating the ligation of Fe-S clusters in the small subunit of [NiFe] hydrogenases [23,24]. These Fe-S clusters are believed to function as a molecular wire guiding electrons out of the active site and to a biological electron carrier molecule (Figure 1) [10,25,26]. Conversely, in a hydrogen-production scenario, these Fe-S clusters would predominantly operate in the reverse direction, channeling electrons into the active site from an electron donor at the enzyme surface.

At the medial Fe-S cluster, uptake [NiFe] hydrogenases have conserved amino acid residues deviating from the canonical 4-cysteine ligation characteristic of Fe-S cubane clusters. In *A. macleodii* and almost all members of the uptake NiFe hydrogenase family, the medial cluster features a proline in the place of a ligating cysteine resulting in the installation of a 3Fe-4S cluster instead of a 4Fe-4S cluster [26]. Interestingly, the closely related small subunits of the H_2-sensing regulatory hydrogenases, which show almost no hydrogen evolution, and the NiFeSe hydrogenases, which show more hydrogen evolution activity [29], both feature cysteinyl ligations at the homologous amino acid position [24,30]. The unrelated small subunit of the F420-reducing hydrogenase, which also features 3 4Fe-4S clusters, has a 4Fe-4S cluster for its medial cluster [31]. In *A. macleodii* hydrogenase, and uptake hydrogenases in general, the cluster distal to the active site is a 4Fe-4S cluster featuring an unusual iron-histidinyl ligation that is conserved in most [NiFe] hydrogenases. While its function is unknown, the unusual ligation may be related to intramolecular electron transfer [23,32].

In *Desulfovibrio fructosovorans*, substitution of the proline residue associated with the medial cluster to cysteine (*Df* HynA P238C) resulted in approximately 30% reduction of hydrogen uptake activity and a nearly two-fold increase in evolution activity [24]. Separately, when the histidine

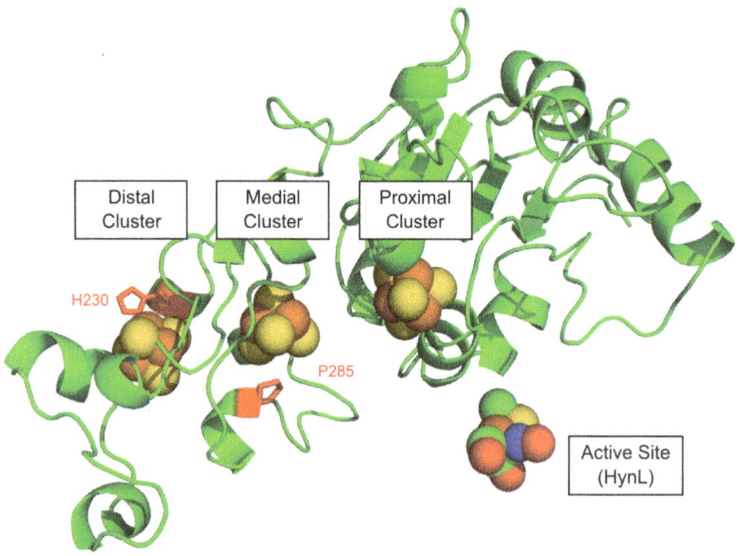

Figure 1 Structural model of the *A. macleodii* [NiFe] uptake hydrogenase small subunit HynS, highlighting the three Fe-S clusters, and residues His230 and Pro285, which were targeted for substitution to cysteine residues. For reference, the location of the active site is shown, but the rest of the large subunit structure was omitted to simplify viewing of the small subunit. The structural model was generated by Phyre using structure PDB: 1h2a as a model [27,28].

residue at the distal cluster was changed to cysteine (*Df* HynA H184C), hydrogen uptake activity was nearly abolished while evolution activity was only 50% reduced [23]. We produced enzyme variants featuring substitutions in the *A. macleodii* hydrogenase at sites homologous to the *D. fructosovorans* uptake hydrogenase substitutions. The variants of the *A. macleodii* enzyme (HynSL - NCBI accession codes: YP_004425481.1, YP_004425482.1) featuring the corresponding substitutions based on homology (HynS H230C, P285C[1]) were created both individually and combined in an effort to find an enzyme with improved hydrogen evolution activity for production of hydrogen in cyanobacteria.

Taking advantage of our previously reported dual-species system, these modified enzymes were first screened in *E. coli* to assess their enzymatic properties. Initial screens with *E. coli* extracts indicated that the doubly-substituted enzyme had the highest hydrogen evolving activity. Purification of the enzyme indicated that this activity was intrinsic to the enzyme variant, and not due to enzyme titer or host factors persistent in crude lysate. Finally, we moved the expression system to the cyanobacterium *S. elongatus*, which recapitulated the relative activities of the enzymes, validating our general strategy of using *E. coli* as a rapid-turnover model system.

Results and discussion

Heterologous expression for [NiFe] hydrogenase in E. coli

Heterologous expression of the *A. macleodii* hydrogenase has been previously reported in both *E. coli* [15] and the cyanobacterium *S. elongatus* [16]. These reports address general issues of aerobic expression of the *A. macleodii* hydrogenase, as in both these reports, aerobic handling was used throughout, established as safe in the original report on the enzyme [14]. We have previously determined that the minimal gene set required for optimal activity in *E. coli* consists of the hydrogenase structural genes, *hynS* and *hynL*, the six pleiotropic maturation factors, *hypABCDEF*, the endoprotease *hynD*, and two additional genes, *hupH* and *orf2*. Additionally, four promoter sites can be used to add additional transcriptional regulation throughout the gene cluster (Figure 2). Expression plasmids used initially consisted of these essential 11 genes driven by a single P_{TRC} promoter installed in "promoter site 1" (Figure 2). The P_{TRC} promoter functions well in both *E. coli* and cyanobacteria [33]. Expression plasmids also contain a spectinomycin resistance gene and *lacI* gene that function in both *E. coli* and cyanobacteria. Variants containing a P_{T7} promoter or P_{TRC} promoters at all four "promoter site"s were also generated. For integration into cyanobacteria, these features are flanked by 1-kb of *S. elongatus* DNA on either side for homologous recombination into neutral site 1 (NS1) [34]. Amino acid substitutions were generated by PCR of the operon using primers which encode for mutations corresponding to the substitutions, followed by assembly with an empty vector. Amino acid substitutions of HynS can be transferred by restriction digest and reassembly process. In this process, the wild type sequence is excised using restriction enzymes whose sites flank the region of interest and the new sequence is installed by reassembly of the vector with PCR products featuring the corresponding nucleic acid sequence mutations (see Methods).

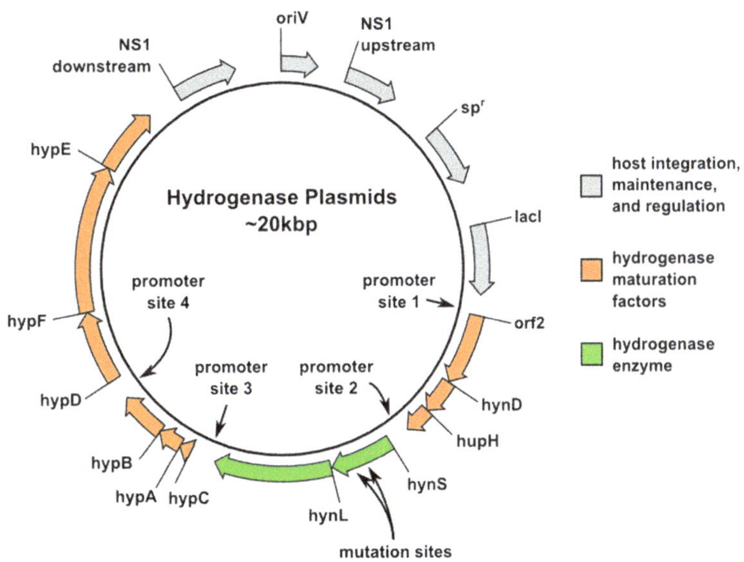

Figure 2 A schematic diagram of the *A. macleodii* hydrogenase expression vectors used in our study. Detailed maps and sequences for all vectors are available in the supplemental information.

Evolution activity assay

To test the effect of our substitutions on the hydrogen evolution activity, we performed methyl viologen hydrogen production assays on lysates of *E. coli* cultures expressing the wild type *A. macleodii* hydrogenase (WT), the *A. macleodii* hydrogenase bearing the individual substitutions H230C, P285C, or the doubly-substituted HynS H230C/P285C. In this assay, electrons are transferred to methyl viologen by oxidation of sodium dithionite, then from methyl viologen to hydrogenase. At the hydrogenase active site, these electrons are recombined with protons to generate molecular hydrogen, which undergoes a phase separation from solution and is detectable by gas chromatographic analysis (GC) of the assay vial headspace. Compared to wild type, lysates containing the singly-substituted enzymes had slightly lower H_2 evolution activity. The H230C lysate had approximately half the activity of WT, and the P285C lysate had approximately one third the WT activity (Figure 3A). Lysates containing the H230C/P285C doubly-substituted enzyme, by contrast, exhibited a 3-4-fold increase in evolution activity compared to WT (Figure 3A).

Uptake activity assay

To test the effect of our substitutions on hydrogen uptake, we performed a benzyl viologen hydrogen uptake assay on *E. coli* lysates expressing the wild type *A. macleodii* [NiFe] hydrogenase, and variants featuring the substitutions H230C, P285C, and H230C/P285C. In this reaction the hydrogenase catalyzes the dissociation of hydrogen to protons and electrons. The electrons are transferred to benzyl viologen, causing a color change detectable by spectrophotometry at 555 nm. Lysates containing the H230C and P285C enzymes each had slightly increased uptake activity relative to WT; H230C had nearly twice the uptake activity as WT while P285C had slightly less activity than H230C. The hydrogen uptake activity of lysate containing the doubly-substituted enzyme was similar to WT and within the range of experimental error (Figure 3B).

Western blot

Because our activities are normalized against total protein from crude *E. coli* lysate, it was necessary to rule out the possibility that differences in hydrogenase expression levels were responsible for the observed differences in activity. Proteins in crude extracts were separated by SDS-PAGE and subjected to western blotting to estimate the relative expression of hydrogenase. Because the primary sequence of HynL is unaffected by the substitutions we made in HynS, we expect the relative intensity of resulting HynL bands to be a good indicator of enzyme titer. Lysates were diluted to 5 mg mL^{-1} total protein and enzyme activity was measured to ensure activities agreed with those measured previously (Additional file 1: Figure S2). To confirm approximately equivalent loading, a gel slice was stained with Sypro Ruby to illustrate protein content of loaded samples (Figure 4, top). Anti-HynL western blot (Figure 4, bottom) of the lysates shows two bands, presumably corresponding to the immature and the C-terminally processed mature forms of the enzyme subunit [35]. Identically-handled lysate was subjected to brief thermal denaturation (65°C, 10 min) and a second clearing step. Lysate treated in this manner retained nearly all activity (Additional file 1: Figure S2) but lost the larger HynL band in the western blot, lending support to the top band being the unprocessed, immature form of HynL. These western blots suggest that all variants were expressed at approximately equivalent levels and that there were no major defects in expression.

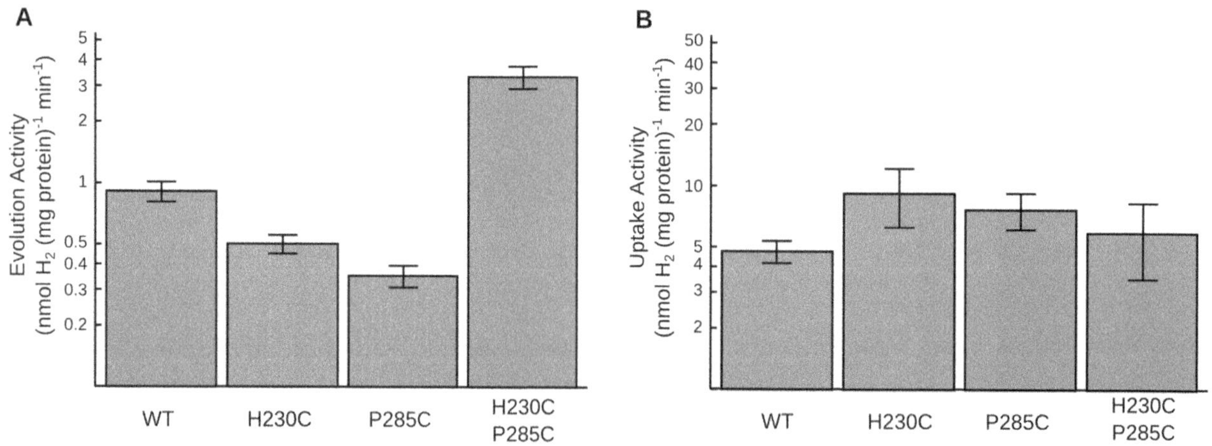

Figure 3 Hydrogen evolution (A) and uptake (B) activities for lysates of *E. coli* strain FTD147 over-expressing wild type *A. macleodii* hydrogenase or substitutions H230C, P285C, and H230C/P285C. Activities on both graphs are plotted on matching log scales over a 50-fold range to allow comparison of equivalent fold-changes. Error bars represent one standard error of the mean.

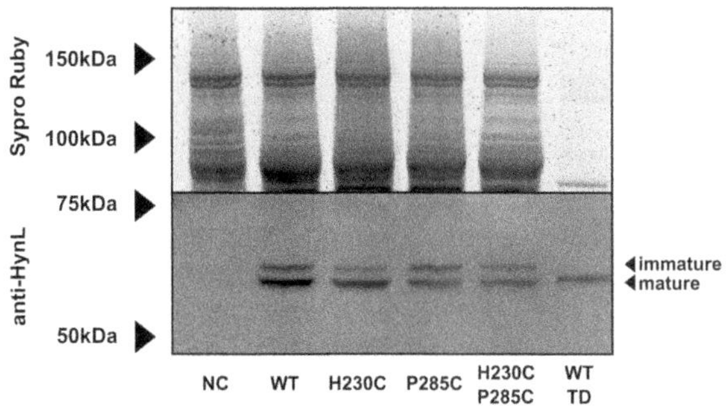

Figure 4 Expression analysis for lysates of *E. coli* strain FTD147 with no hydrogenase construct (NC), over-expressing wild type *A. macleodii* hydrogenase enzyme (WT), or H230C, P285C, and H230C/P285C substitutions, and wild type hydrogenase subjected to thermal denaturation (WT TD). Before loading, proteins were normalized for protein content at 5 mg mL^{-1}. Associated enzyme activites can be found in Figure S1. Top: Sypro-Ruby stain of the gel, indicating approximately equivalent amounts of protein loaded per lane. Bottom: anti-HynL western blot of samples.

Evolution assay on purified enzyme

To further test if the observed improvement in evolution activity found in the doubly-substituted enzyme is the result of an intrinsic change in enzymatic activity, we generated WT and doubly-substituted enzymes with strep-tags at the C-terminus of HynS. This tag was used in an affinity-purification strategy (Figure 5), and the resulting purified fraction was tested by the hydrogen evolution assay (Additional file 1: Table S1). Some impurities persist in the streptactin-mediated purification; we believe the major contamination in the SDS-PAGE of the strep-tagged protein to be immature small subunit lacking TAT-mediated cleavage, as well as trace amounts of high-molecular weight proteins that persist in the partially purified fractions.

Nonetheless, we found that purification afforded a more than 100-fold improved specific activity for both wild type and doubly-substituted enzymes. Furthermore, after parallel purification the doubly-substituted, strep-tagged enzyme still exhibited three-fold more activity than the wild type, strep-tagged enzyme, consistent with our observations using unpurified *E. coli* lysates in the hydrogen evolution assay (Figure 2A).

Thermal and oxygen tolerance assays

To test that the doubly-substituted enzyme did not gain improved evolution activity at the expense of other desirable intrinsic properties, we subjected the *E. coli* lysates to a thermal denaturation assay and an assay testing function in the presence of oxygen. We chose thermal denaturation (85°C for 1h) as a qualitative measure of structural perturbation and hypothesized that the doubly-substituted enzyme would exhibit less thermotolerance than the wild type enzyme due to its modifications. Other treatments were tested; for simplicity we report the harshest treatment we tested that exhibited > 10% activity. We also tested if the doubly-substituted enzymes would retain function in partial oxygen atmospheres as we had previously observed for the wild-type enzyme. We found that the doubly-substituted enzyme was not less thermotolerant than the WT enzyme within the range of experimental error. Substituting the two residues did not result in a significant defect in the ability to function in the presence of 1% oxygen (Figure 6).

Relative performance in other bacterial strains

To test the robustness of our observation that the doubly-substituted enzyme has improved hydrogen evolution activity, we expressed the enzyme with different promoters and in different host species. Plasmid pIY003 expresses the hydrogenase behind a single P_{TRC} promoter, with pIY007 as the corresponding plasmid bearing the doubly-substituted enzyme. Plasmid pIY033 expresses the hydrogenase gene cluster behind a single P_{T7} promoter with pIY038 as the corresponding plasmid bearing the doubly-substituted enzyme; pRC41-4 is a plasmid featuring the hydrogenase gene cluster supported by 4 P_{TRC} promoters with pCM012 as the corresponding plasmid bearing the doubly-substituted enzyme. While pIY003 and pIY007 were tested in the FTD147 strain, the pIY033/pIY038 pair was tested in an *E. coli* strain in which all of its hydrogenases were knocked out (BL21(DE3)Δ*hycG*Δ*hyaB*Δ*hybC*Δ*hyfG*, hereafter referred to as BL21(DE3)ΔH_4) [18]. The pRC41-4 and pCM012 plasmids were tested in the cyanobacterial strain *S. elongatus* PCC 7942 PW416 (Δ*hoxYH*:: Em^r) [16]. Our data indicate that *E. coli* strain FTD147 expressing the doubly-substituted enzyme from pIY007 showed 3.7±0.6-fold higher activity than the same strain expressing the wild-type enzyme from pIY003; *E. coli* strain BL21 (DE3)ΔH_4 expressing the doubly-substituted enzyme from pIY038 showed 2.3±0.8-fold higher activity than the same strain expressing the wild-type enzyme, and an *S. elongatus* strain expressing the doubly-substituted enzyme from pCM012 showed 5.1±0.7-fold higher activity than a strain expressing the wild-type enzyme (Figure 7). Thus, relative to the WT enzyme, the doubly-substituted enzyme consistently shows increased activity -- in both *E. coli* strains as well as in a cyanobacterial expression system.

By substituting two residues in the small subunit of *A. macleodii* [NiFe] uptake hydrogenase, we have increased its hydrogen evolution activity. To our knowledge this is the first reported instance of this doubly-substituted enzyme analyzed in the context of hydrogen evolution. In the homologous doubly-substituted enzyme from *D. fructosovorans* constructed by Dementin and co-workers [36], activity in the evolving direction was not reported. Comparing the homologous substitutions in *A. macleodii* and *D. fructosovorans* hydrogenases, we observe qualitatively different structure-activity relationships. Thus, the

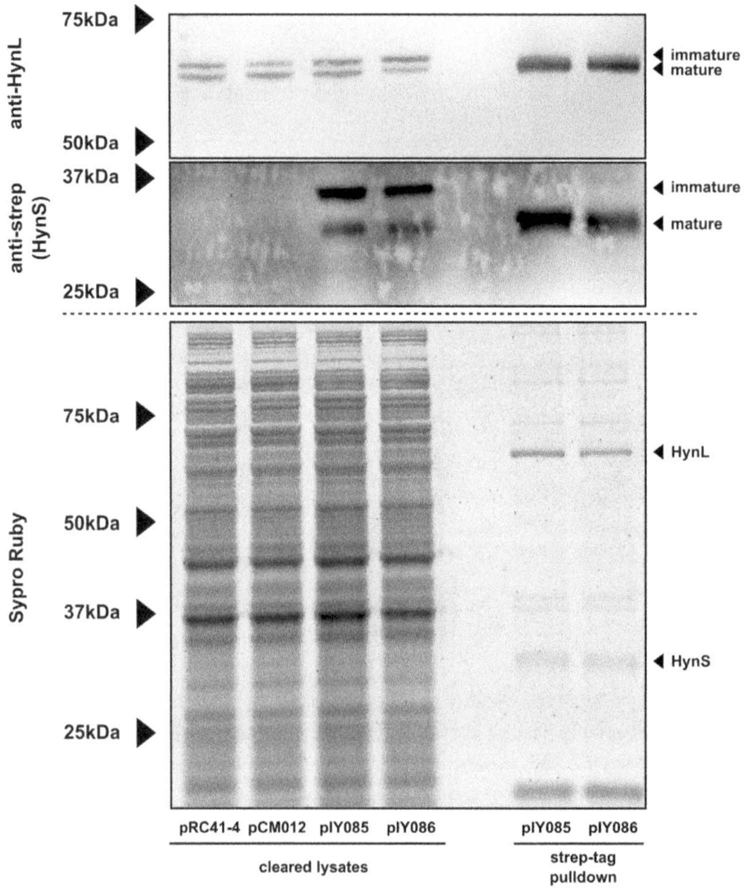

Figure 5 Purification of hydrogenase from *E. coli* over-expressing wild type *A. macleodii* hydrogenase (pIY085) or H230C/P285C doubly-substituted enzyme (pIY086). Associated enzyme activities can be found in Table S1. Samples are either cleared lysates (left) or Streptactin-purified proteins (right) as indicated at the bottom of the figure. Top: anti-HynL western blot. Middle: anti-Strep western blot, detecting HynS-strep protein. Bottom: Sypro-Ruby stain of a parallel gel. pRC41-4 and pCM012 are constructs expressing wild type and doubly-substituted enzymes, respectively, and are included as negative controls for the strep-tag blot. pIY085 and pIY086, respectively, are the equivalent constructs, except bearing a C-terminal strep-tag on the HynS protein.

effect of these substitutions may not be generalizable across different members of the enzyme family. Indeed, the inverse experiment, changing a medial 4Fe-4S cluster to a 3Fe-4S cluster in the *Methanococcus janaschii* F_{420}-reducing hydrogenase, resulted in severely impaired uptake, inconsistent with the loose pattern otherwise observed in uptake hydrogenases [31]. As such, others seeking to modify [NiFe] hydrogenase Fe-S cluster ligations should consider testing both single substitutions as well as the double substitution that are homologous to those discussed here when seeking an enzyme suitable for their needs.

The doubly-substituted *A. macleodii* hydrogenase we present here is an attractive launching point for future engineering of hydrogenases redesigned for use in photoautotrophs. The two substitutions have minimal effects on the thermotolerance of the enzyme, as measured by loss of activity after the harsh treatment of 85°C for 1 hour. Treatment at lower temperatures or shorter times did not cause appreciable loss of activity in either doubly-substituted or wild type enzymes (data not shown). Our observations suggest that the substitutions have not created severe structural changes, likely allowing further modification of the enzyme without disrupting function. Furthermore, the ability to function in the presence of partial oxygen, a critical property of photoautotrophic hydrogen production systems, appears to be unaffected and may be able to be improved by further modification.

Finally, because *A. macleodii* hydrogenase can be expressed as active enzyme in both *E. coli* [15] and in the photoautotrophic cyanobacterium *S. elongatus* [16], our strategy moving forward is to use expression in *E. coli* as a preliminary model system to rapidly generate and test promising modifications of the enzyme prior to experimentation in *S. elongatus*, which has a longer turnaround time for genetic manipulations. The H230C/P285C enzyme demonstrated the same qualitative performance over the wild type enzyme in extracts prepared from both *S. elongatus* and *E. coli*, validating

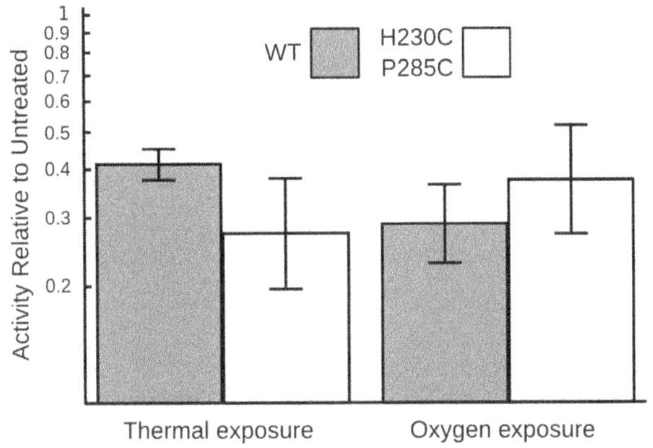

Figure 6 Response of wild type and doubly-substituted *A. macleodii* hydrogenase to various environmental conditions. Thermotolerance was measured as hydrogen evolution activity after denaturation at 85°C for 1 hour. Sensitivity to oxygen was measured by relative H-D exchange rate in the presence of 1% oxygen. Results are expressed as the ratio of activity remaining in the treated samples relative to the untreated controls, and plotted on a log scale over one decade. Error bars represent one geometric standard error of the geometric mean.

this strategy. Initial attempts at in vivo activity were performed in both host species aerobically and anaerobically, but did not yield measurable hydrogen production. To that end, further modifications we are pursuing to achieve this important result include: improving the oxygen tolerance by engineering a substitution leading to a 4Fe-3S proximal Fe-S cluster that is not natively present in our enzyme [37-40], and connecting the hydrogenase to cyanobacterial electron carriers, such as ferredoxin.

Conclusion

In summary, we have used an *E. coli* - cyanobacteria shuttle system to rapidly test candidate modifications of a heterologously-expressed *A. macleodii* [NiFe] hydrogenase in *E. coli* before transferring the most promising constructs to the slower-growing cyanobacteria. As a test of this strategy, we substituted the *A. macleodii* hydrogenase with modifications that in other systems have been shown to change the ligations of the medial and distal Fe-S clusters. When expressed in *E. coli*, we found that neither substitution alone improved enzyme activity; however, when combined, both substitutions increased hydrogen evolution activity approximately 4-fold relative to wild type, while hydrogen uptake remained unchanged. As we did not directly verify that these substitutions resulted in the corresponding alterations to the Fe-S clusters, we cannot completely rule out the possibility of more subtle effects on electron-transfer kinetics in the absence of the expected, new cysteinyl-iron bonds or the possibility that the native enzyme did not feature the expected Fe-S clusters. The effect of this combined modification was recapitulated with the enzymes in enriched form and with the enzymes expressed in cyanobacteria, suggesting that the enzymatic improvement is intrinsic to the enzyme and independent of extrinsic host factors. These results together validate our strategy of piloting hydrogenase modifications in the rapid-turnaround *E. coli* model system for use in cyanobacterial hosts.

Methods

Molecular biology

Primers, full plasmid maps and sequences are provided in supplementary materials. Plasmid pIY003 was constructed by PCR-amplification of the hydrogenase operon from *A. macleodii* genomic DNA. Two amplicons spanning the operon were produced where the sequences had overlapping ends with each other and to the plasmid pTRC-NS1 (the backbone plasmid from [16]), which was linearized by double digest with the restriction enzymes BamHI and EcoRI. These three DNA fragments were assembled using Gibson isothermal assembly [41], and transformed into DH10B (Invitrogen) cells by electroporation. After screening clones by PCR, candidate constructs were sequence-verified by Sanger DNA sequencing. Plasmids pIY004, bearing the H230C substitution and pIY006, bearing the P285C substitution were constructed similarly, except using PCR products generated from primers containing

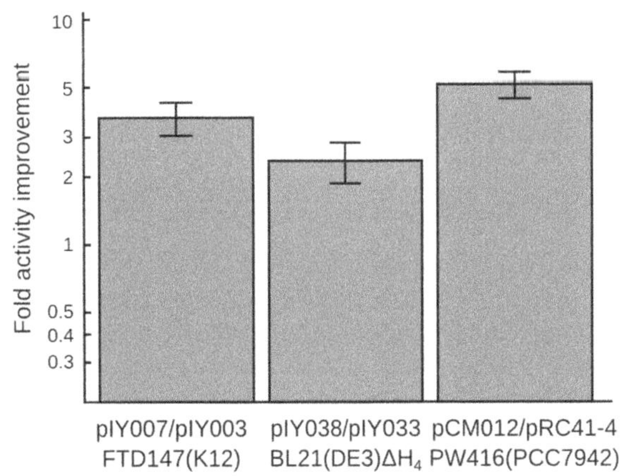

Figure 7 Relative hydrogenase evolution activities of heterologous expression constructs bearing the wild type vs. doubly-substituted *A. macleodii* hydrogenase. pIY007/pIY003 compares the doubly-substituted enzyme (DS) and the wild type (WT) enzyme produced by the hydrogenase operon as driven by P_{TRC} and transformed into the hydrogenase-free *E. coli* K12 strain FTD147. pIY038/pIY033 compares DS and WT enzymes produced by the hydrogenase operons as driven by P_{T7} and transformed into the hydrogenase-free *E. coli* BL21(DE3)ΔH_4 derivative. pCM012/pRC41-4 compares the hydrogenase operons driven by 4 P_{TRC} promoters interspersed in the operon and integrated into an otherwise hydrogenase-free *S. elongatus* cyanobacterial strain. Relative activities are plotted on a log scale over a 50-fold range. Error bars represent one geometric standard error of the geometric mean.

the appropriate DNA nucleotide substitutions; plasmid pIY007, bearing the doubly-substituted enzyme, was constructed by repeating the procedure used to generate pIY006 except using pIY004 plasmid as a template.

Plasmid pIY033 was constructed by moving the hydrogenase operon into a pDEST-23 vector backbone using Gibson isothermal assembly followed by re-amplification of this operon (now including T7 promoter and terminator) and recloning into the pTRC-NS1 plasmid by Gibson isothermal assembly. Successful construction was verified by Sanger sequencing. pIY038 was constructed by PCR amplification of a segment of pIY007 around the *hynS* ORF followed by Gibson isothermal assembly into plasmid pIY033 which was linearized by double digest with the BamHI and AgeI restriction enzymes. Construction of plasmid pRC41-4 will be discussed elsewhere but the sequence is provided in the Supplemental Information; pCM012 was constructed by PCR amplification of the second half of the *hynS* ORF from pIY038, followed by Gibson isothermal assembly into pRC41-4 which was linearized by double digest with BamHI and AgeI restriction enzymes. Successful construction of pIY038 and pCM012 was verified by Sanger sequencing initiated with PCR primers used to amplify the modified region. Plasmids pIY085 and pIY086 were constructed similarly to pCM012, with pIY003 and pIY007 as templates, respectively, except using two sequential rounds of amplification with two reverse primers together generating sequence encoding the strep-tag epitope –RSAWSHPQFEK and the sequence bridging the C-terminal stop codon of *hynS* and the beginning of *hynL* up to the AgeI restriction site. Successful construction of pIY085 and pIY086 was verified by Sanger sequencing initiated with the PCR primers used to amplify the modified region.

A. macleodii Hydrogenase Expression in *E. coli*

All four plasmids pIY003, pIY004, pIY006, pIY007 were transformed into *E. coli* strain FTD147 cells, an MC4100 (DE3) strain with the three active *E coli* hydrogenases removed [42]. Plasmids pIY033 and pIY038 were transformed into BL21(DE3)ΔH_4 cells [18]. To grow *E. coli* cultures used to prepare extracts, 25 mL of autoinducer media [43] supplemented with 0.1 mM $NiCl_2$ and 40 µg mL^{-1} spectinomycin was inoculated with 100 µL of a starter culture of the appropriate strain. Following inoculation, the expression culture was grown in an unbaffled 250 mL Erlenmeyer flask, rotating at 200 rpm, under ambient room atmosphere at 30°C overnight (~10-14 hrs). Under these conditions, the culture grew to a density in the approximate range of OD_{600} 6–7. Piloting experiments showed that growth in baffled flasks resulted in higher specific activity but was avoided to maintain comparability with previous experiments [15]. This was then harvested by centrifugation; the pellet was separated from the supernatant media and resuspended after addition of 0.8 mL lysis buffer (10 mM Tris, pH 7; 1 mM DTT; 0.5 mM EDTA). Lysis was performed by sonication on wet ice for one minute using a microprobe sonicator (Branson, Sonifier Model 250) at setting "4", 40% duty cycle. The lysate was cleared by centrifugation for 10 minutes at 16,800 x g, at 4°C, and the supernatant was reserved as the "cleared lysate". Protein content in the cleared lysate was measured with Bradford reagent (Bio-Rad) and comparison against a BSA (NEB) standard curve.

A. macleodii Hydrogenase Expression in *S. elongatus*

Plasmids pRC41-4 and pCM012 were mobilized into *S. elongatus* PCC 7942 strain PW416 by conjugative transfer from *E. coli* as previously described [16,44]. Genomic integration into neutral site I (NSI) was verified by PCR. Cyanobacterial cultures were grown in 50 mL Bg11 medium in baffled 250 mL flasks to an OD_{730} of approximately 0.5. The cultures were induced by addition of IPTG to a final concentration of 0.25 mM, and additionally 0.5 µM $NiCl_2$ was supplemented. Cells were grown for an additional 24 h, centrifuged, resuspended in 800 µL sonication buffer and sonicated as for *E. coli*. The lysates were cleared as described above for *E. coli* and the supernatant was reserved as the "cleared lysate". *In vitro* hydrogenase evolution assays were conducted as described below.

Evolution activity assay

Cleared lysate (0.2 mL) was added to 1.7 mL of methyl viologen assay buffer containing 1.5 mL deionized water, 0.1 mL 40 mg mL^{-1} methyl viologen (Aldrich), and 0.1 mL of 500 mM potassium phosphate, pH 7.0 (Gibco). The combined samples in 13 mL gastight vials was sealed using rubber septa (Aldrich), and sparged under argon (Westair) for 20 minutes to remove oxygen. After sparging, 0.1 mL of 2 M sodium dithionite (Aldrich) was added by syringe anoxically. The resulting assay solution was a 1:10 dilution of lysate containing 25 mM potassium phosphate, 8 mM methyl viologen, and 100 mM sodium dithionite, pH ~7.0. *E. coli* samples were incubated for 2 hours at 30°C, followed by gas chromatography (CP-3800, Varian) using a Fused Silica Molsieve 5A column (CP7537, Varian) of 100 µL samples taken from the vial headspace. Previous experiments [13,14] and piloting experiments for this experiment demonstrated time-dependent linearity of hydrogen evolution over the course of ~20 hours, suggesting that enzyme quality is unlikely to have been affected by the reaction conditions, including damage by oxidized dithionite products. *S. elongatus* samples were prepared similarly but were incubated at 30°C overnight before GC analysis. Hydrogen peaks were identified and integrated to quantify hydrogenase yield by comparison to a standard curve prepared with pure hydrogen from which a specific activity was calculated based on total protein in the lysate. For plasmids pIY003, pIY004, pIY006, and pIY007, the evolution activity was measured for four biological replicates with three technical replicates for each expression. For plasmids pIY033, pIY038, the evolution activity was measured for three technical replicates, and for pRC41-4, and pCM012, the evolution activity was measured for three biological replicates.

Uptake activity assay

Cleared lysate (0.1 mL) was added to 0.9 mL of benzyl viologen assay buffer containing 0.05 mL of benzyl viologen (65 mg mL^{-1}) and 0.05 mL of 0.5 M potassium phosphate buffer, pH 7.0, and 0.8 mL deionized water. The combined assay sample was prepared in a 2 mL round-rim disposable UV cuvette (Brandtech) and sealed with a rubber septum (Aldrich), coated with a small amount of vacuum grease (Fisher). The cuvette was sparged for 20 minutes under 10% hydrogen balanced by nitrogen (Westair). Cuvettes were then transferred to a multichannel spectrophotometer and monitored at 555 nm at 30°C. For all four variants, four instances of expression were conducted with four replicates for each instance, such that three of these replicate experiments appeared to have self-consistent progress curves without prejudice to the data processing. Reaction rates were identified by fitting a linear regression to the data between OD_{555} = 1 to OD_{555} = 3.

Thermal exposure assay

The assay was conducted as the evolution assay, except cleared lysate was further incubated at 85°C for one hour and centrifuged a second time for 10 minutes at 16,800 x g and 4°C to clear additional insoluble debris prior to addition to MV buffer. Equivalent volumes of heat-treated and control-treated cleared lysate were added to assay buffer, and total hydrogen evolution activities without normalizing to protein concentration were used for subsequent calculations. Tolerance to thermal exposure was expressed by dividing the mean of three technical replicates of the denatured sample by the mean of three technical replicates of the undenatured sample. This process was subjected to three biological replicates; the geometric mean and geometric standard error are reported.

Oxygen exposure assay

The H-D exchange assay to test sensitivity to oxygen exposure was performed as follows: cleared lysate (0.1 mL) was added to 0.8 mL of mass spectrometry buffer containing 0.05 mL of 0.5 M potassium phosphate buffer, pH 7.0 (Gibco), and 0.75 mL deionized water. The samples were placed in 10 mL scintillation vials, and capped with new rubber septa. Vials were sparged for 10 minutes under pressurized argon. A polyimide-coated fused silica capillary, ID 50 μm OD 220 μm (Scientific Instrument Services) adapted for mass spectrometry was then inserted into the vial headspace followed by mass spectrometric monitoring (Omnistar, Pfeiffer). The microcapillary substituted for the standard metal probe supplied with this instrument samples very low quantities of total gas making up an inconsequential amount of total gas volume for the duration of the experiment. Hydrogen (100%, Westair) was injected to a final concentration of 10% hydrogen headspace and verified by mass spectrometry. For microaerobic samples, air was injected to create a final 1% oxygen mixture. Anaerobic samples were incubated for 2h at 30°C to ensure quantitative enzyme recovery, followed by injection of 0.1 mL D_2O, followed by monitoring of the developing m/z = 3, until $\chi_{HD} > 1000$ ppm. Microaerobic samples, prepared in parallel, were then injected with 0.1 mL D_2O followed by overnight mass spectrometric monitoring at m/z = 3. Oxygen tolerance was measured by filtering spikes from the progress curve, calculating enzyme rate and normalizing the slope of the microaerobic activity against the slope of the anaerobic activity. This process was subjected to three biological replicates; the geometric mean and geometric standard errors are reported.

Protein purification

Cleared lysate from *E. coli* strains containing plasmid pIY085 and pIY086 was further purified using Streptactin magnetic bead resin (Qiagen). Cleared lysate was prepared as above but in NP buffer (50 mM sodium phosphate, pH 8.0, 300 mM NaCl) with 1 mM dithiothreitol and NeutrAvidin (Thermo) at 10 μg mL^{-1} final concentration. Streptactin resin (100 μl) was combined with cleared lysate (400 μl) and incubated at 4°C for 1 h with end-over-end rotation. The supernatant was removed and the resin was washed three times for five minutes each at room temperature in NP buffer containing 0.05% sodium dodecyl sulfate (SDS). This was followed by two washes in NP buffer supplemented with 0.01% Tween 20 as above. Final elution was performed at room temperature in 100 μl NP buffer supplemented with 0.01% Tween 20 and 10 mM biotin.

Western blot

Lysates were diluted to 5 mg mL^{-1} total protein. One lysate from an overexpression of WT hydrogenase was further treated by denaturation at 65°C for 10 minutes and clearing by centrifugation, at 20,000 x g for 10 minutes. 25 μL of 4x NuPAGE SDS-PAGE loading buffer (Invitrogen) was added to 75 μL aliquots, followed by boiling for 10 minutes. These samples were frozen at −80°C and on a subsequent day thawed and 10 μL was separated onto a 10% NuPAGE Bis-Tris gel with the NuPAGE MOPS-SDS running buffer system (Invitrogen). Gel electrophoresis was performed at 150 V for 5 hours at 4°C, and the gel was cut separating the gel into two halves at approximately 75 kDa. The top gel was stained using the Sypro Ruby kit (Invitrogen) as a total protein loading control; the bottom gel was transferred onto nitrocellulose using NuPage transfer buffer (Invitrogen) with 30% MeOH for 10 hours at 10 mA constant current. These samples were then hybridized using rabbit-anti-HynL primary antibody [13] and goat-anti-rabbit-DyLight secondary antibody (Thermo), and imaged using a Typhoon fluorescence scanning imager (GE).

For gels associated with protein purification, samples were prepared as described above, but for cleared lysate, approximately 5 µg was loaded per lane and for purified proteins, approximately 0.5 µg protein was loaded per lane. Electrophoresis of the gels for western blot analysis of Strep-tagged HynS was only performed for 1 h at 150 V. Monoclonal antibody to the strep-tag antigen (Qiagen) was used with goat-anti-mouse-DyLight secondary antibody (Thermo) and imaged as above.

Endnotes

[1]In this paper the amino acid sequence numbers used for *A. macleodii* HynS are relative to the translational start methionine as annotated (NCBI: YP_006975190.1), not the putative starting residue after N-terminal processing by the TAT system, as this site has not been experimentally verified.

Additional file

Additional file 1: Table S1. Specific activities of enzyme preparations. **Table S2,** Primers used in this study. **Table S3,** (22pp) Maps and sequences for vectors used in this study. **Figure S1**, A representative enzyme activity progress curve.

Abbreviation

WT: Wild type.

Competing interests

The authors declare that they have no competing interests.

Authors' contributions

ITY conceived and designed the experiments, created plasmid constructs, collected all data pertaining to *E. coli*, and wrote the paper. CWM designed and created plasmid constructs, and helped draft the manuscript. TAN designed and created plasmids, and helped draft the manuscript. HOS designed the experiments, participated in its coordination, and helped draft the manuscript. PDW conceived and designed the experiments, collected all data pertaining to *S. elongatus*, performed enzyme purifications, and wrote the paper. All authors read and approved the final manuscript.

Acknowledgements

We thank Pin-Ching Maness for advice on experimental procedures. We thank Phillip Wright for kindly providing us with the BL21ΔH_4 strain. Funding for all aspects of this work was provided by the U.S. Department of Energy, Fuel Cell Technologies Office (DE-FG36-05GO15027).

References

1. Muellerlanger F, Tzimas E, Kaltschmitt M, Peteves S: **Techno-economic assessment of hydrogen production processes for the hydrogen economy for the short and medium term.** *International Journal of Hydrogen Energy* 2007, **32**:3797–3810.
2. Beer LL, Boyd ES, Peters JW, Posewitz MC: **Engineering algae for biohydrogen and biofuel production.** *Current opinion in biotechnology* 2009, **20**:264–271.
3. Angermayr SA, Hellingwerf KJ, Lindblad P, De Mattos MJT: **Energy biotechnology with cyanobacteria.** *Current opinion in biotechnology* 2009, **20**:257–263.
4. Ghirardi ML, Dubini A, Yu J, Maness P-C: **Photobiological hydrogen-producing systems.** *Chemical Society reviews* 2009, **38**:52–61.
5. Vignais PM, Billoud B: **Occurrence, classification, and biological function of hydrogenases: an overview.** *Chemical reviews* 2007, **107**:4206–4272.
6. Tamagnini P, Leitão E, Oliveira P, Ferreira D, Pinto F, Harris DJ, Heidorn T, Lindblad P: **Cyanobacterial hydrogenases: diversity, regulation and applications.** *FEMS microbiology reviews* 2007, **31**:692–720.
7. Houchins JP, Burris RH: **Occurrence and localization of two distinct hydrogenases in the heterocystous cyanobacterium Anabaena sp. strain 7120.** *J. Bacteriol* 1981, **146**:209–214.
8. Cournac L, Guedeney G, Peltier G, Vignais PM, De BL: **Sustained Photoevolution of Molecular Hydrogen in a Mutant of Synechocystis sp. Strain PCC 6803 Deficient in the Type I NADPH-Dehydrogenase Complex.** *Journal of Bacteriology* 2004, **186**:1737–1746.
9. Ghirardi ML, Posewitz MC, Maness P-C, Dubini A, Yu J, Seibert M: **Hydrogenases and hydrogen photoproduction in oxygenic photosynthetic organisms.** *Annual review of plant biology* 2007, **58**:71–91.
10. Parkin A, Sargent F: **The hows and whys of aerobic H2 metabolism.** *Current opinion in chemical biology* 2012, **16**:26–34.
11. Fritsch J, Lenz O, Friedrich B: **Structure, function and biosynthesis of O(2)-tolerant hydrogenases.** *Nature reviews. Microbiology* 2013, **11**:106–114.
12. McIntosh CL, Germer F, Schulz R, Appel J, Jones AK: **The [NiFe]-hydrogenase of the cyanobacterium Synechocystis sp. PCC 6803 works bidirectionally with a bias to H2 production.** *Journal of the American Chemical Society* 2011, **133**:11308–11319.
13. Maróti G, Tong Y, Yooseph S, Baden-Tillson H, Smith HO, Kovács KL, Frazier M, Venter JC, Xu Q: **Discovery of [NiFe] hydrogenase genes in metagenomic DNA: cloning and heterologous expression in Thiocapsa roseopersicina.** *Applied and environmental microbiology* 2009, **75**:5821–5830.
14. Vargas WA, Weyman PD, Tong Y, Smith HO, Xu Q: **[NiFe] hydrogenase from Alteromonas macleodii with unusual stability in the presence of oxygen and high temperature.** *Applied and environmental microbiology* 2011, **77**:1990–1998.
15. Weyman PD, Vargas WA, Chuang R-Y, Chang Y, Smith HO, Xu Q: **Heterologous expression of Alteromonas macleodii and Thiocapsa roseopersicina [NiFe] hydrogenases in Escherichia coli.** *Microbiology (Reading, England)* 2011, **157**:74–1363.
16. Weyman PD, Vargas WA, Tong Y, Yu J, Maness P-C, Smith HO, Xu Q: **Heterologous expression of Alteromonas macleodii and Thiocapsa roseopersicina [NiFe] hydrogenases in Synechococcus elongatus.** *PloS one* 2011, **6**:e20126.
17. Bock A, King P, Blokesch M, Posewitz M: **Maturation of Hydrogenases.** *Advances in Microbial Physiology* 2006, **51**:1–225.
18. Wells MA, Mercer J, Mott RA, Pereira-Medrano AG, Burja AM, Radianingtyas H, Wright PC: **Engineering a non-native hydrogen production pathway into Escherichia coli via a cyanobacterial [NiFe] hydrogenase.** *Metabolic engineering* 2011, **13**:445–453.
19. Sun J, Hopkins RC, Jenney FE, McTernan PM, Adams MWW: **Heterologous expression and maturation of an NADP-dependent [NiFe]-hydrogenase: a key enzyme in biofuel production.** *PloS one* 2010, **5**:e10526.
20. Kim JYH, Jo BH, Cha HJ: **Production of biohydrogen by heterologous expression of oxygen-tolerant Hydrogenovibrio marinus [NiFe]-hydrogenase in Escherichia coli.** *Journal of biotechnology* 2011, **155**:312–319.
21. Lukey MJ, Parkin A, Roessler MM, Murphy BJ, Harmer J, Palmer T, Sargent F, Armstrong FA: **How Escherichia coli is equipped to oxidize hydrogen under different redox conditions.** *The Journal of biological chemistry* 2010, **285**:3928–3938.
22. Abou Hamdan A, Dementin S, Liebgott P-P, Gutierrez-Sanz O, Richaud P, De Lacey AL, Rousset M, Bertrand P, Cournac L, Léger C: **Understanding and tuning the catalytic bias of hydrogenase.** *Journal of the American Chemical Society* 2012, **134**:8368–8371.
23. Dementin S, Belle V, Bertrand P, Guigliarelli B, Adryanczyk-Perrier G, De Lacey AL, Fernandez VM, Rousset M, Léger C: **Changing the ligation of the**

distal [4Fe4S] cluster in NiFe hydrogenase impairs inter- and intramolecular electron transfers. *Journal of the American Chemical Society* 2006, **128:**5209–5218.

24. Rousset M, Montet Y, Guigliarelli B, Forget N, Asso M, Bertrand P, Fontecilla-Camps JC, Hatchikian EC: **[3Fe-4S] to [4Fe-4S] cluster conversion in Desulfovibrio fructosovorans [NiFe] hydrogenase by site-directed mutagenesis.** *Proceedings of the National Academy of Sciences of the United States of America* 1998, **95:**11625–11630.
25. Fontecilla-Camps JC, Volbeda A, Cavazza C, Nicolet Y: **Structure/function relationships of [NiFe]- and [FeFe]-hydrogenases.** *Chemical reviews* 2007, **107:**4273–4303.
26. Volbeda A, Charon MH, Piras C, Hatchikian EC, Frey M, Fontecilla-Camps JC: **Crystal structure of the nickel-iron hydrogenase from Desulfovibrio gigas.** *Nature* 1995, **373:**580–587.
27. Higuchi Y, Yagi T, Yasuoka N: **Unusual ligand structure in Ni-Fe active center and an additional Mg site in hydrogenase revealed by high resolution X-ray structure analysis.** *Structure* 1997, **5:**1671–1680.
28. Kelley LA, Sternberg MJE: **Protein structure prediction on the Web: a case study using the Phyre server.** *Nature protocols* 2009, **4:**363–371.
29. Parkin A, Goldet G, Cavazza C, Fontecilla-Camps JC, Armstrong FA: **The difference a Se makes? Oxygen-tolerant hydrogen production by the [NiFeSe]-hydrogenase from Desulfomicrobium baculatum.** *Journal of the American Chemical Society* 2008, **130:**13410–13416.
30. Garcin E, Vernede X, Hatchikian E, Volbeda A, Frey M, Fontecilla-Camps J: **The crystal structure of a reduced [NiFeSe] hydrogenase provides an image of the activated catalytic center.** *Structure* 1999, **7:**557–566.
31. Bingemann R, Klein A: **Conversion of the central [4Fe-4S] cluster into a [3Fe-4S] cluster leads to reduced hydrogen-uptake activity of the F420-reducing hydrogenase of Methanococcus voltae.** *European Journal of Biochemistry* 2000, **267:**6612–6618.
32. Volbeda A, Darnault C, Parkin A, Sargent F, Armstrong FA, Fontecilla-Camps JC: **Crystal Structure of the O(2)-Tolerant Membrane-Bound Hydrogenase 1 from Escherichia coli in Complex with Its Cognate Cytochrome b.** *Structure* 2013, **21:**184–190.
33. Huang H-H, Camsund D, Lindblad P, Heidorn T: **Design and characterization of molecular tools for a Synthetic Biology approach towards developing cyanobacterial biotechnology.** *Nucleic acids research* 2010, **38:**2577–2593.
34. Bustos SA, Golden SS: **Light-regulated expression of the psbD gene family in Synechococcus sp. strain PCC 7942: evidence for the role of duplicated psbD genes in cyanobacteria.** *Molecular & general genetics* 1992, **232:**221–230.
35. Theodoratou E, Paschos A, Magalon A, Fritsche E, Huber R, Böck A: **Nickel serves as a substrate recognition motif for the endopeptidase involved in hydrogenase maturation.** *European Journal of Biochemistry* 2000, **267:**1995–1999.
36. Dementin S, Burlat B, Fourmond V, Leroux F, Liebgott P-P, Abou Hamdan A, Léger C, Rousset M, Guigliarelli B, Bertrand P: **Rates of intra- and intermolecular electron transfers in hydrogenase deduced from steady-state activity measurements.** *Journal of the American Chemical Society* 2011, **133:**10211–10221.
37. Goris T, Wait AF, Saggu M, Fritsch J, Heidary N, Stein M, Zebger I, Lendzian F, Armstrong FA, Friedrich B: **others: A unique iron-sulfur cluster is crucial for oxygen tolerance of a [NiFe]-hydrogenase.** *Nature Chemical Biology* 2011, **1:**1–9.
38. Pandelia M-E, Nitschke W, Infossi P, Giudici-Orticoni M-T, Bill E, Lubitz W: **Characterization of a unique [FeS] cluster in the electron transfer chain of the oxygen tolerant [NiFe] hydrogenase from Aquifex aeolicus.** *Proceedings of the National Academy of Sciences of the United States of America* 2011, **108:**6097–6102.
39. Shomura Y, Yoon K-S, Nishihara H, Higuchi Y: **Structural basis for a [4Fe-3S] cluster in the oxygen-tolerant membrane-bound [NiFe]-hydrogenase.** *Nature* 2011, **479:**253–256.
40. Fritsch J, Scheerer P, Frielingsdorf S, Kroschinsky S, Friedrich B, Lenz O, Spahn CMT: **The crystal structure of an oxygen-tolerant hydrogenase uncovers a novel iron-sulphur centre.** *Nature* 2011, **479:**249–252.
41. Gibson DG, Young L, Chuang R-Y, Venter JC, Hutchison CA, Smith HO: **Enzymatic assembly of DNA molecules up to several hundred kilobases.** *Nature methods* 2009, **6:**343–345.
42. Redwood MD, Mikheenko IP, Sargent F, Macaskie LE: **Dissecting the roles of Escherichia coli hydrogenases in biohydrogen production.** *FEMS microbiology letters* 2008, **278:**48–55.
43. Studier FW: **Protein production by auto-induction in high density shaking cultures.** *Protein expression and purification* 2005, **41:**207–234.
44. Elhai J, Wolk CP: **Conjugal transfer of DNA to cyanobacteria.** *Methods in enzymology* 1988, **167:**747–754.

3

Polylactic acid composites incorporating casein functionalized cellulose nanowhiskers

Jin Gu[1] and Jeffrey M Catchmark[1,2*]

Abstract

Background: Polylactic acid (PLA) is considered to be a sustainable alternative to petroleum-based polymers for many applications. Using cellulose fiber to reinforce PLA is of great interest recently due to its complete biodegradability and potential improvement of the mechanical performance. However, the dispersion of hydrophilic cellulose fibers in the hydrophobic polymer matrix is usually poor without using hazardous surfactants. The goal of this study was to develop homogenously dispersed cellulose nanowhisker (CNW) reinforced PLA composites using whole milk casein protein, which is an environmentally compatible dispersant.

Results: In this study, whole milk casein was chosen as a dispersant in the PLA-CNW system because of its potential to interact with the PLA matrix and cellulose. The affinity of casein to PLA was studied by surface plasmon resonance (SPR) imaging. CNWs were functionalized with casein and used as reinforcements to make PLA composites. Fluorescent staining of CNWs in the PLA matrix was implemented as a novel and simple way to analyze the dispersion of the reinforcements. The dispersion of CNWs in PLA was improved when casein was present. The mechanical properties of the composites were studied experimentally. Compared to pure PLA, the PLA composites had higher Young's modulus. Casein (CS) functionalized CNW reinforced PLA (PLA-CS-CNW) at 2 wt% filler content maintained higher strain at break compared to normal CNW reinforced PLA (PLA-CNW). The Young's modulus of PLA-CS-CNW composites was also higher than that of PLA-CNW composites at higher filler content. However, all composites exhibited lower strain at break and tensile strength at high filler content.

Conclusions: The presence of whole milk casein improved the dispersion of CNWs in the PLA matrix. The improved dispersion of CNWs provided higher modulus of the PLA composites at higher reinforcement loading and maintained the strain and stress at break of the composites at relatively low reinforcement loading. The affinity of the dispersant to PLA is important for the ultimate strength and stiffness of the composites.

Keywords: Cellulose nanowhiskers, Whole milk casein, Polylactic acid, Surface plasmon resonance, Fluorescence microscopy

Introduction

The development of biodegradable polymer composites, especially composites combining biodegradable matrices and structural biodegradable reinforcements, is of great importance currently. Among these polymer matrices, polylactic acid (PLA) is being used widely in automobile, food packaging and pharmaceutical industries due to its high mechanical performance [1]. It is a commercially available polyester polymerized from lactic acid, which is produced from biorenewable sugar based sources, such as corn starch and sugarcane.

Nature fiber such as lignocellulose or purified cellulose derived from a variety of nature feedstocks (e.g. woods, cotton, and bacterial cellulose) has great potential to replace synthetic reinforcements such as glass fiber or fiber produced using petroleum derived polymers used in composite materials. To this end, various forms of cellulose including microcrystalline cellulose (MCC) or nanodimensional cellulose such as cellulose nanowhiskers (CNWs) and cellulose nanofibrils (CNFs) have been explored as reinforcements for different composite materials [2]. These celluloses are typically derived from controlled acid hydrolysis or mechanical homogenization of

* Correspondence: jcatchmark@engr.psu.edu
[1]Intercollege Graduate Degree Program in Plant Biology, The Pennsylvania State University, University Park, PA 16802, USA
[2]Department of Agricultural and Biological Engineering, The Pennsylvania State University, 109 Ag. Eng. Bldg., University Park, PA 16802, USA

plant materials [3]. During the acid hydrolysis, the amorphous regions of the CNWs were preferentially removed. They are of particular interest because of their desirable mechanical properties (Young's modulus of crystal cellulose is in the range of 100–160 GPa [4]) and bio-availability.

Natural fiber additives are challenging to incorporate into non-polar matrix materials due to their polar surface chemistries. This situation results in poor dispersion of the fiber in the matrix and poor interfacial adhesion limiting improvements in mechanical properties. This has been found to be true for PLA-nanocellulose composites [1,5].

To address these issues, different processing techniques have been implemented including solvent casting [5], premixing the materials before processing [1,6], addition of surfactant/plasticizer to improve the dispersion [7-10] and chemical modification of the nanofibers [11,12]. Generally speaking, solvent casting of cellulose nanocomposites results in more homogeneous materials than other industrially used processing techniques such as melt compounding [2]. However, the above processing techniques have used hazardous chemicals or could not achieve good dispersion. Such approaches would retract significantly from the ecological benefits of using PLA/cellulose composites.

Low cost agricultural products such as proteins from soy bean, corn and milk have been used to make plastic since the first half of the last century [13]. The milk proteins, named caseins, consist of four protein subunits (α_{s1}, α_{s2}, β and κ). Caseins are amphiphilic containing both hydrophobic and hydrophilic domains. They are distinct proteins which are able to disperse, solubilize and compatibilize the insoluble components (colloidal calcium phosphate, CCP) into an incompatible solution (water) by a soluble micelle [14-16]. Casein was extensively used as a wood adhesive before 1930's. Its affinity to cellulose has been proposed to be associated with the interaction between the hydroxyl groups of cellulose and the peptide bonds of casein [17,18]. The proposed binding mechanism also involves carbohydrate-π interactions [19,20].

Inspired by the unique amphiphilic nature of casein proteins, this work aimed to explore the possibility of using ecologically compatible caseins as dispersants of cellulose into PLA. Surface plasmon resonance (SPR) imaging was used to characterize the interaction of PLA and the whole milk casein. The affinity of PLA to casein coated surfaces was studied in order to demonstrate the functionality of casein as a dispersant in the PLA matrix. Subsequently, casein functionalized CNWs-PLA composites were prepared using solvent casting method. The dispersion of CNWs in PLA was studied by staining the CNWs with Calcofluor White, which is a fluorescence dye known to bind cellulose, in a simple and effective manner compared to previous characterization methods such as electron microscopy [8]. The effect of fillers on the mechanical properties and crystallinity of the composites was evaluated by tensile testing and X-ray diffraction, respectively.

Results and discussion

Affinity of PLA to whole milk casein

Figure 1a depicts a typical SPR result for casein adsorption onto gold surfaces and PLA adsorption onto casein coated surfaces. Arrows on the figure indicate the times when the solutions were changed. This result is repeatable [21]. When the casein solution was flowed across the surface, an increase in SPR intensity was observed due to the adsorption of casein molecules onto the gold surface. Flow of SPR buffer on the casein coated surface caused a decrease of SPR intensity which indicated some reversibly bound casein left the surface. The coverage of casein reached equilibrium after the surface was exposed to the running buffer for about 1 minute.

The mechanism of casein adsorption onto gold surfaces has been studied previously. Liu et al. [22] proposed that the carboxylate or amine groups of the caseins exhibited an affinity to gold particles. In previous works [23-25], it was demonstrated that casein was able to form a bilayer on hydrophilic surfaces. Caseins behave like copolymers consisting of blocks containing hydrophobic and hydrophilic domains. The hydrophilic domains of caseins interacted with the hydrophilic gold surface causing the hydrophobic domains to orient away from the surface. The hydrophobic domains were then exposed to the solution allowing the adsorption of a second layer of casein through hydrophobic interactions (Figure 2). Using quartz crystal microbalance measurements, Ozeki et al. [24] stated that the first layer of casein on the hydrophilic surface was tightly bound and the second layer was reversibly bound. The decrease of SPR intensity when SPR buffer was flowed across the casein coated surface could be explained by desorption of the reversibly bound casein layer.

To study the interaction between immobilized casein and PLA, a PLA solution was flowed across the casein coated surface. As indicated in Figure 1a, the SPR intensity was increased by 70 units (arbitrary intensity units) in about two hundreds seconds. The initial decrease of SPR intensity was due to a change in refractive index of the PLA solution. After adsorption of PLA to the casein, the SPR intensity began to increase. The curve became flat after a steep increase indicating saturation of the binding. The adsorption of PLA to casein reached equilibrium after about three hundred seconds. Finally, the SPR buffer was again flowed across the surface. The increase of SPR intensity further confirmed that the refractive index of the buffer was higher than that of the PLA solution. To demonstrate that PLA did not interact with the gold surface and the overall increase of SPR intensity in Figure 1a was caused by casein-PLA interaction, a control experiment

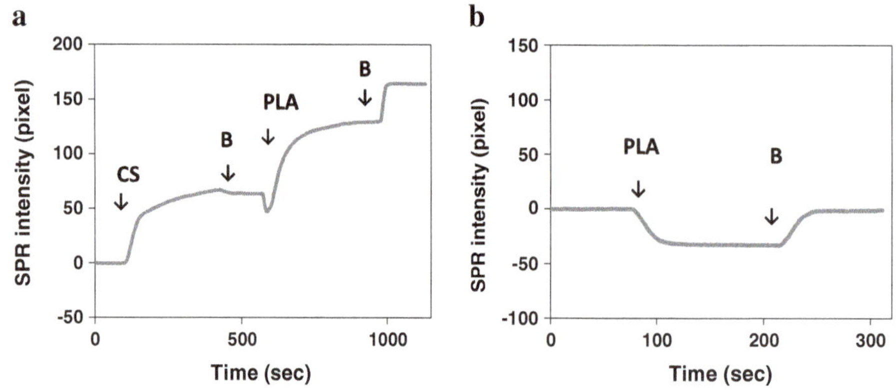

Figure 1 SPR results at a fixed angle. (a) The change of SPR intensity when the casein solution was flowed across the gold surface, and followed by 1 mM PLA solution. **(b)** Control experiment showing little change of final SPR intensity when PLA was flowed across the gold surface. The gold film was first exposed to buffer **(B)**, then casein (CS) or PLA solution. Arrows indicate the times when the solutions were changed.

was performed. A PLA solution was flowed across the gold surface without a casein coating. Almost no net change of SPR intensity was observed when SPR buffer was flowed after exposure of the PLA solution to the gold surface (Figure 1b). The change of intensity from PLA to buffer in Figure 1b was also comparable to that in Figure 1a.

PLA is a non-polar and hydrophobic polymer. A hydrophobic interaction between PLA and casein occurred when PLA solution was flowed across this surface. Figure 2 shows a schematic illustration of the SPR imaging experiments. Casein was able to bind to gold through its hydrophilic domains. A reversibly bound casein layer was deposited on the first casein layer through hydrophobic interactions. After flowing in the casein-free buffer, some casein in the second layer left the surface and the hydrophobic domain of the casein was exposed. PLA was then able to interact with the hydrophobic domain of casein.

To study the strength of interaction between casein and PLA, PLA solutions with different concentrations (from 0.01-0.5 mM) were flowed across casein coated surfaces in sequence. The SPR intensity was recorded at equilibrium and converted to a change in relative reflectivity value as shown in Figure 3 (See Equation 1 in Materials and methods). The trend reached a saturation point at 0.5 mM PLA. The data was fitted to a Langmuir isotherm. The adsorption coefficient and the maximum change of relative reflectivity were calculated by Equation 2. The adsorption coefficient for PLA binding to a casein coated surface was determined to be $5.40 \pm 0.51 \times 10^4$ M^{-1}. The maximum change of relative reflectivity value was $9.64 \pm 0.20\%$. The R^2 of the fitting was as high as 0.98 which indicated Langmuir isotherm could best describe the adsorption behavior. This also suggested that PLA only formed a monolayer on the casein coated surface. The adsorption coefficient indicated a weak interaction between PLA and casein (compared to other protein carbohydrate interaction [26], where K is around 10^6-10^7 M^{-1}).

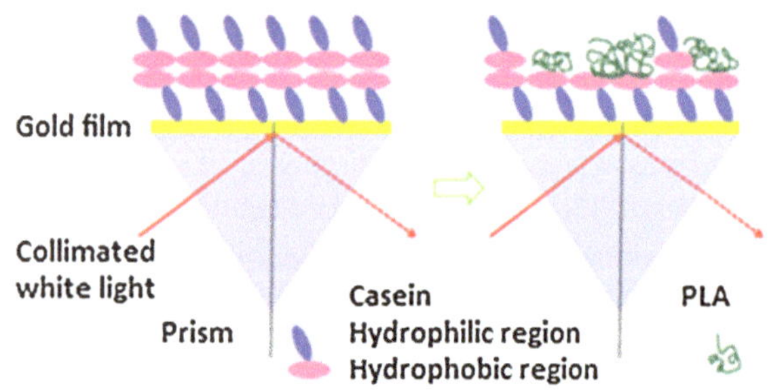

Figure 2 Schematic illustration of the SPR experiment showing deposition of casein onto the gold surface (left) and adsorption of PLA onto the casein coated surface (right).

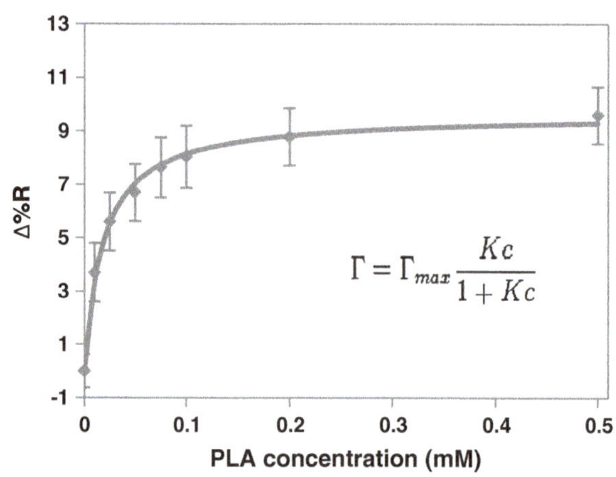

Figure 3 Change in relative reflectivity value when 0.01, 0.025, 0.05, 0.075, 0.1, 0.2, 0.25, 0.5 mM PLA solutions were flowed across the casein coated surface in sequence. Data is presented as mean standard deviation of five determinations and fit to the Langmuir isotherm (solid line).

Distribution of fillers in composites

CNWs were highly crystalline nanofibers obtained by the acid hydrolysis of cotton cellulose. Stable colloidal suspensions were obtained because sulfuric acid treated cellulose carried negative charged sulfate groups [3]. Most CNWs were found to have lengths of approximately 200 nm to 300 nm with diameters of 15–30 nm (Figure 4, also determined previously [27]). Larger particles measuring several micrometers long which were not completely hydrolyzed were also observed in small amounts. The crystallinity of CNWs was as high as approximately 90% as studied previously [28]. Both freeze-dried CNWs and casein (CS) functionalized CNWs (CS-CNWs) were able to suspend in dimethylformamide (DMF) solution. DMF was selected as the solvent in this study due to its low toxicity (compared to solvents used in other studies, such as chloroform), ability to suspend CNWs, and dissolve PLA. Samir et al. [29] pointed out that the high value of the dielectric constant of DMF and its wettability on CNWs resulted in stable CNW suspensions. This method increased the dispersion of CNWs in the PLA matrix as compared to other solvents such as acetone or chloroform.

Photographs of PLA films with different reinforcement concentrations are presented in Figure 5a. The pure PLA control and PLA with 2% CNW (CNW2) or with 2% CS-CNW (CS-CNW2) were transparent. PLA films with 5% or 10% CNW (CNW5 and CNW10) exhibited some white spots which may be the CNWs aggregates. The transparency of the PLA films containing CS-CNW was decreased with 5% CS-CNW (CS-CNW5), and even more with 10% CS-CNW(CS-CNW10) filler content, but few white spots were observed.

In order to further understand the distributions of fillers in the PLA composites, the optical microscope images of PLA, PLA-CNW2 and PLA-CS-CNW2 were obtained as shown in Figure 5b,c,d. When the DMF solvent evaporated, single PLA crystals were formed which subsequently arranged themselves into large, sphere-like structures, called spherulites. Spherulites exhibiting diameters of approximately 50 microns formed in pure PLA films. In PLA-CNW2, large spherulites with similar size to those observed in pure PLA films were found together with many additional smaller spherulites. Medium size spherulites with 20–30 μm diameters were observed in PLA-CS-CNW2 composites. Since nucleation of PLA on the reinforcement surface was observed previously in composites [30], the change of spherulite size in the PLA-CS-CNW2 composite could possibly result from the presence of the fillers. The tiny spherulites in PLA-CNW2 may grow on the surface of CNW aggregates. In PLA-CS-CNW2, since PLA had higher affinity to the casein-coated fillers, the CS-CNWs may be better nucleating agents. Also, the distribution of filler was more uniform in the PLA-CS-CNW2, resulting in a larger number of smaller spherulites.

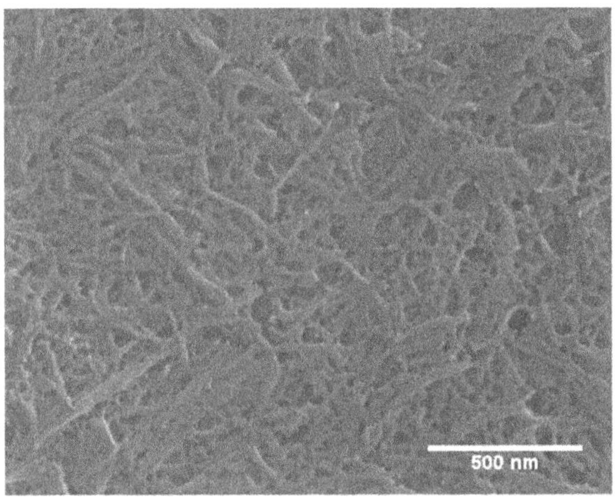

Figure 4 SEM image of CNWs prepared with sulfuric acid.

To further understand the distribution of the filler, the composite films with Calcofluor staining (see Materials and methods section) were investigated as shown in Figure 5e, f. Agglomerates of CNWs were discernible by their higher local fluorescence intensity. PLA-CNW2 composite showed randomly distributed bright aggregation spots. The aggregations of CNWs are highlighted with circles in Figure 5e. PLA-CS-CNW2 composite, however, showed evenly distribution of spots (Figure 5f). It suggests that the CS-CNWs were better dispersed in the PLA matrix.

The sodium hydroxide in the stain dissolved the PLA polymer allowing the surface of CNWs to adsorb Calcofluor and became fluorescent. In alkaline condition, the crystallinity of the PLA was first increased and then decreased [31]. The hydrolytic degradation was a two-stage process. It indicated that the amorphous part of the PLA was hydrolyzed first. During our short experimental time, no significant erosion of the composite was observed. This is a simple method to observe the dispersion of CNWs in PLA matrix compared to other methods, such as electron microscopy and atomic force microscopy [8]. Prior work successfully developed a method for fluorescent labeling of CNWs [32], but the chemical modification of CNWs may have an impact on the filler-matrix or filler-filler interaction.

Mechanical properties of PLA composites

Figure 6a,b shows typical tensile curves for pure PLA and PLA composites. Data extracted from these plots are presented in Figure 6c,d,e and Table 1 with mean and standard deviations. Values of the Young's modulus which are different based on statistical analysis ($p < 0.05$) are identified (small letters) in Figure 6c. It is clearly showed that the addition of CNWs had a positive effect on the Young's modulus of the composites. The effect of reinforcement was achieved even with only 2 wt%

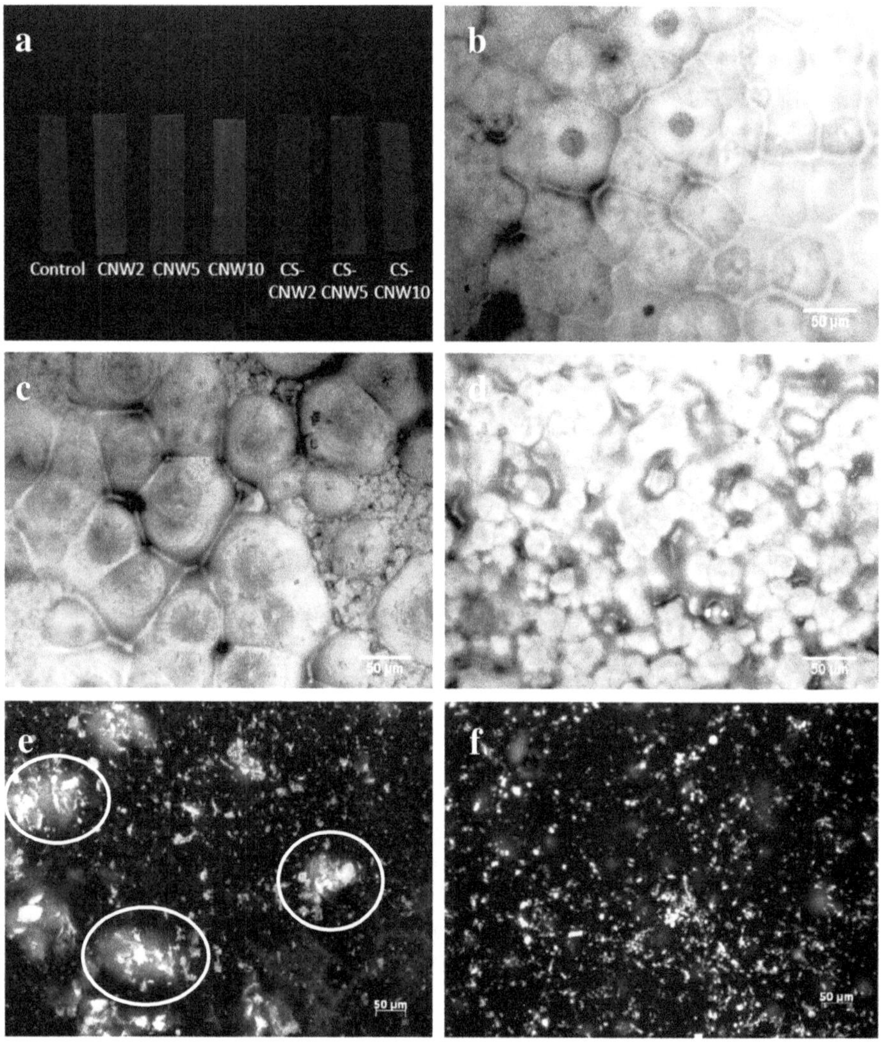

Figure 5 Visual comparison of PLA and its composites. (a) Image showing the appearance of different samples. **(b-d)** Optical micrographs of the materials: **(b)** PLA; **(c)** PLA-CNW2; and **(d)** PLA-CS-CNW2. **(e-f)** Fluorescent images of the composites: **(e)** PLA-CNW2; **(f)** PLA-CS-CNW2. The aggregations of CNWs are circled in **(e)**.

CNWs. The Young's modulus of PLA composites was increased from 2.5 GPa to roughly 3.0 GPa at 2 wt% filler loading. Compared to PLA-CNW composites, PLA-CS-CNW composites showed higher Young's modulus with increasing filler contents. The Young's modulus was 3.5 GPa for PLA-CS-CNW10 compared to 2.9 GPa for PLA-CNW10. The standard deviations were higher for PLA-CNW composites which may result from the inhomogeneous distribution of the CNW filler in the matrix. As shown in Figure 5, the distribution of fillers was better in PLA-CS-CNW composites, which may result in improved reinforcement. Moreover, the improved modulus may also result from changes in spherulite formation. The CS-CNW composite exhibited more numerous, smaller spherulites.

The strain at break of the composites was generally lower than that of pure PLA with one exception which was PLA-CS-CNW2. It is a common observation that the addition of fibers to the polymer reduces the elongation at break in thermoplastic composites [1,33]. At low CS-CNW content, the better adhesion of CS-CNW to the matrix may reduce the voids at the fiber matrix interface and thus the matrix could elongate as pure PLA. Also, with decreased spherulite size in PLA-CS-CNW2, the strain at failure, tensile strength and toughness could increase as shown in other polymer [34]. At higher CS-CNW content, the strain at break decreased. The decrease was more significant when compared to the PLA-CNW composites, suggesting that the casein was effective at improving the bonding of the CNWs to PLA. Since the filler only contained 0.1% casein (data shown in Materials and methods), the CNW surface may not be fully covered by the casein, possibly suggesting that improved coating of the casein on the CNWs or

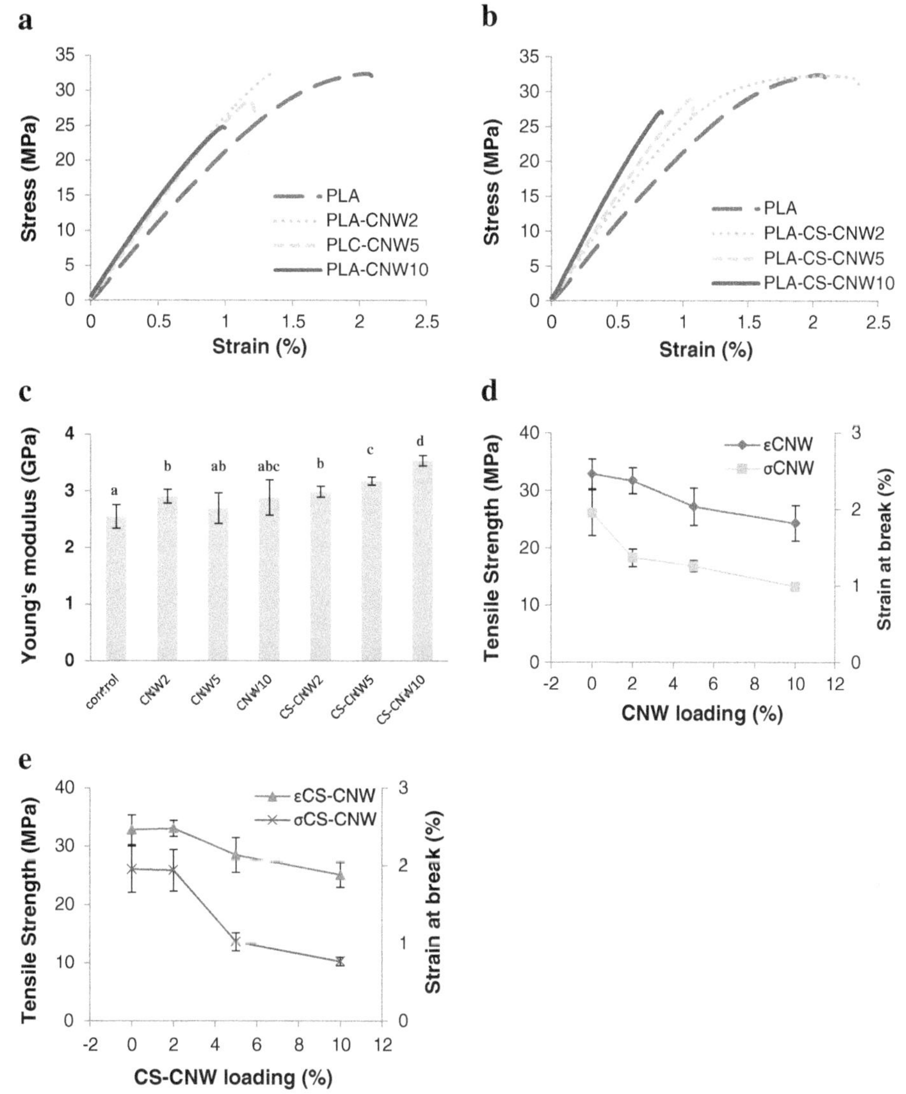

Figure 6 Mechanical properties of PLA and its composites. (a-b) Typical strain–stress curves of PLA and PLA composites: **(a)** PLA-CNW composites and **(b)** PLA-CS-CNW composites; **(c)** Young's modulus of all samples. T-test results are displayed as letters on the top of columns. A column with the same letter represents no significant difference ($p > 0.05$). **(d)** Tensile strength and strain at break of PLA-CNW composites; **(e)** Tensile strength and strain at break of PLA-CS-CNW composites. ε represents stress. σ represents strain.

Table 1 Mechanical properties and crystallinity of PLA and PLA composites

Samples	Young's modulus (GPa)	Tensile strength (MPa)	Strain at break (%)	Crystallinity (%) (XRD)
PLA	2.5 ± 0.2	32.8 ± 2.6	2.0 ± 0.3	47.8 ± 4.1
PLA-CNW2	2.9 ± 0.1	31.6 ± 2.3	1.4 ± 0.1	48.6 ± 0.5
PLA-CNW5	2.7 ± 0.3	27.1 ± 3.2	1.3 ± 0.1	43.2 ± 0.7
PLA-CNW10	2.9 ± 0.3	24.3 ± 3.0	1.0 ± 0.1	37.1 ± 1.8
PLA-CS-CNW2	3.0 ± 0.1	33.1 ± 1.4	1.9 ± 0.3	41.7 ± 1.3
PLA-CS-CNW5	3.2 ± 0.1	28.4 ± 3.0	1.0 ± 0.1	44.8 ± 4.4
PLA-CS-CNW10	3.5 ± 0.1	25.1 ± 2.1	0.8 ± 0.1	40.1 ± 1.3

other fillers may further improve performance. As the distribution of CS-CNW was better than CNW, more interface area between filler and matrix may be expected and thus the fracture initiating points may increase and strongly influence the mechanical properties of the composites. Another possibility is that a small amount of free casein protein was present in the matrix causing brittleness of the composites, since protein itself can be brittle [35]. This is unlikely in the current study, however, since the total amount of casein in the composite did not exceed 0.01% w/w (CS:PLA) in the case of highest CNW loading (10%).

The tensile strength was comparable for pure PLA and PLA composites with 2% filler loading. At higher filler loading, the tensile strength was decreased for both fillers types. It decreased even more for PLA-CNW composites. The decrease of strength in the composites is likely due to poor stress transfer across the interface [33].

Effect of fillers on the crystallinity of polylactic acid

Figure 7 shows the X-ray diffraction pattern of PLA and PLA composites. Four crystalline PLA peaks at 14.8, 16.4, 19.0 and 22.3° 2θ are obvious. A broad peak at around 17° 2θ showed the amorphous portion of the PLA. In PLA composites, a cellulose peak at around 22.7° 2θ could be found and its intensity increased with increase of CNW fillers. The intensity of the peak at 16.4 ° 2θ was decreased for all the composites compared to PLA which suggested a decrease of crystallinity in the composites. The crystallinity of PLA and its composites is summarized in Table 1. Generally, the crystallinity of PLA composites was lower than that of pure PLA (except PLA-CNW2, which was statistically not different from the control, $p > 0.05$), but even the highest decrease was only about 10%. Although it is shown that the incorporation of a CNW filler impacts the crystallization of PLA, both specific changes in crystal formation and how these changes impact mechanical performance of CNW-PLA composites require further study.

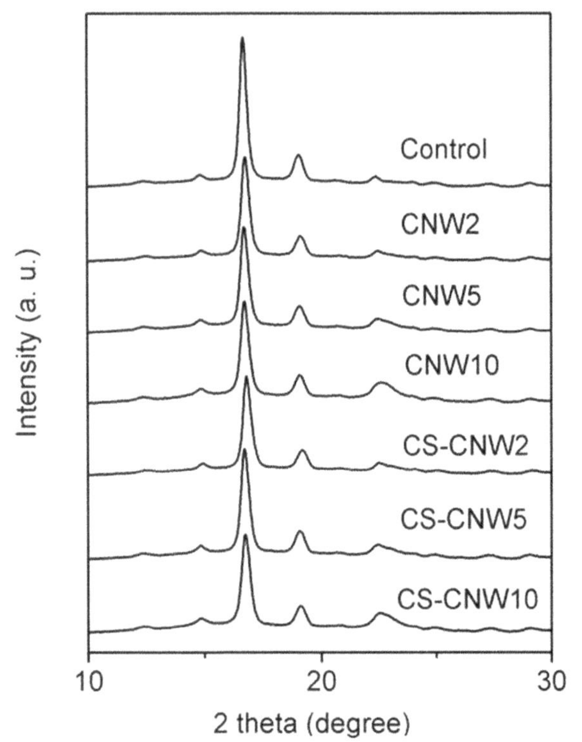

Figure 7 X-ray diffraction patterns of PLA and its composites.

Conclusions

In this study, amphiphilic whole milk casein was explored as a dispersant in PLA-CNW composites. The SPR study demonstrates the binding of casein to PLA. The binding affinity was 5.40×10^4 M^{-1}. The pure PLA and PLA composite samples were prepared using a solvent casting method. The dispersion of CNWs in PLA was studied by Calcofluor staining. The results show the dispersion of CS-CNWs in PLA matrix was better than unmodified CNWs. The evaluation of the mechanical properties of pure PLA and its composites shows the reinforcement effect of CNWs. The Young's modulus of PLA composites with 10% CS-CNW was 40% higher than that of the pure PLA. CS-CNW composites with 2% filler loading maintained high strain and stress at break. However, the tensile strength and strain at break were decreased with 5% or 10% CS-CNW loading. Most PLA composites exhibited lower crystallinity than pure PLA. Smaller and more numerous spherulites were also observed in PLA-CS-CNW composites demonstrating the improved CNW dispersion. Based on the results of this study, casein may be applied as a dispersant and binder in different composite systems such as those involving clays and minerals, as well as fibers, due to the binding ability of casein to such materials.

Materials and methods

Materials

PLA oligomer (OLYGOs Bioresin 120, MW: 2500-3500 Da, NatureWorks™ LLC) was used for SPR studies. High molecular weight PLA (NatureWorks™ 4032D, MW: 195,000-205,000 Da) was used to make the composites. Whole milk casein proteins (from bovine milk, Sigma) were used as CNW dispersants in the composites. Dimethylsulfoxide (DMSO, Burdick & Jackson), dimethylformamide (DMF, Macron Chemical), sulfuric acid (Mallinckrodt Baker) and sodium hydroxide (EMD chemicals Inc) were all A.C.S. grade and used as received. The SPR chips (18 mm × 18 mm × 1 mm with 10 angstroms of Cr followed by 500 angstroms of Au on standard float glass) were from EMF Corporation. Cellulose from cotton (Whatman™ CF11) was used to prepare CNWs.

Calcofluor White Stain (Sigma) was used to stain CNWs. It consisted of Calcofluor White M2R (1 g/L) and Evans Blue (0.5 g/L).

SPR imaging

The SPR chips were soaked in Piranha solution (sulfuric acid: 30% hydrogen peroxide = 3:1, v/v) for 10–15 minutes to remove any surface contamination before use. Caseins were dissolved in the SPR running buffer (PBS: DMSO = 4:1, v/v, pH = 7.3) until saturation (~0.1 mg/mL). 1 mM PLA solution was prepared in buffer and diluted to 0.01, 0.025, 0.05, 0.075, 0.1, 0.2, 0.5 mM to study the binding of PLA to a casein coated SPR chip.

The SPR imaging system (GWC Technologies, SPRimager®) was used to detect the binding between PLA and casein. The SPR buffer was first flowed across the surface followed by the casein solution. Subsequently, the surface was exposed to the SPR buffer, PLA solution, and then the SPR buffer to study the affinity of PLA to casein.

SPR data analysis

An SPR data analysis software (V++, Digital Optics) was used to collect all SPR images and process the data. The software was able to average the signal intensity from selected regions of interest on the chip. The percent reflectivity values were subsequently obtained as follows (from GWC Technologies Technical Note 125):

$$\Delta\%R = \frac{100 \times \text{SPR value} \times 0.85 \times (1\text{-d})}{\text{s-pol pixel intensity}} \qquad (1)$$

SPR value was obtained under p-polarized light. In order to convert the SPR value to the reflectivity, the intensity for s-polarized light (s-pol pixel intensity) was obtained for regions of interest. d is the density of the filter used when obtaining s-pol pixel intensity (in this system, d = 0.75). A correction factor (0.85) adjusts for the differential reflectivity of p-polarized and s-polarized light as it traverses the prism and reflects off the gold. Adsorption coefficient for the binding of PLA to casein surfaces was determined by Langmuir isotherm:

$$\Gamma = \Gamma \max \frac{Kc}{1 + Kc} \qquad (2)$$

where K = Langmuir equilibrium constant or adsorption coefficient, c = PLA concentration, Γ = surface coverage of adsorbed molecules and Γ_{max} = maximum surface coverage as c increases. Change of reflectivity value (Δ% R) reflects the change of surface coverage of adsorbed molecules (Γ). The curve fitting was performed with Origin 8.0 (Origin Lab Corporation) using a hyperbola function.

CNW production

The method by Bondeson et al. [36] was used to prepare CNWs. Generally, CF11 cellulose of 55 g was hydrolyzed in 500 mL 63.5% w/w sulfuric acid at 45°C for 90 min. The reaction was stopped by putting the container on ice until the temperature dropped to 15°C. The excess sulfuric acid was removed by centrifugation (10 min, 16,000 g). The supernatant was removed and replaced by DI water. Centrifugation and washing steps were repeated until the pH of the solution reached 3. The pH of the suspension was then increased to 7 using 0.1 M sodium hydroxide. Finally, the CNW suspension was dialyzed against deionized (DI) water and sonicated to help dispersion of the CNWs (10 min, Branson Model 5510, Danbury). The suspension formed two layers after sedimentation over a few days. The top layer in which CNWs were suspended was used for this study.

Field emission scanning electron microscopy

CNWs were air dried and sputter coated with a thin gold film (approximately 5 nm) to avoid charging. A LEO 1530 field emission scanning electron microscope (Leo Co., Oberkochen, Germany) was operated at 5 kV to study the morphology of the CNWs.

Adsorption of casein to CNWs

The amount of adsorption of casein to CNWs was studied using the Bradford method [37]. A standard protein concentration curve was generated by plotting absorbance at 595 nm versus known bovine serum albumin protein solution concentration (UV-2250 UV–VIS spectrophotometer, Shimadzu Corp., Japan). Whole milk casein was dissolved in sodium hydroxide (pH = 10) to make a 1 mg/mL solution. Approximately 50 µg/mL and 100 µg/mL casein solution was obtained by diluting a 1 mg/mL solution. It was shown that the suspension of casein in alkaline solutions can disrupt the micelle [38]. 100 mg freeze dried CNWs were mixed with 4 mL casein solutions overnight with mechanical stirring. Subsequently, the CNWs suspension was centrifuged at 16,000 g for 10 min. The unbound casein was left in the supernatant and the protein concentration of supernatant was determined. The difference of protein concentration before and after mixing to CNWs was converted to the amount of casein adsorbed to CNWs. For 50 µg/mL casein solution, the adsorption of casein to CNWs was 0.92 ± 0.07 $\mu g_{casein}/mg_{CNW}$. For 100 µg/mL casein solution, the adsorption was 1.20 ± 0.13 $\mu g_{casein}/mg_{CNW}$. Casein solution with higher concentration was also used and the amount of adsorption did not increase significantly. The adsorption of casein to CNWs should approach the saturation point at roughly 0.1% w/w casein/CNW.

To make casein functionalized CNWs (CS-CNW), freeze dried CNWs were mixed with casein solution (pH = 10) overnight with mechanical stirring. The concentration of whole milk casein was selected to ensure saturation of the binding sites on CNWs as shown above. The CNWs suspension was centrifuged at 16,000 g for 10 min and the supernatant was replaced with DI water three times to remove any unbound casein.

Preparation of PLA-CNW composites

Freeze-dried CNWs and CS-CNWs were re-suspended in DMF by mechanical stirring and sonication. The suspensions were stable and no precipitation was observed over three days. High molecular weight PLA was dissolved in the suspension at 100°C forming a 5 wt% solution. Pure PLA samples were also dissolved in DMF. Composites with 2%, 5% and 10% CNW or CS-CNW were casted on a flat glass plate, allowing the DMF to evaporate at 100°C overnight and then dry in a vacuum oven for 24 h at 50°C. The resulting samples were used for all the studies below. Typical film thicknesses were approximately 50 μm. The samples were stored in vacuum oven at room temperature until testing.

Fluorescence microscopy

The composite samples were characterized using an optical microscope (Zeiss Axio imager A1m). The samples were examined on a clean glass slide. For fluorescence image, a drop of Calcofluor White Stain and a drop of 10% sodium hydroxide were added to the sample and a cover slide was placed over the specimen. The sample was undisturbed for 20 min and then examined under UV light. A 350 nm bandpass optical filter was used to filter only the Calcoflour fluorescence for imaging of the cellulose. The sodium hydroxide was used to slowly dissolve the PLA to expose the cellulose for staining.

Tensile testing

A dynamic mechanical analyzer (Model Q800, TA Instruments) were used to study the mechanical properties of pure PLA and PLA composites. The typical sample dimension for evaluation of tensile modulus was 25 mm × 5 mm × 50 μm. The actual dimension of the samples was measured by caliper and recorded. The test was performed at a strain ramp speed of 5 mm/min until break at a temperature of 35°C. Tensile strength (MPa), strain at break (%), and the Young's modulus (GPa), which was calculated from the data of stress and strain in the initial linear region, were recorded. For each material, at least five samples were used to obtain the average value and standard deviation. The significant difference of the data was evaluated using T-tests (with $p < 0.05$ confident interval).

X-ray diffraction

X-ray diffraction (XRD) experiments (PANalytical Empyrean) with CuKα radiation generated at 45 kV and 40 mA was used to study the crystallinity of PLA and PLA composites. XRD data were obtained at a rate of 2 degrees per minute from 5 to 50 degrees 2θ. Four major crystalline peaks and one amorphous peak (fixed at 17 degrees 2θ) of PLA were deconvoluted using Peakfit (www.systat.com). In the PLA composites, the cellulose peak was assigned at 22.7 degrees 2θ. Gaussian function was used to fit the peaks. The crystallinity of PLA was calculated as the area of PLA crystalline peaks divided by the total area of PLA peaks (the area under the cellulose peak was not included).

Competing interests

The authors declare that they have no competing interests.

Authors' contributions

JG carried out the experiments and data analysis. JC conceived of the study and participated its design. Both authors read and approved the final manuscript.

Acknowledgments

XRD and SEM were supported by the Pennsylvania State University Materials Research Institute Nanofabrication Lab and the National Science Foundation Cooperative Agreement No. ECS-0335765. We acknowledge Dr. Xuepei Yuan and Dr. T.C. Mike Chung at PSU for help with the preparation of composites.

References

1. Jonoobi M, Harun J, Mathew AP, Oksman K: **Mechanical properties of cellulose nanofiber (CNF) reinforced polylactic acid (PLA) prepared by twin screw extrusion.** *Compos Sci Technol* 2010, **70**:1742–1747.
2. Samir MASA, Alloin F, Dufresne A: **Review of recent research into cellulosic whiskers, their properties and their application in nanocomposite field.** *Biomacromolecules* 2005, **6**:612–626.
3. Moon RJ, Martini A, Nairn J, Simonsen J, Youngblood J: **Cellulose nanomaterials review: structure, properties and nanocomposites.** *Chem Soc Rev* 2011, **40**:3941–3994.
4. Eichhorn SJ, Dufresne A, Aranguren M, Marcovich NE, Capadona JR, Rowan SJ, Weder C, Thielemans W, Roman M, Renneckar S, *et al*: **Review: current international research into cellulose nanofibres and nanocomposites.** *J Mater Sci* 2010, **45**:1–33.
5. Iwatake A, Nogi M, Yano H: **Cellulose nanofiber-reinforced polylactic acid.** *Compos Sci Technol* 2008, **68**:2103–2106.
6. Kowalczyk M, Piorkowska E, Kulpinski P, Pracella M: **Mechanical and thermal properties of PLA composites with cellulose nanofibers and standard size fibers.** *Compos Part a-Appl S* 2011, **42**:1509–1514.
7. Bondeson D, Oksman K: **Dispersion and characteristics of surfactant modified cellulose whiskers nanocomposites.** *Compos Interface* 2007, **14**:617–630.
8. Kvien I, Tanem BS, Oksman K: **Characterization of cellulose whiskers and their nanocomposites by atomic force and electron microscopy.** *Biomacromolecules* 2005, **6**:3160–3165.
9. Oksman K, Mathew AP, Bondeson D, Kvien I: **Manufacturing process of cellulose whiskers/polylactic acid nanocomposites.** *Compos Sci Technol* 2006, **66**:2776–2784.
10. Petersson L, Kvien I, Oksman K: **Structure and thermal properties of poly (lactic acid)/cellulose whiskers nanocomposite materials.** *Compos Sci Technol* 2007, **67**:2535–2544.
11. Frone AN, Berlioz S, Chailan JF, Panaitescu DM, Donescu D: **Cellulose fiber-reinforced polylactic acid.** *Polym Composite* 2011, **32**:976–985.
12. Wang B, Sain M: **The effect of chemically coated nanofiber reinforcement on biopolymer based nanocomposites.** *Bioresources* 2007, **2**:371–388.

13. Onwulate CI, Thomas AE, Cooke PH: **Effects of biomass in polyethylene or polylactic acid composites.** *J Biobased Mater Bio* 2009, **3**:172–180.
14. McKenzie HA: *Milk proteins: Chemistry and Molecular Biology.* New York, London: Academic Press; 1970.
15. Horne DS: **Casein structure, self-assembly and gelation.** *Curr Opin Colloid In* 2002, **7**:456–461.
16. Fox PF, McSweeney PLH: *Advanced dairy chemistry.* 3rd edition. New York, NY: Kluwer/Plenum; 2003.
17. Yang G, Zhang LN, Han HW, Zhou JP: **Cellulose/casein blend membranes from NaOH/urea solution.** *J Appl Polym Sci* 2001, **81**:3260–3267.
18. Zhang LN, Yang G, Xiao L: **Blend Membranes of Cellulose Cuoxam/Casein.** *J Membrane Sci* 1995, **103**:65–71.
19. Mears D, Catchmark JM: **Quantification of casein and cellulose binding affinity.** In *ASABE annual meeting.* Pittsburg, PA: ASABE; 2010.
20. Kiehna SE, Laughrey ZR, Waters ML: **Evaluation of a carbohydrate-pi interaction in a peptide model system.** *Chem Commun* 2007, **43**:4026–4028.
21. Gu J, Catchmark JM: **Interaction of casein functionalized cellulose whiskers with poly(lactic acid).** In *ASABE annual meeting.* Louisville, KY: ASABE; 2011.
22. Liu Y, Guo R: **The interaction between casein micelles and gold nanoparticles.** *J Colloid Interf Sci* 2009, **332**:265–269.
23. Nylander T, Wahlgren NM: **Forces between adsorbed layers of beta-casein.** *Langmuir* 1997, **13**:6219–6225.
24. Ozeki T, Verma V, Uppalapati M, Suzuki Y, Nakamura M, Catchmark JM, Hancock WO: **Surface-bound casein modulates the adsorption and activity of kinesin on SiO(2) surfaces.** *Biophys J* 2009, **96**:3305–3318.
25. Verma V, Hancock WO, Catchmark JM: **The role of casein in supporting the operation of surface bound kinesin.** *J Biol Eng* 2008, **2**:14.
26. Smith EA, Thomas WD, Kiessling LL, Corn RM: **Surface plasmon resonance imaging studies of protein-carbohydrate interactions.** *J Am Chem Soc* 2003, **125**:6140–6148.
27. Gu J, Catchmark JM, Kaiser EQ, Archibald DD: **Quantification of cellulose nanowhiskers sulfate esterification levels.** *Carbohyd Polym* 2013, **92**:1809–1816.
28. Guo J, Catchmark JM: **Surface area and porosity of acid hydrolyzed cellulose nanowhiskers and cellulose produced by Gluconacetobacter xylinus.** *Carbohyd Polym* 2012, **87**:1026–1037.
29. Samir MASA, Alloln F, Sanchez JY, El Kissi N, Dufresne A: **Preparation of cellulose whiskers reinforced nanocomposites from an organic medium suspension.** *Macromolecules* 2004, **37**:1386–1393.
30. Mathew AP, Oksman K, Sain M: **The effect of morphology and chemical characteristics of cellulose reinforcements on the crystallinity of polylactic acid.** *J Appl Polym Sci* 2006, **101**:300–310.
31. Zhang XQ, Espiritu M, Bilyk A, Kurniawan L: **Morphological behaviour of poly(lactic acid) during hydrolytic degradation.** *Polym Degrad Stabil* 2008, **93**:1964–1970.
32. Dong SP, Roman M: **Fluorescently labeled cellulose nanocrystals for bioimaging applications.** *J Am Chem Soc* 2007, **129**:13810–13811.
33. Mathew AP, Oksman K, Sain M: **Mechanical properties of biodegradable composites from poly lactic acid (PLA) and microcrystalline cellulose (MCC).** *J Appl Polym Sci* 2005, **97**:2014–2025.
34. Ehrenstein G: *Polymeric materials: structure-properties-applications.* Cincinnati, Ohio: Hanser Publishers; 2001.
35. Cao N, Fu YH, He JH: **Preparation and physical properties of soy protein isolate and gelatin composite films.** *Food Hydrocolloid* 2007, **21**:1153–1162.
36. Bondeson D, Mathew A, Oksman K: **Optimization of the isolation of nanocrystals from microcrystalline cellulose by acid hydrolysis.** *Cellulose* 2006, **13**:171–180.
37. Bradford MM: **A rapid and sensitive method for the quantitation of microgram quantities of protein utilizing the principle of protein-dye binding.** *Anal Biochem* 1976, **72**:248–254.
38. Vaia B, Smiddy MA, Kelly AL, Huppertz T: **Solvent-mediated disruption of bovine casein micelles at alkaline pH.** *J Agr Food Chem* 2006, **54**:8288–8293.

4

Model-driven elucidation of the inherent capacity of *Geobacter sulfurreducens* for electricity generation

Longfei Mao* and Wynand S Verwoerd

Abstract

Background: *G. sulfurreducens* is one of the commonest microbes used in microbial fuel cells (MFCs) for organic-to -electricity biotransformation. In MFCs based on this microorganism, electrons can be conveyed to the anode via three ways: 1) direct electron transfer (DET) mode, in which electrons of reduced c-type cytochromes in the microbial outer membrane are directly oxidized by the anode; 2) mediated electron transfer (MET) mode, in which the reducing potential available from cell metabolism in the form of NADH is targeted as an electron source for electricity generation with the aid of exogenous mediators; and 3) a putative mixed operation mode involving both electron transfer mechanisms described above (DET and MET). However, the potential of *G. sulfurreducens* for current output in these three operation modes and the metabolic mechanisms underlying the extraction of the reducing equivalents are still unknown.

Results: In this study, we performed flux balance analysis (FBA) of the genome-scale metabolic network to compute the fundamental metabolic potential of *G. sulfurreducens* for current output that is compatible with reaction stoichiometry, given a realistic nutrient uptake rate. We also developed a method, **f**lux variability **a**nalysis with **t**arget flux **min**imization (FATMIN) to eliminate futile NADH cycles. Our study elucidates the possible metabolic strategies to sustain the NADH for current production under the MET and Mixed modes. The results showed that *G. sulfurreducens* had a potential to output current at up to 3.710 A/gDW for DET mode, 2.711 A/gDW for MET mode and 3.272 A/gDW for a putative mixed MET and DET mode. Compared with DET, which relies on only one contributing reaction, MET and Mixed mode were more resilient with ten and four reactions respectively for high current production.

Conclusions: The DET mode can achieve a higher maximum limit of the current output than the MET mode, but the MET has an advantage of higher power output and more flexible metabolic choices to sustain the electric current. The MET and DET modes compete with each other for the metabolic resource for the electricity generation.

Keywords: MFC, Microbial fuel cell, Geobacter sulfurreducens, Bioelectricity, Flux balance analysis, Flux variability analysis, Flux minimization, FATMIN

* Correspondence: longfei.mao@lincolnuni.ac.nz
Centre for Advanced Computational Solutions, Wine, Food & Molecular Bioscience Department, Lincoln University, Ellesmere Junction Road, Lincoln 7647, New Zealand

Background

Geobacter sulfurreducens has been extensively studied for microbial fuel cell (MFC) applications since discovered in 1994 [1]. The electrons in the metabolism of *G. sulfurreducens* can be converted to a comparatively large amount of electrical energy through the direct electron transfer (DET) established between reducing processes and the MFC electrode. DET relies on membrane bound c-type cytochromes and electrically conducting nanowires [2], and allows MFCs based on this microbe to obviate the need to replenish exogenous mediators. DET has been suggested to resemble mechanisms observed for electron transport to Fe(III) citrate [3].

Although *G. sulfurreducens* do not excrete redox shuttles [4,5], exogenous redox shuttles can be added to pull electrons from the reduced intracellular carriers (e.g., NADH) to sustain power generation in MFCs in a mediated way [5]. The electron shuttle compounds allow bacteria to utilise a remote electron acceptor (anode) that is not directly accessible to the cells, enhancing the electron transfer efficiency between the cellular metabolism and the electrode [6]. Therefore, the mediated electron transfer (MET) mode could also be advantageous for some MFCs in which the microbes are not grown on the anode.

For the mediated electron generation process, the intracellular electron-shuttling compound NADH has received considerable attention, since it is used to accept the electrons released from an organic substrate and donates them to various oxidoreductase enzymes and electron acceptors (such as quinone and ferricyanide) in the respiratory electron transfer chain of all living cells [7-12]. Exogenous mediators, such as bromocresol green (BG) and neutral red (NR), may target any steps of NADH regeneration in the intracellular electron-transfer pathways in cells and convey the electrons to an extracellular electrode (anode) [13-15]. Besides, it has been proposed that the theoretical limit of MFC voltage output is the potential difference between NADH and the reaction in the cathode if microorganisms are used as a biological catalyst [10,16]. Therefore, targeting NADH as the electron source in MFCs of MET mode can liberate the maximum power achievable for a microorganism.

Furthermore, a putative mixed operation mode (Mixed mode) involving both the DET and MET may exist when the cells are grown on the anode and the redox shuttles are added to the anodic reactor of the MFC. The putative Mixed mode may not be practical since different anode potentials could be required to accept the electrons from c-type cytochrome and the reduced mediators, but study of the Mixed mode can help elucidate the metabolic connection between the MET and DET modes.

Scanning the previous literature regarding MFCs based on *G. sulfurreducens* indicates that there are two questions needed to be answered: 1) What is the maximum potential of *G. sulfurreducens* for current output in MFCs under the three aforementioned operation modes? 2) Since a variety of reactions inside a cell are associated with NADH regeneration, what is the best metabolic strategy to choose NADH supply reactions, to maximize the current output in the MET and Mixed modes?

To address these two issues, in this study, we employed flux balance analysis (FBA) to examine the maximum amperage output of metabolic-driven electricity generation based on *G. sulfurreducens* in three electricity generation modes, namely, the c-type cytochrome dependent DET mode, the NADH-dependent MET mode and a Mixed mode combining the DET and MET modes (Table 1). We then demonstrated theoretical trade-offs between amperage yield and the biomass production (growth) rate. For elucidation of the metabolic strategies for sustaining free NADH flux for electricity generation in the MET and Mixed modes, we developed a method, **F**lux variability **A**nalysis with **T**arget flux **Min**imization (FATMIN), to identify the enzymatic mechanisms that could be used by the cell to regenerate NADH at a high rate, subject to the network stoichiometry and substrate uptake. Finally, we analysed the effect of varying substrate uptake rate on the growth rate and the amperage output in the three modes.

Results and discussion

Impact of the redox perturbation on the biomass production

Figure 1 shows how production of reducing equivalent competes with biomass production for metabolic resources. The increase in COI drove the fluxes through electron transfer reactions in the three MFC operating modes towards their maximum allowable values and corresponding biomass formation rates to their minimum values. This reflects that at a high COI (>1000),

Table 1 The metabolism and electron transfer types investigated in the present metabolic modelling

Organism	Transfer type	Electron source (Terminal bacterial electron shuttle)	Metabolic type
G.sulfurreducens	DET (Membrane-driven)	c-type cytochrome	Anaerobic & Heterotrophic
	MET (Mediator-driven)	NADH	
	The Mixed mode (Membrane-driven and Mediator-driven at the same time)	c-type cytochrome & NADH	

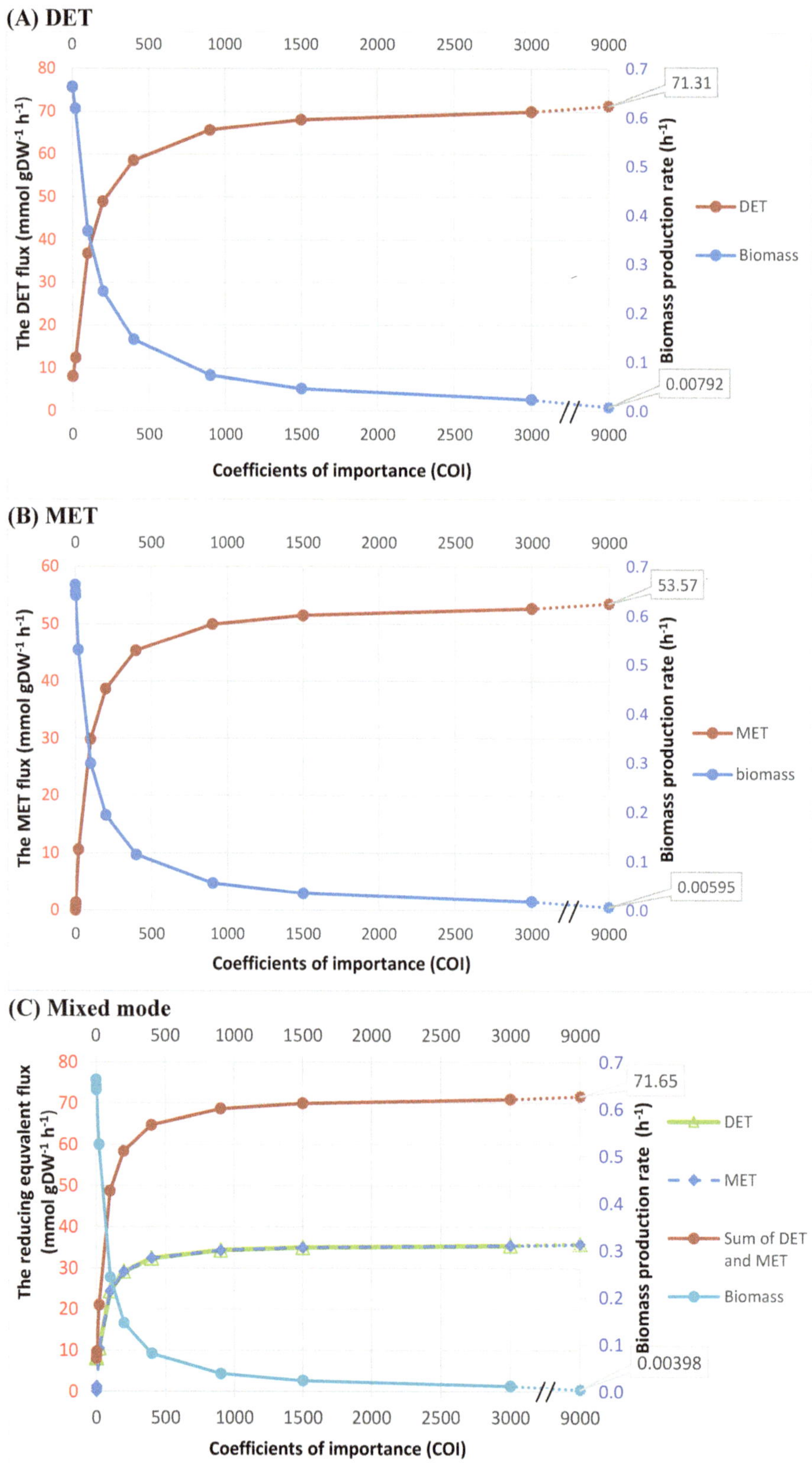

Figure 1 (See legend on next page.)

(See figure on previous page.)
Figure 1 The effect of varying COI on the biomass production and electron transfer rates. The reducing equivalent supplying rates in the (**A**) DET, (**B**) MET or (**C**) Mixed mode, and the biomass yields in the cases of (**A-C**), as functions of weighting ratios of the electron transfer rate versus biomass in *G. sulfurreducens* metabolic model. A series of coefficients of importance (COI), 1,2,20, 200, 400, 900, 1500 and 3000 were applied for the electron transfer flux. The changing trends of electron transfer rates in the tree modes are compared to those of the corresponding biomass yields. COI is the stoichiometry representation used in the compound objective function. The higher the COI means the priority for associated electron transfer reaction is higher. The reducing equivalent represents NADH in the MET mode, whereas it denotes an assumed product of the reaction catalysed by cytochrome c reductase for the DET mode.

only a small part of the energy is available to microorganisms in MFCs for their growth, as a large part is converted to electricity. The increase in COI was able to decrease the production rate of the biomass yield and increase the electron transfer much more significantly when the COI range was below 400 than above 400. This suggests that the metabolic states modelled by COI higher than 400 would be much more difficult to achieve for the microorganisms than those metabolic states with COI lower than 400. In other words, it is increasingly difficult for the cell to direct metabolic resources to sustain the increase in electricity generation. Since the COI of 400 can produce about 85%, 82% and 81% of the maximum electron transfer rates (at the COI of 9000) in the MET, DET and Mixed modes respectively, a major portion of the maximum theoretical MFC current yield should be achievable in practice for *G. sulfurreducens*.

There was a plateau in the electron transfer rates when the COI was higher than around 1000, which was indicated by the observation that a large increase in COI only resulted in little change in electron transfer and biomass production rates. This indicates that the metabolic states modelled by COI higher than 1000 have captured the highly perturbed metabolic behaviours under the three electron transfer modes and they were no different from one another for studying the metabolic basis of these modes.

The COI can be as large as 9000, because a large portion of this number is contributed by the high molecular weight of the biomass of the microorganisms. The coefficient for each component in the biomass equation is not a pure stoichiometric coefficient, which is different from other reactions in the model. These coefficients can convert the unit of the resulting flux through the biomass equation from mmol/gDW/h to g/gDW/h (h^{-1}). An introductory exposition of the formulation of the biomass equation can be found in Feist *et al.* 2010 [17].

The present modelling allowed free selection of acetate uptake rate in a proper range (0–18 mmol/gDW/h) to maximize the objectives, i.e., biomass production rate, and NADH or c-type cytochrome involved electron transfer rates. Since acetate is the main source for producing biomass, NADH and c-type cytochrome, the acetate uptake rate stayed at the maximum (18 mmol/gDW/h) in all the simulation results. This suggests that the increased fluxes through the NADH and reduced c-type cytochrome regenerating reactions, accompanied with the raised COI, did not result from the increased availability of acetate but were associated with the redirection of electron flux for biomass formation towards mediators in the MET mode or the electrode in the DET mode.

In the Mixed mode, the total electron flux available for current production was the sum of the fluxes through the c-type cytochrome dependent and NADH-targeted pathways, which are individually exploited by the DET and MET modes respectively. The COIs influenced the combined production of the NADH and the reduced form of the c-type cytochrome in the same way as observed for the cases of MET and DET modes. Since COIs were equally assigned for DET and MET, the two electron transfer rates were nearly the same. Equal weighting may not necessarily represent actual physiological priorities, but is appropriate to evaluate how DET and MET mutually influence each other. Figure 1 shows that the Mixed mode did not produce significantly higher electron flux than either MET or DET alone. This result suggests that the DET competed with the MET for the metabolic resources.

Figure 2 shows how reduction equivalent is allocated to internal use (i.e. maintenance of the cell metabolism) and supply to the external circuit, for various cell growth rates.

Generally, the increase in the electron transfer rates is linearly correlated with a decrease in biomass yield in the three operation modes. For DET mode (Figure 2(A)), the optimal metabolic state that resulted in the maximum growth rate of 0.6622 h^{-1} necessitates a flux of 8.092 mmol/gDW/h through reaction CYOR1m. This indicates that disposal of some electrons to exogenous Fe(III) can benefit the growth of *G.sulfurreducens*. When the COI associated with CYOR1m increased to 3000, which resulted in a highly perturbed metabolic state corresponding to around the upper limit of current output, the capability of the metabolic network for the electron transfer in DET mode increased by 8.65 fold (765%) to 69.69 mmol/gdW/h (not specified on the plot) compared with the optimal state.

Unlike the DET mode, which relies on one single enzymatic contribution (c-type cytochrome involved

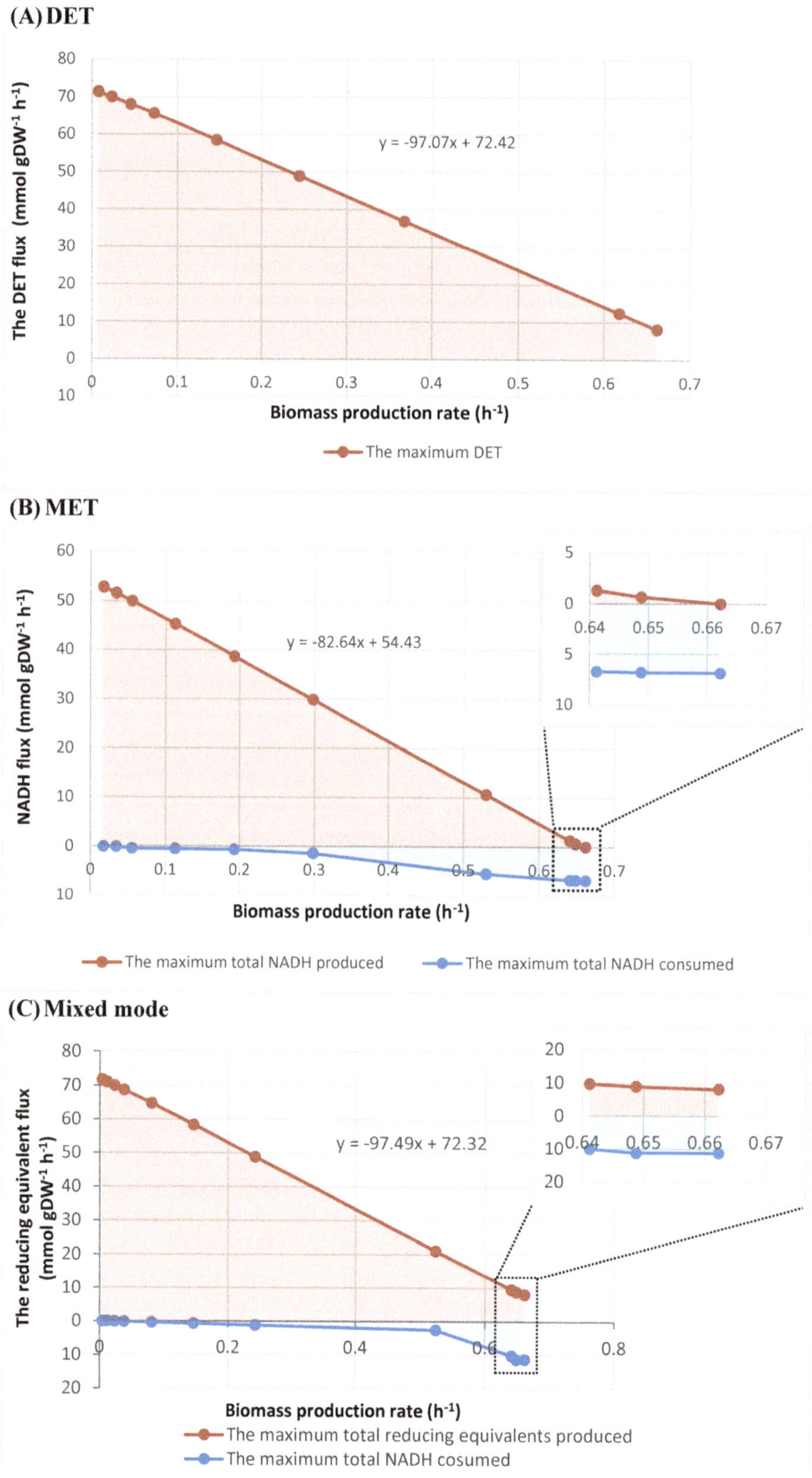

Figure 2 (See legend on next page.)

(See figure on previous page.)
Figure 2 The relationship of the biomass production and electron transfer rates. The reducing equivalent supplying rates in the (**A**) DET, (**B**) MET and (**C**) Mixed modes, and the NADH consumption rate for cellular use in the (**B**) MET and (**C**) Mixed modes, as functions of biomass yield. The line represents the maximal electron transfer rates and biomass production rate, while any point within the pink area represents all allowable electron transfer rates and biomass production rates. The blue area represents the total NADH-consuming flux for normal cellular function. The distance between the two lines across the pink and blue areas represents the total available NADH flux in the cell at a metabolic state related to a specific biomass production rate; inset, enlargement of boxed area. The reducing equivalent represents NADH in the MET mode, whereas it denotes an assumed product of the reaction catalysed by cytochrome c reductase for the DET mode.

reaction), there were 51 internal reactions with potential to generate or consume NADH for the MET mode in the network model. Sum of the fluxes of all NADH producing reactions gives the total NADH production rate at a particular metabolic state, whereas sum of the fluxes of all NADH consuming reactions is the total NADH consumption rate for maintenance. The net yield of NADH reactions is the total NADH production rate available for MFCs (i.e., NADH_mfc).

For MET mode (Figure 2(B)), the total NADH production rate (sum of the fluxes of the NADH_mfc and NADH_consumed) increased from 6.865 mmol/gDW/h at the optimal growth rate of 0.6622 h^{-1} to 52.80 mmol/gDW/h at the growth rate of 0.01757 h^{-1} (COI =3000). In the metabolic state simulated by COI of 3000, the metabolism was capable of increasing the NADH_mfc yield by up to 7.5 fold (653%) compared with the base state optimised for growth. The increase in the flux of NADH_mfc was predicted to result from a combination of two mechanisms, i.e., increasing the NADH production and decreasing the NADH needed for biomass production and maintenance.

In addition, the rise in NADH regeneration rate by increasing COI accidentally activated some NADH-consuming futile cycles, leading to very high flux values (larger than 100 mmol/gdW/h) through some NADH-consuming reactions. Increasing flux through these futile cycle reactions did not change the growth rate and the net NADH production rate. Since the total NADH flux for normal cellular function was taken as the sum of all NADH-consuming fluxes, the contained futile values could make the calculated total NADH flux much higher than the actual ones. Because the task to eliminate all futile cycles in the whole network diverges from the purpose of the present study, we did not carry out the analysis of all futile pathways using methods such as elementary mode or extreme pathway, which can only be performed on a small-scale network [18].

Instead, in an attempt to eliminate flux contributions of those futile cycle reactions to the total NADH consumption rate associated with growth and non-growth maintenance, several sub-sets of constraints were used for the simulations with different COI according to a simple method (Additional file 1). This was used instead of the relatively complex pipeline in the FATMIN algorithm, because the present section simply uses a single FBA rather than elucidation of the detailed flux range of individual reactions for which FATMIN was developed.

Unlike DET and MET modes, the Mixed mode relied on the metabolic capability for concurrent regeneration of NADH and reduced c-type cytochrome. The analysis showed that the metabolism could increase the regeneration capability of the two metabolites by 3.7 fold (267.1%), compared with its base state, to 71 mmol/gDW/h (COI of 3000). The increase was attributed to a combination of three mechanisms, i.e., increasing the NADH and reduced c-type cytochrome regeneration rate and decreasing the NADH needed for biomass production and maintenance.

The mechanisms by which the NADH is additionally produced for current output are elucidated in the next section.

Metabolic strategies for sustaining a high flux of NADH under the MET and Mixed modes

Since the further increase in COI can barely increase the electron transfer rates when the COI was above 3000, we chose the metabolic states modelled with COI of 3000 as the reference states to investigate the metabolic behaviours underlying the high electron transfer rates in the MET and Mixed modes. In fact, for identifying the metabolic behaviours under high electron transfer rates, it does not matter whether the metabolism is extremely perturbed (such as the metabolic state modelled by COI of 3000 or higher) or just a little bit less extremely perturbed (COI of 1000), as these metabolic states were all situated on the plateau of the lines (Figure 1).

The MET mode

When the mediator deprives NADH of electrons forming NAD^+, the NAD^+/NADH ratio in the metabolism is changed and consequently the cell has to adjust its metabolism to regenerate NADH from NAD^+ in order to maintain a proper redox balance for survival. To elucidate the metabolic adjustments induced by heavy NADH loss (COI=3000), the range of variability of intracellular fluxes were calculated. The results show that although many reactions in the network can produce NADH, under heavy NADH demand, only a few reactions can sustain large NADH drain, while other

NADH producing reactions become deactivated to leave the metabolic resource for the highly efficient NADH producers.

In total, ten out of fifty-one NADH involved reactions were identified as capable of regenerating NADH at a high rate under the redox perturbation (Table 2). Among the ten reactions, one reaction (AKGD) (EC 2.3.1.61) had a potential to produce NADH at up to a flux rate of 17.73 mmol/gDW/h. Each of the other nine reactions possessed capability to solely supply almost all of the 52.80 mmol/gDW/h, which was the maximum total NADH flux that could be achieved for the metabolism of *G. sulfurreducens* at COI of 3000. Therefore, the probed maximum NADH production rate could result from any combination of lesser fluxes of the nine reactions that made up that value. This indicates that the number of optimal solutions will be unlimited. The non-uniqueness of the obtained solution corresponds to the alternative optima in the linear optimization and the existence of alternative pathways that result in equivalent mutant phenotypes regarding the required metabolic adjustment.

Among the nine reactions, three reactions (i.e., AKGD, ME1x and PDH) were part of the central metabolism pathway (TCA cycle). Four reactions (i.e., ALAD_L, GLUDx, GSADH and PGCD) were related to the amino acid metabolism. Another two reactions (i.e., TDMAT6 and HDMAT7) were located in the fatty acid metabolism. Therefore, this result extends the previous notion that the TCA cycle is the main source of NADH for *G. sulfurreducens* when acetate is fed as the electron donor [19].

Except the nine reactions, all other NADH involved reactions were found to have very low or no variability ($v_{i\ min}/v_{i\ max}$>0.99) under high current output. The rigid variability may be attributed to the fact that NADH is a highly connected metabolite ensuring a degree of network connectivity sufficient to limit the availability of alternative flux distributions because of cascading effects on other metabolites.

The analysis presented does not fully specify the solution space. The Max/Min limits in Table 2 describes a rectangular cuboid in 10 dimensions, that encloses the feasible flux points but would in general still contain flux value combinations that are not feasible even if the values do add up to the 52.80 mmol/gDW/h value that was determined as the combined maximum. A full specification requires determination of all vertices of the solution space. Elucidation of these vertices is the task targeted by techniques such as elementary mode and extreme pathway analysis, but applications of these tools are restricted to small- and medium-scale networks of limited connectivity [20] and not suitable for the network size of *G. sulfurreducens* used here.

The Mixed mode

Similar to the case of the MET, a continuous range of solutions for achieving the high electron transfer rate in the Mixed mode involving a combination of six reactions for NADH regeneration plus one c-type cytochrome dependent DET reaction (Table 3) was obtained. It is shown that the electrons leaking from the site of c-type cytochrome influenced the possible cellular strategies to supply high NADH flux. The metabolic strategies probed for the Mixed and MET modes shared one common reaction, AKGD (EC: 2.3.1.61), to sustain high NADH flux. Nonetheless, the other eight reactions that could regenerate NADH at high rates in the MET mode were not active in the Mixed mode, which chose two new reactions, ALDD2x (EC: 1.2.1.3) and GCCc (EC: 1.8.1.4) located in the amino acid metabolism as the NADH suppliers. Besides, in the Mixed mode, another three reactions were identified capable of consuming NADH at high rates. These NADH consuming reactions were deregulated to balance the excess NADH produced by the identified three producing reactions, making up a net sum flux of NADH of 70.96 mmol/gDW/h (obtained when COI=3000). Since a range of reactions (nine for MET and three for Mixed mode) in different biological sub-systems could be targeted as the efficient NADH suppliers, NADH-dependent electricity generation should be more robust against momentary shifts in environmental conditions than the pure DET mode.

In the case of MET, NADH production was increased by directing the electrons towards the reactions in the Krebs cycle and fatty acid biosynthesis, so fewer electrons are available for the downstream pathway of oxidative phosphorylation, which is associated with DET, to produce energy for the cell maintenance and growth. This accounts for the phenomenon shown here that MET and DET competed with each other for electron resources in the Mixed mode and a combination of both did not double the current output of MET or DET alone.

To demonstrate how nutrient uptake is channelled to biomass growth and current yield respectively, Figure 3 compares the corresponding fluxes at different uptake rates.

Effect of varying acetate uptake rate on the predicted biomass and reducing equivalent production rates

The biomass production rates or the electron transfer rates were all linearly proportional to the acetate uptake rate in the three operation modes. This suggests that acetate uptake rate is the major limiting factor for both biomass and current yields. In addition, as seen from the different Y axis scales (i.e., the slopes of the trend lines), the priority set for electricity generation in the three modes lead to the suppression of biomass production

Table 2 The identified reactions that contribute significantly to the predicted maximum NADH production rate

Reaction ID	NADH flux (mmol/gDW/h)		Enzyme	EC. No.	Reaction	Subsystems
	Min	Max				
AKGD	0	17.73	2-oxoglutarate dehydrogenase	2.3.1.61	[c] : akg + coa + nad --> co2 + nadh + succoa	Central Metabolism
ALAD_L	0	52.77	L-alanine dehydrogenase	1.4.1.1	[c] : ala-L + h2o + nad --> h + nadh + nh4 + pyr	Amino Acid Metabolism
GLUDx	0	52.72	glutamate dehydrogenase (NAD)	1.4.1.3	[c] : glu-L + h2o + nad <==> akg + h + nadh + nh4	Amino Acid Metabolism
GSADH	0	52.77	L-glutamate 5-semialdehyde dehydrogenase	1.2.1.38	[c] : glu5sa + h2o + nad --> glu-L + (2) h + nadh	Amino Acid Metabolism
HDMAT7	−0.00007	52.77	Hexadecanoyl-[acyl-carrier protein]:malonyl-CoA C-acyltransferase	1.3.1.9	[c] : nad + palmACP <==> h + hdeACP + nadh	Fatty Acid Synthesis
ME1x	0	52.77	malic enzyme (NAD)	1.1.1.37	[c] : mal-L + nad --> co2 + nadh + pyr	Central Metabolism
P5CD	0	52.77	1-pyrroline-5-carboxylate dehydrogenase	1.5.1.12	[c] : 1pyr5c + (2) h2o + nad --> glu-L + h + nadh	Amino Acid Metabolism
PDH	0	52.77	pyruvate dehydrogenase	1.2.4.1	[c] : coa + nad + pyr --> accoa + co2 + nadh	Central Metabolism
PGCD	0.0196	52.79	phosphoglycerate dehydrogenase	1.1.1.95	[c] : 3pg + nad --> 3php + h + nadh	Amino Acid Metabolism
TDMAT6	−0.00404	52.77	Tetradecanoyl-[acyl-carrier protein]:malonyl-CoA C-acyltransferase	1.3.1.9	[c] : myrsACP + nad <==> h + nadh + tdeACP	Fatty Acid Synthesis

(The reaction and metabolite abbreviations are listed in the Additional file 2).

Table 3 The identified reactions that contribute significantly to the predicted maximum NADH production rate

Reaction ID	NADH flux (mmol/gDW/h)		Enzyme	EC. No.	Reaction	Subsystems
	Min	Max				
NADH producing reactions						
AKGD	0	35.48	2-oxoglutarate dehydrogenase	2.3.1.61	[c] : akg + coa + nad --> co2 + nadh + succoa	Central Metabolism
ALDD2x	8.855	35.32	aldehyde dehydrogenase (acetaldehyde, NAD)	1.2.1.3	[c] : acald + h2o + nad --> ac + (2) h + nadh	Amino Acid Metabolism
GCCc	4.427	17.66	part of glycine-cleavage complex	1.8.1.4	[c] : dhlpro + nad --> h + lpro + nadh	Amino Acid Metabolism
NADH consuming reactions						
GAPD	−52.96	−0.04026	glyceraldehyde-3-phosphate dehydrogenase (NAD)	1.2.1.12	[c] : g3p + nad + pi <==> 13dpg + h + nadh	Central Metabolism
GLUSx	−52.92	0	glutamate synthase (NADH2)	1.4.1.14	[c] : akg + gln-L + h + nadh --> (2) glu-L + nad	Amino Acid Metabolism
NADH5	−10.58	0	NADH dehydrogenase (Menaquinone 7 & 2 protons)	1.6.5.3	(3) h[c] + mqn7[c] + nadh[c] --> (2) h[e] + mql7[c] + nad[c]	Energy Metabolism
DET						
CYOR1m	35.48	35.48	cytochrome-c reductase (menaquinol 7: 1 protons)	1.7.2.2	(2) ficytcc[c] + mql7[c] --> (2) focytcc[c] + h[e] + h[c] + mqn7[c]	Energy Metabolism

(The reaction and metabolite abbreviations are listed in the Additional file 2).

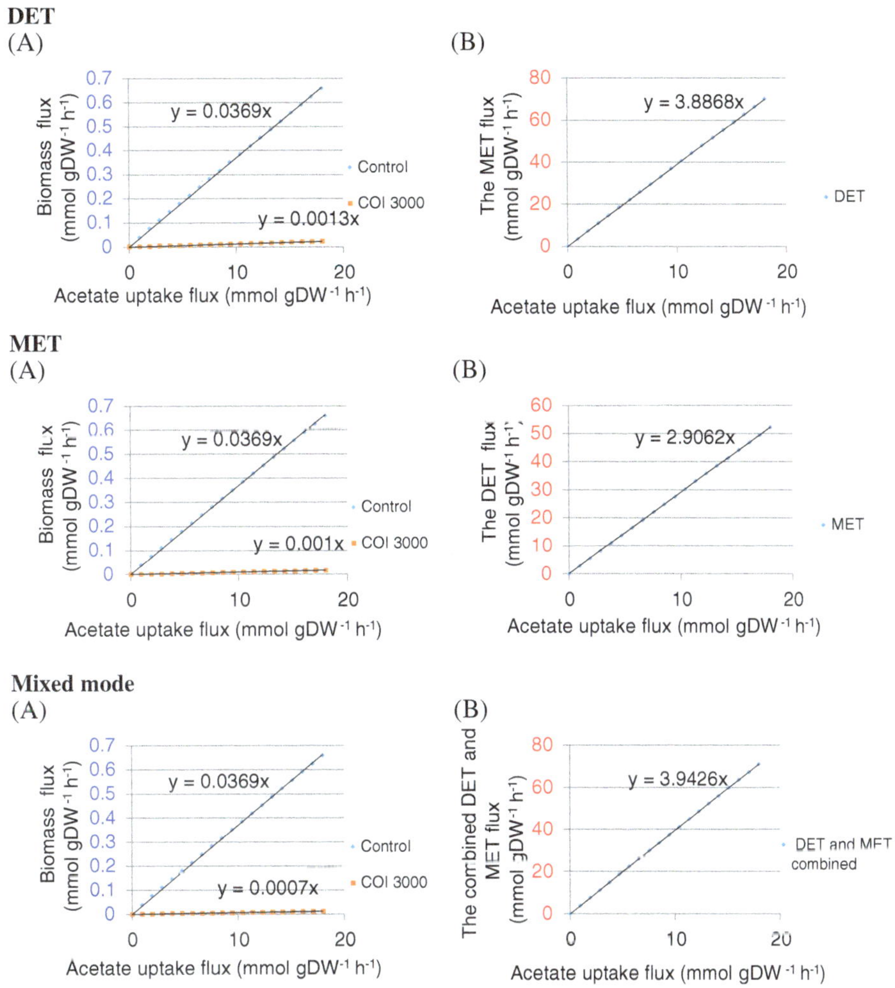

Figure 3 The effect of varying glucose uptake rate on the biomass and reducing equivalent production rates. (A) The effect of varying acetate uptake rate on the biomass growth rates when biomass production is set as biological priority for the organisms (use of only biomass maximization as objective function) and when the electron transfer in the three modes is set as the priority (use of a COI of 1 for biomass production and a COI of 3000 for the reducing equivalent production in the objective function); (**B**) The effect of varying acetate uptake rate on reducing equivalent production rate flux when a COI of 3000 is used for the reducing equivalent production and a COI of 1 for biomass production.

and current production was much less acetate-costly than the biomass production.

The Mixed mode (slope=3.9426) can achieve higher efficiency of converting substrate into electricity in MFCs than the DET (slope=3.8868) and MET (slope=2.9062) modes. The high conversion efficiency reached by the Mixed mode is ascribed to the fact that two by-products, reduced c-type cytochrome and NADH, were both acceptable for electricity generation, leaving more freedom for the metabolism to adjust to maximize the objective than only one byproduct, such as in the either case of DET or MET.

Finally, the three operation modes are compared for their theoretically maximum current output in Figure 4 and Table 4.

Comparison of amperage output of the three operations modes

A growth rate of 0.06 h^{-1}, which is achievable in the experiment [21], was also chosen as a reference value in addition to the 5% optimal growth rate. Among the three tested electron transfer modes, DET delivered the highest maximum limits of the current output, whereas MET achieved the lowest maximum limit. This could be attributed to the fact that *G. sulfurreducens* can conserve energy for cellular growth during the process of disposing electrons to exogenous electron acceptor [22] and consequently less additional energy resources (i.e., acetate) are required for DET mode to achieve the same level of current output as the MET. Therefore, these organisms can gain an energetic advantage by growth on or near the

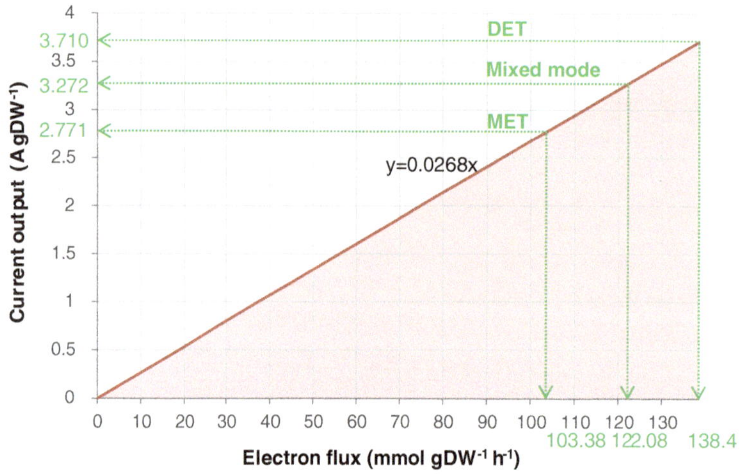

Figure 4 The current output (A/g) as a function of electron flux. The dark red line denotes the maximal current outputs and NADH_mfc production rate, while the area represents all allowable current outputs and electron production rates. The round dotted arrow line indicates the maximal current output and corresponding electron production rate when the growth rate is set to 5% of the predicted maximum growth rate ($0.033\ h^{-1}$).

electrode. Nonetheless, despite its higher current output, the maximum power output of the DET mode was lower than those achievable by the other two modes, due to the low standard potential of c-type chromosome used by the DET mode.

For the MET, mediators are needed to extract the redox potential out of cells and thus the cells that are located far away from electrode surface can still contribute to the current production. On the other hand, the DET depends on the 2–3 μm long pili of *G. sulfurreducens* to deliver the electron from the outer membrane to the anode [23,24]. Consequently, compared with the MET, the DET requires a much closer distance between the cell and the anode surface, and thus the actual current output will be restricted by the number of electrochemically active cells that near the surface area of the electrode. This may account for the observations that the DET based systems usually deliver quite low current [6].

The result shows that the maximum output of the Mixed mode was a little bit higher than MET. However, based on this, it cannot be concluded that Mixed mode is more competitive than MET. In effect, the Mixed mode may require independent circuits to collect the power output from the DET and MET concurrently. Furthermore, the Mixed mode requires involvement of mediators, large surface contact between the cells and special electrode design that can accept electrons from the two reducing agents with different potentials, which creates more demands for the engineering design and has impacts on power density and the long-term stability of the MFCs. The extra complication probably does not justify the design effort needed for mixed mode operation.

The DET and Mixed mode involved higher coulombic efficiencies than MET mode, which implies that more of electron flows were directed to current production rather than cellular synthesis. In a previous MFC review

Table 4 Comparison of predicted amperages and power outputs of the three modes

	Conditions	Biomass growth rate (h^{-1})	Electron (mmol $gDW^{-1}\ h^{-1}$)	Amperage (A gDW^{-1})	Coulombic efficiency (CE%)	Theoretical limit of the power output (W gDW^{-1})
MET	The growth rate achieved in the acetate limiting condition)	0.06	98.94	2.652	68.71%	2.201
DET			133.2	3.570	92.50%	0.9139
Mixed mode			116.8	3.131	81.13%	2.599
MET	5% of the maximum growth rate	0.033	103.4	2.771	71.79%	2.300
DET			138.4	3.710	96.12%	0.950
Mixed mode			122.1	3.272	84.78%	2.716

Note: c-type cytochrome c and NADH have different standard potentials. The power density for the Mixed mode was calculated by sum of the power contributions from the DET and the MET, i.e., the current output of DET and MET were multiplied with their appropriate voltages respectively.

article, the MET system was suggested to produce higher current and power densities, but have lower coulombic efficiency than the DET [6]. This is because part of the electrons from the substrate are used by the cell to form some electrochemically inactive side products [6].

The present study computed the theoretical maximum CE% achievable for the DET as 79.34% based on the stoichiometry in the network, indicating the acetate-fed MFCs using *Geobacter* species can effectively oxidize acetate with electron transfer to electrodes. However, the computed value is still much lower than the values reported by several previous studies, which claimed that *G. sulfurreducens* in MFCs could convert acetate to current with CEs of 89% [25] and above 90% [26,27]. And some studies assumed a CE of 100% for the DET via c-type cytochromes [28,29].

The reasons for such a discrepancy between the computed CE in this study and previous reported ones are speculated to be as follows: 1) The acetate consumed by the cell was calculated by measuring the change of concentration of acetate in the culture media. The measured point is always far away from the cells. A gradient of acetate concentration exists in the distance between the measured point and the cell surface, i.e., the acetate concentration is lower at the cell surface and much higher at the point where the measurement takes place, because starting with a homogenous acetate concentration, after MFC operation the concentration will drop more at the cell surface than at the measurement point. Thus the measured acetate uptake rate is lower than the actual uptake rate. According to the formula: $CE\% = \frac{\text{Acetate used for current production}}{\text{Total acetate consumed by the cell}}$, a underestimated total acetate uptake rate could increase the resulting CE%. 2) The whole experiment period did not last long enough for the microbes to form a steady state. 3) Microbes can consume other nutrients in the media or reserved in the cells, such as other dead cells, for electron production, but this factor was not taken into account for calculating the total substrate consumed by the cell.

Microorganisms have a strong propensity to aggregate and proliferate into biofilm on the solid surfaces [30]. A previous MFC study reported a current output of 1.4 A/m^2 and a power density of 0.5 W/m^2 by a biofilm of wild-type *G. sulfurreducens* [27]. The cell protein density of the biofilm was measured as 0.571 g protein/m^2 [27]. With an assumed protein content of 46% of cell dry weight [31], the cell density of the biofilm is estimated to be 1.241 g/m^2 and the previously reported current flow (1.4 A/m^2) and power density (0.5 W/m^2) can be converted into 1.128 A/gDW and 0.4028 W/gDW respectively. Clearly, these values are much lower than the computed ones (3.710 A/gDW and 0.95 W/gDW) for the DET mode in the present study. The difference between the *in silico* predicted maximum electric outputs and observed ones may be attributed to the fact that some electrons and potential would be lost during the electron transfer between the biofilm and the electrode, and consequently the actual current and power density will be lower than these computed values. Understanding and optimizing the micro-electrode interaction is another challenge to overcome for the MFC research.

It is commonly observed that mixed cultures usually produce much higher current densities at high coulombic efficiency compared with pure cultures, due to many advantageous features arising from the mutualism among microbial diversity [12,32]. However, by improving the engineering design, the MFC based on *G.sulfurreducens* can produce a power density similar to those reported with various mixed cultures while maintaining high coulombic efficiency [26].

Conclusions

Analysis of the metabolic flux models predicted that *G. sulfurreducens* in the DET mode achieved the highest current output (3.710 A/gDW), followed by the Mixed mode (3.272 A/gDW), and MET the lowest (2.771 A/gDW). In the DET mode, the enzymatic activity of one reaction determined the current generation in MFCs. On the contrary, in the MET mode, the reducing molecule, NADH, could be efficiently regenerated and supplied from ten identified reactions, while in the Mixed mode the number of the possible reactions supplying NADH reduced to three. The decreased reaction number indicates that the DET and MET compete with each other for the metabolic resources. The increases in the electron transfer rates in the three modes were linearly correlated with the drop in growth rate, which was, however, achieved by the increasingly rising metabolic 'cost' (modelled by COI).

The multiplicity of reactions capable of individually supplying the maximal current output indicates a range of feasible solutions, rather than just a single idealised metabolic state. This leaves room for good performance to be obtained from different cellular phenotypes and suggests robustness to changes in environmental or metabolic conditions.

While this work identified the maximum capability of *G. sulfurreducens* for current production, this leaves open the question of practical strategies to achieve this potential. One possibility may be a metabolic engineering strategy such as adaptive evolution. In this way, if the right selection pressure is applied, there is a continuous selection for beneficial mutations that can attain the high predicted current output. As an alternative approach, gene (reaction) deletion can also be a fast way to generate beneficial mutant strains, facilitated with the reaction knockout identifying algorithms such as OptKnock [33-35] and OptGene [36,37].

Methods

Flux variability analysis with target flux minimization (FATMIN)

Background

FBA relies on linear programming to find the optimal solution for reaction fluxes, which maximizes or minimizes a given objective function. However, the identified solution may not be unique and an infinite number of different flux vectors may exist producing an identical optimal objective value. In the context of metabolic models, these flux vectors are called alternate optimal solutions (AOS) or equivalent phenotypic states [38-41]. Previous methods for the determination of AOS include 1) vertex enumeration such as extreme pathway or elementary mode analysis [40,42,43], and 2) flux variability analysis (FVA) [44]. However, the computation of extreme pathways and elementary modes are too time-consuming and sometimes impossible, even for a medium-sized network model (up to one hundred reactions), and FVA results may contain large futile values due to the presence of loops in the network.

In constraint-based modelling, the constraints mainly used are mass balance and metabolite uptake rates. However, thermodynamic limitations associated with the reactions are difficult to implement and commonly overlooked. Without thermodynamic constraints, non-physical fluxes can be computed for some metabolic reactions if they form an internal cycle (Figure 5(A)). Such a cycle of reactions violate a "loop law" that is analogous to Kirchhoff's second law for electrical circuits, as discussed previously by Beard *et al.* [45].

In practical FBA calculations, the presence of a futile loop is often signalled by implausibly high flux values for some reactions. It is common practice to set an artificial upper limit of e.g. 10000 on all fluxes, simply to avoid numerical problems in the optimization algorithm. Fluxes of loop reactions are often observed to reach values of this magnitude, but simple strategies such as blocking these reactions by constraining their fluxes to zero only work in the simplest cases.

One reason for this is that individual loop reactions may simultaneously also participate in productive (non-futile) fluxes carried by active pathways. In this case flux values along the loop may all be high, but with different values, because the observed fluxes result from superimposing productive and futile fluxes. This has to be disentangled without affecting productive fluxes. This problem is particularly common when loops involve a currency metabolite such as NADH.

Another reason is that as loop fluxes do not contribute to the objective value, it is conceivable that some such

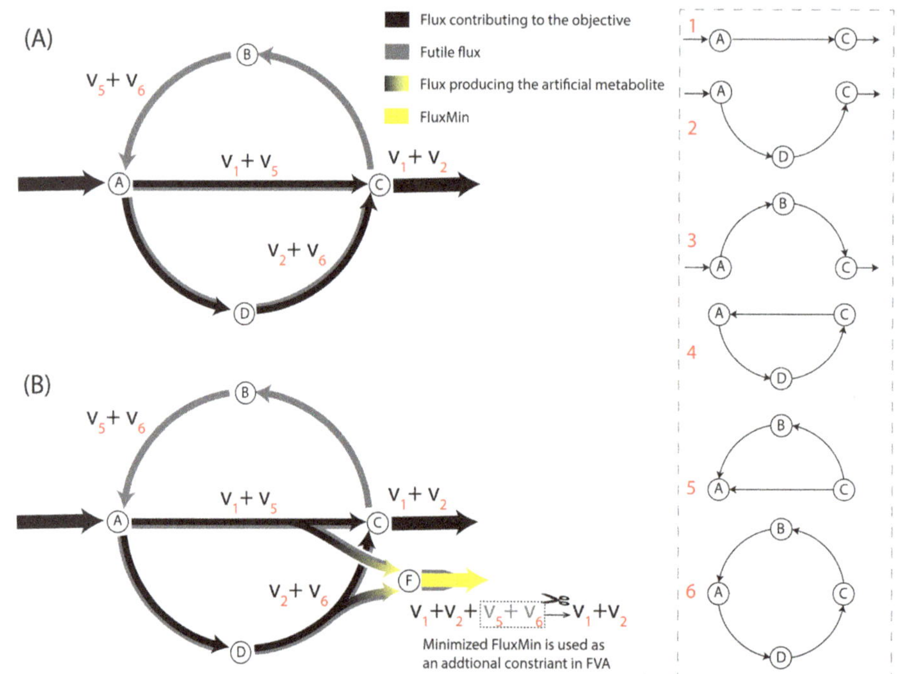

Figure 5 The schematic diagram of elimination of the futile cycle in an imaginary network by the FATMIN. (**A**) Flux distribution at steady state without employment of the FATMIN; (**B**) The minimization in the FATMIN to identify another constraint that can remove the futile values associated with cycles 5 and 6. The inset shows alternate path through the network, with numbers used as flux indices in the main figure. Path 3 and 4 carry no flux with reaction directions as shown, but are examples of paths that become active if reactions are made reversible.

fluxes may have another value, even zero, in the particular flux distribution that is selected from the solution space by the optimization algorithm.

More elaborate strategies that have been proposed to remove these thermodynamically infeasible futile cycles, include known flux directionality [46], energy-balance equations [45], known [47-52] or predicted [53] thermodynamic parameters, and nonlinear constraints to eliminate flux distributions that utilize reactions which cannot be thermodynamically feasible under physiological conditions [54]. But these thermodynamic constraints are difficult to obtain and nonlinear constraints can make application to large-scale systems computationally challenging [48]. Therefore, the listed methods above cannot be applied on genome scale systems [48].

Alternatively, loops can be removed by minimizing network flux [47,55-57]. The application of flux minimization can produce a most likely stationary flux distribution. The notion behind this method is that the flux through any reaction of a metabolic network requires some 'effort' and a metabolic network is inclined to fulfil the same biological objective, such as growth, with minimum metabolic effort [47]. A drawback of this approach is that the use of the minimum sum of all fluxes in the network as an auxiliary criterion in FBA by previous methods, can forestall variability within optimal flux distribution.

Below, we present a new method based on FVA and FBA to determine AOS and efficiently eliminate flux loops in large metabolic networks.

The FATMIN algorithm

The basis of FATMIN is FVA, which is used to probe the feasible flux ranges of desired reactions, such as NADH involved reactions in the present study. Importantly, the method can constrain the pointlessly high fluxes values of the loop reactions to a biologically meaningful value. It is intended to deal with such loops involving one or more *target metabolites* that are of specific interest; here this is NADH.

All reactions that produce the target metabolite and carry a flux higher than a threshold value (that is chosen to be higher than any realistic flux value) are identified as *target reactions*. To identify which portion of the high value is compulsory to achieve the value of the FBA objective function, we add an artificial metabolite to each target reaction, and add another reaction (e.g., reaction ID: FluxMin) that only comprises the artificial metabolite. FluxMin acts as a drain reaction for the artificial metabolite, and its flux is the sum of all production fluxes. Minimizing this flux while maintaining the objective value, in effect eliminates futile fluxes.

The idea is demonstrated in Figure 5(A). Metabolite C is taken as the target metabolite that this simple network produces. The network contains two alternative routes (1 and 2) for producing C and two cycles (5 and 6). All reactions are irreversible, so with paths through the network chosen as indicated, all flux values are nonnegative. A new metabolite labelled "F" is added to both reactions that produce C, as well as the associated drain reaction, to give the augmented network in Figure 5(B). Balancing fluxes at all nodes of the augmented network shows that although the net production of C (made up of the contributions v_1 and v_2) is independent of the futile fluxes, the flux of the added reaction FluxMin is $v_1 + v_2 + v_5 + v_6$. Since the value of $v_1 + v_2$ remains fixed for a chosen objective value and negative values for v_5 and v_6 are not allowed, minimization of FluxMin reduces both of the cyclic fluxes to zero. The resulting minimal value is subsequently used as a constraint on FluxMin in a repeated FVA calculation for the augmented network.

Although the explanation above referred to fluxes along individual pathways – in effect, elementary modes – the strength of the FATMIN method is that these do not in fact have to be identified to apply the method. Simply adding FluxMin to the network and minimizing its flux ensures that the cyclic components implicitly contained in it are eliminated.

This result relies on the fact that negative flux values are excluded. Realistic networks containing reversible reactions are usually handled in FBA by allowing positive and negative flux values. To maintain the exclusion of negative fluxes, in the FATMIN minimization step reversible reactions involving the target metabolite are conceptually split into two counter directional reactions, and the artificial metabolite only entered into the branch that produces the target metabolite. In practice, because the other branch does not contribute to FluxMin, it can be omitted. So the procedure followed is that where the FVA range of a target reaction exceeds the threshold on both negative and positive fluxes, the negative lower limit constraint is replaced by zero for finding the minimal FluxMin flux value. We stress that for the subsequent FluxMin-constrained FVA calculations the lower limit is restored to its original value.

Another point illustrated by the example is that the FATMIN strategy does not restrict the optimal solution space; e.g. all v_1 and v_2 value pairs allowed by the original objective value remain viable in the augmented network. Finally, we remark that the strategy described here is easily extended for multiple target metabolites by simply introducing a unique artificial metabolite and drain reaction for each.

Overall, FATMIN can be summarised as a modelling sequence FBA+ FVA +FBA +FVA, which is implemented as a computational pipeline consisting of the following steps, where we take NADH as the target metabolite:

1. Perform FBA to calculate the optimal objective $Z_{optimum}$. Then set the obtained optimal $Z_{optimum}$ as the constraint and perform flux variability analysis (FVA) to calculate a feasible flux range for each reaction.
2. Extract NADH reactions and calculate the feasible range of NADH flux values associated with each reaction by multiplying reaction rate with NADH stoichiometry coefficient.
3. Any reaction with an absolute value of the NADH flux boundaries higher than a threshold T (here, we chose T = 100 mmol/gDW/h) is identified as a target reaction. In such a reaction, if the absolute value of the lower limit is higher than T, we decompose the reversible reaction into two reactions in forward and backward direction respectively.
4. Extend the network by adding an artificial metabolite F as a product to each target reaction, and an F drain reaction FluxMin to the network model.
5. Use the FBA optimum ($Z_{optimum}$) as the constraint and minimize the flux of reaction FluxMin.
6. Perform FVA on the augmented network while using the obtained minimal flux value obtained at step 5 as a constraint for the flux of FluxMin.
7. Re-calculate the numerical range of NADH flux values using the method in step 2.

This method shares some ideas with [58], in which the solution space is described in terms of three key characteristics: linealities which are the reversible (bi-directional) infinite reactions, rays which are the irreversible infinite reactions, and vertices which are the corner points of the shape formed by interception of the polyhedral cone representing the convex constraint space with the objective plane. However, there could be millions of vertices, from which one cannot identify biological significance. Therefore, FATMIN intends to remove only the rays and linealities, which matched irreversible and reversible cycles. The rays and linealities are independent of the growth medium [58]. Since FATMIN is based on FVA, the method inherits the function of the FVA in elucidating the phenomena of equivalent optima in the network. Nonetheless, unlike the results of FVA that could contain infeasible, futile values, the result of FATMIN are the feasible flux ranges for the reactions.

The drawback of FVA is that this tool only produces the outline of a rectangular cuboids or 'box' encompassing the polygon of the solutions, but cannot illustrate the real shape of the polygon, which indicates that FVA results cannot reflect the relationships between different reaction fluxes, e.g., how the increase in the fluxes of some reactions influence the fluxes of the other reactions. The solution 'box' is discussed in the case of the MET (The MET mode section of the Results and discussion).

Assumptions for present modelling

To model the metabolisms under electricity generation, a few assumptions were made: 1) the MET mode relies on a putative ideal mediator (e.g., Neutral red [59]) that can enter cytoplasm and transfer electrons from NADH/NAD^+ cycle to anode. 2) Only pure culture is used for electricity generation. 3) The MFC reactor is a chemostat that can provide optimum conditions to meet all the needs of microbial growth. Subsequently, in order to ensure that the cells are not limited by substrate uptake rates, the model is allowed to adjust substrate uptake rate within a proper range as necessary for optimal growth yield. The range constraints on the reactions used in the modelling are specified in the model (Additional file 3). In addition, since the carbon and energy source were limited to a proper range, there is no necessity to use a two-step optimization strategy, i.e., the first is to maximize biomass with no constraint and the second is to minimize energy usage while fixing the biomass, as described in [31,60,61].

Modelling electrode interactions

The interactions with an electrode were captured by introducing three reactions into the model reconstruction (Table 5). These reactions represent the net reaction that takes place between the reducing equivalents and the mediators (or electrode) in MFCs. Introduction of these reactions create additional channels for electrons to escape from the metabolism and the fluxes of these reactions are subject to the mass balance rule in the FBA modelling. Because NADH and NAD are native metabolites of the microorganisms, these added equations did not lead to non-native by-product production. The reducing equivalent represents NADH in the MET mode, whereas it denotes an assumed product of the reaction catalysed by cytochrome c reductase for the DET mode. In both cases, one cycle of the electron transfer yields two electrons. The process is schematically shown in Figure 6.

Table 5 The added reactions for modelling the interaction of the microorganisms and the electrode in MFCs

Operation mode of the MFC	Reaction ID	Reaction
MET	1NADHmfc	nadh --> nadh_mfc
	2NADHmfc	nadh_mfc ->nad+h_emm
DET	3Cytcmfc	h_edm --> h_edm_out

nadh_mfc: the NADH available for MET mode of MFCs;
h_emm: the H ions as the by-product released from the reaction 2NADHmfc;
h_edm: the H ions as the by-product released from cytochrome-c reductase. (CROR1m) catalysed reaction; h_edm_out: the extracellular form of h_edm.

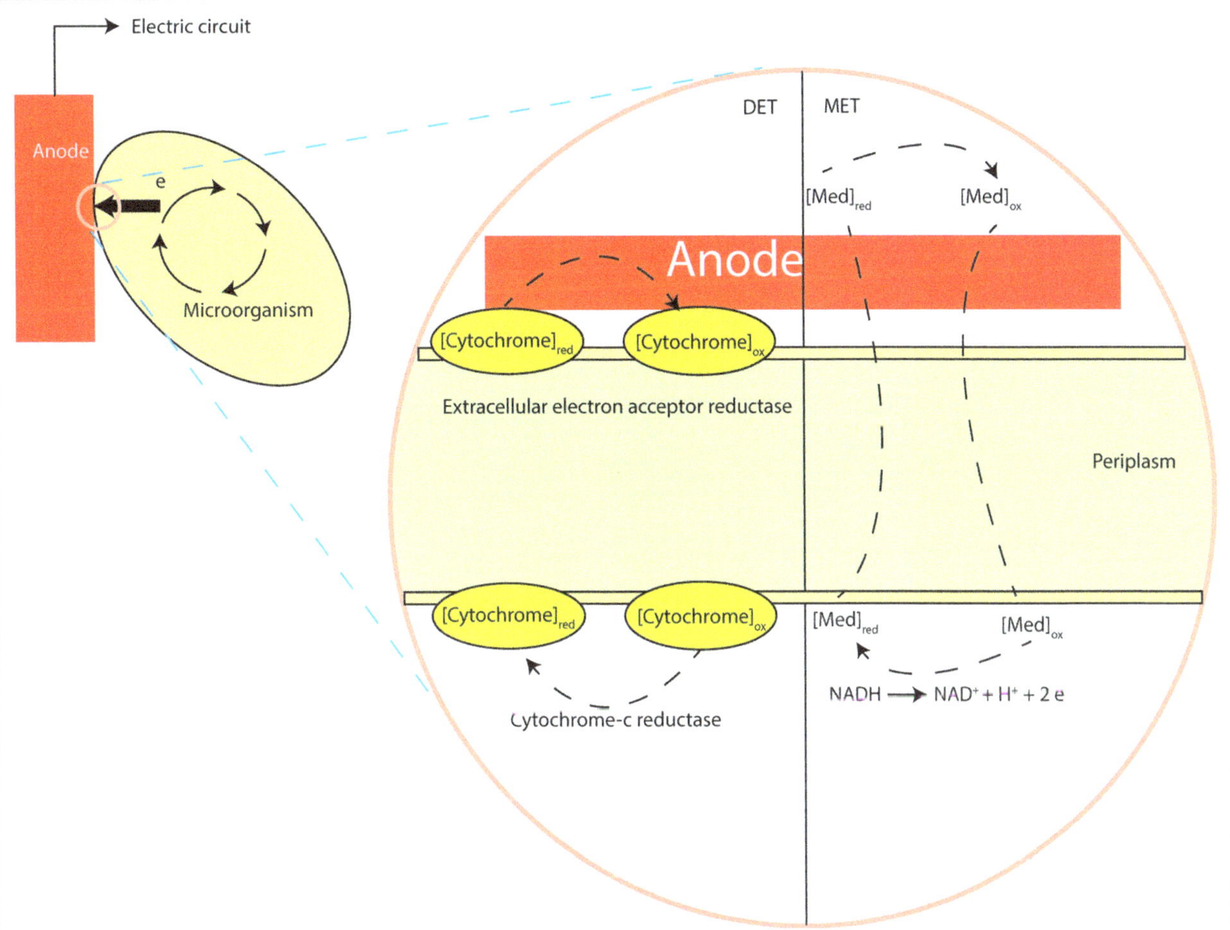

Figure 6 A schematic of the anodic mechanisms in MFCs. (i) DET, the electron transfer involving c-type cytochrome. (ii) MET, the electron transfer driven by a mediator used process; red, reduced form; ox, oxidized form. Microbes take up substrate generating carbon dioxide and protons. This process yield electrons for metabolic benefit, i.e., growth, and reduce Med_{ox} in the cytosol into Medred. Med_{red} diffuse into contact with the electrode, where Med_{red} reduce the electrode generating electrical current. The oxidized form, Med_{ox}, diffuses back through anolyte for reuse by the microbes. Periplasm (only exist in gram-negative bacteria such as G. sulfurreducens).

Scanning previous MFC studies indicates that the electron-transfer property of mediators can be specific to each cell design and vary between cells and operational conditions even for the same reaction. This indicates that there is no such a certain mediator 'supreme' for all MFCs. Therefore, to avoid limiting the scope of the present study to a particular mediator and leave room for researchers to refine the electron-mediating properties of mediators, we choose to model the function of mediators, skipping their properties such as electric potential and the ability to penetrate the cellular membrane.

Locating the target site in the DET mode

The c-type cytochrome is the electron source for the DET mode (Figure 7) [62]. In the network model, the reaction that captures the c-type cytochrome production and the direct electron transfer to the exogenous electron acceptor are retrieved and listed in Table 6.

The reaction CYOR1m makes the model suitable for studying direct electron transfer in the case of MFC. The availability of the c-type cytochrome for extracellular electron transfer is determined by the flux of this reaction. It is assumed that one round of this reaction cycle produces two quotients of the reduced c-type cytochrome for the extracellular electron transfer reaction (FERCYT), equivalent of two quotients of electrons for the electric circuit.

Objective equations

For modelling the maximum metabolic potential for the current production and the pertinent metabolic behaviours in the MET mode, the cells were tested for their capabilities to convert low energetic co-substrates (NAD) to their highly energetic counterparts (NADH). For the DET mode, the oxidation rate of c-type cytochrome by the exogenous electron acceptor (i.e., the anode) was

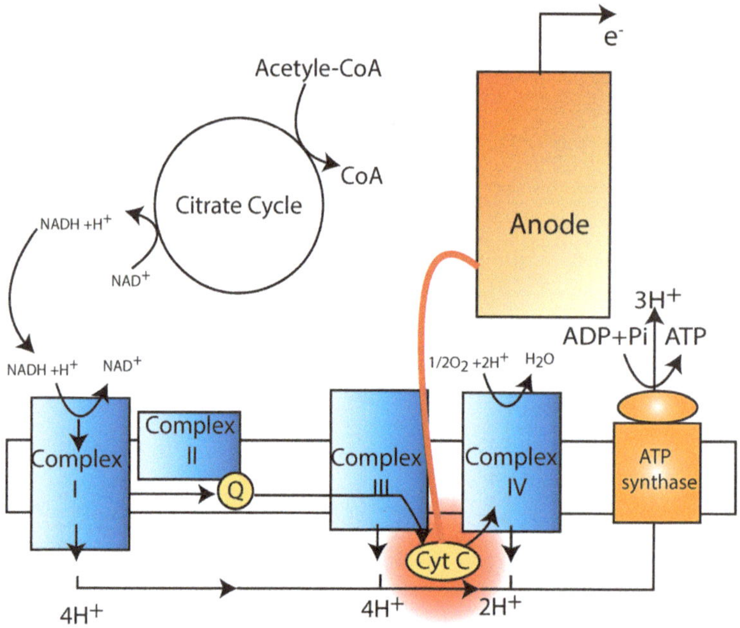

Figure 7 A schematic illustration of the electron source in direct electron transfer mode. Cyt C, Cytochrome c, which is not part of an enzyme complex and moves between complex III and IV as a freely soluble protein.

maximized. For the Mixed mode, the maximization of the rates of the MET and DET were equally combined. Hence, the Mixed mode was just a special case of the MET mode with the reduced type c cytochrome as an additional electron supplier.

The present study used an optimization based framework based on a hypothesized objective function produced by a weighted combination of fluxes (Table 7). Characterizing changes to NADH yield and biomass growth rate can aid in understanding the impact of the enhanced current extraction on cellular metabolism. We used a compound objective function comprising maximizations of NADH and biomass concurrently. The weighting for each component objective is termed as the coefficient of importance (COI), which was used to define the competing priority of one objective with respect to the other one. Thus, the objective functions were defined by setting weightings on multiple reactions (i.e., a set of COIs). COI can also serve as a tool to quantify the fraction of the additive contribution of a given flux to a fitness (objective) function [35]. In this way, the metabolic 'cost' for achieving a desired objective was quantitatively evaluated.

A high COI implies that the corresponding flux is preferentially maximized by the network, whereas a low COI indicates the converse (i.e., a low priority for the maximization). In the present modelling, COI can also be considered as the degree of NADH perturbation caused by the mediators of MFCs. A high COI magnitude associated with NADH is used to model NADH drain in MFCs, while the COI assigned to the biomass production by default is 1 in the network model.

The use of COI is also a way to represent regulatory mechanisms of the cell to increase the electron transfer rate in DET and MET for MFCs while trying to maximize the cellular growth rate. The mediator in MFCs deprives the cell of the electrons, and the cell metabolism reacts to the deprivation by regulating the metabolism to increase NADH production because the cell requires NADH for growth. In other words, we use the growth objective to model the normal cell metabolism. Then we introduce the COI to model the regulation process of the cell (Table 8).

The growth-coupled product flux ensures that the metabolic network is forced to produce the desired reaction flux while still supporting a functional state of the

Table 6 The reactions of the extracellular electron transfer in *G. sulfurreducens*

Reaction ID	Enzyme	Reaction
CYOR1m	cytochrome-c reductase	(2) ficytcc[c] + mql7[c] -->(2)focytcc[c]+h_edm [e] +h[c] + mqn7[c]
FERCYT	Fe (III) Reductase	fe3[e] + focytcc[c] --> fe2[e] + ficytcc[c]

(The reaction and metabolite abbreviations were described in Sun, *et al.* (2009) [63] and also in the Additional file 2).

Table 7 Formulation of objective functions for modelling perturbed metabolisms under electricity generation

MET	Biomass production + (COI_{DET}) NADH_mfc
DET	Biomass production + (COI_{MET}) h_edm_out
Mixed mode	Biomass production+(COI_{MET})NADH_mfc+(COI_{DET}) h_edm_out

Note: NADH_mfc denotes the NADH flow that contributes to the electricity generation; h_edm_out represents the flux of the DET reaction that involves the cytochrome c reductase.

network for a biologically meaningful goal, i.e., to grow (to survive). The predictions of these models on growth rate have been verified by the experimental studies where the original network was published [31].

The values of the COI's in this study should not be directly interpreted as specifications of the relative priorities of the two objective functions, because due to the different units of measurement of biomass and reduction equivalent objectives, they also contain a unit conversion factor. The value of this factor is not easily determined because it is related to the "molecular weight" of the biomass, but it is not clear what meaning to attach to one "molecule" of biomass.

Conversion of units of flux and current

Current (in amperes) was integrated over time and converted to electrons recovered by using the following conversions: 1 C = 1 A × 1 s, 1 C = 6.24 × 10^{18} electrons, and 1 mol = 6.02 × 10^{23} electrons (Faraday's constant 96485 C/mol). Therefore, one flux unit (mmol/g/h) can be converted into A/g as follows:

$$1\ \text{mmol/g/h} = \frac{1\ \text{mol}}{1000\text{g} \times 3600\text{s}} \times \frac{96485\text{C}}{\text{mol}} = 0.0268\text{A/g}$$

Coulombic efficiency (CE)

One of the parameters commonly used to quantify the performance of MFCs is the Coulombic efficiency (CE); the CE is defined as the ratio of electrons transferred to the anode to that in the starting substrate. For the characterization of the energy efficiency of the metabolism, the full oxidation of acetate with oxygen as the oxidant (see as follows) will be used as the reference reaction.

$$\text{Acetate oxidation reaction} : CH_3COOH \cdot + \cdot 2 \cdot H_2O \cdot \rightarrow \cdot 2 \cdot CO_2 \cdot + \cdot 8 \cdot H^+ + \cdot 8 \cdot e^- .$$

Based on such stoichiometric information, 1 mol of acetate was converted into 8 mol of electrons (1 mol of acetate = 8 mol e^-) [64].

$$CE\% = \frac{C_{output}}{C_{substrate}} \times 100\% = \frac{\text{reducing equivelent produced (mmol/gDW/h)}}{\text{acetate uptake rate (mmol/gDW/h)} \times 8}\%$$

Table 8 The principle of the bi-objective optimization in modelling regulatory mechanisms under energy extraction

Objective one	←	Regulation	→	Objective two
(Maximization of growth rate)		(modelled by way of COI)		(Maximization of election transfer in DET and/or MET, which mimics the metabolic perturbation caused by DET and/or MET for electricity generation)

Table 9 The standard potential of the anodic and catholic reactions funneling electrons to the electrode (measured at pH 7)

		Redox couple	$E^{o'}$ (V)
Anode	MET	$NAD^+ + H^+ + 2e^- \rightarrow NADH$	−0.320 [66]
	DET	Cytochrome c(Fe^{3+})+e^-→Cytochrome c(Fe^{2+})	+0.254 [66]
Cathode		$O_2 + 4H^+ + 4e^- \rightarrow 2H_2O$	+0.51 [6,67,68]

Calculation of theoretical power outputs of the three tested modes

The calculations in the present study are only based on a flux balance viewpoint and thus do not take into account ohmic resistances, concentration polarization and kinetic constraints during a MFC operation. For the sake of simplification the calculations of cell voltage (i.e., cell potential) are based on formal potentials (at pH 7) of the involved biological and electrochemical redox processes. Therefore, this calculated Electromotive force (EMF) provides an upper limit for the cell voltage and the actual potential derived from the MFC will be lower due to various potential losses [65]:

$$\Delta E^{\circ'}_{cell} = E^{\circ'}_{cathode} - E^{\circ'}_{anode}$$

Where $\Delta E^{\circ'}_{cell}$ is the standard cell potential (aka., electromotive force); $E^{\circ'}_{cathode}$ is the standard potential of cathode oxidation; $E^{\circ'}_{anode}$ is the standard potential of anode reduction.

Table 10 The theoretical limit of standard anode potentials of MFC based on *G sulfurreducens*

Electron transfer mode	$E^{\circ'}_{anode}$	$E^{\circ'}_{cathode}$	$\Delta E^{\circ'}_{cell}$
MET	−0.32	0.51	0.83
DET	0.254	0.51	0.256

The formal potentials of the anode and cathode used for calculation of power density in the three operation modes were summarized in Table 9.

The calculation of standard cell potential of MFC (Table 10) shows that as long as the same electron donors are used as fuel there is little that can be done in terms of discovering bacteria capable of creating stronger potentials.

Calculation of the viable growth rate

Several modelling works assumed 1% of the maximum theoretical biomass yield as the viability threshold for computational identification of lowest growth rate [69,70]. Nevertheless, we chose 5% of the maximum theoretical biomass yield in this study since this value is more conservative and more likely to be practically viable.

Simulating the growth of *G. sulfurreducens*

The flux values in the simulations were in units of mmol/gDW (dry weight)/h. The maximum acetate uptake rate was constrained to 18 mmol/gDW/h [71]. Non-growth associated ATP maintenance demand was set to 0.45 mmol/gDW/h [31,72]. The maximum Fe3+ and NH_4^+ uptake rates were set to 568 mmol/g/h and 0.468 mmol/g/h respectively [71]. The exchange rate of other external metabolites such as CO_2, H_2O, K^+, Mg^{2+}, PO_4^{3-}, and SO_4^{2-} were allowed to freely enter and leave the network [31].

Analysis technique

An updated metabolic network model of *G sulfurreducens* PCA strain [31,63] was used as the backbone for all *in silico* analysis. The original network model in Excel format was parsed into SBML using the converter integrated in the NetSplitter [73,74]. The growth rates and electron production were computationally determined using Flux Balance Analysis (FBA) [75,76]. Computations were performed with COBRA Toolbox [77] in MATLAB (The Math-Works Inc., Natick, MA, USA) and OptFlux [78]. FATMIN is a pipeline comprising FBA and FVA and thus can be performed with either COBRA or OptFlux according to users' preference.

Additional files

Additional file 1: A simple method to eliminate encountered futile cycle reactions during FBA modelling.

Additional file 2: Flux profiles and FATMIN results.

Additional file 3: SBML file of the *G sulfurreducens* network model.

Abbreviations

MFC: Microbial fuel cell; NAD: Nicotinamide adenine dinucleotide; NADH: Nicotinamide adenine dinucleotide (reduced form); DET: Direct electron transfer; MET: Mediated electron transfer; BG: Bromocresol green; NR: Neutral red; COI: coefficient of importance; AOS: alternate optimal solutions; FATMIN: Flux variability analysis with target flux minimization; FBA: Flux balance analysis; FVA: Flux variability analysis.

Competing interests

The authors declare that they have no competing interests.

Authors' contributions

LM and WV conceived of the study, developed the method and wrote the manuscript. LM performed *in silico* modelling and modelling analysis. WV supervised the work. Both authors read and approved the final manuscript.

References

1. Caccavo F Jr, Lonergan DJ, Lovley DR, Davis M, Stolz JF, McInerney MJ: **Geobacter sulfurreducens sp. nov., a hydrogen- and acetate-oxidizing dissimilatory metal-reducing micro organism.** *Appl Environ Microbiol* 1994, **60**:3752–3759.
2. Nevin KP, Kim BC, Glaven RH, Johnson JP, Woodard TL, Methe BA, Didonato RJ, Covalla SF, Franks AE, Liu A, Lovley DR: **Anode biofilm transcriptomics reveals outer surface components essential for high density current production in Geobacter sulfurreducens fuel cells.** *PLoS One* 2009, **4**:e5628.
3. Bond DR, Lovley DR: **Electricity production by geobacter sulfurreducens attached to electrodes.** *Appl Environ Microbiol* 2003, **69**:1548–1555.
4. Nevin KP, Lovley DR: **Mechanisms for accessing insoluble Fe(III) oxide during dissimilatory Fe(III) reduction by Geothrix fermentans.** *Appl Environ Microbiol* 2002, **68**:2294–2299.
5. Straub KL, Schink B: **Evaluation of electron-shuttling compounds in microbial ferric iron reduction.** *FEMS Microbiol Lett* 2003, **220**:229–233.
6. Schröder U: **Anodic electron transfer mechanisms in microbial fuel cells and their energy efficiency.** *Phys Chem Chem Phys* 2007, **9**:2619–2629.
7. Kim BH, Park HS, Kim HJ, Kim GT, Chang IS, Lee J, Phung NT: **Enrichment of microbial community generating electricity using a fuel-cell-type electrochemical cell.** *Appl Microbiol Biotechnology* 2004, **63**:672–681.
8. Du Z, Li H, Gu T: **A state of the art review on microbial fuel cells: a promising technology for wastewater treatment and bio energy.** *Biotechnol Adv* 2007, **25**:464–482.
9. Wang K, Liu Y, Chen S: **Improved microbial electro catalysis with neutral red immobilized electrode.** *J Power Sources* 2011, **196**:164–168.
10. Rabaey K, Verstraete W: **Microbial fuel cells: Novel biotechnology for energy generation.** *Trends Biotechnol* 2005, **23**:291–298.
11. Rosenbaum M, Schröder U: **Photo microbial solar and fuel cells.** *Electro analysis* 2010, **22**:844–855.
12. Yang Y, Xu M, Guo J, Sun G: **Bacterial extracellular electron transfer in bio electrochemical systems.** *Process Biochem* 2012, **47**:1707–1714.
13. Babanova S, Hubenova Y, Mitov M: **Influence of artificial mediators on yeast-based fuel cell performance.** *J Biosci Bioeng* 2011, **112**:379–387.
14. McKinlay JB, Zeikus JG: **Extracellular iron reduction is mediated in part by neutral red and hydrogenase in Escherichia coli.** *Appl Environ Microbiol* 2004, **70**:3467–3474.
15. Park DH, Zeikus JG: **Electricity generation in microbial fuel cells using neutral red as an electronophore.** *Appl Environ Microbiol* 2000, **66**:1292–1297.
16. Logan BE: **Exoelectrogenic bacteria that power microbial fuel cells.** *Nat Rev Micro* 2009, **7**:375–381.
17. Feist AM, Palsson BO: **The biomass objective function.** *Curr Opin Micro bio* 2010, **13**:344–349.
18. Yeung M, Thiele I, Palsson B: **Estimation of the number of extreme pathways for metabolic networks.** *BMC Bioinforma* 2007, **8**:363.
19. Galushko AS, Schink B: **Oxidation of acetate through reactions of the citric acid cycle by Geobacter sulfurreducens in pure culture and in syntrophic coculture.** *Arch Microbiol* 2000, **174**:314–321.
20. Terzer M, Stelling J: **Large-scale computation of elementary flux modes with bit pattern trees.** *Bioinformatics* 2008, **24**:2229–2235.
21. Esteve-Nunez A, Rothermich M, Sharma M, Lovley D: **Growth of Geobacter sulfurreducens under nutrient-limiting conditions in continuous culture.** *Environ Micro bio* 2005, **7**:641–648.
22. Lovley DR: **Extracellular electron transfer: wires, capacitors, iron lungs, and more.** *Geobiology* 2008, **6**:225–231.

23. Reguera G, McCarthy KD, Mehta T, Nicoll JS, Tuominen MT, Lovley DR: **Extracellular electron transfer via microbial nanowires.** *Nature* 2005, **435**:1098–1101.
24. Reguera G, Nevin KP, Nicoll JS, Covalla SF, Woodard TL, Lovley DR: **Biofilm and nanowire production leads to increased current in Geobacter sulfurreducens fuel cells.** *Appl Environ Microbiol* 2006, **72**:7345–7348.
25. Rabaey K, Lissens G, Siciliano SD, Verstraete W: **A microbial fuel cell capable of converting glucose to electricity at high rate and efficiency.** *Biotechnol Lett* 2003, **25**:1531–1535.
26. Nevin KP, Richter H, Covalla SF, Johnson JP, Woodard TL, Orloff AL, Jia H, Zhang M, Lovley DR: **Power output and columbic efficiencies from biofilms of Geobacter sulfurreducens comparable to mixed community microbial fuel cells.** *Environ Microbiol* 2008, **10**:2505–2514.
27. Yi H, Nevin KP, Kim BC, Franks AE, Klimes A, Tender LM, Lovley DR: **Selection of a variant of Geobacter sulfurreducens with enhanced capacity for current production in microbial fuel cells.** *Biosens Bio electron* 2009, **24**:3498–3503.
28. Lovley DR: **Microbial fuel cells: novel microbial physiologies and engineering approaches.** *Curr Opin Biotechnol* 2006, **17**:327–332.
29. Lovley DR: **Bug juice: harvesting electricity with microorganisms.** *Nat Rev Microbiol* 2006, **4**:497–508.
30. Stewart PS, Franklin MJ: **Physiological heterogeneity in biofilms.** *Nat Rev Micro* 2008, **6**:199–210.
31. Mahadevan R, Bond DR, Butler JE, Esteve-Nunez A, Coppi MV, Palsson BO, Schilling CH, Lovley DR: **Characterization of metabolism in the Fe(III)-reducing organism Geobacter sulfurreducens by constraint-based modeling.** *Appl Environ Microbiol* 2006, **72**:1558–1568.
32. Lovley DR: **Electro microbiology.** *Annu Rev Microbiol* 2012, **66**:391–409.
33. Pharkya P, Burgard AP, Maranas CD: **Exploring the overproduction of amino acids using the bilevel optimization framework Opt Knock.** *Biotechnol Bioeng* 2003, **84**:887–899.
34. Reed J, Vo T, Schilling C, Palsson B: **An expanded genome-scale model of Escherichia coli K-12 (iJR904 GSM/GPR).** *Genome Biol* 2003, **4**:R54.
35. Burgard AP, Maranas CD: **Optimization-based framework for inferring and testing hypothesized metabolic objective functions.** *Biotechnology Bioeng* 2003, **82**:670–677.
36. Rocha M, Maia P, Mendes R, Pinto JP, Ferreira EC, Nielsen J, Patil KR, Rocha I: **Natural computation meta heuristics for the in silico optimization of microbial strains.** *BMC Bioinforma* 2008, **9**:499.
37. Patil K, Rocha I, Forster J, Nielsen J: **Evolutionary programming as a platform for in silico metabolic engineering.** *BMC Bioinforma* 2005, **6**:308.
38. Lee K, Berthiaume F, Stephanopoulos GN, Yarmush DM, Yarmush ML: **Metabolic flux analysis of post burn hepatic hyper metabolism.** *Metab Eng* 2000, **2**:312–327.
39. Mahadevan R, Schilling CH: **The effects of alternate optimal solutions in constraint-based genome-scale metabolic models.** *Metab Eng* 2003, **5**:264–276.
40. Reed JL, Palsson BØ: **Genome-scale in silico models of E. coli have multiple equivalent phenotypic states: assessment of correlated reaction subsets that comprise network states.** *Genome Res* 2004, **14**:1797–1805.
41. Thiele I, Fleming RMT, Bordbar A, Schellenberger J, Palsson BØ: **Functional characterization of alternate optimal solutions of Escherichia coli's transcriptional and translational machinery.** *Biophys J* 2010, **98**:2072–2081.
42. Price ND, Famili I, Beard DA, Palsson BO: **Extreme pathways and Kirchhoff's second law.** *Biophys J* 2002, **83**:2879–2882.
43. Schuster S, Hilgetag C: **On elementary flux modes in biochemical reaction systems at steady state.** *J Biol Syst* 1994, **2**:165–182.
44. Duarte N, Herrgard M, Palsson B: **Reconstruction and validation of Saccharomyces cerevisiae iND750, a fully compartmentalized genome-scale metabolic model.** *Genome Res* 2004, **14**:1298–1309.
45. Beard DA, Liang SD, Qian H: **Energy balance for analysis of complex metabolic networks.** *Biophys J* 2002, **83**:79–86.
46. Varma A, Palsson BO: **Metabolic capabilities of Escherichia coli: I. synthesis of biosynthetic precursors and cofactors.** *J Theor Biol* 1993, **165**:477–502.
47. Holzhütter HG: **The principle of flux minimization and its application to estimate stationary fluxes in metabolic networks.** *Eur J Biochem* 2004, **271**:2905–2922.
48. Henry CS, Broadbelt LJ, Hatzimanikatis V: **Thermodynamics-based metabolic flux analysis.** *Biophys J* 2007, **92**:1792–1805.
49. Fleming RMT, Thiele I, Provan G, Nasheuer HP: **Integrated stoichiometric, thermodynamic and kinetic modelling of steady state metabolism.** *J Theor Biol* 2010, **264**:683–692.
50. Fleming RMT, Thiele I: **von Bertalanffy 1.0: a COBRA toolbox extension to thermodynamically constrain metabolic models.** *Bioinformatics* 2011, **27**:142–143.
51. Cogne G, Rügen M, Bockmayr A, Titica M, Dussap C-G, Cornet J-F, Legrand J: **A model-based method for investigating bioenergetic processes in autotrophically growing eukaryotic microalgae: Application to the green algae Chlamydomonas reinhardtii.** *Biotechnol Prog* 2011, **27**:631–640.
52. Noor E, Bar-Even A, Flamholz A, Lubling Y, Davidi D, Milo R: **An integrated open framework for thermodynamics of reactions that combines accuracy and coverage.** *Bioinformatics* 2012, **28**:2037–2044.
53. De Martino D, Figliuzzi M, De Martino A, Marinari E: **A scalable algorithm to explore the Gibbs energy landscape of genome-scale metabolic networks.** *PLoS Comput Biol* 2012, **8**:e1002562.
54. Beard DA, Qian H: **Thermodynamic-based computational profiling of cellular regulatory control in hepatocyte metabolism.** *Am J Physiol Endocrinol Metab* 2005, **288**:E633–E644.
55. Lewis NE, Hixson KK, Conrad TM, Lerman JA, Charusanti P, Polpitiya AD, Adkins JN, Schramm G, Purvine SO, Lopez-Ferrer D, *et al*: **Omic data from evolved E. coli are consistent with computed optimal growth from genome-scale models.** *Mol Syst Biol* 2010, **6**:390.
56. Schuetz R, Zamboni N, Zampieri M, Heinemann M, Sauer U: **Multidimensional optimality of microbial metabolism.** *Science* 2012, **336**:601–604.
57. Klitgord N, Segrè D: **The importance of compartmentalization in metabolic flux models: yeast as an ecosystem of organelles.** *Genome Inform* 2010, **22**:41–55.
58. Kelk SM, Olivier BG, Stougie L, Bruggeman FJ: **Optimal flux spaces of genome-scale stoichiometric models are determined by a few subnetworks.** *Sci Rep* 2012, **2**:580.
59. Park DH, Kim SK, Shin IH, Jeong YJ: **Electricity production in bio fuel cell using modified graphite electrode with Neutral Red.** *Biotechnol Lett* 2000, **22**:1301–1304.
60. Boyle NR, Morgan JA: **Flux balance analysis of primary metabolism in Chlamydomonas reinhardtii.** *BMC Syst Biol* 2009, **3**:4.
61. Shastri AA, Morgan JA: **Flux balance analysis of photoautotrophic metabolism.** *Biotechnol Prog* 2005, **21**:1617–1626.
62. Aklujkar M, Coppi MV, Leang C, Kim BC, Chavan MA, Perpetua LA, Giloteaux L, Liu A, Holmes D: **Proteins involved in electron transfer to Fe(III) and Mn (IV) oxides by Geobacter sulfurreducens and Geobacter uraniireducens.** *Microbiology* 2013, **159**:515–535.
63. Sun J, Sayyar B, Butler JE, Pharkya P, Fahland TR, Famili I, Schilling CH, Lovley DR, Mahadevan R: **Genome-scale constraint-based modeling of Geobacter metallireducens.** *BMC Syst Biol* 2009, **3**:15.
64. Kim K-Y, Chae K-J, Choi M-J, Ajayi FF, Jang A, Kim C-W, Kim IS: **Enhanced coulombic efficiency in glucose-fed microbial fuel cells by reducing metabolite electron losses using dual-anode electrodes.** *Bioresour Technol* 2011, **102**:4144–4149.
65. Logan BE, Hamelers B, Rozendal R, Schröder U, Keller J, Freguia S, Aelterman P, Verstraete W, Rabaey K: **Microbial fuel cells: methodology and technology.** *Environ Sci Technol* 2006, **40**:5181–5192.
66. Virdis B, Freguia S, Rozendal RA, Rabaey K, Yuan Z, Keller J: **4.18 - Microbial Fuel Cells.** In *Treatise on Water Science*. Edited by Peter W. Oxford: Elsevier; 2011:641–666.
67. Cheng S, Liu H, Logan BE: **Increased power generation in a continuous flow MFC with advective flow through the porous anode and reduced electrode spacing.** *Environ Sci Technol* 2006, **40**:2426–2432.
68. Zhao F, Harnisch F, Schröder U, Scholz F, Bogdanoff P, Herrmann I: **Challenges and constraints of using oxygen cathodes in microbial fuel cells.** *Environ Sci Technol* 2006, **40**:5193–5199.
69. Suthers PF, Zomorrodi A, Maranas CD: **Genome-scale gene/reaction essentiality and synthetic lethality analysis.** *Mol Syst Biol* 2009, **5**:301.
70. Zomorrodi A, Maranas C: **Improving the iMM904 S. cerevisiae metabolic model using essentiality and synthetic lethality data.** *BMC Syst Biol* 2010, **4**:178.
71. Zhuang K, Izallalen M, Mouser P, Richter H, Risso C, Mahadevan R, Lovley DR: **Genome-scale dynamic modeling of the competition between Rhodoferax and Geobacter in anoxic subsurface environments.** *ISME J* 2011, **5**:305–316.

72. Segura D, Mahadevan R, Juarez K, Lovley DR: **Computational and experimental analysis of redundancy in the central metabolism of Geobacter sulfurreducens.** *PLoS Comput Biol* 2008, **4**:e36.
73. Verwoerd W: **A new computational method to split large biochemical networks into coherent subnets.** *BMC Syst Biol* 2011, **5**:25.
74. Verwoerd W: **Interactive extraction of metabolic subnets – the netsplitter software implementation.** *Journal of Molecular Engineering and Systems Biology* 2012, **1**:2–2.
75. Varma A, Palsson BO: **Metabolic flux balancing: basic concepts, scientific and practical Use.** *Nat Biotech* 1994, **12**:994–998.
76. Orth JD, Thiele I, Palsson BO: **What is flux balance analysis?** *Nat Biotech* 2010, **28**:245–248.
77. Schellenberger J, Que R, Fleming RM, Thiele I, Orth JD, Feist AM, Zielinski DC, Bordbar A, Lewis NE, Rahmanian S, *et al*: **Quantitative prediction of cellular metabolism with constraint-based models: the COBRA Toolbox v2.0.** *Nat Protoc* 2011, **6**:1290–1307.
78. Rocha I, Maia P, Evangelista P, Vilaca P, Soares S, Pinto JP, Nielsen J, Patil KR, Ferreira EC, Rocha M: **OptFlux: an open-source software platform for in silico metabolic engineering.** *BMC Syst Biol* 2010, **4**:45.

Large area micropatterning of cells on polydimethylsiloxane surfaces

Mahmoud E Moustafa, Venkat S Gadepalli, Ahmed A Elmak, Woomin Lee, Raj R Rao and Vamsi K Yadavalli*

Abstract

Background: Precise spatial control and patterning of cells is an important area of research with numerous applications in tissue engineering, as well as advancing an understanding of fundamental cellular processes. Poly (dimethyl siloxane) (PDMS) has long been used as a flexible, biocompatible substrate for cell culture with tunable mechanical characteristics. However, fabrication of suitable physico-chemical barriers for cells on PDMS substrates over large areas is still a challenge.

Results: Here, we present an improved technique which integrates photolithography and cell culture on PDMS substrates wherein the barriers to cell adhesion are formed using the photo-activated graft polymerization of polyethylene glycol diacrylate (PEG-DA). PDMS substrates with varying stiffness were prepared by varying the base to crosslinker ratio from 5:1 to 20:1. All substrates show controlled cell attachment confined to fibronectin coated PDMS microchannels with a resistance to non-specific adhesion provided by the covalently immobilized, hydrophilic PEG-DA.

Conclusions: Using photolithography, it is possible to form patterns of high resolution stable at 37°C over 2 weeks, and microstructural complexity over large areas of a few cm^2. As a robust and scalable patterning method, this technique showing homogenous and stable cell adhesion and growth over macroscales can bring microfabrication a step closer to mass production for biomedical applications.

Keywords: Cell micropatterning, Poly (dimethyl siloxane), Photolithography, Poly (ethylene glycol)

Background

The development of methods to spatially direct cell growth in two and three dimensions is a fundamental challenge for *in vitro* research and simulating *in vivo* cellular microenvironments [1,2]. Beyond applications in tissue engineering and microarray technologies, precisely controlling the location of cells has potential in furthering our understanding of fundamental cellular processes [3-5]. Precise regulation of cell response and fate can reveal insights into intercellular interactions and cues [6]. By integrating microfabrication strategies, it is possible to form efficiently controlled cell cultures, or lead to hierarchical organization as tissues and organs [7]. Typically, spatial control has been achieved by creating well-defined physical or biochemical barriers, or cell-adhesive regions to encourage specific attachment. Conversely, controlling non-specific adsorption in regions where cell growth is not desired provides a similar effect. Nevertheless, maintaining microscale precision and uniformity over large areas (cm), a characteristic of live tissues, remains a challenge for translation to application.

Various synthetic and natural materials are used as substrates for cell growth and differentiation [8]. As an alternative to rigid polystyrene and glass surfaces, poly (dimethylsiloxane) (PDMS) is a versatile polymer that has been widely used as an elastic, stretchable, cellular substrate in the form of microfluidic channels, microwells, and micro- and nano-pillars [3,9,10]. PDMS possesses unique advantages including optical transparency, biocompatibility, flexibility, tunable mechanical properties, oxygen permeability, durability and low cost [11]. However, the surface of PDMS is highly hydrophobic (contact angle ~105°) which tends to result in the non-specific adsorption of proteins and other biomolecules required for cell attachment and growth [12,13]. Surface modification is therefore required for effective spatial regulation of cells. Typically, cells have been grown on

* Correspondence: vyadavalli@vcu.edu
Department of Chemical and Life Science Engineering, Virginia Commonwealth University, Richmond, VA 23284, USA

PDMS substrates without specific spatial control [11,14]. On the other hand, micropatterning strategies on PDMS have usually involved microcontact printing (μCP) to form high resolution cell-adhesive patterns [15,16]. Surface modification using plasma oxidation of PDMS to increase hydrophilicity followed by surface functionalization has also been reported [17]. However, these physisorption approaches are typically non-covalent in nature, confined to small areas (mm), or do not present adequate physical barriers for cellular growth, making them short-lived [18]. Forming physico-chemical barriers can confine cells to adhesive regions, while allowing growth over extended periods of time and over large areas. Adapting photolithography to form geometrically distinct barriers for specific cell attachment provides the ability to easily fabricate high resolution patterns over large areas [5,19].

Surface modification using hydrophilic and neutrally charged polymers, in particular homo and hetero-functional polyethylene glycol (PEG) hydrogels, has been extensively used to repel non-specific protein adsorption and guide cell attachment [20-24]. Using PEG on PDMS therefore provides a means to precisely direct cell adhesion. However, covalently attaching PEG to PDMS has been difficult. Photo-induced grafting for surface modification of PDMS was first demonstrated using acrylic acid (PAA), acrylamide and polyethylene glycol methacrylate (PEG-MA) monomers [25]. However, both PEG-MA and PAA are not optimal owing to surface charges and gradual loss of hydrophilicity resulting in eventual cell adhesion. Further, PEG-MA yields fragile patterning and cannot withstand physiological or microfluidic shear stresses [26]. Photo-induced graft polymerization using polyethylene glycol diacrylate (PEG-DA) was used to micropattern PDMS, which is effective with long-lasting hydrophilic properties and stable patterns over 2 months [10]. The micropatterned-PEGDA-grafted PDMS was applied to protein adsorption and cell adhesion. However, a reliable strategy to form stable, micro and macroscale patterns over large areas on different PDMS compositions is still challenging.

One of the advantages of using PDMS is its tunable mechanical nature. By controlling the ratio of monomer to crosslinker, the stiffness of the underlying substrate can be altered. This in turn has a great influence on the cell growth on the surface [27]. For instance, our group and others have previously shown that human embryonic stem cell proliferation can be affected by varying the stiffness [28,29]. Characteristically, exposure of cell-binding motifs differ based on the nanoscale surface stiffness, reflecting changes in cell behavior [30]. In this work, we investigate the photopatterning of PEG-DA hydrogels that can be used on PDMS surfaces with different substrate stiffness. We show a facile strategy that allows the fabrication of stable, high resolution patterns for microfluidics and culture of cells over large areas (several cm). Areas covered by PEG-DA are used to prevent non-specific adhesion and confine the cells to spatially defined microstructural features. We demonstrate fibroblast attachment to these patterns and show homogenous and stable cell adhesion and growth over macroscales that can bring microfabrication a step closer to mass production over larger scales for biomedical applications.

Results and discussion

Non-specific adsorption of proteins on surfaces is a common problem with various biomedical devices such as biosensors, microfluidic devices, and microarrays. Despite a host of favorable properties including flexibility, tunable mechanical properties and oxygen permeability, PDMS surfaces have required surface modification owing to a hydrophobic nature and such non-specific adsorption [11,31]. A commonly used method for blocking the adsorption of proteins involves immobilizing hydrophilic and neutrally charged polymers to protect the surface [31]. Strategies to prevent protein adsorption can also be used to spatially corral cells on to modified PDMS surfaces [13]. In particular, immobilization of poly (ethylene glycol) (PEG) on surfaces has been widely adopted [21]. Using acrylate functionalized PEGs further allows the integration of microfabrication via photolithography on such surfaces [32]. In this work, micropatterns of PEG-DA were covalently attached to PDMS surfaces as a means to physically control the spatial positioning of cells at the microscale. Since the PEG regions are resistant to the non-specific adhesion of cells, cell growth is confined to exposed regions of the PDMS which in turn, can be functionalized as desired.

Microchannels of PEG-DA on PDMS surfaces

PDMS consists of a precursor containing dimethylsiloxane oligomers with vinyl-terminated end groups, mixed with a curing agent containing a crosslinking agent and an inhibitor. Upon crosslinking, the oligomers undergo hydrosilylation and form a Si-C bond [11]. Several studies have reported on microcontact printing (μCP) as a means to pattern cell adhesive moieties on PDMS down to micro and even nanoscales [19,33,34]. However questions regarding resolution over large areas, or stability under complex environmental or mechanical cues indicate that these methods may not be optimal for long-term cellular or high-throughput studies. Here PEG polymers which behave as a negative photoresist are covalently grafted on PDMS via UV photopolymerization [25,35]. This is achieved by first creating active locations for grafting by forming free radicals, followed by covalent attachment of the photocrosslinkable PEG chains to the activated locations. Benzophenone has been previously mixed into the PDMS pre-polymer to form a photoreactive version of PDMS [36]. Here the benzophenone acts as a photoinitiator to aid in the grafting of the monomer. Uncrosslinked PEG-DA following UV exposure through a photomask can then be washed away (developing step) (Figure 1).

Figure 2A shows large scale patterning of 25 and 50 μm channel features on PDMS, reaching an overall area of 2.25 cm^2 while maintaining their fidelity uniformly throughout the substrate. Squares with sides of 150 μm are regions of PEG-DA hydrogel to limit protein and cell adhesion [20,24,37]. The reproducibility of such qualities is vital towards testing selective cell adhesion and growth. This technique can be easily extended to patterns of varying complexity and feature resolutions down to a few micron (Figure 2C). To show that the patterned channels were well developed and reaching the underlying PDMS surface, protein adsorption was tested using fluorescein-labeled albumin. As shown in Figure 2B, the green fluorescence in the channels shows patterned protein adhesion to the channels but not on the PEG-DA squares, signifying specific protein adsorption to the PDMS. Further, it verifies the absence of PEG-DA (due

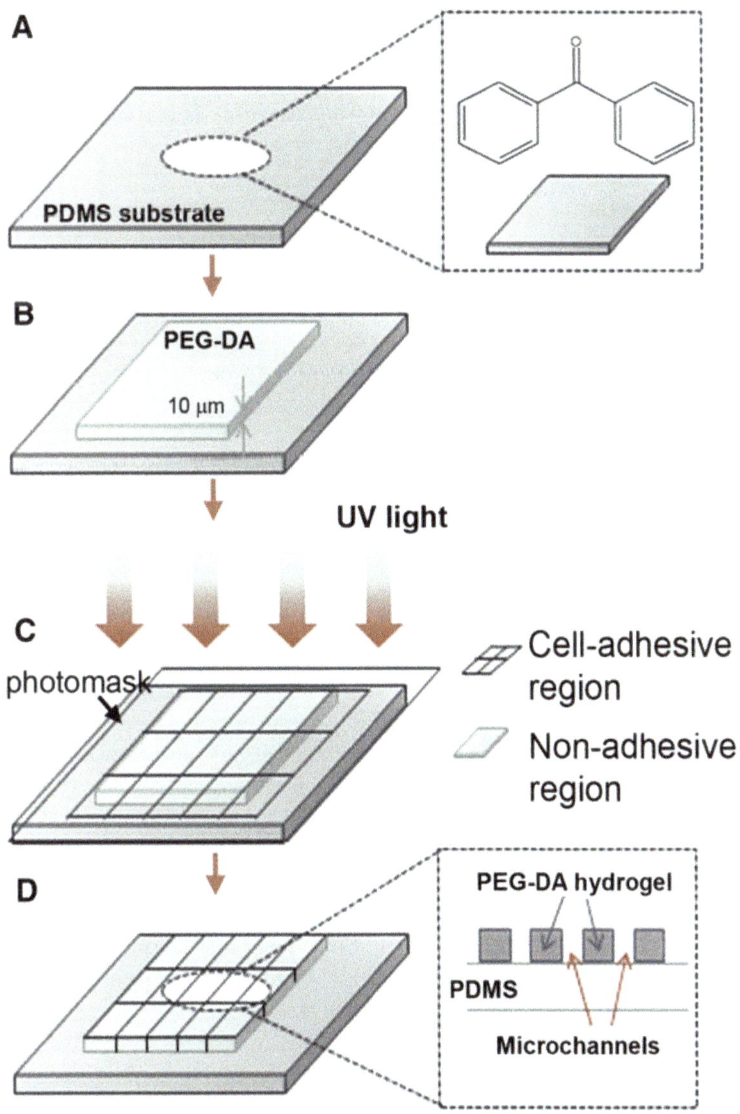

Figure 1 Schematic of micropatterning on PDMS substrates. (A) Benzophenone diffusion on to PDMS surface. **(B)** A PEG-DA layer is initially spin coated on the surface (shown to different scale for clarity). **(C)** Exposure of UV through bright field mask. **(D)** Hydrophilic hydrogel micropatterns constructed on the substrate for cells.

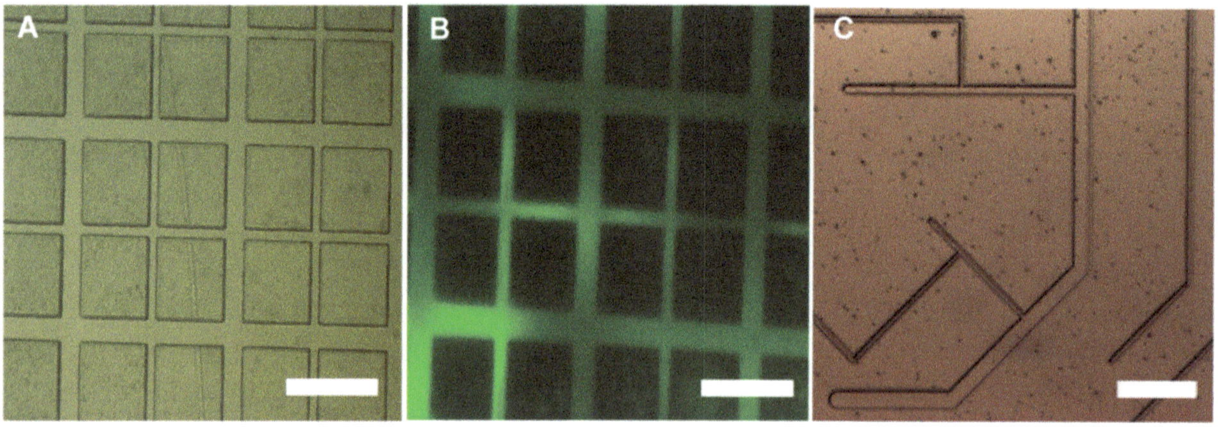

Figure 2 Micropatterns of PEG on PDMS substrates. (A) Optical image of micropatterned channels on 10:1 PDMS with 25 and 50 μm width with 150 μm PEG-DA squares. Selective FITC-BSA adhesion to the channels signified by the green fluorescence and resistance to adhesion by the dark PEG-DA 150 μm squares in **(B) (C)** Different patterns with features down to 10 μm can be easily patterned via graft polymerization and photolithography (Scale bar = 200 μm).

to potential overexposure at the bottom of the channel network) that might hinder protein attachment. This is vital to ensure cell adhesion which would be otherwise impeded by PEG-DA residue in the channels.

Optimization of micropatterning and stability

The overall strategy focused on two primary objectives – i) high fidelity of micro-architectures over a large (cm) area and, ii) stability of the covalently bound PEG microstructures to permit long-term cell culture. The covalent attachment of PEG to the underlying PDMS was developed using modifications of the methods described earlier [10,35]. In order to form stable patterns over large areas, several parameters had to be optimized including intensity of the UV light source and time of exposure, which contribute to the amount of light energy transferred to the substrate. Over or under-development during photolithography can result in the formation of cloudy hydrogels, or microstructures that delaminate from the PDMS within a few hours. One successful modification used involves brief intervals between UV exposures that allows heat to dissipate and prevent possible thermal polymerization. In addition, the concentration of the photoinitiator benzophenone, and the chain transfer agent benzyl alcohol are also modulated. Benzyl alcohol aids in the diffusion of the reactive monomers to PDMS surface by decreasing the solution viscosity [31,35]. This facilitates a stable attachment of the hydrogel to the underlying PDMS which can otherwise delaminate given the large areas. Hydrogels were examined for stability by incubating in a water bath at 37°C over a period of 2 weeks. No delamination was observed and the hydrogels remained intact with the channels maintaining their width. Samples were also observed to be stable over a month when stored dry and rehydrated, showing that this is a robust method. (Image shown in the Additional file 1: Figure S1).

Removal of benzophenone post-patterning

A potential concern for cell toxicity is the presence of residual photoinitiator benzophenone following UV polymerization [9,38]. Further, it was observed that increasing benzophenone exposure beyond 2 minutes results in cloudy suspensions and decreases the stability of the hydrogel. In an earlier work, it was shown that the concentration of benzophenone could be increased to 5% but excess causes the mixture to crystallize. This was also observed in our studies with decreased hydrogel stability [39]. Our reported approach using a low starting concentration of benzophenone ensured that residual photoinitiator was removed prior to cell culture. Immersion of PDMS in acetone was previously suggested as a means to remove benzophenone [10]. However, such treatment resulted in dry hydrogels that delaminate instantly. In our experiments, immersion in a 50 wt. % acetone solution was used, followed by incubation in water overnight and extensive washing. To confirm that the photoinitiator was completely removed, ATR-IR spectroscopy was used to verify the extent of removal. Figure 3 shows surface analyses of PDMS samples for benzophenone following photopolymerization and both conventional and modified acetone treatments. This is monitored by observing peaks near 3000 cm^{-1} representing the C-H stretching of an aromatic ring and the carbonyl group (bridging the two phenyl rings) peak at 1720 cm^{-1}. Upon rinsing the samples with the 50% acetone solution and storing them in water overnight, an attenuation of these peaks shows the benzophenone is completely removed. The untreated PDMS is shown as a comparison in the lower trace.

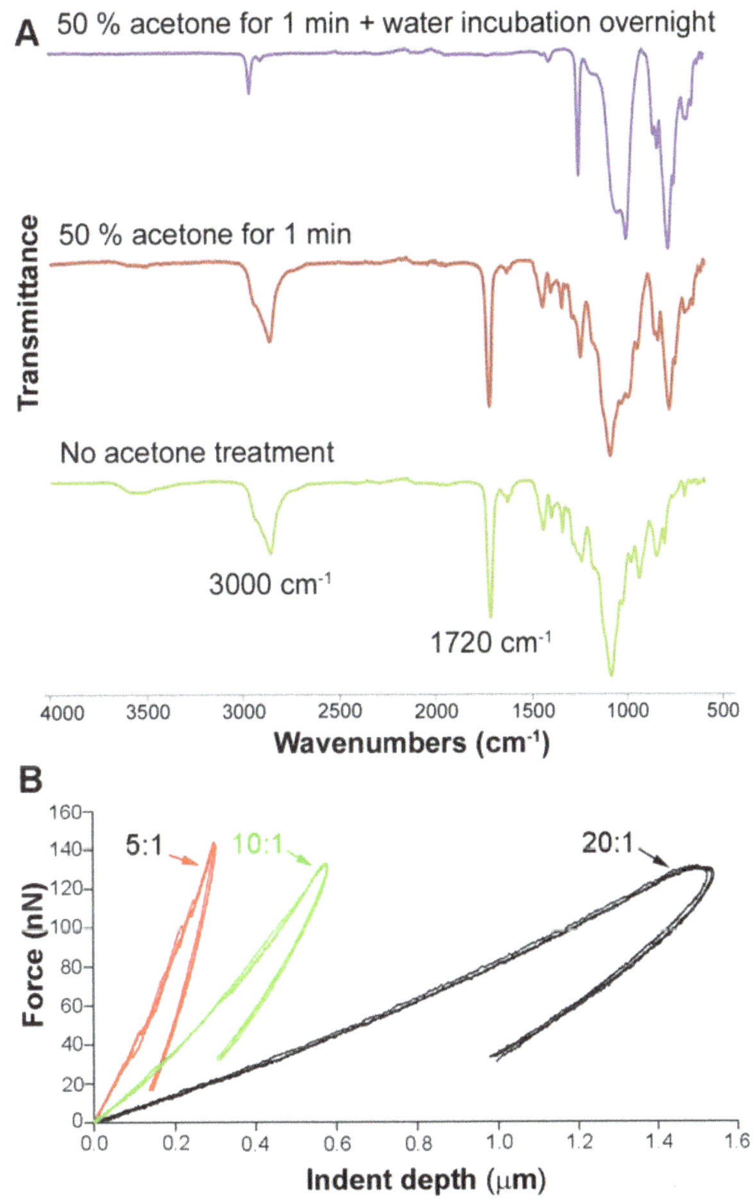

Figure 3 Surface characterization and mechanical property measurement. (A) Observation of peaks at 3000 and 1720 cm^{-1} indicates that benzophenone is washed away upon rinsing PDMS with 50 wt% acetone solution for 1 minute and overnight incubation in water (top). Washing with acetone for 1 minute (middle) and untreated PDMS (lower) are shown for comparison **(B)** Sample loading curves of nanoindentation measurements on PDMS. The 20:1 (base:crosslinker ratio) shows more compliant surfaces in comparison to the stiffer 10:1 and 5:1 samples.

Effect of substrate stiffness

Cell behavior in terms of proliferation, spreading and attachment can be regulated by altering the stiffness of the substrate. Acting as the *in vitro* extra cellular matrix (ECM), the mechanical properties influence the chemical and physical cues responsible for cell fate. Diverse cell types vary in terms of adhesion and proliferation to changes in stiffness of the substrate. For instance, neural progenitor cells were found to favor neuron and astrocyte differentiation on softer surfaces, but oligodendrocyte differentiation on stiffer substrates [40]. This indicates that such change in the mechanical properties of the substrate can also influence lineage specification. One of the primary advantages of PDMS over substrates such as glass and polystyrene is its tunable mechanical nature and flexibility. However, to date, most studies involving changes in stiffness of PDMS used un-patterned surfaces [11,28,30]. Here, in order to optimize cell micropatterning for different cell types, underlying substrate stiffness was further tested as an additional cue.

The stiffness of PDMS can be altered by manipulating the base precursor to cross-linker ratio which dictates the quantity of un-crosslinked oligomers [1,28]. Typically, optimal PDMS, in terms of cross-linking density and amount of un-crosslinked oligomers, is standardized as 10:1 (base: crosslinker). Here, three ratios were tested – 5:1, 10:1 and 20:1 PDMS, where an increase in the amount of crosslinker results in stiffer substrates [41]. Uncrosslinked, low molecular weight oligomers that can influence the surface chemistry and therefore cell behavior on the substrate. Excess curing agent, as in the 5:1 PDMS, means a stiffer substrate since further crosslinking is promoted [42]. Surplus of precursor, as in the 20:1 PDMS, leads to a softer substrate due to insufficient curing agent and the presence of unlinked vinyl terminated oligomers [11]. While temperature and curing time also contribute to the rigidity of PDMS, they were kept constant at 62°C for 24 hours. Table 1 demonstrates the variance in the Young's modulus as obtained using AFM-based nanoindentation. As expected, increasing the ratio of crosslinker to precursor causes an increase in the modulus and stiffness [43] (Figure 3B). The micropatterning strategy developed above is versatile enough to be adaptable to all stiffness of PDMS without any significant change in protocol. Typically the different PDMS substrates required adjustments in only the time of UV exposure. Images showing protein adsorption to channels of 5:1 and 20:1 PDMS is presented in the Additional file 1: Figure S2.

Cell culture on patterned PDMS surfaces

The patterned surfaces were studied for cell adhesion and growth with a focus on cell avoidance to the PEG regions and confinement to PDMS. Since native PDMS does not promote cell attachment due to its high hydrophobicity [12], coating the PDMS with fibronectin allowed cell adhesion and proliferation. In our experiments, two cell lines - 3T3 mice fibroblasts and human dermal fibroblasts (HDFs) were tested at a density of 5×10^4 cells/well on fibronectin coated PDMS samples. After 6 days, the PEG-DA hydrogel regions showed no cell attachment as expected. The cells were shown to be fully proliferated and interconnected at high density throughout the microchannel patterns to form a network over a large area (Figure 4). Here we show representative images of each culture. The 50 μm channels showed the aggregation of several cells compared to the smaller 25 μm ones which had only a few cells aligned. This indicated that not only was the adhesion confined to the PDMS channels, but that the cells responded to spatial cues and distributed accordingly. Consistent proliferation of the cells throughout the pattern was also achieved over a large area (typically 1–2.5 cm) indicating that this is a scalable method that can be used to cover an area such as a tissue culture well or petridish. The cells were fixed using 4% paraformaldehyde in PBS and stained with both DAPI and phalloidin. Figure 5B shows fluorescence imaging of the cells with the nuclear and cytoskeleton stains overlaid.

Table 1 Elastic moduli and stiffness of PDMS with varying base to curing agent ratios

Base: CA	Modulus (MPa)	Stiffness (N/m)
5:1	6.10 ± 0.11	1.90 ± 0.04
10:1	2.95 ± 0.05	1.35 ± 0.02
20:1	1.38 ± 0.05	1.10 ± 0.04

Earlier, a higher adhesion density of 3T3 mouse fibroblasts on un-patterned 20:1 PDMS than the stiffer 5:1 and 10:1 surfaces had been reported. Despite being less stiff at the macroscale, 20:1 PDMS is more pliable at the nanoscale resulting in exposed cell-binding motifs [30]. On the other hand, a different study found transformed fibroblasts to grow at similar rates on PDMS irrespective of stiffness [11]. In our experiments, expectedly, the morphology and slow proliferation of the adhered cells is an indication that substrate stiffness is an important contributor in micropatterned samples as well. This is observed in Figure 3 which shows consistent fidelity of the features across samples but lower cell density. Both 5:1 and 20:1 PDMS maintained precise cell adhesion to the channels but at lower densities compared to the 10:1 samples (Figure 5c, d). Favoring 10:1 PDMS over 20:1 supports the claim that fibroblasts may migrate preferentially to stiffer substrates and exhibit stronger traction forces [44]. Similarly, the cell proliferation on stiffer 5:1 PDMS was higher in comparison to 20:1 PDMS. Overall this could be attributed to the optimal cross-linking density of 10:1 PDMS and the higher amount of un-crosslinked components when increasing the cross-linker or based beyond the normal ratio [11]. These components can be either mobile affecting the nutrients in growth media or stationary on the PDMS surface influencing cell attachment and growth. However, since softer substrates are desirable for certain cultures, it is important to note that overall, the formation of high-resolution PEG microstructures on PDMS of different stiffness demonstrates that this strategy can be adapted to substrates of tunable mechanical properties to control cell growth.

Conclusions

The integration of microfabrication and cell culture can result in efficient and directed cellular responses with uniform cell patterning over large length scales. In this study, we demonstrate a technique for the fabrication of stable and long-lasting PEG-DA hydrogels that are photografted on PDMS of varying stiffness. The resultant micropatterns can be formed at a high resolution down to 5–10 μm. The patterns are robust over a period of weeks and maintain function in cell culture medium,

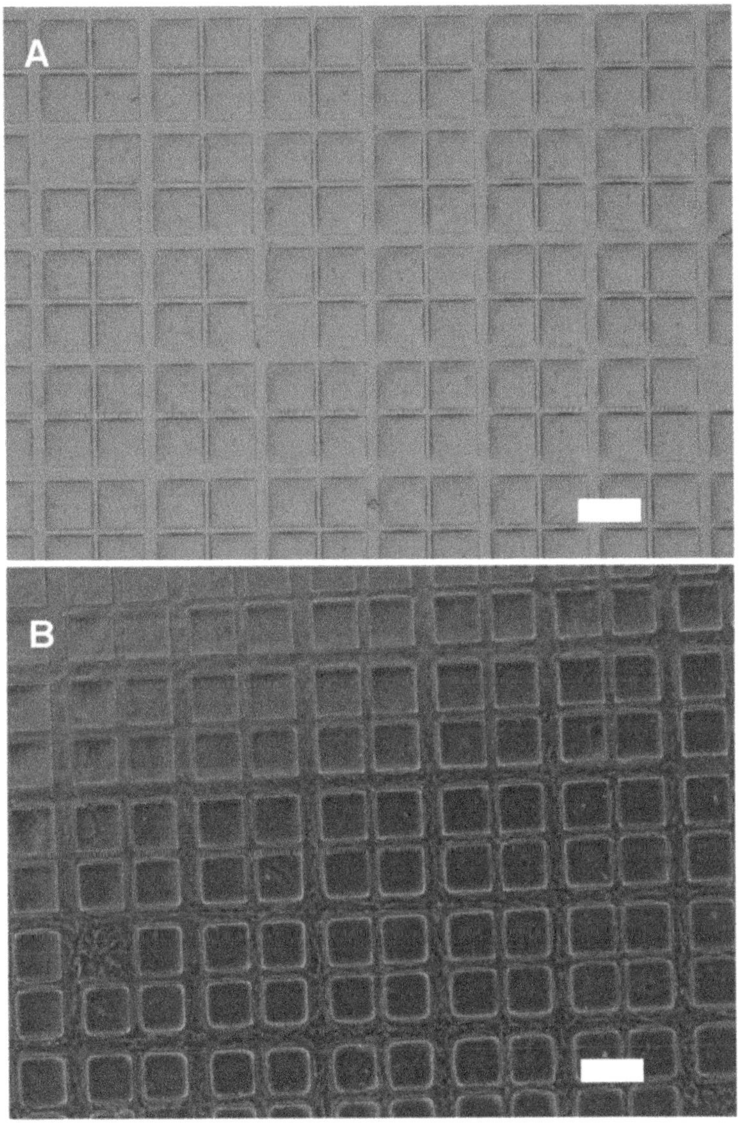

Figure 4 Patterning of fibroblasts. (A) Large scale micropatterning of PEG-DA by photografting on PDMS. **(B)** Mouse 3T3 fibroblasts grown on the patterns. Features had high fidelity over the total area of the micropatterns ~1.5 × 1.5 cm.

specifically resisting cell adhesion and directing cell growth. By increasing benzyl alcohol concentration in the PEG-DA monomer, decreasing benzophenone treatment and using intermittent UV exposure, both hydrogel attachment and resolution were optimized. Such measures further lead to specific, high density protein and cell adhesion on PDMS that was uniformly micropatterned up to an area of 2–4 cm^2. The effect of altering the mechanical properties of the patterned substrate, through stiffness, on cell behavior was examined on 5:1, 10:1 and 20:1 PDMS samples. Controlled cell adhesion and proliferation was achieved for all three ratios with the cells migrating specifically to the PDMS channels surrounded by PEG-DA. 10:1 PDMS, which is very commonly used, was also shown to be a versatile and flexible surface in terms of growing fibroblasts on micropatterned regions. The ability to further functionalize and tune the PDMS surface and its underlying stiffness, provides a method to tailor the attachment and culture of various cells in different geometries. This study shows potential to further increase the micropatterning areas while reaching smaller widths on the microscale. More complex features can also be adapted that mimic the *in vivo* microenvironment and be used for directing cells into specific lineages and controlling their fate.

Methods

Materials

Poly (dimethyl siloxane) (PDMS) prepolymer and curing agent (Sylgard 184) were obtained from Dow Corning

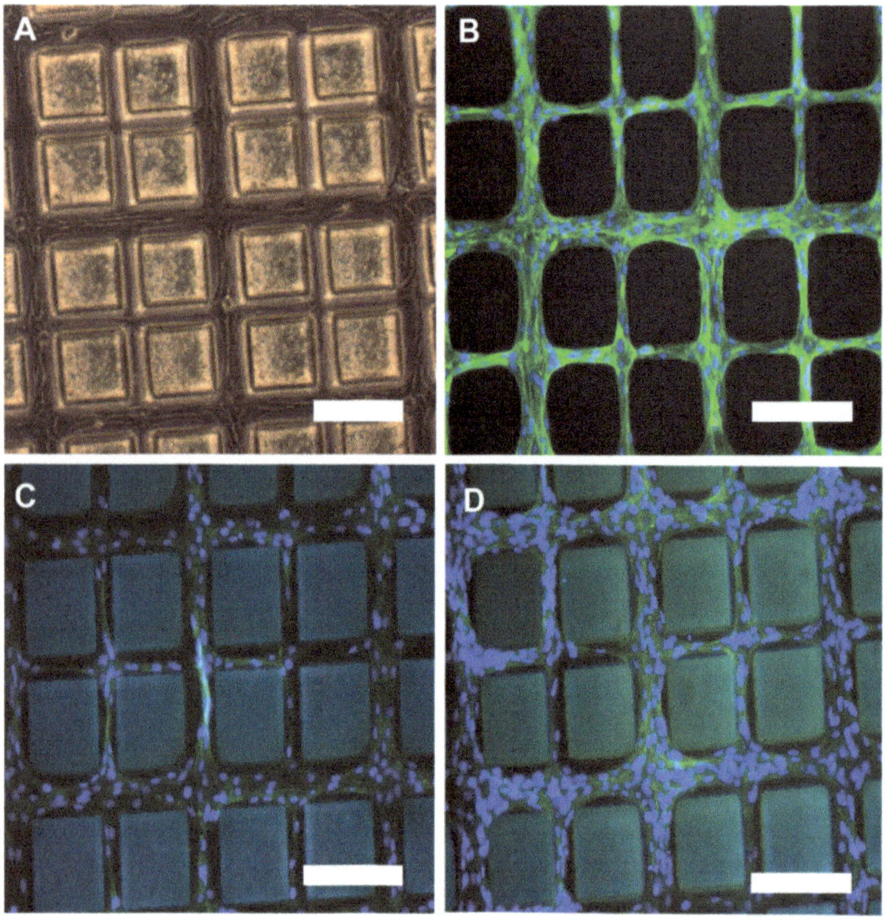

Figure 5 Effect of substrate stiffness. (A) A bright field optical image of human dermal fibroblasts after 6 days on 10:1 PDMS forming an interconnected network around 150 μm squares of PEG-DA. Phalloidin and DAPI stained cells on **(B)** 10:1 PDMS **(C)** 20:1 PDMS and **(D)** 5:1 PDMS shown with the nuclei and cytoskeleton overlaid. (Scale bar on all images = 200 μm).

(Midland, MI). PEG-DA (number average molecular weight, 575) and albumin-fluorescein isothiocyanate conjugate were purchased from Sigma-Aldrich (St. Louis, MO). Benzophenone, acetone, Benzyl alcohol, Sodium periodate ($NaIO_4$), methanol, paraformaldehyde powder and Phosphate buffered saline (PBS) solution (10X) were obtained from Fisher Scientific (Fair Lawn, NJ) and used as received. 3T3 mice fibroblasts and human dermal fibroblasts were used to test cell adhesion and cell growth. 4′, 6-diamidino-2-phenylindole (DAPI) and Alexa Fluor Phalloidin 488 (Life Technologies, Grand Island, NY) were used for staining. A bright-field reflective chrome photomask was designed using CleWin and custom fabricated to form large area grids consisting of 50 and 25 μm lines.

UV photografting of PEG-DA on PDMS

To prepare the PDMS substrates of varying stiffness, the mass ratio of base to curing agent was varied to form (in order of decreasing stiffness) 5:1, 10:1 and 20:1 samples. 12.5 g of pre-polymer was mixed with 1.25 g, 0.625 g and 2.5 g of curing agent respectively and added to a 60 mm plastic petridish. After overnight curing at 62°C, the PDMS was peeled off and diced into squares (~1-4 cm^2). PDMS slabs were then immersed in a 10 wt. % benzophenone solution in acetone for 2 minutes. The samples were rinsed with methanol and air dried.

1 ml of 40wt% PEG-DA solution was prepared by dissolving 400 μl of PEG-DA, 10 μl of 100mM $NaIO_4$ (1 mM) and 50 μl of benzyl alcohol (5 wt. %) in water. 65 μl of the reaction solution was cast on the PDMS surfaces. PEG-DA was photopolymerized through a photomask using a 365 nm, 500 mW/cm^2 light source (OmniCure S1000, Lumen Dynamics). PEG-DA behaves as a negative photoresist in the presence of UV light and is crosslinked owing to the photoinitiator benzophenone [32,45]. The schematic for this reaction is shown in the (Additional file 1: Figure S4). The exposure conditions were optimized by varying intensity and time of exposure to obtain large patterned areas. Figure 1 shows a schematic of the steps involved in obtaining a stable, micropatterned PEG-DA hydrogel on PDMS.

Micropatterned PDMS samples were stored in 1x PBS and sterilized by UV exposure for 30 minutes (<0.01 W/cm^2) prior to cell studies. To promote cell attachment, samples were subsequently immersed in 3 ml of 6 μg/ml fibronectin solution for one hour prior to cell culture.

Cell culture on micropatterned samples

3T3 mice fibroblasts were cultured at 37°C in 5% humidified environment in Dulbecco's modified eagle's medium (DMEM) with 10% fetal bovine serum. On reaching 90% confluence in four days, the cells were trypsinized (0.25% trypsin solution) and passaged at a density of 5×10^4 cells per PDMS sample. Human dermal fibroblasts (HDFs) were maintained in minimum essential medium supplemented with 2 mM l-glutamine, 1% penicillin/streptomycin, 15% fetal bovine serum, non-essential and essential amino acids, sodium pyruvate and vitamins [46]. Cells were cultured on the 5:1 and 20:1 samples at the same density. For cell staining, PDMS samples were each covered with 1 mg/ml of albumin-fluorescein isothiocyanate (FITC-BSA) conjugate for 30 minutes before being examined under a fluorescence microscope (Nikon ECLIPSE TE2000-U). Following 6 days of cell culture, the samples were fixed by adding 4% paraformaldehyde solution for 30 minutes. Phalloidin 488 was added as a cytoskeletal stain and DAPI as a nuclear stain for 30 minutes each. The samples were washed 3 times for 5 minutes each using PBS wash buffer following each of the fixing and staining steps.

Nanomechanical measurements of PDMS substrates

Mechanical properties of crosslinked films of PDMS were measured using AFM-based nanoindentation (MFP-3D, Asylum Research, Santa Barbara, CA). All samples were indented using an AC160 TS cantilever (Olympus Research, Tokyo, Japan) with nominal spring constants varying from 30–40 N/m. The actual spring constants were determined prior to each experiment using the thermal fluctuation method on a hard mica surface [47]. Different PDMS samples were indented in air with ~30 indents at different areas on the surface using constant force mode (100 nN). The Young's modulus and stiffness were obtained via the Oliver-Pharr model in Igor Pro 6.22 A (Wavemetrics Inc., OR) [48] .

Additional file

Additional file 1: Figure S1. Micropatterned PEG-DA squares on PDMS after 2 months storage. Samples could be rehydrated and used for cell culture. **Figure S2.** Large scale grids formed on PDMS. Squares are 150 μm PEG-DA surrounded by 25 and 50 μm channels. Adsorption experiment with FITC-BSA on the different substrates showing the non-specific adsorption to the PEG is minimal with the protein attaching to the PDMS. **Figure S3.** Large area micropatterning of cells on PDMS. Area on the left shows a PEG-DA square that had delaminated. Cells reach the underlying PDMS substrate and begin to proliferate. **Figure S4.** Schematic showing the chemistry of photo-induced polymerization of PEG-DA in the presence of benzophenone as the photoinitiator and UV light irradiation. The PEG-DA behaves as a negative photoresist and crosslinks in the presence of UV light turning from liquid to solid in the process. (*adapted from C. Decker'– "Photoinitiated crosslinking polymerization", Prog. Polym. Sci. 21, 593-650, 1996*).

Competing interests

The authors declare that they have no competing interests.

Authors' contributions

MM, AAE and WL synthesized the microfabricated substrates. MM and VG conducted the cell culture experiments. RRR and VKY supervised the project and experimental design. MM, VKY and RR drafted the manuscript. All authors have read and approved the final manuscript.

Acknowledgements

This research was supported by a grant from the National Science Foundation (CBET - 1144611). Photomasks were custom fabricated in the Wright Virginia Microelectronics Center in the VCU School of Engineering. The authors thank Congzhou Wang for help with nanoindentation experiments and Dr. Xuejun Wen for the 3T3 mouse fibroblasts used in this study.

References

1. Wang L, Sun B, Ziemer KS, Barabino GA, Carrier RL: **Chemical and physical modifications to poly (dimethylsiloxane) surfaces affect adhesion of Caco-2 cells.** *J Biomed Mater Res A* 2009, **93A:**1260–1271.
2. Walker GM, Zeringue HC, Beebe DJ: **Microenvironment design considerations for cellular scale studies.** *Lab Chip* 2004, **4:**91–97.
3. Choi JH, Lee H, Jin HK, Bae J-s, Kim GM: **Micropatterning of neural stem cells and Purkinje neurons using a polydimethylsiloxane (PDMS) stencil.** *Lab Chip* 2012, **12:**5045–5050.
4. Goudar VS, Suran S, Varma MM: **Photoresist functionalisation method for high-density protein microarrays using photolithography.** *Micro Nano Lett* 2012, **7:**549–553.
5. Ross AM, Lahann J: **Surface engineering the cellular microenvironment via patterning and gradients.** *J Polym Sci Pol Phys* 2013, **51:**775–794.
6. Folch A, Toner M: **Microengineering of cellular interactions.** *Annu Rev Biomed Eng* 2000, **2:**227–256.
7. Weibel DB, DiLuzio WR, Whitesides GM: **Microfabrication meets microbiology.** *Nat Rev Microbiol* 2007, **5:**209–218.
8. Ross AM, Jiang ZX, Bastmeyer M, Lahann J: **Physical aspects of cell culture substrates: topography, roughness, and elasticity.** *Small* 2012, **8:**336–355.
9. Chen W, Lam RHW, Fu J: **Photolithographic surface micromachining of polydimethylsiloxane (PDMS).** *Lab Chip* 2012, **12:**391–395.
10. Sugiura S, Edahiro JI, Sumaru K, Kanamori T: **Surface modification of polydimethylsiloxane with photo-grafted poly (ethylene glycol) for micropatterned protein adsorption and cell adhesion.** *Colloid Surf B* 2008, **63:**301–305.
11. Lee JN, Jiang X, Ryan D, Whitesides GM: **Compatibility of mammalian cells on surfaces of poly (dimethylsiloxane).** *Langmuir* 2004, **20:**11684–11691.
12. Park JY, Ahn D, Choi YY, Hwang CM, Takayama S, Lee SH, Lee S-H: **Surface chemistry modification of PDMS elastomers with boiling water improves cellular adhesion.** *Sensor Actuat B Chem* 2012, **173:**765–771.
13. Wu MH: **Simple poly (dimethylsiloxane) surface modification to control cell adhesion.** *Surf Interface Anal* 2009, **41:**11–16.
14. Zhang WJ, Choi DS, Nguyen YH, Chang J, Qin LD: **Studying cancer stem cell dynamics on PDMS surfaces for microfluidics device design.** *Sci Rep* 2013, **3:**2322.
15. Sugaya S, Kakegawa S, Fukushima S, Yamada M, Seki M: **Micropatterning of hydrogels on locally hydrophilized regions on PDMS by stepwise solution dipping and in situ gelation.** *Langmuir* 2012, **28:**14073–14080.
16. Beduer A, Vieu C, Arnauduc F, Sol J-C, Loubinoux I, Vaysse L: **Engineering of adult human neural stem cells differentiation through surface micropatterning.** *Biomaterials* 2011, **33:**504–514.

17. Bodas D, Khan-Malek C: **Formation of more stable hydrophilic surfaces of PDMS by plasma and chemical treatments.** *Microelectron Eng* 2006, **83**:1277–1279.
18. de Silva M, Desai R, Odde D: **Micro-patterning of animal cells on PDMS substrates in the presence of serum without use of adhesion inhibitors.** *Biomed Microdevices* 2004, **6**:219–222.
19. Kane RS, Takayama S, Ostuni E, Ingber DE, Whitesides GM: **Patterning proteins and cells using soft lithography.** *Biomaterials* 1999, **20**:2363–2376.
20. Kang K, Kang G, Lee BS, Choi IS, Nam Y: **Generation of patterned neuronal networks on cell-repellant poly (oligo(ethylene glycol) methacrylate) films.** *Chem Asian J* 2010, **5**:1804–1809.
21. Rogers CI, Pagaduan JV, Nordin GP, Woolley AT: **Single-monomer formulation of polymerized polyethylene glycol diacrylate as a nonadsorptive material for microfluidics.** *Anal Chem* 2011, **83**:6418–6425.
22. Harris JM, Zalipsky S: *Poly (ethylene glycol).* Washington DC: American Chemical Society; 1997.
23. Koh W-G, Revzin A, Simonian A, Reeves T, Pishko M: **Control of mammalian cell and bacteria adhesion on substrates micropatterned with poly (ethylene glycol) hydrogels.** *Biomed Microdevices* 2003, **5**:11–19.
24. Revzin A, Tompkins RG, Toner M: **Surface engineering with poly (ethylene glycol) photolithography to create high-density cell arrays on glass.** *Langmuir* 2003, **19**:9855–9862.
25. Hu SW, Ren XQ, Bachman M, Sims CE, Li GP, Allbritton N: **Surface modification of poly (dimethylsiloxane) microfluidic devices by ultraviolet polymer grafting.** *Anal Chem* 2002, **74**:4117–4123.
26. Patrito N, McCague C, Norton PR, Petersen NO: **Spatially controlled cell adhesion via micropatterned surface modification of poly (dimethylsiloxane).** *Langmuir* 2007, **23**:715–719.
27. Stevens MM, George JH: **Exploring and engineering the cell surface interface.** *Science* 2005, **310**:1135–1138.
28. Eroshenko N, Ramachandran R, Yadavalli VK, Rao RR: **Effect of substrate stiffness on early human embryonic stem cell differentiation.** *J Biol Eng* 2013, **7**:7.
29. Engler AJ, Sen S, Sweeney HL, Discher DE: **Matrix elasticity directs stem cell lineage specification.** *Cell* 2006, **126**:677–689.
30. Seo J-H, Sakai K, Yui N: **Adsorption state of fibronectin on poly (dimethylsiloxane) surfaces with varied stiffness can dominate adhesion density of fibroblasts.** *Acta Biomater* 2013, **9**:5493–5501.
31. Almutairi Z, Ren CL, Simon L: **Evaluation of polydimethylsiloxane (PDMS) surface modification approaches for microfluidic applications.** *Colloid Surface A* 2012, **415**:406–412.
32. Revzin A, Russell RJ, Yadavalli VK, Koh W-G, Deister C, Hile DD, Mellott MB, Pishko MV: **Fabrication of poly (ethylene glycol) hydrogel microstructures using photolithography.** *Langmuir* 2001, **17**:5440–5447.
33. Whitesides GM, Ostuni E, Takayama S, Jiang X, Ingber DE: **Soft lithography in biology and biochemistry.** *Ann Rev Biomed Eng* 2001, **3**:335–373.
34. Ruiz SA, Chen CS: **Microcontact printing: a tool to pattern.** *Soft Matter* 2007, **3**:168–177.
35. Wang YL, Lai HH, Bachman M, Sims CE, Li GP, Allbritton NL: **Covalent micropatterning of poly (dimethylsiloxane) by photografting through a mask.** *Anal Chem* 2005, **77**:7539–7546.
36. Jothimuthu P, Carroll A, Bhagat AAS, Lin G, Mark JE, Papautsky I: **Photodefinable PDMS thin films for microfabrication applications.** *J Micromech Microeng* 2009, **19**:045024.
37. Schlapak R, Pammer P, Armitage D, Zhu R, Hinterdorfer P, Vaupel M, Frühwirth T, Howorka S: **Glass surfaces grafted with high-density poly (ethylene glycol) as substrates for DNA oligonucleotide microarrays.** *Langmuir* 2005, **22**:277–285.
38. Anadon A: **Toxicological evaluation of benzophenone.** *EFSA J* 2009, **1104**:1–30.
39. Bhagat AAS, Jothimuthu P, Papautsky I: **Photodefinable polydimethylsiloxane (PDMS) for rapid lab-on-a-chip prototyping.** *Lab Chip* 2007, **7**:1192–1197.
40. Leipzig ND, Shoichet MS: **The effect of substrate stiffness on adult neural stem cell behavior.** *Biomaterials* 2009, **30**:6867–6878.
41. Brown XQ, Ookawa K, Wong JY: **Evaluation of polydimethylsiloxane scaffolds with physiologically-relevant elastic moduli: interplay of substrate mechanics and surface chemistry effects on vascular smooth muscle cell response.** *Biomaterials* 2005, **26**:3123–3129.
42. Mata A, Fleischman AJ, Roy S: **Characterization of polydimethylsiloxane (PDMS) properties for biomedical micro/nanosystems.** *Biomed Microdevices* 2005, **7**:281–293.
43. Carrillo F, Gupta S, Balooch M, Marshall SJ, Marshall GW, Pruitt L, Puttlitz CM: **Nanoindentation of polydimethylsiloxane elastomers: effect of crosslinking, work of adhesion, and fluid environment on elastic modulus.** *J Mater Res* 2006, **21**:535–537.
44. Lo CM, Wang HB, Dembo M, Wang YL: **Cell movement is guided by the rigidity of the substrate.** *Biophys J* 2000, **79**:144–152.
45. Decker C: **Photoinitiated crosslinking polymerization.** *Prog Polym Sci* 1996, **21**:593–650.
46. Abraham S, Riggs MJ, Nelson K, Lee V, Rao RR: **Characterization of human fibroblast-derived extracellular matrix components for human pluripotent stem cell propagation.** *Acta Biomater* 2010, **6**:4622–4633.
47. Hutter JL, Bechhoefer J: **Calibration of atomic-force microscope tips.** *Rev Sci Instrum* 1993, **64**:1868–1873.
48. Oliver WC, Pharr GM: **Measurement of hardness and elastic modulus by instrumented indentation: advances in understanding and refinements to methodology.** *J Mater Res* 2004, **19**:3–20.

6

Engineered protein A ligands, derived from a histidine-scanning library, facilitate the affinity purification of IgG under mild acidic conditions

Masayuki Tsukamoto[1,3], Hideki Watanabe[2], Ayako Ooishi[2] and Shinya Honda[1,2,3*]

Abstract

Background: In antibody purification processes, the acidic buffer commonly used to elute the bound antibodies during conventional affinity chromatograph, can damage the antibody. Herein we describe the development of several types of affinity ligands which enable the purification of antibodies under much milder conditions.

Results: *Staphylococcal* protein A variants were engineered by using both structure-based design and combinatorial screening methods. The frequency of amino acid residue substitutions was statistically analyzed using the sequences isolated from a histidine-scanning library screening. The positions where the frequency of occurrence of a histidine residue was more than 70% were thought to be effective histidine-mutation sites. Consequently, we identified PAB variants with a D36H mutation whose binding of IgG was highly sensitive to pH change.

Conclusion: The affinity column elution chromatograms demonstrated that antibodies could be eluted at a higher pH (ΔpH**≧2.0) than ever reported ($\Delta pH = 1.4$) when the *Staphylococcal* protein A variants developed in this study were used as affinity ligands. The interactions between *Staphylococcal* protein A and IgG-Fab were shown to be important for the behavior of IgG bound on a SpA affinity column, and alterations in the affinity of the ligands for IgG-Fab clearly affected the conditions for eluting the bound IgG. Thus, a histidine-scanning library combined with a structure-based design was shown to be effective in engineering novel pH-sensitive proteins.

Keywords: Antibody, Affinity purification, *Staphylococcal* protein A, Histidine-scanning Library, Combinatorial screening

Background

Monoclonal antibodies are currently one of the most important applications in biotechnology and as a promising class of drugs in the biopharmaceutical industry [1,2]. Affinity chromatography is commonly used for antibody purification on both the laboratory and industrial scale because it provides easy, fast and selective purification [3]. However, conditions for efficiently eluting monoclonal antibodies from affinity columns require improvement. Conventional affinity chromatography for antibodies uses an acidic buffer solution for elution, which would cause antibodies to denature and/or to form aggregates, leading to immunogenicity and other adverse reactions [4,5]. To solve these problems, several types of affinity ligands have been developed for the elution of antibodies under milder conditions [6-11].

The affinity ligands commonly used for antibody purification are *Staphylococcal* protein A (SpA) and *Streptococcal* protein G (SpG). Optimizing the pH-sensitive interactions of these affinity ligands with the target antibodies could improve elution conditions and guarantee the quality of purified antibody. A number of studies using various strategies [7,9-19] to modulate the pH sensitivity of protein-protein interactions have been carried out. Most of pH-sensitive* proteins in these studies were produced by introducing histidine-mutations based on structural designs, such as that by Brown *et al.* [7] on engineered pH-sensitive SpA B-domain (PAB) variants. In previous studies, we developed rational strategies for histidine-mutations to produce pH-sensitive IgG-binding SpG [9] and SpA [11].

* Correspondence: s.honda@aist.go.jp
[1]Department of Medical Genome Sciences, Graduate School of Frontier Sciences, The University of Tokyo, Kashiwa, Japan
[2]Biomedical Research Institute National Institute of Advanced Industrial Science and Technology (AIST), Higashi, Tsukuba, Ibaraki 305-8566, Japan
Full list of author information is available at the end of the article

To obtain pH-sensitive proteins, it is important to explore the most effective positions for histidine-substitution. Gera *et al.* [10] performed a histidine-scanning mutagenesis study where a number of Sso7d protein variants with individual single histidine-substitutions were systematically examined; they identified a pH-sensitive mutant containing a single histidine substitution. In addition to these structure-based designs, combinatorial screening is also a promising approach. Murtaugh *et al.* [19] reported the engineering of pH-sensitive proteins using a histidine-scanning library of anti-RNase A single domain VHH antibodies, and identified pH-sensitive antibodies with multiple histidine substitutions that retained near wild-type affinity but with high sensitivity to pH change for elution.

In this paper, we describe the design and generation of pH-sensitive PAB variants by introducing a histidine-mutation using both structure-based design and combinatorial screening. First, a histidine-scanning library for PAB was created based on structural information. The effective histidine-mutation positions were evaluated from statistical analysis of the frequency of occurrence of amino acids in the sequences isolated using a T7 phage display screening. We also investigated the pH value at which IgG peaks eluted from PAB variant-immobilized affinity columns. To better understand the effects of histidine-mutation at molecular level, we evaluated the structural stability of each histidine-substituted PAB variant and its binding affinity to IgG by Surface Plasmon Resonance (SPR) analysis.

Results and discussion

Generation of histidine-scanning library

To design the histidine-scanning library, we focused on the IgG-Fc binding interface of PAB for the mutation positions to be randomized. Our previous study [9] revealed that positions near positively charged residues on a target protein were effective sites for histidine-mutation. First, positively charged residues (histidine, lysine and arginine) on IgG1-Fc under acidic conditions were searched using the structure of PAB complexed with IgG1-Fc [pdb: 1FC2] [20]. Distances from the positively charged residues in IgG1-Fc to all the PAB residues were calculated (Table 1). Next, the solvent exposure of each residue in PAB was calculated in order to select the residues on the protein surface using the subunit structure of PAB in the complex (Table 1).

Several residues important in the interaction between PAB and IgG1-Fc have been previously identified by mutational experiments [7,11,21-24] and by molecular simulation calculations [25,26]. However, we did not take these positions into particular consideration for our library design in order to test a wide range of mutation positions in PAB.

The theoretical molecular diversity of a histidine-scanning library can be controlled by the number of mutation positions. In order to restrict the library to fit within the size range of T7 phage-display screening, we determined the maximum number of mutation positions. The seventeen mutation positions were finally selected (Table 1).

Screening of histidine-scanning library

The histidine-scanning library was screened using a T7 phage-display method. The initial library size (5.4×10^7 pfu) of T7 phages estimated by plaque assays was larger than the theoretical value (2.5×10^7 pfu) of the objective library, so a highly diverse library was produced.

After each round of selection, the T7 phages were cloned and several dozen of the sequences were determined. The ratios of unique sequences out of the sequences determined after each round were as follows (the number of unique sequences/the number of total sequences determined): 100%

Table 1 Mutation positions for histidine-scanning library

Number	5	6	9	10	11	13	14	15	17	24	25	27	28	31	32	35	36
Aminoacid residue of wild type	F	N	Q	Q	N	F	Y	E	L	E	E	R	N	I	Q	K	D
Distance (Å)	5.7	11	6.4	4.2	6.5	3.8	3.4	7.9	4.0	8.4	8.7	8.6	6.9	5.8	8.1	4.5	9.8
fSASA (%)	78	94	36	47	64	40	63	50	46	78	96	23	71	17	60	64	68
Mixed codon used in the library	YWT	MAY	CAW	CAW	MAY	YWT	YAT	SAW	CWK	SAW	SAW	CRY	MAY	MWT	CAW	MAW	SAT
Substituted amino acids in the library	H					H		H	H	H	H			H		H	
	Y	H	H	H	H	Y	H	D	Q	D	D	H	H	N	H	N	H
	L					L		Q		Q	Q			L		Q	

Distance in the table indicates the length between the PAB residue and the nearest positively charged residue in IgG-Fc. fSASA indicates the extent of solvent exposure of the side chain in wild-type amino acid residues of PAB. The seventeen mutation positions finally selected were classified into two groups: (1) the PAB residues were located within 9 Å from the positively charged IgG-Fc residues and (2) the PAB residues were located between 9 and 11 Å from the IgG-Fc residues, and with a fractional solvent-accessible surface area (fSASA) ((fSASA) = (SASA) (native)/SASA (denatured)) value of more than 68%. The following abbreviations are used for mixed bases: R = (A or G), Y = (C or T), M = (A or C), K = (G or T), S = (G or C) and W = (A or T). Substituted amino acid residues in the library indicate non-wild-type residues encoded by a mixed codon.

(the initial library: 84/84), 100% (after the 1st round: 50/50), 97.5% (after the 2nd round: 78/84), 51.7% (after the 3rd round: 30/58), 25.0% (after the 4th round: 20/88), and 19.3% (after the 5th round: 18/93). Sixty one out of 93 clones examined after the final round had identical sequences (Additional file 1: Table S2). It should be mentioned that the frequency of occurrence of the wild-type or histidine residue at each mutation position fluctuated (Additional file 2: Figure S1, Additional file 3: Figure S2 and Additional file 4: Figure S3).

Frequency of occurrence of amino acid residues of isolated PAB variants

The frequency of occurrenc of the amino acids was calculated from 18 unique sequences determined from the final round (Figure 1). The positions at which the frequency of the histidine residue was less than 20% were: F13, Y14, L17, E24 and I31, which were previously shown to be essential for the interaction with IgG-Fc [6,7,21-26]. For example, Brown *et al.* [7] used ELISA to show that the binding affinity of rabbit IgG with a PAB mutant (F13H) at pH 8.0 was more than 100-fold weaker than with wild-type PAB.

The positions at which the frequency of occurrence of a histidine residue was more than 70% were expected to be effective histidine-mutation sites. They were: Q9, Q10, R27, K35 and D36. The frequency of a histidine residue at D36 was the highest, although D36 has been little investigated in previous studies [7,21-26]. Hence, a single histidine-substituted (D36H) PAB variant and four double histidine-substituted PAB ones containing the D36H mutation and either the Q9H, Q10H, R27H or K35H mutation were produced.

Elution profile of bound IgG from the PAB variant columns

The affinity chromatograms with the PAB variants described above indicated that the pH values at which IgG eluted from all the PAB variant columns were higher than the elution pH from the PAB01 (Wild-type (WT)) column (Table 2, Additional file 5: Figure S4). First, the pH sensitivity of PAB02 (D36H) was clearly higher. For the double substitution PAB variants, the pH values of the IgG elution peaks from the PAB03 (Q9H, D36H) and PAB04 (Q10H, D36H) columns were significantly higher than that of the PAB02 (D36H) column. On the other hand, the pH required to elute IgG from the PAB05 (R27H, D36H) and PAB06 (K35H, D36H) column was similar to that of the PAB02 column. Therefore, the Q9, Q10 and D36 histidine-mutations showed much better pH sensitivity. The *ΔpH* of the IgG elution peaks from the PAB03 and PAB04 columns was more than ever reported (Table 2, Additional file 6: Table S3).

Next, we evaluated the effects of combinations of the histidine-mutations. The pH values for eluting IgG from a single histidine-substituted PAB variant (PAB07 (Q9H) or PAB08 (Q10H)) column was slightly higher than that required with PAB01. Thus, the large effects of Q9H and Q10H on the pH sensitivity of the PAB03 and PAB04 columns were likely due to the combination effect with D36H.

Structural stability of PAB variants

We attempted to evaluate the effect of introducing a histidine-mutation on the structure and thermal stability of PAB variants. The circular dichroism (CD) spectra of all PAB variants at room temperature were similar to

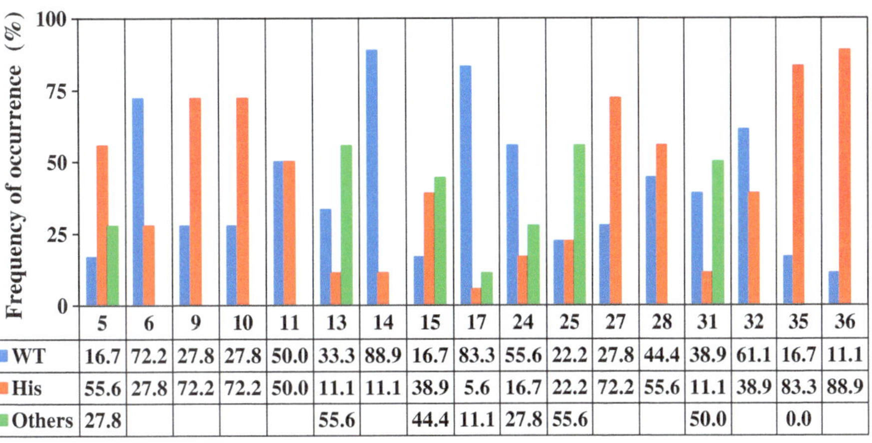

	5	6	9	10	11	13	14	15	17	24	25	27	28	31	32	35	36
WT	16.7	72.2	27.8	27.8	50.0	33.3	88.9	16.7	83.3	55.6	22.2	27.8	44.4	38.9	61.1	16.7	11.1
His	55.6	27.8	72.2	72.2	50.0	11.1	11.1	38.9	5.6	16.7	22.2	72.2	55.6	11.1	38.9	83.3	88.9
Others	27.8					55.6		44.4	11.1	27.8	55.6			50.0		0.0	

amino acid residue number

Figure 1 Frequency of occurrence of amino acid residues after the final round. The frequency of occurrence of amino acid residues was calculated from 18 unique determined sequences of PAB variants after the final round. The ordinate indicates the percentage of the frequency of occurrence. The abscissa indicates the amino acid residue number. WT, wild-type residue: His, histidine residue: Others, non-wild-type and non-histidine residues. The number in the lower three rows indicate percentage.

Table 2 The pH values of the IgG elution peaks and the thermal stability of PAB variants

Protein	Elution Peak pH	*ΔpH*	Stability T_m (K)	ΔT_m
PAB01 (Wild Type)	3.5	-	346.4	-
PAB02 (D36H)	4.6	1.1	345.8	−0.6
PAB03 (Q09H, D36H)	5.5	2.0	343.7	−2.7
PAB04 (Q10H, D36H)	6.8	3.3	343.1	−3.3
PAB05 (R27H, D36H)	4.7	1.2	316.2	−30.2
PAB06 (K35H, D36H)	4.4	0.9	342.5	−3.9
PAB07 (Q09H)	3.9	0.4	344.9	−1.5
PAB08 (Q10H)	3.7	0.2	343.6	−2.8
PAB10 (Q32H)	4.2	0.7	341.4	−5.0

Affinity columns were prepared using the PAB variants. The captured IgG on each column was eluted with a decreasing pH gradient. The pH value in the table indicates the peak position of the eluted IgG on the affinity chromatography. The midpoint of the transition during thermal denaturation (T_m) was determined from circular dichroism melting measurements.
ΔpH = (the pH value at which IgG eluted from mutant column) - (the pH value at which IgG eluted from wild-type column).
ΔT_m = (the thermal stability of PAB mutant expressed as T_m) - (that of the thermal stability of PAB wild-type, PAB01).

Table 3 Binding affinity of IgG for PAB variants

Protein	pH = 7.4		pH = 5.0	
	K_D (×10^{-9}) (M^{-1})	K_D ratio	K_D (×10^{-9}) (M^{-1})	K_D ratio
PAB01 (WT)	5.1 ± 3.2	-	7.6 ± 4.3	-
PAB02 (D36H)	33.6 ± 19.2	6.6	832 ± 443	110
PAB03 (Q9H, D36H)	288 ± 146	56	ND	-
PAB04 (Q10H, D36H)	6830 ± 3600	1300	ND	-
PAB05 (R27H, D36H)	80.7	16	2090	270
PAB06 (K35H, D36H)	550 ± 143	110	3860	510
PAB07 (Q9H)	13.9 ± 6.3	2.7	30.6	4
PAB08 (Q10H)	50.8 ± 20.5	10	951	120
PAB09 (R27H)	5.6 ± 1.3	1.1	7.0	0.92
PAB10 (Q32H)	2.8 ± 1.3	0.55	96.8	13

The binding assay was performed under neutral (pH = 7.4) and acidic (pH = 5.0) conditions using SPR. The IgG was immobilized on the surface of a CM5 sensor chip (GE Healthcare).
K_D, equilibrium dissociation constant: K_D ratio, (K_D of histidine-substituted PAB variant)/(K_D of Wild-type PAB): ND, not determined.

that of PAB01 (WT). The midpoint of the transition during thermal denaturation (T_m) was determined from CD melting measurements (Table 2, Additional file 7: Figure S5). The pH-sensitive PAB variants with effective histidine-mutations (Q9H, Q10H and D36H) showed a slight decrease in thermal stability. An increase in the number of histidine-mutations led to a decrease in the T_m value. Thus, histidine-mutations apparently have additive effects on thermal stability. In summary, however, the decrease in thermal stability due to histidine-mutations was rather limited and would not prevent utilization of these mutants as affinity ligands.

Binding assay of PAB variants

Interaction under neutral conditions

We examined the binding affinity of IgG for each of the four single histidine-substituted PAB variants under neutral conditions (pH = 7.4) (Table 3). Of these, PAB08 (Q10H) showed the weakest binding affinity, about 10-fold weaker than that of PAB01 (WT). The binding affinities of IgG for the double histidine-substituted PAB variants containing D36H were lower than those of their single substitution PAB variant counterparts. PAB03 (Q9H, D36H) was exhibited 56-fold weaker binding affinity than PAB01. Thus, the IgG binding of PAB variant columns in this study was decreased (Additional file 8: Figure S6). Still, except for PAB04, the interaction between IgG and each of the PAB variants with single or double histidine substitutions would be strong enough to retain IgG on an affinity column using these variants as ligands.

Interaction under acidic conditions

The binding affinity of IgG for PAB01 under the acidic condition (pH = 5.0) (Table 3) was about 1.5-fold weaker than at neutral pH, still quite strong. PAB07 (Q9H) and PAB09 (R27H) also retained strong affinity. Brown *et al.* [7] showed that the K35H PAB variant exhibited strong affinity under both neutral (pH = 8.0) and acidic (pH = 5.0) conditions.

On the other hand, the binding affinity of IgG for PAB02 (D36H) and PAB08 (Q10H) was more than 100-fold weaker than that for PAB01. The binding affinities for PAB03 (Q9H, D36H) and PAB04 (Q10H, D36H) were too weak to be determined under the acidic condition. Thus, histidine-mutations at Q9, Q10 and D36, as well as the double-mutations, Q9H plus D36H and Q10H plus D36H, appeared to have favorable pH-sensitivity. It should also be noted that histidine-mutation at D36 improved the pH sensitivity of all the four histidine-mutations under both neutral and acidic conditions.

The effects of histidine-mutations

The observed effects of the histidine-mutations could be explained based on crystal structure and biophysical data. Q9 and Q10 of PAB probably contribute to the interaction with IgG-Fc. The binding affinity of PAB08 (Q10H) under the neutral condition was about 10-fold weaker than that of PAB01 (WT). Five residues in IgG-Fc are found within 4 Å of Q10 of PAB. Therefore, the steric hindrance generated by the Q10H mutation likely affects the interaction with IgG-Fc. IgG elution from the affinity columns and SPR measurement results under the acidic condition demonstrated that the effects of histidine-mutation at Q10 are more significant than mutation at Q9. Because the positively charged residue (His435) in IgG-Fc resided near Q10 of PAB (<4.2 Å), electrostatic repulsion would increase when the pH value changes from neutral to acidic.

As described below, however, the decrease in binding affinity due to the D36H mutation (PAB02) was unlikely to be affected by modulation of the interaction of the IgG-Fc binding interface of PAB.

Effect of the D36H mutation

Ile253 of IgG-Fc resides 6.89 Å from D36 of PAB in the PAB-IgG1-Fc complex [20]. Since this is not a short distance, the effects of the histidine-mutation at D36 obtained from the screening experiment are difficult to explain if the effect were caused by modulation of the interaction between PAB and IgG-Fc. In fact, it was previously observed that SpA binds not only to IgG-Fc but also to IgG-Fab [27-30]. We focused on the IgG-Fab binding interface of PAB.

We evaluated the binding affinity of IgG-Fc and IgG-Fab for PAB01 (WT) and PAB02 (D36H) (Tables 3 and 4) using SPR. The interaction between PAB02 and IgG-Fc was found to be similar to that of PAB01 under both neutral and acidic conditions. On the other hand, the binding affinity of PAB02 for IgG-Fab under the neutral condition was 2.3-fold weaker than that of PAB01, and the binding affinity of PAB02 under the acidic condition was too weak to be determined. These results suggested that the interaction of IgG-Fab with PAB would significantly contribute to the effects of histidine-mutation at D36.

Next, we analyzed the effects of D36H using the structure of SpA D-domain complexed with IgG-Fab [30]. D36 is conserved in all the SpA domains (A, B, C, D and E). The crystal structure indicates that six residues in IgG-Fab are within 4 Å of D36 in PAB. Therefore, the steric hindrance due to the histidine-mutation of D36 would affect the interaction with IgG-Fab under neutral conditions. Because a positively charged residue (Arg2519) in IgG-Fab is present near D36 in PAB (<3.0 Å), electrostatic repulsion under acidic conditions should be stronger than under neutral conditions.

Table 4 Binding affinity of IgG-Fc (upper rows) or IgG-Fab (lower rows) for PAB01 (WT) and PAB02 (D36H)

Protein	pH = 7.4		pH = 5.0	
	K_D (×10^{-9}) (M^{-1})	K_D ratio	K_D (×10^{-9}) (M^{-1})	K_D ratio
PAB01 (WT)	12.8	-	1080	-
PAB02 (D36H)	15.5	1.2	1350	1.3
	pH = 7.4		pH = 5.0	
Protein	K_D (×10^{-5}) (M^{-1})	K_D ratio	K_D (×10^{-5}) (M^{-1})	K_D ratio
PAB01 (WT)	2.9	-	ND	-
PAB02 (D36H)	6.7	2.3	ND	-

The binding assay was performed under neutral (pH = 7.4) and acidic (pH = 5.0) conditions using SPR. IgG-Fc or IgG-Fab was immobilized on the surface of a CM5 sensor chip (GE Healthcare).

K_D, equilibrium dissociation constant: K_D ratio, (K_D of histidine-substituted PAB variant)/(K_D of Wild-type PAB or PAZ): ND, not determined.

We also focused on Arg2519 in IgG-Fab as it would be involved in the effects of histidine-mutation of D36. Q32 of PAB is located within 4 Å of Arg2519 of IgG-Fab. The PAB Q32H variant (PAB10) was produced, and the pH value of the IgG elution peaks from the affinity column with PAB10 was found to be significantly higher than those values for PAB01 (WT), PAB07 (Q9H) and PAB08 (Q10H) columns. This suggested that the introduction of histidine-mutations into the IgG-Fab binding interface of PAB was as effective as introducing them into the IgG-Fc binding interface. In addition, the majority of amino acid residues corresponding to Arg2519 of the top 10 therapeutic antibodies are positively charged ones (Additional file 9: Table S4). Therefore, the PAB affinity ligands with D36H mutation should be effective for therapeutic antibody purification.

Although the binding affinity of IgG-Fab for PAB01 (WT) was more than 1000-fold weaker than that of IgG-Fc (Table 4), a possible reason for the remarkable effects of the D36H mutation would be a change in avidity. IgG has four SpA binding sites: two are strong binding sites in the IgG-Fc region and two are weak binding sites in the IgG-Fab region. On an affinity column, since a plurality of interactions for these four sites would bind the IgG molecule, the cooperative effect of interactions between SpA and IgG would be strong. Since the D36H mutation would result in just 2 binding sites instead of 4 in the interaction with IgG, the effect of avidity would likely decrease significantly, thus promoting dissociation under moderately acidic conditions.

Elution profile of bound IgG from 4 × PAZ variant columns

An affinity column of tandem SpA (4 × PAZ) consisting of four SpA Z-domains was fabricated using pH-sensitive double substituted PAB variant (PAB03) and its performance was compared with commercially available affinity columns (Figure 2).

IgG was eluted with a stepwise pH decrease. The relative amounts of IgG eluted under mild acidic conditions (pH = 5 ~ 6) was 67% at pH = 6.0, 82% at pH = 5.5 and 90% at pH = 5.0. In contrast, IgG bound to the 4 × PAZ01 column or to a commercially available SpA column (rProtein A) (GE Healthcare) did not elute at pH = 5.0. Although a part of IgG injected onto the 4 × PAZ03 column seemed to flow through, more than the case of the 4 × PAZ01 column, with pH 7.5 binding buffer, the amounts of IgG leaking out were very little, if any, with pH 9.0 binding buffer (Additional file 10: Figure S7). Thus, the 4 × PAZ03 (Q9H, D36H) column would allow the retention of IgG by optimizing the affinity column condition and the elution from

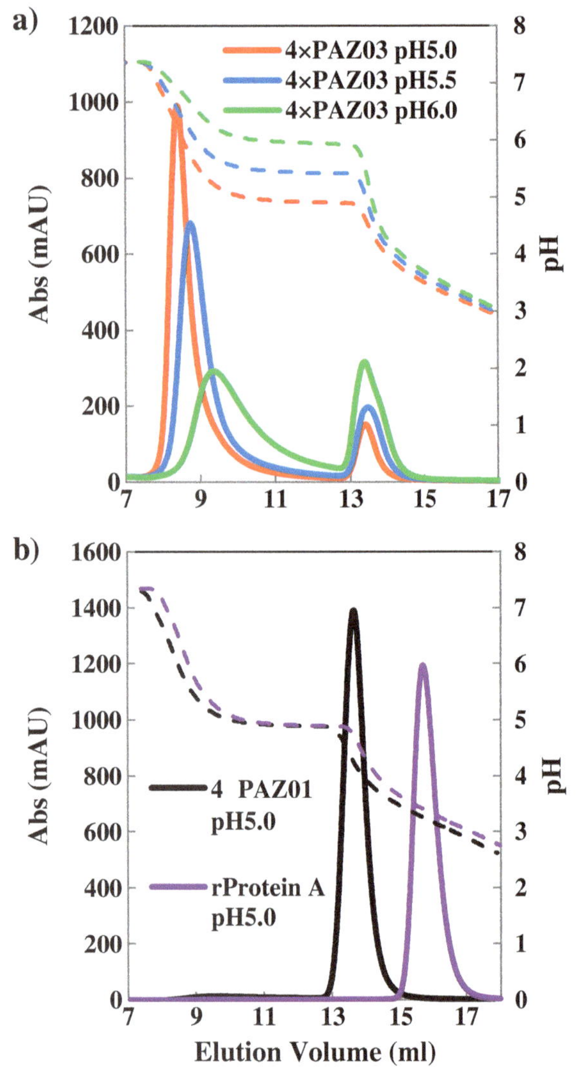

Figure 2 Elution profile of IgG on affinity columns with immobilized 4 × PAZ variants. (a) Elution profile of IgG from affinity columns prepared with immobilized 4 × PAZ03. IgG was eluted with pH 5.0, pH 5.5 or pH 6.0 buffer, **(b)** Elution profile of IgG from affinity columns prepared with immobilized 4 × PAZ01 or rProteinA (GE Healthcare). Elution was performed using pH 5.0 buffer. Affinity columns were prepared using the 4 × PAZ variants. The captured IgG on the column was eluted using a buffer with the pH indicated in the Figure. Afterwards, the residual IgG was eluted with a decreasing pH gradient. The ordinate on the right indicates the pH of the solution. The ordinate on the left indicates the absorbance at 280 nm (mAU: milliabsorbance units). The abscissa indicates the elution volume. The solid and dashed lines show the elution patterns of IgG on 4 × PAZ variants or rProteinA (GE Healthcare) columns, as shown in the Figure, and the pH value of the elution buffer, respectively.

the column under much milder pH conditions than previously reported [7,8,11].

Conclusion

We have shown that a histidine-scanning library combined with a structure-based design was effective in engineering novel pH-sensitive proteins. The positions at which histidine residues occurred with high frequency should be suitable sites for introducing histidine-mutations. Large effects of PAB mutations such as Q9H and Q10H, combined with D36H, were observed. The chromatograms demonstrated that antibodies could be eluted at a higher pH than ever reported previously when the protein A variants developed in this study were used as affinity ligands. The interactions between SpA and IgG-Fab are apparently important for the behavior of IgG bound on a SpA affinity column, and alterations in the affinity of the ligands for IgG-Fab affected the elution condition for bound IgG. Therefore, not only the interactions with IgG-Fc but also with IgG-Fab should be considered when designing a new affinity ligand for IgG purification.

Materials and methods

Histidine-scanning library design

Distances from positively charged residues (histidine, lysine and arginine residues) on IgG1-Fc to all residues of PAB in the crystal structure of the complex of PAB bound to IgG1-Fc [pdb: 1FC2] (Table 1) [20] were calculated using the program CONTACT in CCP4 suite [31]. The solvent-accessible surface area (SASA) was calculated for the subunit structure of PAB in the complex. The extent of solvent exposure of all residues of PAB was determined by the equation (Table 1) fSASA (a fractional solvent-accessible surface area) = SASA (native)/SASA (denatured), where a Gly-Xaa-Gly tripeptide model was assumed to adopt the denatured structure. The calculation was performed using surface Racer 3.0 [32].

Gene fragments for constructing a mutated PAB library were prepared with chemically synthesized oligonucleotides (Additional file 11: Table S1), purchased from Rikaken, by the overlap extension method using combinations of these oligonucleotides. At the mutation position of the oligonucleotide, we introduced mixed codons comprising three mixed nucleotides, encoding either the wild-type residue or a histidine residue. However, genes encoding neither the wild-type nor a histidine residue at that position were also produced, since the codon was a random mixture of three nucleotides. Consequently, genes encoding three or four different kinds of residues were obtained for several mutation positions.

The gene fragments thus prepared were treated with restriction enzymes *EcoR* I and *Hind* III and linked to the 3′-terminal side of the g10 gene on the T7 phage genome through ligation reaction (16°C, 16 hr) with T7 phage vector (T7Select1-1b) (Novagen). The linked T7 phage genome was subjected to an *in vitro* T7 phage particle packaging reaction (22°C, 2 hr) to prepare the initial library.

An *E. coli* BLT5403 strain cultured in 200 mL LB medium to O.D.600 = 1.0 was infected with this initial library, and a T7 phage library was collected by amplification. About 4 hours after infection, the amplified phages were recovered (T7 phages have bacteriolytic ability and are thus released from the bacterial body) in the supernatant after centrifugation. To the supernatant, 1/6 volume of 50% polyethylene glycol (PEG, molecular weight: 8000) and 1/10 volume of 5 M NaCl were added, and the mixture was stirred at 4°C overnight. The PEG-precipitated phages were partially purified by centrifugation. Then, the phages were lysed by repeated pipetting in TBST buffer (10 mM Tris–HCl (pH = 7.5), 150 mM NaCl, and 0.1% Tween 20) and the resultant suspension was filtered through a 0.22 μm filter to obtain the phage library displaying the mutated PAB. This is referred to as the initial phage library.

Phage panning and selection

Human monoclonal IgG1 antibody (Chugai) was biotinylated using NHS-biotin (Roche) according to the manufacturer's protocol. The biotinylated monoclonal antibody was mixed with streptavidin magnetic beads (Promega) to immobilize the antibodies on the magnetic beads.

For the binding step, the suspension of the phage library ($10^9 \sim 10^{11}$ phage/mL titer) displaying the mutated PAB prepared above was added to the biotinylated antibody-immobilized magnetic beads, and the mixture was shaken at 25°C for 1 hour. Afterwards, the beads with bound phages were pulled down by using a magnetic stand (Promega). The beads were washed 10 to 20 times with TBST buffer to remove antibody-unbound phages.

For the elution step, phages specifically bound with the antibodies were recovered by adding a 50 mM sodium acetate solution (pH = 5.0). The eluted phages were subjected to the infection/amplification/PEG-precipitation operation described in the preceding paragraph to prepare the phage library displaying the mutated PAB. This is referred to as the phage library after the first round.

The operation described above, i.e., the screening experiment, was repeated 5 times in total. The eluting solution used in the screening experiment from the second round onwards was 50 mM sodium acetate (pH = 6.0). The selective pressure of screening was changed at each stage by gradually decreasing the amount of biotinylated antibody-immobilized magnetic beads.

Protein preparation

Expression vectors for single or double histidine-substituted PAB variants were prepared using a QuickChange™ site-directed mutagenesis kit (Stratagene). The expression vector for wild-type PAB (pET16-b) was used as the template. The primer sequences (Rikaken) used for this procedure are shown in Additional file 11: Table S1.

The expression vectors for the tandem SpA Z-domain (4 × PAZ) variants comprised artificial pET16-b vectors containing four sets of the SpA Z-domain gene and the C-terminal cysteine codon and were purchased from Genscript.

E. coli strain BL21 (DE3) was transformed with these plasmid vectors and cultured in 2 × YT medium containing 100 μg/ml ampicillin to O.D.600 = 0.8 to 1.0 at 37°C. Recombinant gene expression of PAB and the 4 × PAZ variants was induced by the addition of isopropylthio-β-D-galactopyranoside (IPTG) to a final concentration of 1 mM, then the transformants were further cultured at 37°C for 2 hours. The bacteria harvested were suspended in TBST buffer and homogenized by ultrasonication. PAB and 4 × PAZ variants were purified from the cell lysate by IgG Sepharose 6 Fast Flow chromatography (GE Healthcare). The purity of the sample was checked with Tricine-SDS-PAGE.

Affinity chromatography

To prepare a PAB variant-immobilized affinity column, a purified PAB variant was applied to a HiTrap NHS HP column (GE Healthcare), incubated at room temperature (overnight), and then washed with 0.5 M acetic acid (pH 2.5). Affinity chromatography was performed with an AKTA purifier (GE Healthcare) using the columns as above. Human IgG1 was injected onto the column equilibrated with TBST buffer. The captured IgG was eluted with a decreasing pH gradient by using 50 mM sodium citrate (pH = 7.0) and 0.5 M acetic acid (pH = 2.5) (Table 2, Additional file 5: Figure S4). The chromatograms obtained were processed using UNICORN version 4.12 (GE Healthcare).

To prepare a tandem SpA Z-domain (4 × PAZ) variant-immobilized affinity column, a purified 4 × PAZ variant was chemically coupled to sepharose resin through the sulfhydryl group of the C-terminal cysteine using a SulfoLink Immobilization kit (Pierce). Affinity chromatography was performed as above. The captured IgG was eluted using 50 mM sodium citrate (pH = 5.0, 5.5 or 6.0). Afterwards, the residual IgG on the 4 × PAZ column was eluted with a decreasing pH gradient by using 50 mM sodium citrate (pH = 5.0, 5.5, or 6.0) and 0.5 M acetic acid (pH = 2.5) (Figure 2).

Stability analysis

PAB variants were dissolved in 50 mM sodium phosphate (pH = 7.0). Circular dichroism (CD) spectra were obtained on a J-805 spectropolarimeter (JASCO) at various temperatures. CD melting curves were obtained by monitoring the ellipticity at 222 nm with increasing temperature from 278 to 373 K at a heating rate of 1.0 K/min. The thermodynamic parameters of proteins for equilibrium unfolding were obtained through a fitting calculation using a two-state transition model [33] on the assumption that heat capacity changes (ΔC_p) of all proteins are constant and fixed to the value of the wild-type domain (ΔC_p = 2.643 kJ/mol•K) (Table 2). The value of ΔC_p was estimated using the linear relationship between molecular weight and ΔC_p displayed by various proteins [34]. Numerical fitting calculations were carried out using IGOR software (Wavemetrics). Other procedures were the same as described previously [35].

Binding affinity analysis

Binding analyses of a PAB variant to IgG, IgG-Fc and IgG-Fab were performed using a Biacore T100 (GE Healthcare). Either 4000 resonance units (RU) of human monoclonal IgG1, 1250 RU of the Fab region of human polyclonal antibody (Jackson Immunoresearch), or 1250 RU of the Fc region of human polyclonal antibody (Jackson Immunoresearch) was immobilized on the surface of a CM5 sensor chip (GE Healthcare) using amine coupling chemistry. A binding assay was performed under neutral and acidic conditions as follows: (1) 10 mM HEPES (pH = 7.4), 150 mM NaCl and Tween 20 (0.05%); (2) 10 mM sodium acetate (pH = 5.0), 150 mM NaCl and Tween 20 (0.05%). The binding data was fitted to a 1:1 binding model to determine the dissociation constant (K_D) using Biacore T100 Evaluation software (Tables 3 and 4).

Endnote

*In this study, ligand proteins with favorable pH sensitivities are referred to as "pH-sensitive". **ΔpH = (the pH value at which IgG eluted from mutant column) - (the pH value at which IgG eluted from wild-type column).

Additional files

Additional file 1: Table S2. Eighteen unique sequences of PAB variants after the final round. Amino acid residues at the indicated position number are shown for $PAB_{wild\text{-}type}$ and for 18 unique sequences from the final round, $PAB_{T7phage}$01-18.

Additional file 2: Figure S1. Frequency of occurrence of amino acid residues for each mutation position (5F, 6N, 9Q, 10Q, 11N and 13F). The frequency of occurrence of amino acid residues was calculated from the determined sequences of PAB variants after each round. The title of each section of the Figure, such as Position 5F, indicates the wild-type residue (in this case, F) of the mutation position. The ordinate indicates the percentage of the frequency of occurrence. The abscissa indicates the round number. Wild-type residue (blue), Histidine residue (red), Non-wild-type and non-histidine residues (green or purple).

Additional file 3: Figure S2. Frequency of occurrence of amino acid residues for each mutation position (14Y, 15E, 17L, 24E, 25E and 27R). see the caption in Additional file 2: Figure S1.

Additional file 4: Figure S3. Frequency of occurrence of amino acid residues for each mutation position (28N, 31I, Q32, 35K and 36D). see the caption in Additional file 2: Figure S1.

Additional file 5: Figure S4. Elution profiles of IgG on affinity columns with immobilized PAB variants. Affinity columns were prepared using the PAB variants. The captured IgG on the column was eluted with a decreasing pH gradient. The ordinate on the right indicates the pH of the solution. The ordinate on the left indicates the absorbance at 280 nm (mAU: milliabsorbance units). The abscissa indicates the elution volume. The solid and dashed lines show the elution patterns of IgG on affinity columns with immobilized PAB variants as shown in the Figure and the pH value of the elution buffer, respectively.

Additional file 6: Table S3. PAB variants data by mutational experiments [7, 11, 21-24] and by molecular simulation calculations [25-26]. The binding affinity of IgG for PAB variants was determined using ELISA and SPR. None: Almost the same binding affinity, Very small: 2-fold to < 5-fold decrease, Small: 5-fold to < 10-fold decrease, Large: 10-fold to < 100-fold decrease, Very large: 100-fold decrease <.

Additional file 7: Figure S5. Circular dichroism melting curves of PAB variants. (a) The curves of the single histidine substituted PAB variants. (b) The curves of the double histidine substituted PAB variants. Circular dichroism melting curves were obtained by monitoring the ellipticity at 222 nm with increasing temperature on a J-805 spectropolarimeter (JASCO). The mole fractions of the proteins in an unfolded state (thick lines) are shown as a function of temperature. Theoretical curves (thin lines) were calculated using a two-state equilibrium transition model.

Additional file 8: Figure S6. Binding efficiency of PAB variants. The PAB variant (about 72 μg) was immobilized on NHS-activated agarose gel. The IgG (about 1mg) solution was added to the PAB variant immobilized agarose gel, and the mixture was shaken at 25°C for 30 min in pH 9.0 buffer (25 mM Tris–HCl (pH=9.0), 2.5 M NaCl, and 0.1% Tween 20). The amounts of IgG in supernatant after centrifugation were determined and the binding efficiency was calculated. The ordinate indicates the binding efficiency.

Additional file 9: Table S4. Amino acid residues of therapeutic antibodies related to the effects of D36H. In this study, Arg2519 of IgG-Fab was found to be related to the effects of D36H mutation. Of top 10 therapeutic antibodies in 2010, 80% of the amino acid residues corresponding to Arg2519 are positively charged residues. Sales rank of therapeutic antibodies in 2010 are cited from the article (John G. Elvin *et al.* International Journal of Pharmaceutics 440 (2013) 83– 98).

Additional file 10: Figure S7. Elution profile of IgG on affinity columns with immobilized 4xPAZ variants. (a) Elution profile of IgG from affinity columns prepared with immobilized 4xPAZ01 or 4xPAZ03. IgG was bound with pH 7.5 buffer (25 mM Tris-HCl (pH=7.5), 150 mM NaCl, and 0.1% Tween 20) at a flow rate of 0.5 mL/min, (b) Elution profile of IgG from affinity columns prepared with immobilized 4xPAZ01 or 4xPAZ03. IgG was bound with pH 7.5 buffer (25 mM Tris-HCl (pH=7.5), 150 mM NaCl, and 0.1% Tween 20) at a flow rate of 0.1 mL/min, (c) Elution profile of IgG from affinity columns prepared with immobilized 4xPAZ01 or 4xPAZ03. IgG was bound with pH 9.0 buffer (25 mM Tris-HCl (pH=9.0), 2.5 M NaCl, and 0.1% Tween 20) at a flow rate of 0.1 mL/min. Affinity columns were prepared using the 4xPAZ variants. The captured IgG on the column was eluted using a buffer with the pH indicated in the Figure. Afterwards, the residual IgG was eluted with a decreasing pH gradient. The ordinate on the right indicates the pH of the solution. The ordinate on the left indicates the absorbance at 280 nm (mAU: milliabsorbance units). The abscissa indicates the elution volume. The solid and dashed lines show the elution patterns of IgG on 4xPAZ variant columns, as shown in the Figure, and the pH value of the elution buffer, respectively. The inlets show the elution patterns with magnified scale from 0 to 8 mL.

Additional file 11: Table S1. List of primers. Primers 1~4 indicate the oligonucleotides for constructing a histidine-scanning library. Primers 5~12 indicate the oligonucleotides for introducing histidine-mutations. The following abbreviations are used for mixed bases: R=(A or G), Y=(C or T), M=(A or C), K=(G or T), S=(G or C) and W=(A or T).

Abbreviations

SpA: *Staphylococcal* protein A; Ig: Immunoglobulin; PAB: *Staphylococcal* protein A B-domain; PAZ: *Staphylococcal* protein A Z-domain.

Competing interests
We declare that all authors are inventors on a pending patent using these *Staphylococcal* protein A variants.

Authors' contributions
MT conducted experiments, analyzed data, and co-wrote the manuscript. HW and AO conducted experiments and analyzed data. SH designed the study, analyzed data and co-wrote the manuscript. All authors read and approved the final manuscript.

Acknowledgements
We thank Dr. Takashi Shimizu for critical reading of the manuscript. This study was partly supported by a grant for the project to build infrastructure for creating next-generation drugs for personalized medicine from the Ministry of Economy, Trade and Industry, Japan.

Author details
[1]Department of Medical Genome Sciences, Graduate School of Frontier Sciences, The University of Tokyo, Kashiwa, Japan. [2]Biomedical Research Institute National Institute of Advanced Industrial Science and Technology (AIST), Higashi, Tsukuba, Ibaraki 305-8566, Japan. [3]Manufacturing Technology Association of Biologics, Chuo-ku, Kobe, Japan.

References

1. Brekke O, Sandlie I: **Therapeutic antibodies for human diseases at the dawn of the twenty-first century.** *Nat Rev Drug Discov* 2003, **2**:52–62.
2. Maggon K: **Monoclonal antibody "gold rush".** *Curr Med Chem* 2007, **14**:1978–1987.
3. Shukla A, Hubbard B, Tressel T, Guhan S, Low D: **Downstream processing of monoclonal antibodies–application of platform approaches.** *J Chromatogr B* 2007, **848**:28–39.
4. Hermeling S, Crommelin D, Schellekens H, Jiskoot W: **Structure-immunogenicity relationships of therapeutic proteins.** *Pharm Res* 2004, **21**:897–903.
5. Schellekens H: **How to predict and prevent the immunogenicity of therapeutic proteins.** *Biotechnol Annu Rev* 2008, **14**:191–202.
6. Li R, Dowd V, Stewart D, Burton S, Lowe C: **Design, synthesis, and application of aprotein A mimetic.** *Nat Biotechnol* 1998, **16**:190–195.
7. Brown N, Bottomley S, Scawen M, Gore M: **A study of the interactions between an IgG-binding domain based on the B domain of staphylococcal protein A and rabbit IgG.** *Mol Biotechnol* 1998, **10**:9–16.
8. Gulich S, Uhlen M, Hober S: **Protein engineering of an IgG-binding domain allows milder elution conditions during affinity chromatography.** *J Biotechnol* 2000, **76**:233–243.
9. Watanabe H, Matsumaru H, Ooishi A, Feng Y, Odahara T, Suto K, Honda S: **Optimizing pH response of affinity between protein G and IgG Fc.** *J Biol Chem* 2009, **284**:12373–12383.
10. Gera N, Hill A, White D, Carbonell R, Rao B: **Design of pH Sensitive Binding Proteins from the Hyperthermophilic Sso7d Scaffold.** *PLoS One* 2012, **7**(11):e48928.
11. Watanabe H, Matsumaru H, Ooishi A, Honda S: **Structure-based histidine substitution for optimizing pH-sensitive *Staphylococcus* protein A.** *J Chromatogr B* 2013, **929**:155–160.
12. Tawfik D, Chap R, Eshhar Z, Green B: **pH on-off switching of antibody-hapten binding by site-specific chemical modification of tyrosine.** *Protein Eng* 1994, **7**:431–434.
13. Morag E, Bayer E, Wilchek M: **Reversibility of biotin-binding by selective modification of tyrosine in avidin.** *Biochem J* 1996, **316**:193–199.
14. Linder M, Nevanen T, Teeri TT: **Design of a pH-dependent cellulose-binding domain.** *FEBS Lett* 1999, **447**:13–16.
15. Sarkar C, Lowenhaupt K, Horan T, Boone T, Tidor B, Lauffenburger D: **Rational cytokine design for increased lifetime and enhanced potency using pH-activated "histidine switching".** *Nat Biotechnol* 2002, **20**:908–913.
16. Marttila A, Hytonen V, Latinen O, Bayer E, Wilchek M, Kulomaa M: **Mutation of the important Tyr-33 residue of chicken avidin: functional and structural consequences.** *Biochem J* 2003, **369**:249–254.
17. Nordlund HR, Hytonen VP, Laitinen OH, Uotila ST, Niskanen EA, Savolainen J, Porkka E, Kulomaa MS: **Introduction of histidine residues into avidin subunit interfaces allows pH-dependent regulation of quaternary structure and biotin binding.** *FEBS Lett* 2003, **555**:449–454.
18. Igawa T, Ishii S, Tachibana T, Maeda A, Higuchi Y, Shimaoka S, Moriyama C, Watanabe T, Takubo R, Doi Y, Wakabayashi T, Hayasaka A, Kadono S, Miyazaki T, Haraya K, Sekimori Y, Kojima T, Nabuchi Y, Aso Y, Kawabe Y, Hattori K: **Antibody recycling by engineered pH-dependent antigen binding improves the duration of antigen neutralization.** *Nat Biotechnol* 2010, **28**:1203–1207.
19. Murtaugh M, Fanning S, Sharma T, Terry A, Horn J: **A combinatorial histidine scanning library approach to engineer highly pH-dependent protein switches.** *Protein Sci* 2011, **20**:1619–1631.
20. Deisenhofer J: **Crystallographic Refinement and Atomic Models of a Human Fc Fragment and Its Complex with Fragment B of Protein A from Staphylococcus aureus at 2.9- and 2.8-A Resolution.** *Biochemistry* 1981, **20**:2361–2396.
21. Gore M, Ferris W, Popplewell A, Scawen M, Atkinson T: **pH-sensitive interactions between IgG and a mutated IgG-binding protein based upon two B domains of Protein A from Staphylococcus aureus.** *Protein Eng* 1992, **5**:577–582.
22. Cedergren L, Andersson R, Jansson B, Uhlen M, Nilsson B: **Mutational analysis of the interaction between staphylococcal protein A and human IgG1.** *Protein Eng* 1993, **6**:441–448.
23. Tashiro M, Montelione G: **Structures of bacterial immunoglobulin-binding domains and their complexes with immunoglobulins.** *Curr Opin Struct Biol* 1995, **5**:471–481.
24. Jendeberg L, Persson B, Andersson R, Karlsson R, Uhlen M, Nilsson B: **Kinetic analysis of the interaction between protein A domain variants and human Fc using plasmon resonance detection.** *J Mol Recognit* 1995, **9**:270–278.
25. Salvalaglio M, Zamolo L, Busini V, Moscatelli D, Cavallotti C: **Molecular modeling of protein A affinity chromatography.** *J Chromatogr A* 2009, **1216**:8678–8686.
26. Huang B, Liu F, Dong X, Sun Y: **Molecular mechanism of the affinity interactions between protein A and human immunoglobulin G1 revealed by molecular simulations.** *J Phys Chem B* 2011, **115**(14):4168–4176.
27. Hillson J, Karr N, Oppliger I, Mannik M, Sasso E: **The structural basis of germline-encoded VH3 immunoglobulin binding to staphylococcal protein A.** *J Exp Med* 1993, **178**:331–336.
28. Jansson B, Uhlen M, Nygren P: **All individual domains of staphylococcal protein A show Fab binding.** *FEMS Immunol Med Microbiol* 1998, **20**:69–78.
29. Starovasnik M, O'connell M, Fairbrother W: **and F. Kelley, R.: Antibody variable region binding by Staphylococcal protein A: thermodynamic analysis and location of the Fv binding site on E-domain.** *Protein Sci* 1998, **8**:1423–1431.
30. Graille M, Stura E, Corper A, Sutton B, Taussig M, Charbonnier J, Silverman GJ: **Crystal structure of a Staphylococcus aureus protein A domain complexed with the Fab fragment of a human IgM antibody: Structural basis for recognition of B-cell receptors and superantigen activity.** *Proc Natl Acad Sci U S A* 2000, **97**:5399.
31. Winn MD, Ballard CC, Cowtan KD, Dodson EJ, Emsley P, Evans PR, Keegan RM, Krissinel EB, Leslie AGW, McCoy A, McNicholas SJ, Murshudov GN, Pannu NS, Potterton EA, Powell HR, Read RJ, Vagin A, Wilson KS: **Overview of the CCP4 suite and current developments.** *Acta Crystallogr D Biol Crystallogr* 2011, **67**:235–242.
32. Tsodikov O, Record MT Jr, Sergeev TV: **Novel computer program for fast exact calculation of accessible and molecular surface areas and average surface curvature.** *J Comput Chem* 2002, **23**:600–609.
33. Honda S, Kobayashi N, Munekata E: **Thermodynamics of a beta-hairpin structure: evidence for cooperative formation of folding nucleus.** *J Mol Biol* 2000, **295**:269–278.
34. Makhatadze G, Privalov PL: **Energetics of protein structure.** *Adv Protein Chem* 1995, **47**:307–417.
35. Honda S, Yamasaki K, Sawada Y, Morii H: **10 residue folded peptide designed by segment statistics.** *Structure* 2004, **12**:1507–1518.

7

Half-life measurements of chemical inducers for recombinant gene expression

Nicolo' Politi[1,2], Lorenzo Pasotti[1,2], Susanna Zucca[1,2], Michela Casanova[1,2], Giuseppina Micoli[3], Maria Gabriella Cusella De Angelis[2] and Paolo Magni[1,2*]

Abstract

Background: Inducible promoters are widely spread genetic tools for triggering, tuning and optimizing the expression of recombinant genes in engineered biological systems. Most of them are controlled by the addition of a specific exogenous chemical inducer that indirectly regulates the promoter transcription rate in a concentration-dependent fashion. In order to have a robust and predictable degree of control on promoter activity, the degradation rate of such chemicals should be considered in many applications like recombinant protein production.

Results: In this work, we use whole-cell biosensors to assess the half-life of three commonly used chemical inducers for recombinant *Escherichia coli*: Isopropyl β-D-1-thiogalactopyranoside (IPTG), anhydrotetracycline (ATc) and N-(3-oxohexanoyl)-L-homoserine lactone (HSL). A factorial study was conducted to investigate the conditions that significantly contribute to the decay rate of these inducers. Temperature has been found to be the major factor affecting ATc, while medium and pH have been found to highly affect HSL. Finally, no significant degradation was observed for IPTG among the tested conditions.

Conclusions: We have quantified the decay rate of IPTG, ATc and HSL in many conditions, some of which were not previously tested in the literature, and the main effects affecting their degradation were identified via a statistics-based framework. Whole-cell biosensors were successfully used to conduct this study, yielding reproducible measurements via simple multiwell-compatible assays. The knowledge of inducer degradation rate in several contexts has to be considered in the rational design of synthetic biological systems for improving the predictability of induction effects, especially for prolonged experiments.

Keywords: Degradation rate, Chemical inducers, Synthetic biology, Whole-cell biosensors, BioBrick™, IPTG, ATc, HSL

Background

Isopropyl β-D-1-thiogalactopyranoside (IPTG), anhydrotetracycline (ATc) and N-(3-oxohexanoyl)-L-homoserine lactone (HSL) are popular chemical inducers that can be used to exogenously control the transcriptional activity of wild-type or synthetic LacI-, TetR- and LuxR-regulated promoters, respectively [1,2]. They have been used in a wide number of engineered biological systems to manipulate the production of recombinant proteins in bacteria [3-7], yeasts [8,9] and also in mammalian cells [10]. Synthetic biology advances have provided several examples of finely tuned highly complex gene networks that can be predictably controlled by means of IPTG, ATc and HSL [11,12]. Although such chemicals have been successfully used in biotechnological research, problems could occur when experiments are performed for prolonged times, because inducers may degrade and gene expression could change accordingly. This is a common situation when inducible promoters are used to drive or optimize the production process of industrially-relevant genes in recombinant cultures that are to be grown for long times (e.g., days) before achieving the optimal amount of product (e.g., a protein, a drug or a biofuel). Although the use of chemical inducers in a large-scale industrial context is sometimes undesirable, as they can account for non-negligible costs, inducible promoters can be used to explore the optimal gene dosage in laboratory-scale experiments, before being substituted with constitutive promoters with optimal

* Correspondence: paolo.magni@unipv.it
[1]Dipartimento di Ingegneria Industriale e dell'Informazione, Università degli Studi di Pavia, via Ferrata 3, Pavia, Italy
[2]Centro di Ingegneria Tissutale, Università degli Studi di Pavia, via Ferrata 3, Pavia, Italy
Full list of author information is available at the end of the article

activity [13] or promoters that can be induced by cheaper signals, such as temperature or pH [14]. In this framework, degradation of inducer molecules must be taken into account to predict the actual promoter transcription during the optimization process in the lab.

In the present work, we face these problems by assessing the degradation rate of IPTG, ATc and HSL in *Escherichia coli* culture or sterile broth in two commonly used growth media, which are popular in synthetic biology studies: L-broth (LB) and M9 supplemented with thiamine and casamino acids with glycerol as carbon source (M9). Acidity- and temperature-dependent effects are also studied.

Whole-cell biosensors were used as measurement tools for this purpose. Their essential design principles are reviewed in [15] and can be easily generalized and adapted for the detection of the chemical inducers of interest in this work. Other more precise and sensitive methods could also be used to detect these inducers, such as mass spectrometry, although more expensive in terms of instrumentation [16,17].

IPTG is known to be a highly stable molecule in solution at room temperature [18] and in bacterial cultures under commonly used conditions for recombinant protein production [19,20]. IPTG degradation in cultures and its uptake by different *E. coli* strains was also studied [19], demonstrating substantial differences between strains with or without lactose permease, which is actively involved in IPTG transportation into the cells. However, no reports exist on IPTG stability as a function of temperature, media, pH or combinations of such factors. To our knowledge, no degradation assay has been performed for ATc in media or cultures and, in other modelling works, its degradation rate was assumed to be equal to the one of tetracycline in water [21]. Degradation of HSL and other N-acyl homoserine lactones has been mainly studied as a function of pH [22,23], temperature [22], acyl chain length [22] and of specific lactonase enzymes produced by other microbes [24-26], but no extensive study has been carried out on its degradation in lactonase-free cultures in different *E. coli* growth media. Enzyme-independent lactonolysis was found to be caused by ring opening of the molecule in a pH-dependent manner, with an increasing rate for longer acyl chains and at higher temperature [22]. Such process was found to be reversible only by acidifying the solution at pH < 3.0, thus making this process non-reversible under the standard growth conditions of *E. coli* [22].

Since the advances of synthetic biology are enabling the construction of increasingly complex controllable systems [12] and industrially relevant solutions [27], predictability has become a crucial point for the realization and debugging of customized technologies. With this work, we intend to provide useful data to support the predictable tuning of gene expression in recombinant *E. coli* cultures.

Methods

Reagents and media

IPTG (I1284, Sigma Aldrich) and ATc (631310, Clontech) were purchased as 200 mM and 2 mg/ml liquid stocks, respectively, and were routinely stored at −20°C. HSL (K3007, Sigma Aldrich) powder was dissolved in deionized water and filter-sterilized (0.2 μm) to prepare a 2 mM stock solution, routinely stored at −20°C.

LB composition is: 10 g/l NaCl, 5 g/l yeast extract and 10 g/l tryptone. M9 composition is: 11.28 g/l M9 salts (M6030, Sigma), 2 mM $MgSO_4$, 0.1 mM $CaCl_2$, 2 g/l casamino acids, 1 mM thiamine hydrochloride and 4 ml/l glycerol. They were prepared as described in [28]. When appropriate, the pH was adjusted by adding hydrochloric acid or sodium hydroxide.

Biosensors

Figure 1 describes the recombinant *E. coli* bearing the biosensors. The IPTG-, ATc- and HSL-biosensor genetic devices, called BBa_J107010, BBa_I13521 and BBa_J107053, respectively, are BioBrick™ parts from the MIT Registry of Standard Biological Parts (Registry) [29].

They have been assembled into the pSB3K3 medium-copy plasmid (with kanamycin-resistance marker and p15A replication origin) by using the BioBrick™ Standard Assembly procedure [30] and conventional molecular biology techniques [28]. Detailed description about plasmid construction, as well as full DNA sequences, can be found in the individual parts web pages of the Registry [29]. TOP10 (Invitrogen) *E. coli* strain was used as a host for cloning. MG1655-Z1 *E. coli* strain [31] was used as the final host for the biosensing plasmids, as its genome encodes constitutive LacI and TetR over-expression cassettes. Chemically competent TOP10 were transformed according to manufacturer's instructions, while MG1655-Z1 competent cells were prepared as described in [28] and were heat-shock transformed at 42°C. The described whole-cell biosensors were routinely stored at −80°C in 20% glycerol stocks and they were always grown in selective media with 20 mg/l of kanamycin.

Sampling

1 mM of IPTG, 50 ng/ml of ATc or 50 nM of HSL were added to 3 ml of sterile medium (*sterile broth*) or to 3 ml of an MG1655-Z1 *E. coli* culture harbouring no plasmids (*cultured broth*), prepared by inoculating 3 ml of sterile medium (LB or M9) with a bacterial colony isolated from a streaked LB agar plate. Sterile and cultured broths, in 15-ml tubes, were incubated with shaking at 220 rpm at 30°C or 37°C. The pH of media was set at 6.0 or 7.0. As a result, each inducer was tested in 16 conditions (cultured/sterile broth; pH 6.0/7.0; 30/37°C; LB/M9 medium). At appropriate time points, 100 μl were taken from the cultured broth and

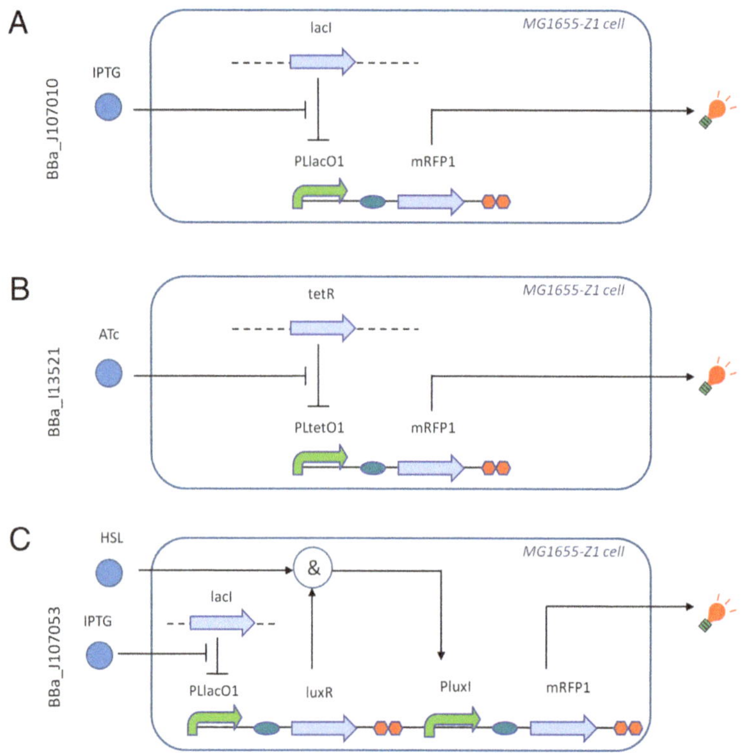

Figure 1 Biosensors used in this study. IPTG- **(A)**, ATc- **(B)** and HSL- **(C)** biosensor constructs and functioning. The BioBrick™ code (BBa_code) of the biosensor-encoding sequences and the microbial host are reported for each construct. All of them were used in the pSB3K3 medium-copy BioBrick™ vector backbone with the p15A replication origin and kanamycin resistance marker. Symbols: curved arrows represent promoters, straight arrows represent coding sequences, ovals represent RBSs, octagons represent transcriptional terminators, "&" means that both inducer and gene product are required to activate the promoter, circles represent chemical inducers, bulbs represent a red fluorescence output and finally dashed lines represent MG1655-Z1 genomic DNA. PLlacO1 and PLtetO1 are synthetic promoters that can be repressed by LacI and TetR, respectively. PluxI is the wild type luxI promoter from *Vibrio fischeri* that can be activated by the LuxR-HSL complex. In the HSL-biosensor **(panel C)**, luxR expression was always induced by 500 μM of IPTG, as gene expression is driven by the PLlacO1 promoter.

centrifuged (13,000 rpm, 1 min), bacteria-free supernatant was taken and stored at −20°C before analysis. For sterile broth, 100 μl were taken and directly stored at −20°C. For each investigated condition, at least two independent tubes were prepared and sampled in different days. When required, cultured broths were prepared by inoculating a colony of the lacY-deficient strain DB3.1 [32,33] or MC1061 [34] instead of MG1655-Z1.

The growth of cultures was measured by optical density at 600 nm (OD600) in a 96-well microplate (Greiner) in the Infinite F200 reader (Tecan). The presented OD600 measurements are background (sterile media)-subtracted absorbance values, relative to the pathlength of 200 μl of liquid in microplate.

Biosensor-based fluorescence assays

0.5 ml of selective M9 were inoculated with 1 μl of the appropriate biosensor glycerol stock and incubated for 16 h at 37°C, 220 rpm. Bacteria were 50-fold diluted in fresh selective M9 and grown under the same conditions as before. If using the HSL-biosensor, IPTG (500 μM) was added after 1 h to induce LuxR expression. After 2–3 h from dilution, 190-μl aliquots were transferred into a 96-well microplate (Greiner). Cells at this point are at an OD600 of about 0.03 (exponential growth phase). Wells were induced with 10 μl of properly diluted samples. Standard calibration curves were obtained by inducing the wells with 10 μl of properly diluted inducer amounts. The microplate was incubated in the Infinite F200 reader and it was assayed with the following kinetic cycle, programmed via the i-control™ software (Tecan): linear shaking (3-mm amplitude, 15 s), wait (5 s), OD600 measurement, fluorescence measurement (excitation at 535 nm, emission at 620 nm, gain = 50, to detect RFP signal), repeat cycle every 5 min.

Each calibration curve dilution sample was prepared in the same growth medium of the sampled cultured/sterile broth. A non-fluorescent culture and sterile medium were always included in each experiment to estimate the background fluorescence and absorbance. Calibration curve points were assayed in duplicate, while a single well was used for the unknown samples points (technical replicates of the same sample give highly reproducible measurements with an average CV lower than 7%, data not shown).

When measured values were under the detection limit of the standard curve, experiments were repeated by adding 20 μl of sample to 180 μl of biosensor. When measured values were still under the detection limit of the standard curve (with this dilution, usually 100 μM for IPTG, 10 ng/ml for ATc and 2 nM for HSL), they were set at the detection limit value itself. These detection limits correspond to 10%, 20% and 4% of the IPTG, ATc and HSL initial concentrations.

Data analysis

The acquired time series were processed as described previously [35,36] to obtain a value proportional to the average RFP protein synthesis rate per cell (S_{cell}), which is related to the inducer concentration in the wells. Briefly, raw absorbance and fluorescence time series were background-subtracted over time by using sterile medium and non-fluorescent culture as background to obtain the actual fluorescence (F) and absorbance (OD600) signals in the well, proportional to the per-well RFP molecules and number of bacteria, respectively. The (dF/dt)/OD600 signal was averaged at its steady-state to obtain S_{cell}. The steady-state is reached after a time that depends on the dynamics of each specific biosensor. S_{cell} is measured as arbitrary units (AU) of RFP per minute per cell.

The obtained standard calibration curve was fitted with the Hill function $S_{cell}(I) = \delta + \alpha/(1 + (K_m/I)^n)$, where $\delta + \alpha$ is the S_{cell} at maximum induction, K_m is the inducer concentration that yields $S_{cell} = \delta + \alpha/2$, n is the Hill coefficient, δ is the basic activity of the biosensor in absence of inducer and I is the inducer concentration. The inducer concentration of the unknown samples was computed from their S_{cell} value via the identified Hill function representing the calibration curve. When required, the resulting value was multiplied by the applied dilution factor of the sample.

Each inducer decay curve was expressed as percentage of its value at t = 0 and, for each tested condition, the independently measured time series were pooled in a single time series c(t).

Assuming a first-order exponential decay for c(t), its natural logarithm was fitted with the linear function $\ln(c(t)) = \ln(c(0)) - k * t$, where k is the degradation rate of the inducer. Fitting was performed with MATLAB R2011b (MathWorks, Natick, MA) via the *regress* routine, which also computes the 95% confidence intervals of k. When required, inducer half-life was computed as ln(2)/k. The measurement unit of k is h^{-1}.

For each inducer, the following linear model with interactions was used to determine the experimental factors that significantly affect the decay rate:

$$\begin{aligned}\ln(c(t)) = {} & \alpha + \alpha_{Temp}\cdot Temp + \alpha_{pH}\cdot pH + \alpha_{Med}\cdot Med + \alpha_{Ster}\cdot Ster \\ & + \alpha_{Temp\cdot pH}\cdot Temp\cdot pH + \alpha_{Temp\cdot Med}\cdot Temp\cdot Med \\ & + \alpha_{Temp\cdot Ster}\cdot Temp\cdot Ster + \alpha_{pH\cdot Med}\cdot pH\cdot Med \\ & + \alpha_{pH\cdot Ster}\cdot pH\cdot Ster + \alpha_{Med\cdot Ster}\cdot Med\cdot Ster \\ & -(\beta_t + \beta_{Temp}\cdot Temp + \beta_{pH}\cdot pH + \beta_{Med}\cdot Med + \beta_{Ster}\cdot Ster \\ & + \beta_{Temp\cdot pH}\cdot Temp\cdot pH + \beta_{Temp\cdot Med}\cdot Temp\cdot Med \\ & + \beta_{Temp\cdot Ster}\cdot Temp\cdot Ster + \beta_{pH\cdot Med}\cdot pH\cdot Med \\ & + \beta_{pH\cdot Ster}\cdot pH\cdot Ster + \beta_{Med\cdot Ster}\cdot Med\cdot Ster)\cdot t \end{aligned} \quad (1)$$

This model describes ln(c(t)) behaviour by using time (t) as continuous independent variable and pH, temperature (Temp), sterility (Ster) and medium (Med) as Boolean independent variables assuming the −1 or +1 value, where −1 corresponds to pH 6.0, 30°C, sterile broth and LB, while +1 corresponds to pH 7.0, 37°C, cultured broth and M9, respectively.

The MATLAB *regstats* routine was used to estimate the regression coefficients and their p-value (P). To test if an experimental factor (or interaction between two factors) significantly contributes to inducer degradation, its β coefficient was considered, i.e., the coefficient of the factor*time term. When $P < 0.05$ for a given coefficient, the related factor (or interaction) is considered to significantly affect the inducer decay rate.

The variability of ln(c(t)) explained by the j-th factor (SS_j) or an interaction term was quantified and compared with the residual unexplained variability of the following (reference) null model (SS_{tot}):

$$\begin{aligned}\ln(c(t)) = {} & \alpha + \alpha_{Temp}\cdot Temp + \alpha_{pH}\cdot pH + \alpha_{Med}\cdot Med + \alpha_{Ster}\cdot Ster \\ & + \alpha_{Temp\cdot pH}\cdot Temp\cdot pH + \alpha_{Temp\cdot Med}\cdot Temp\cdot Med \\ & + \alpha_{Temp\cdot Ster}\cdot Temp\cdot Ster + \alpha_{pH\cdot Med}\cdot pH\cdot Med \\ & + \alpha_{pH\cdot Ster}\cdot pH\cdot Ster + \alpha_{Med\cdot Ster}\cdot Med\cdot Ster\text{-}\beta_t\cdot t \end{aligned} \quad (2)$$

In particular, SS_{tot}, SS_j and the residual unexplained variability SSE can be calculated as follows:

$$SS_{tot} = \sum_i \left(y_i - y_{null,i}\right)^2 \quad (3)$$

$$SS_j = SS_{tot} - \sum_i \left(y_i - y_{null,i} + \beta_j\cdot X_{ji}\cdot t_i\right)^2 \quad (4)$$

$$SSE = \sum_i \left(y_i - y_{full,i}\right)^2 \quad (5)$$

where y_i are the experimental data, $y_{full,i}$ is the full model (Eq.1) output prediction computed in the i-th experimental point, $y_{null,i}$ is the null model (Eq.2) output prediction computed in the i-th experimental point, β_j is the regression coefficient of the j-th factor or interaction term and X_{ji} is the value (−1 or +1) of the j-th factor or interaction term in the i-th experimental point. When X_{ji} is an interaction term, it is computed as the product of the individual factors values (−1 or +1) in the i-th point. As a result, SS_j is the variability of the output that is explained by the $\beta_j \cdot X_j \cdot t$ term alone, thus quantifying the importance of the j-th factor or interaction term. SSE represents the output variability that remains unexplained by the full model (which includes all the considered factors and interaction terms).

When appropriate, the linear model was used without considering cultured broth data and, in this case, sterility was not included as a factor.

Measurement of ATc with High Performance Liquid Chromatography (HPLC)

HPLC is known to enable ATc measurements via UV detection [37]. Here, a 10 AD/vp HPLC (Shimadzu) was used with a Discovery C18 HPLC column 150 × 4.6 mm, 5 μm (Supelco) and a Diode Array UV Detector SPD-M10AVP (Shimadzu). The column was kept at 25°C. The

Table 1 Degradation rate measurements, expressed as h^{-1}, for IPTG, ATc and HSL in all the tested conditions

pH	Temperature	Sterility	Medium	k (L; U), IPTG	k (L; U), ATc	k (L; U), HSL
6	30°C	Sterile	LB	0.001	0.021	0.001
				(−0.002; 0.005)	(0.018; 0.024)	(−0.001; 0.003)
6	30°C	Sterile	M9	−0.005	0.029	0.021
				(−0.017; 0.007)	(0.022; 0.035)	(0.012; 0.029)
6	37°C	Sterile	LB	0.001	0.039	0.01
				(−0.002; 0.004)	(0.03; 0.049)	(0.007; 0.014)
6	37°C	Sterile	M9	−0.002	0.044	0.034
				(−0.01; 0.006)	(0.042; 0.047)	(0.024; 0.044)
7	30°C	Sterile	LB	0.005	0.017	0.019
				(0; 0.01)	(0.016; 0.019)	(0.013; 0.024)
7	30°C	Sterile	M9	−0.001	0.021	0.085
				(−0.005; 0.004)	(0.012; 0.031)	(0.069; 0.102)
7	37°C	Sterile	LB	−0.004	0.029	0.039
				(−0.011; 0.002)	(0.019; 0.038)	(0.037; 0.042)
7	37°C	Sterile	M9	−0.003	0.041	0.076
				(−0.01; 0.003)	(0.036; 0.046)	(0.037; 0.114)
6	30°C	Cultured	LB	0.116	0.021	−0.006
				(0.09; 0.143)	(0.007; 0.035)	(−0.009; -0.003)
6	30°C	Cultured	M9	0.037	0.026	0.007
				(0.027; 0.048)	(0.018; 0.033)	(0.002; 0.012)
6	37°C	Cultured	LB	0.115	0.051	0.027
				(0.089; 0.141)	(0.035; 0.067)	(0.013; 0.04)
6	37°C	Cultured	M9	0.114	0.053	0.013
				(0.07; 0.158)	(0.032; 0.073)	(0.003; 0.024)
7	30°C	Cultured	LB	0.096	0.02	0.063
				(0.072; 0.12)	(0.006; 0.033)	(0.032; 0.094)
7	30°C	Cultured	M9	0.096	0.031	0.062
				(0.069; 0.122)	(0.011; 0.052)	(0.045; 0.079)
7	37°C	Cultured	LB	0.075	0.042	0.091
				(0.024; 0.126)	(0.021; 0.064)	(0.066; 0.116)
7	37°C	Cultured	M9	0.093	0.058	0.088
				(0.071; 0.114)	(0.038; 0.079)	(0.062; 0.114)

Estimated lower (L) and upper (U) 95% confidence intervals of the measured decay rates (k) are also reported. According to the first-order decay model (see Methods), negative values of k are due to experimental noise and indicate a non-significant decaying trend of the inducer. In this table, cultured broth conditions were tested with the MG1655-Z1 strain.

flow rate of the mobile phase was 0.8 ml/min. Gradient elution was performed with solutions A (0.1% formic acid in water) and B (0.1% formic acid in acetonitrile) following the profile: 0–2 min 10% of B, 2–8 min 10% to 30% of B (linear increase), 8–12 min 30% to 80% of B (linear increase), 12–17 min 80% of B (constant), 17–19 min 80% to 10% of B (linear decrease) and 19–25 min 10% (constant). Each single analysis was run for 25 min with additional 3 min for column conditioning. The injection of 100 μl was performed via automatic injector. A standard calibration curve was prepared in sterile M9 at pH 7.0. The LabSolutions software (Shimadzu) was used to analyze HPLC data. As identified by analyses of standard solutions, ATc retention time is 14.26 min and the maximum absorbance of ATc is observed at a wavelength of 430 nm. Experiments and sampling were performed as described above with the following exceptions: culture broth volume was 7 ml, the initial ATc concentration was 200 ng/ml, at appropriate time points 1.5 ml were taken and filter-sterilized (0.2 μm) for HPLC analysis and 100 μl were taken as described above for analysis via biosensor. All the samples were stored at 20°C before the analyses.

Results

Additional file 1: Figure S1 shows calibration curves of the IPTG, ATc and HSL biosensors in a representative experiment. Each biosensor yields reproducible measurements of the same samples in different days with a CV of 7%, 21% and 11% for IPTG, ATc and HSL, respectively (see Additional file 1: Figure S2). No relevant difference in biosensor activity can be detected when a known inducer amount was diluted in either exhausted or sterile medium (see Additional file 1: Figure S3), thus demonstrating that these biosensors can be successfully used to detect inducer concentration in sterile or cultured broth conditions.

Table 1 reports the degradation rates of IPTG, ATc and HSL estimated from experimental data in all the investigated conditions, while Additional file 1: Figure S4-S6 show all the data and fitted curves.

IPTG is stable over 32 hours in all the sterile broth conditions (the 95% confidence interval of k contains zero in all the conditions, i.e., k is not distinguishable from zero from a statistical point of view), while in all the cultured broth conditions IPTG disappears in the supernatant (see Table 1 and Additional file 1: Figure S4). IPTG uptake by lactose permease (LacY) of the MG1655-Z1 strain could explain the observed phenomenon, as described in [19]. Correlation between OD600 and residual percent IPTG over time (see Figure 2A and Additional file 1: Figure S7) is consistent with this statement.

To further validate this, we measured IPTG in the LB pH 7.0 at 37°C cultured broth condition, inoculated with the DB3.1 or MC1061 strains, both deficient in LacY. Results (see Figure 2B-C) showed that IPTG is not degraded in this condition and the inducer disappearance in the medium of MG1655-Z1 was actually due to uptake by LacY. These results confirm the high stability of the IPTG molecule, previously found in literature in different conditions [16,19,20], and the role of lactose permease in IPTG uptake by LacY + strains. None of the tested environmental factors have been found to affect IPTG stability.

ATc is degraded in all the tested conditions with an average half life of about 20 hours (see Table 1). The factors that give a significant contribution in degradation were identified by fitting a linear regression model (see Methods section). According to Table 2, the statistically significant factors (see p-values) are temperature, sterility and medium. Temperature was found to be the main significant factor affecting degradation, while sterility and medium give a minor contribution (see Figure 3A).

In particular, 42% of the total variability is explained by temperature, while sterility and medium together account for only 11%. The decay rate of ATc is 2-fold higher at 37°C ($\beta + \beta_{Temp} = 0.045\ h^{-1}$) than at 30°C ($\beta - \beta_{Temp} = 0.023\ h^{-1}$),

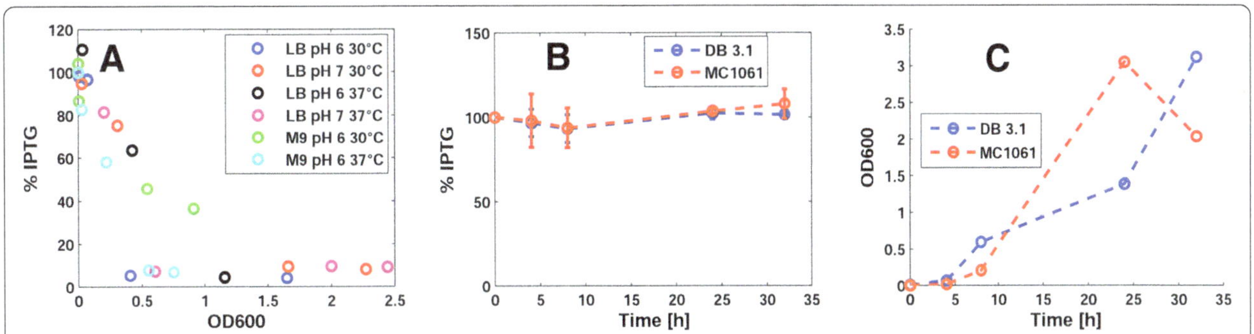

Figure 2 IPTG measured in the supernatant of cultures of strains with or without lactose permease, as a function of cell density. Measured IPTG in supernatants of MG1655-Z1 (lacY+) cultured broth conditions as a function of cell density, expressed as OD600 **(A)**. Measured IPTG in supernatants of DB3.1 and MC1061 (lacY-) cultured broth conditions in LB pH 7.0 at 37°C over time **(B)**. Growth of DB3.1 and MC1061, expressed as OD600, in a representative experiment performed in LB pH 7.0 at 37°C **(C)**. In all the panels, circles represent data points, with colours specified in the legends, and error bars represent the 95% confidence intervals of the average.

Table 2 Parameters of the linear model

Parameter	Identified value	p-value
ATc		
α	4.562	1.2e-165
α_{Temp}	0.0003582	0.99
α_{pH}	0.002346	0.93
α_{Med}	0.02576	0.31
α_{Ster}	0.01122	0.66
$\alpha_{Temp \cdot pH}$	−0.01358	0.59
$\alpha_{Temp \cdot Med}$	0.02222	0.38
$\alpha_{Temp \cdot Ster}$	0.004408	0.86
$\alpha_{pH \cdot Med}$	0.02403	0.34
$\alpha_{pH \cdot Ster}$	0.01493	0.56
$\alpha_{Med \cdot Ster}$	0.04917	5.4e-2
$\boldsymbol{\beta_t}$	**0.03398**	**2.7e-52**
$\boldsymbol{\beta_{Temp}}$	**0.01076**	**1.5e-12**
β_{pH}	−0.001510	0.28
$\boldsymbol{\beta_{Med}}$	**0.003901**	**5.4e-3**
$\boldsymbol{\beta_{Ster}}$	**0.003825**	**6.4e-3**
$\beta_{Temp \cdot pH}$	−0.0006518	0.64
$\beta_{Temp \cdot Med}$	0.0004395	0.75
$\beta_{Temp \cdot Ster}$	0.002527	7.0e-2
$\beta_{pH \cdot Med}$	0.001577	0.26
$\beta_{pH \cdot Ster}$	0.001638	0.24
$\beta_{Med \cdot Ster}$	0.0002971	0.83
HSL		
Parameter	**Identified value**	**p-value**
α	4.441	9.4e-74
α_{Temp}	−0.09555	3.5e-2
α_{pH}	−0.1136	1.3e-2
α_{Med}	−0.1403	2.4e-3
$\alpha_{Temp \cdot pH}$	−0.05275	0.24
$\alpha_{Temp \cdot Med}$	−0.1082	1.8e-2
$\alpha_{pH \cdot Med}$	−0.1057	2.0e-2
$\boldsymbol{\beta_t}$	**0.03560**	**1.8e-22**
β_{Temp}	0.004155	9.1e-2
$\boldsymbol{\beta_{pH}}$	**0.01911**	**4.4e-11**
$\boldsymbol{\beta_{Med}}$	**0.01827**	**1.8e-10**
$\beta_{Temp \cdot pH}$	−0.001483	0.54
$\beta_{Temp \cdot Med}$	−0.003336	0.17
$\boldsymbol{\beta_{pH \cdot Med}}$	**0.007504**	**2.9e-3**

Estimated regression coefficients and p-values are reported for the study of ATc and HSL. Only sterile broth conditions were considered for HSL. The regression coefficients β (i.e., the coefficients that are multiplied by time) shown in bold are significantly different from zero ($P < 0.05$, t-test) and identify the main factors or interaction terms that significantly contribute to inducer decay.

which corresponds to half-lives of 15 h and 30 h, respectively, without considering the other factors. Conversely, degradation is only 1.26-fold higher in cultured broth or M9 ($\beta + \beta_{Ster} = 0.038\ h^{-1}$, $\beta + \beta_{Med} = 0.038\ h^{-1}$) than in sterile broth or LB ($\beta - \beta_{Ster} = 0.03\ h^{-1}$, $\beta - \beta_{Med} = 0.03\ h^{-1}$). Conclusions do not substantially change if cultured broth conditions are excluded from the study (see Additional file 1: Table S1 and Figure S8): temperature is confirmed to be the most important factor, while the other statistically significant factors (in this case pH and medium) give a much lower contribution.

HPLC was used to confirm the obtained results of ATc decay in one representative condition (M9 cultured broth at 30°C, pH 7.0), yielding consistent ATc measurements and degradation rates (see Additional file 1: Figure S9).

HSL spans a wide range of decay rates, from zero (i.e., no detectable degradation in 32 hours) to a half-life of about 8 hours (see Table 1). Degradation was firstly studied in sterile broth conditions by fitting a linear regression model, since pH is already known to affect inducer decay and in cultured broth conditions bacteria can change medium acidity during growth (see Additional file 1: Figure S10). Table 2 shows that the identified significant factors affecting HSL degradation are medium and pH with a significant interaction between them. They explain 38% (pH), 35% (Med) and 6% (pH:Med) of the output variability, respectively (see Figure 3B). In particular, HSL degradation is negligible in the LB pH 6.0 condition ($\beta - \beta_{pH} - \beta_{Med} + \beta_{pH:Med} = \sim 0\ h^{-1}$). Degradation increases in the LB pH 7.0 ($\beta + \beta_{pH} - \beta_{Med} - \beta_{pH:Med} = 0.029\ h^{-1}$) and M9 pH 6.0 ($\beta - \beta_{pH} + \beta_{Med} - \beta_{pH:Med} = 0.029\ h^{-1}$) conditions, while in M9 pH 7.0 degradation further increases because of the individual contributions of pH, medium and their interaction ($\beta + \beta_{pH} + \beta_{Med} + \beta_{pH:Med} = 0.08\ h^{-1}$). Taken together, the results indicate that pH and medium can tune HSL half life from ~9 hours (M9 pH 7.0) to a time that is much longer than the performed experiments, which are 32-hour long (LB pH 6.0).

HSL decay data from cultured broth conditions were then considered. Figure 4 shows the HSL degradation rates (k) reported in Table 1, measured in the eight cultured broth conditions, as a function of the average pH of a culture in those conditions, computed from the data shown in Additional file 1: Figure S10.

As expected, in both the LB and M9 conditions there is a significant pH-dependent trend ($P < 0.05$, t-test). Moreover, k is systematically higher in M9 than in LB for a wide range of pH. These results confirm the significant contribution of pH and growth medium on HSL decay. However, since the linear regression model was fitted on sterile broth data, the precise prediction of HSL degradation in cultured broth may not be inferred by the model, although a similar medium- and pH-dependent trend is qualitatively observed.

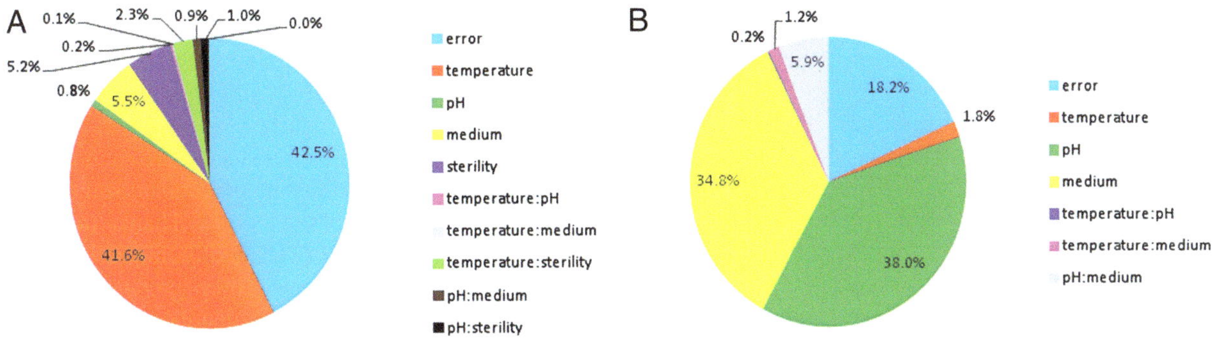

Figure 3 Quantification of the main factors and their interaction terms contributing to the degradation of ATc (A) and HSL (B). The pie charts show the variability of the linear model output explained by each of the considered factors and interaction terms (SS_j, Eq.4). 100% of the pie corresponds to the residual error of the null model (SS_{tot}, Eq.3), while the error term represents the residual error that remains unexplained by the full linear model (SSE, Eq.5). Colon indicates an interaction between two factors. Only sterile broth conditions are considered for HSL and, as a result, the sterility factor and its related interaction terms are not present in panel **B**.

Discussion

The decay rates of the commonly used chemical inducers IPTG, ATc and HSL have been studied in different conditions. Temperature has been found to be the major factor affecting ATc, increasing degradation rate by 2-fold from 30°C to 37°C, while the other significant factors (medium and sterility) gave a minor contribution (<1.3-fold). In light of the reported results for ATc, the measured degradation rate is mainly dependent on non-biological processes as the decay occurs in sterile media, even if the presence of MG1655-Z1 culture slightly increases its degradation by 1.26-fold.

Medium and pH have been found to highly affect HSL in sterile broth, with a significant interaction: while no significant degradation occurs in LB pH 6.0 over a 32-h experiment, the increase of pH from 6.0 to 7.0 or the use of M9 instead of LB yield a significant degradation, which further increases in the M9 pH 7.0 condition. Data from cultured broth experiments qualitatively confirmed the medium and pH contribution. This study of HSL degradation confirms the important contribution of pH previously found in literature [22,23]. Temperature was previously identified as a key factor for the enzyme-independent hydrolysis of HSL, yielding a significantly higher degradation rate at 37°C than at 22°C [22], but in the temperature conditions used here (30°C and 37°C) no difference in degradation was detected.

Finally, no significant degradation was observed for IPTG among the tested conditions, thus confirming the high stability of this molecule, previously reported in high cell density cultures in different conditions [16,19,20].

The two growth media considered in this work, LB and M9, are widely used in biotechnological research and, in particular, in synthetic biology: plasmid propagations are usually performed in LB [28], while many quantitative expression studies are done in LB or M9 [1,2,4,7,38]. The latter is usually preferred when using fluorescent proteins as reporters because of its low background fluorescence. While the aim of our study was to provide useful data to support the predictable tuning of gene expression, our results cannot be directly exploited when different media are used, e.g., in case of high cell density cultures for microbial production of recombinant proteins where defined media are employed [19]. For this reason, additional measurements and linear model update should be performed when dealing with a condition of interest not included in this study.

Here, whole-cell biosensors have been an important tool for the quantification of the molecules of interest, with easy to carry out protocols that are compatible with multi-well microplates. However, it is worth noting that biosensors may provide inaccurate measurements when the culture environment of the samples unpredictably affects the biosensor output. To limit this effect, in this work we always prepared standard curves in the same growth medium of the samples and we also found no significant difference in biosensors activity when sterile media or exhausted media, with a known amount of inducer, were used as samples

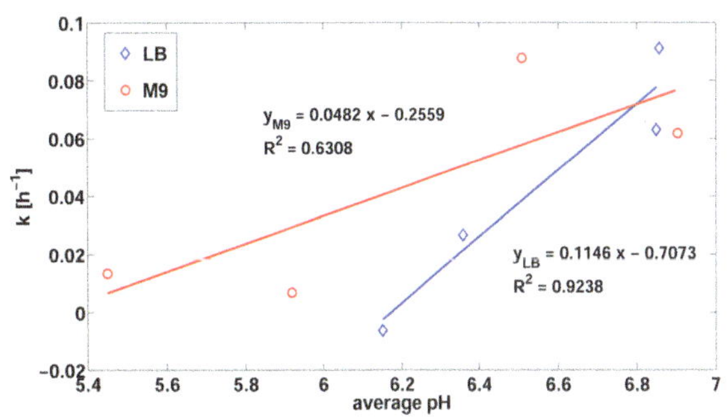

Figure 4 HSL degradation rate (k) as a function of the average pH of bacterial culture in LB or M9. Data points represent the estimated decay rates as a function of pH, taken from Table 1 and Additional file 1: Figure S10, respectively. Solid lines represent data fitting via regression line with the indicated equation and R^2.

(see Additional file 1: Figure S3), thus demonstrating that measurements in the two extreme conditions of the same medium could be correctly performed. Such effects should always be tested when designing new experiments, as they could be dependent on strain, medium, growth conditions, inducer or biosensor. For all these reasons, the results reported in this work are valid under the assumption that only the inducer present in the sample determines the biosensors output, while other factors (e.g., degradation products) are assumed not to significantly affect it. To support the results, HPLC was used to confirm the biosensor measurements of ATc in a representative condition. The results obtained via HPLC were consistent with the measurements performed via the ATc biosensor. Other analytical methods, such as mass spectrometry, could be used to further confirm the obtained results.

The three inducers considered in this work are commonly known to be diffusible through bacterial cell membranes and for this reason intracellular and extracellular concentration of inducers are expected to be the same at equilibrium [19,22,39]. However, the "entrapment" of inducers in cells may affect inducer decay rate measurements in cultured broth conditions. This can occur if intracellular proteins which bind inducers are expressed, thus sequestering free inducer molecules. This effect may lead to overestimation of inducer decay rate and may be important when cells are at a high density, intracellular proteins are abundant or protein-inducer affinity is high. In the experiments carried out in this work in cultured broth conditions, HSL degradation assays are not affected by this phenomenon, as the MG1655-Z1 strain does not express LuxR, which specifically binds HSL. MG1655-Z1 over-expresses LacI, which may contribute, together with LacY, to IPTG entrapment in cells [20]. Since the MC1061 strain does not contain the lacI gene, also IPTG degradation assays in this strain are not affected. According to its genotype, the DB3.1 strain contains the wild-type lacI gene, but no significant concentration decrease is seen from the analysis of supernatants. Measurements may be affected by this phenomenon in ATc decay assays, since MG1655-Z1 over-expresses TetR, which binds ATc molecules. Importantly, inducer molecules could also bind other unspecific intracellular proteins or factors [20]. The quantification of bound inducer is not trivial and cannot be performed via the whole-cell biosensors, HPLC or mass spectrometry methods used or cited above [20]. Other works may be performed to elucidate the possible role of intracellular sequestration of inducers.

Finally, even though IPTG is not degraded and the pH-dependent HSL degradation has been well studied previously, the medium-dependent decay of HSL and ATc, the temperature-dependent decay of ATc, as well as the contribution of *E. coli* cultures on HSL and ATc degradation in cultured broth conditions are not known. Additional studies with different analytical methods could gain insight into the degradation mechanisms, while an explanation of the decay processes or the identification of the degradation products of the inducers of interest is beyond the scope of this work.

In summary, our study has provided a quantification for the inducer decay rates in different conditions, some of which were not tested in published works. Thanks to the factorial study design, a statistics-based framework has been used to find the main effects and their interactions affecting the decay rates. This led to the identification of previously unreported environmental factors and provided insights into the degradation mechanisms of the studied inducers. Despite the empirical nature of the reported results, the knowledge of degradation rate in several contexts can support the rational design of synthetic biological systems by improving the predictability of induction effects, especially for prolonged experiments.

Additional file

Additional file 1: Figure S1. Calibration curve of the IPTG (A), ATc (B) and HSL (C) biosensors in a representative experiment. **Figure S2.** Reproducibility of the IPTG (A), ATc (B) and HSL (C) biosensors measurements in different days. **Figure S3.** Effect of exhausted vs fresh medium on the activity of the IPTG (A), ATc (B) or HSL (C) biosensors. **Figure S4.** Data and fitting in IPTG degradation assays. **Figure S5.** Data and fitting in ATc degradation assays. **Figure S6.** Data and fitting in HSL degradation assays. **Figure S7.** Growth curves in cultured broth conditions. **Figure S8.** Quantification of the main factors and their interaction terms affecting the degradation of ATc in sterile broth. **Figure S9.** ATc measurements via HPLC: peaks of samples with 500 ng/ml, 350 ng/ml, 50 ng/ml and 10 ng/ml of ATc (A); standard calibration curve (B) and comparison between HPLC and biosensor measurements (C). **Figure S10.** pH of cultured broths. **Table S1.** Parameters of the linear model for ATc considering only the sterile broth conditions.

Competing interest
The authors declare that they have no competing interests.

Authors' contributions
NP, LP, SZ and PM conceived the study, designed the experiments and analyzed the data. NP performed the quantitative experiments. NP and SZ constructed the plasmids. SZ, MC, and GM designed and performed the measurements with HPLC. LP, MGCDA and PM wrote the paper. All authors read and approved the final manuscript.

Acknowledgements
This project was partially funded by the Italian Ministero dell'Universita' e della Ricerca through the FIRB ITALBIONET project.

Author details
[1]Dipartimento di Ingegneria Industriale e dell'Informazione, Università degli Studi di Pavia, via Ferrata 3, Pavia, Italy. [2]Centro di Ingegneria Tissutale, Università degli Studi di Pavia, via Ferrata 3, Pavia, Italy. [3]Centro di Ricerche Ambientali, IRCCS Fondazione Salvatore Maugeri, via Salvatore Maugeri 10, Pavia, Italy.

References

1. Lutz R, Bujard H: **Independent and tight regulation of transcriptional units in Escherichia coli via the LacR/O, the TetR/O and AraC/I1-I2 regulatory elements.** *Nucleic Acids Res* 1997, **25**(6):1203–1210.
2. Cox RS 3rd, Surette MG, Elowitz MB: **Programming gene expression with combinatorial promoters.** *Mol Syst Biol* 2007, **3**:145.
3. Gardner TS, Cantor CR, Collins JJ: **Construction of a genetic toggle switch in *Escherichia coli*.** *Nature* 2000, **403**:339–342.
4. Pasotti L, Zucca S, Lupotto M, Cusella De Angelis M, Magni P: **Characterization of a synthetic bacterial self-destruction device for programmed cell death and for recombinant proteins release.** *J Biol Eng* 2011, **5**:8.
5. Kamionka A, Bertram R, Hillen W: **Tetracycline-dependent conditional gene knockout in bacillus subtilis.** *Appl Environ Microbiol* 2005, **71**(2):728–733.
6. Andersen JB, Sternberg C, Poulsen LK, Bjorn SP, Givskov M, Molin S: **New unstable variants of green fluorescent protein for studies of transient gene expression in bacteria.** *Appl Environ Microbiol* 1998, **64**(6):2240–2246.
7. Canton B, Labno A, Endy D: **Refinement and standardization of synthetic biological parts and devices.** *Nat Biotechnol* 2008, **26**:787–793.
8. Ellis T, Wang X, Collins JJ: **Diversity-based, model-guided construction of synthetic gene networks with predicted functions.** *Nat Biotechnol* 2009, **27**:465–471.
9. Blount BA, Weenink T, Vasylechko S, Ellis T: **Rational diversification of a promoter providing fine-tuned expression and orthogonal regulation for synthetic biology.** *PLoS ONE* 2012, **7**(3):e33279.
10. Gossen M, Bujard H: **Anhydrotetracycline, a novel effector for tetracycline controlled gene expression systems in eukaryotic cells.** *Nucleic Acids Res* 1993, **21**(18):4411–4412.
11. Stricker J, Cookson S, Bennett MR, Mather WH, Tsimring LS, Hasty J: **A fast, robust and tunable synthetic gene oscillator.** *Nature* 2008, **456**(7221):516–519.
12. Moon TS, Lou C, Tamsir A, Stanton BC, Voigt CA: **Genetic programs constructed from layered logic gates in single cells.** *Nature* 2012, **491**(7423):249–253.
13. Temme K, Zhao D, Voigt CA: **Refactoring the nitrogen fixation gene cluster from *Klebsiella oxytoca*.** *Proc Natl Acad Sci* 2012, **109**(18):7085–7090.
14. Jana S, Deb JK: **Strategies for efficient production of heterologous proteins in *Escherichia coli*.** *Appl Microbiol Biotechnol* 2005, **67**(3):289–298.
15. Steindler L, Venturi V: **Detection of quorum-sensing N-acyl homoserine lactone signal molecules by bacterial biosensors.** *FEMS Microbiol Lett* 2007, **266**(1):1–9.
16. Fernandez-Castane A, Ruiz J, Caminal G, Lopez-Santin J: **Development and validation of a liquid chromatography-mass spectrometry assay for the quantitation of IPTG in E. Coli fed-batch cultures.** *Anal Chem* 2010, **82**:5728–5734.
17. Gould TA, Herman J, Krank J, Murphy RC, Churchill MEA: **Specificity of acyl-homoserine lactone synthases examined by mass spectrometry.** *J Bacteriol* 2006, **188**:773–783.
18. Sigma Aldrich: *Datasheets of products I5502 and I6758.* [http://www.sigmaaldrich.com/content/dam/sigma-aldrich/docs/Sigma-Aldrich/Product_Information_Sheet/i5502pis.pdf]
19. Fernandez-Castane A, Vine CE, Caminal G, Lopez-Santin J: **Evidencing the role of lactose permease in IPTG uptake by *Escherichia coli* in fed-batch high cell density cultures.** *J Biotechnol* 2011, **157**:391–398.
20. Fernandez-Castane A, Caminal G, Lopez-Santin J: **Direct measurements of IPTG enable analysis of the induction behavior of E. coli in high cell density cultures.** *Microb Cell Fact* 2012, **11**:58.
21. Biliouris K, Daoutidis P, Kaznessis YN: **Stochastic simulations of the tetracycline operon.** *BMC Syst Biol* 2011, **5**:9.
22. Yates EA, Philipp B, Buckley C, Steve A, Chhabra SR, Sockett RE, Goldner M, Dessaux Y, Camara M, Smith H, Williams P: **N-acylhomoserine lactones undergo lactonolysis in a pH-, temperature-, and acyl chain length-dependent manner during growth of *Yersinia pseudotuberculosis* and *Pseudomonas aeruginosa*.** *Infect Immun* 2002, **70**(10):5635–5646.
23. Schaefer AL, Hanzelka BL, Parsek MR, Greenberg EP: **Detection, purification, and structural elucidation of the acylhomoserine lactone inducer of *Vibrio fischeri* luminescence and other related molecules.** *Methods Enzymol* 2000, **305**:288–301.
24. Wang LH, Weng LX, Dong YH, Zhang LH: **Specificity and enzyme kinetics of the quorum-quenching N-acyl homoserine lactone lactonase (AHL-lactonase).** *J Biol Chem* 2004, **279**(14):13645–13651.
25. Dong YH, Wang LH, Xu JL, Zhang HB, Zhang XF, Zhang LH: **Quenching quorum-sensing-dependent bacterial infection by anN-acyl homoserine lactonase.** *Nature* 2001, **411**:813–817.
26. Leadbetter JR, Greenberg EP: **Metabolism of acyl-homoserine lactone quorum-sensing signals by *Variovorax paradoxus*.** *J Bacteriol* 2000, **182**:6921–6926.
27. Kelly J: **The organism is the product.** *ACS Synth Biol* 2012, **1**(1):4–5.
28. Sambrook J, Fritsch EF, Maniatis T: *Molecular cloning: a laboratory manual.* Cold Spring Harbor, N.Y.: Cold Spring Harbor Laboratory Press; 1989
29. MIT: *Registry of standard biological parts.* [http://partsregistry.org].
30. Knight TF: *Idempotent vector design for standard assembly of biobricks. Technical report, MIT DSpace*; 2003 [http://web.mit.edu/synbio/release/docs/biobricks.pdf]
31. Sleight SC, Bartley BA, Lieviant JA, Sauro HM: **Designing and engineering evolutionary robust genetic circuits.** *J Biol Eng* 2010, **4**:12.
32. Bernard P, Couturier M: **Cell killing by the F plasmid CcdB protein involves poisoning of DNA-topoisomerase II complexes.** *J Mol Biol* 1992, **226**(3):735–745.
33. Miki T, Park JA, Nagao K, Murayama N, Horiuchi T: **Control of segregation of chromosomal DNA by sex factor F in Escherichia coli. Mutants of DNA gyrase subunit A suppress letD (ccdB) product growth inhibition.** *J Mol Biol* 1992, **225**:39–52.
34. Durfee T, Nelson R, Baldwin S, Plunkett G, Burland V, Mau B, Petrosino JF, Qin X, Muzny DM, Ayele M, Gibbs RA, Csorgo B, Posfai G, Weinstock GM, Blattner FR: **The complete genome sequence of Escherichia coli DH10B: insights into the biology of a laboratory workhorse.** *J Bacteriol* 2008, **190**:2597–2606.
35. Pasotti L, Quattrocelli M, Galli D, Cusella De Angelis MG, Magni P: **Multiplexing and demultiplexing logic functions for computing signal processing tasks in synthetic biology.** *Biotechnol J* 2011, **6**(7):784–795.
36. Zucca S, Pasotti L, Mazzini G, Cusella De Angelis MG, Magni P: **Characterization of an inducible promoter in different DNA copy number conditions.** *BMC Bioinforma* 2012, **13**:S11.
37. Tsuji K, Goetz JF: **HPLC as a rapid means of monitoring erythromycin and tetracycline fermentation processes.** *J Antibiot* 1978, **31**:302–308.
38. Pasotti L, Politi N, Zucca S, Cusella De Angelis MG, Magni P: **Bottom-up engineering of biological systems through standard bricks: a modularity study on basic parts and devices.** *PLoS One* 2012, **7**:e39407.
39. Batt G, Yordanov B, Weiss R, Belta C: **Robustness analysis and tuning of synthetic gene networks.** *Bioinformatics* 2007, **23**(18):2415–2422.

8

Design and analysis of LacI-repressed promoters and DNA-looping in a cyanobacterium

Daniel Camsund[1], Thorsten Heidorn[2] and Peter Lindblad[1*]

Abstract

Background: Cyanobacteria are solar-powered prokaryotes useful for sustainable production of valuable molecules, but orthogonal and regulated promoters are lacking. The Lac repressor (LacI) from *Escherichia coli* is a well-studied transcription factor that is orthogonal to cyanobacteria and represses transcription by binding a primary *lac* operator (*lacO*), blocking RNA-polymerase. Repression can be enhanced through DNA-looping, when a LacI-tetramer binds two spatially separated *lacO* and loops the DNA. P*trc* is a commonly used LacI-repressed promoter that is inefficiently repressed in the cyanobacterium *Synechocystis* PCC 6803. P*trc*2O, a version of P*trc* with two *lacO*, is more efficiently repressed, indicating DNA-looping. To investigate the inefficient repression of P*trc* and cyanobacterial DNA-looping, we designed a P*trc*-derived promoter library consisting of single *lacO* promoters, including a version of P*trc* with a stronger *lacO* (Ptrc1O-proximal), and dual *lacO* promoters with varying inter-*lacO* distances (the P*trc*2O-library).

Results: We first characterized artificial constitutive promoters and used one for engineering a LacI-expressing strain of *Synechocystis*. Using this strain, we observed that P*trc*1O-proximal is similar to P*trc* in being inefficiently repressed. Further, the P*trc*2O-library displays a periodic repression pattern that remains for both non- and induced conditions and decreases with longer inter-*lacO* distances, in both *E. coli* and *Synechocystis*. Repression of P*trc*2O-library promoters with operators out of phase is less efficient in *Synechocystis* than in *E. coli*, whereas repression of promoters with *lacO* in phase is efficient even under induced conditions in *Synechocystis*. Two well-repressed P*trc*2O promoters were highly active when tested in absence of LacI in *Synechocystis*.

Conclusions: The artificial constitutive promoters herein characterized can be utilized for expression in cyanobacteria, as demonstrated for LacI. The inefficient repression of P*trc* and P*trc*1O-proximal in *Synechocystis*, as compared to *E. coli*, may be due to insufficient LacI expression, or differences in RNAP subunits. DNA-looping works as a transcriptional regulation mechanism similarly as in *E. coli*. DNA-looping contributes strongly to P*trc*2O-library repression in *Synechocystis*, even though they contain the weakly-repressed primary *lacO* of P*trc*1O-proximal and relatively low levels of LacI/cell. Hence, *Synechocystis* RNAP may be more sensitive to DNA-looping than *E. coli* RNAP, and/or the chromatin torsion resistance could be lower. Two strong and highly repressed P*trc*2O promoters could be used without induction, or together with an unstable LacI.

Keywords: Cyanobacteria, Lac repressor, Synthetic biology, Transcriptional systems, P*trc*, *Synechocystis*, LacI, Metabolic engineering tools, Transcriptional regulation

* Correspondence: peter.lindblad@kemi.uu.se
[1]Microbial Chemistry, Department of Chemistry–Ångström Laboratory, Science for Life Laboratory, Uppsala University, P.O. Box 523, SE-75120 Uppsala, Sweden
Full list of author information is available at the end of the article

Background

Cyanobacteria, solar-powered prokaryotes that fix CO_2 from air during photosynthesis, can be engineered to produce a variety of valuable molecules and fuels sustainably [1,2], and hence present a literally green alternative to the use of fossil fuels. The metabolic engineering efforts of even common biotechnological chassis such as *Escherichia coli (E. coli)*, has illustrated the need for synthetic biology to provide new tools, e.g. minimal chassis, vectors and genetic control systems [3]. This is all the more true for cyanobacteria, and even though tools are available for cyanobacterial synthetic biology, more tools, especially promoters, are needed [4,5]. Further, cellular context effects, cross-talk and interactions among genetic components and control systems are some of many often unpredictable causes for failure in engineered biological systems [6]. Orthogonal biological systems, which display minimal cross-talk because their components are decoupled from each other and preferably from the host, could help mitigate these problems [7]. However, most promoters previously used for engineering cyanobacteria are native to the chassis in question [8], so there is a specific lack of orthogonal promoters, constitutive as well as regulated.

The Lac repressor (LacI) from *E. coli* is a well-studied tetrameric (a dimer of dimers) transcription factor [9] orthogonal to cyanobacteria that represents a model system and, in combination with an optimized promoter a tool, for engineering synthetic cyanobacterial transcriptional systems. In wild-type *E. coli*, LacI represses transcription of the *lacZYA* operon mainly by binding its primary *lac* operator (*lacO1*) that is located downstream of and overlapping with 1 bp the transcription start site (TSS) of the promoter [10]. Repression can be relieved by induction using lactose or its stable analog isopropyl β-D-1-thiogalactopyranoside (IPTG), which reduces the affinity of LacI towards its operator approximately 1000-fold [11]. The repression of the *E. coli lac* operon also depends on the auxiliary operators to which LacI binds with differing affinity. Of the three wild-type operators (the primary *lacO1* and the auxiliary operators *O2* and *O3*) that display an approximate dyad symmetry and an artificial perfectly symmetric *lacO* (*lac*Osym) [12], *lac*Osym was found to bind LacI with the greatest affinity, with the binding affinities ordered as *lac*Osym > *lacO1* > *lacO2* > *lacO3* [13]. Early on, it was found that the auxiliary *lac* operators, which are spatially separated from the primary *lacO1*, improve repression through a DNA-looping mechanism that increases the effective local concentration of LacI at the primary repression site [14]. A LacI tetramer bound to separately located *lacO* on the same side of the DNA-helix would be able to bend or loop the DNA (Figure 1). Vice versa, two *lacO* on different sides of the helix would not be able to simultaneously bind LacI by simply looping the DNA, making the formation of LacI-DNA-loops in geometrically unfavorable configurations more difficult. The DNA-looping model of cooperative tetrameric LacI-repression was confirmed in vitro [15], and in vivo [16]. Further, these and subsequent studies showed that DNA-loop-dependent repression varies periodically as a function of the distance between the two operators, with a period of about 11 bp or the approximate numbers of bp per helical turn of supercoiled chromatin DNA in vivo [17-22]. However, it should be noted that the helical repeat of in vivo chromatin DNA (referred to as DNA) will depend on the associated protein factors [20]. In a comprehensive in vivo DNA-looping study using a fixed primary *lacO1* for repression of a *lac*UV5 promoter, the upstream position of an auxiliary *lac*Osym was varied by varying the length of a spacer between the −35 box and the operator, effectively varying the inter-*lacO* distance from 57.5 to about 1500 bp [23]. While confirming previous results regarding DNA-loop mediated repression, this study also found that repression is phase-dependent from the shortest inter-*lacO* distance tested up to approximately 200 bp, that repression is more effective for shorter distances, and that phasing was not observed for distances above 400 bp. The strong repression phase-dependency for short inter-*lacO* distances and the lack of phase-dependency for longer distances was explained by the strong torsion resistance of short segments of DNA, making pure looping more energetically favorable than twisting, and the lower torsion resistance of longer stretches of DNA, removing the barrier for twisted loop formation.

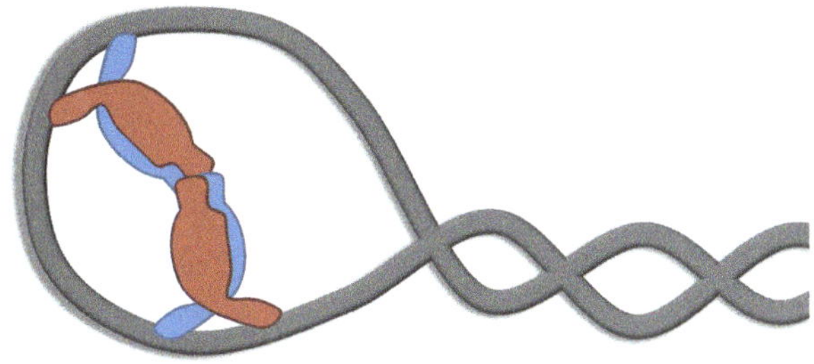

Figure 1 Conceptual image of a Lac repressor tetramer binding inside of an apical loop of plectonemically supercoiled DNA. Inspired by [20].

We previously introduced and characterized two LacI-regulated promoters, P*trc* [24] and P*trc*2O, in the unicellular model cyanobacterium *Synechocystis* PCC 6803 (*Synechocystis*) [25]. Both promoters contain the same P*trc* core promoter, a chimera of the *E. coli trp* operon and *lac*UV5 promoters, including the proximal *lacO1* from the *lac* operon, and proved to be highly active in *Synechocystis* in the absence of LacI. This was expected since P*trc* contains the optimally spaced *E. coli* consensus −35 and −10 elements, which are similar to the consensus −35 and −10 boxes of the cyanobacterial group 1 promoters that are recognized by the primary sigma factor under normal growth conditions [26]. However, we observed that P*trc* is less efficiently repressed by heterologously expressed LacI in *Synechocystis* than in an *E. coli* strain over-expressing LacI. Further, P*trc*2O, a modified P*trc* with an additional *lac*Osym inserted further upstream of the −35 element to enable DNA-loop-enhanced repression, was more efficiently repressed in *Synechocystis*, but could not be efficiently induced [25]. The efficient repression of P*trc*2O, even under full IPTG induction, implies that insufficient levels of heterologously expressed LacI was not the reason for the inefficient repression of P*trc* in *Synechocystis*. In contrast to these results, P*trc* in combination with the *lacI*^{q} gene has proven to be both well-repressed and inducible in the cyanobacterium *Synechococcus elongatus* PCC 7942 [27]. However, when a recent study used a similar P*trc* and *lacI*^{q} setup in *Synechocystis*, no or very little repression of P*trc* could be detected, whereas the contrary was true in *E. coli* [28], in line with our previous results [25]. Further, P*tac* [29], which differs from P*trc* only in that the spacer between the −35 and −10 boxes is 16 bp instead of 17 bp [24], was only partially repressed by LacI when tested in the cyanobacterium *Nostoc* PCC 7120 [30]. The difference in repression of P*trc* between *E. coli* and *Synechocystis* could be due to differences in the cyanobacterial RNA polymerase (RNAP) complex [31,32] and in the interactions between RNAP, promoter and LacI, or simply because of differences in the amount of LacI per cell. Further, P*trc*2O could be efficiently repressed in *Synechocystis*, implying that DNA-looping through the LacI-tetramer may work in a similar way as in *E. coli*. While DNA curvature has been previously recognized to affect gene expression in cyanobacteria [33], as far as we are aware, gene regulation through DNA-looping has not previously been experimentally studied in cyanobacteria.

In the present study, we characterized minimal artificial promoters that can be utilized for stable constitutive expression in cyanobacteria, as demonstrated by using one of them to engineer a LacI-expressing strain of *Synechocystis*. Using this strain, we characterized a version of P*trc* with a stronger *lacO*, P*trc*1O-proximal, which similarly to P*trc* display limited repression as compared to its more efficient repression in *E. coli*. Possible reasons for this were found in the differences between *E. coli* and cyanobacterial RNA polymerases, or in the lower number of LacI/cell in the LacI-expressing strain of *Synechocystis* as compared with the LacI over-expressing strains of *E. coli* herein tested. Further, we characterized the P*trc*2O-library in both *E. coli* and *Synechocystis* and observe several similarities and differences between the two species. While the system displays a similar periodic repression pattern in both, illustrating the presence of DNA-looping in cyanobacteria, differences are found. These differences in repression and induction are attributed to the differing repression efficiency of the proximal *lac* operator of P*trc*1O-proximal, and potential differences between the RNAP sensitivity to DNA-looping and/or the in vivo chromatin DNA torsion-resistance between the two species. These findings will aid in the understanding of native cyanobacterial transcriptional systems and have design implications for engineering synthetic regulated promoters.

Results

Characterization of constitutive promoters in *Synechocystis*

To investigate the activity of minimal and constitutive orthogonal promoters in the unicellular cyanobacterium *Synechocystis* PCC 6803 (*Synechocystis*), eight members from an artificial promoter library spanning a wide range of activities in *Escherichia coli* were selected for testing using a pPMQAK1-carried [25] RBS* [4] EYFP fluorescent protein reporter construct. The artificial BioBrick promoters BBa_J23100, BBa_J23101, BBa_J23102, BBa_J23105, BBa_J23106, BBa_J23109, BBa_J23113 and BBa_J23114 [34] were compared to three promoters native to *Synechocystis*; the nitrate reductase promoter P*nirA*, the plastocyanin promoter P*petE* and the RNase P subunit B promoter P*rnpB*, which have been previously used for engineered expression, using the same reporter construct. The constitutive artificial promoters were found to span a wide range of activities in *Synechocystis*, encompassing and expanding the range covered by the three native promoters (Figure 2).

Construction of a LacI-expressing and a Chloramphenicol resistance-only strain of *Synechocystis*

In our previous study [25] we found that LacI expressed from pPMQAK1 using P*rnpB* and the BBa_B0034 [34] ribosome binding site was sufficient for repression of the original P*trc*2O promoter. For constructing a strain of *Synechocystis* expressing similar levels of LacI from the chromosome instead of from the pPMQAK1 vector, we selected the minimal, constitutive and orthogonal BBa_J23114 promoter (weaker than the P*rnpB* promoter, as characterized in this study) in combination with the RBS* ribosome binding site that was found to be stronger than BBa_B0034 in *Synechocystis* [4]. Since the copy number of pPMQAK1 is expected to be similar to the copy number of the *Synechocystis* chromosome [25] or lower, taking newer data measuring the copy number of the chromosome to about 40 to 220 per cell depending on strain and growth phase into account [35], and assuming that the promoter activities are similar on the pPMQAK1 plasmid and the chromosome, we expected levels of LacI expression to be similar or higher than in our previous study. The LacI-expression cassette, consisting of a forward terminator blocking potentially incoming RNA polymerases from the native part of the chromosome, the

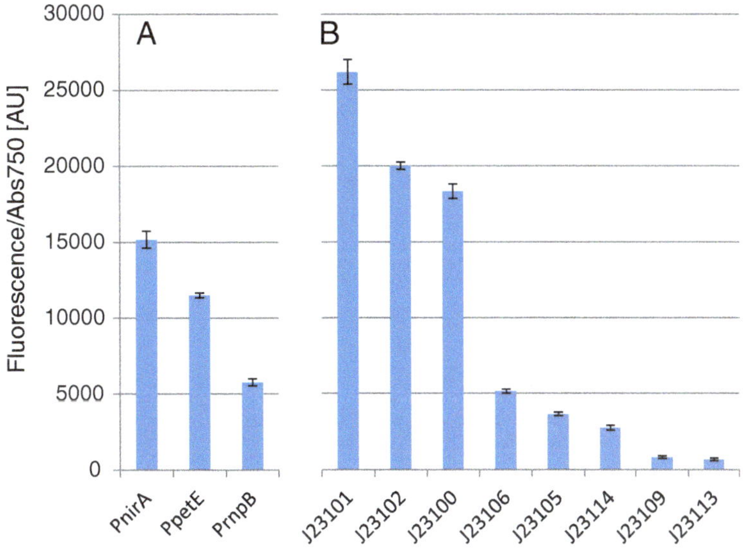

Figure 2 Promoter activities per cell characterized using an EYFP reporter construct carried on pPMQAK1 in *Synechocystis* PCC 6803. **A**. Native promoters from *Synechocystis* included for comparison. P*nirA* is the nitrate reductase promoter, P*petE* is the plastocyanin promoter and P*rnpB* is the RNase P subunit B promoter. **B**. Minimal constitutive promoters from an *Escherichia coli* synthetic promoter library. The J23### promoters correspond to the BBa_J23### BioBrick parts obtained from the iGEM Registry [34]. Promoter activities per cell are given in fluorescence/absorbance at 750 nm in arbitrary units [AU]. Error bars indicate standard error of the mean (n = 6).

BBa_J23114, RBS* and *lacI* CDS expression unit, and another forward terminator to stop BBa_J23114-promoted RNAP from reading out into the subsequent chloramphenicol resistance cassette, was inserted into a previously used neutral site of the chromosome using homologous recombination. The obtained LacI-enhanced strain of *Synechocystis* was confirmed to be fully segregated using PCR with primers flanking the insertion site, and further the areas around the insertion sites were sequence-verified using the same primers. LacI expression was confirmed through western blotting using LacI antibodies (data not shown) and confirmed functional through LacI repression and IPTG induction assays (see following results). Further, as known numbers of cells were used for the western blot of the LacI-enhanced strain of *Synechocystis* and the two *E. coli* strains used herein, NEB5α and DH5αZ1, we were able to quantify the relative levels of LacI/cell among the three strains. NEB5α has about 5X as many LacI/cell as compared to the LacI-enhanced strain of *Synechocystis*, and DH5αZ1 has about 16X as many LacI/cell. Using a previously published value for the number of LacI tetramers/cell in DH5αZ1, 3000/cell [36], as a standard, we estimated the number of LacI tetramers/cell in the LacI-enhanced strain of *Synechocystis* to be 190, and the number of LacI/cell in NEB5α to be about 880.

Finally, to enable the characterization of promoters in a control strain grown under the same conditions as the LacI-enhanced strain of *Synechocystis* but without LacI, we constructed an engineered strain with the chloramphenicol resistance-only inserted into the neutral site. This chloramphenicol resistance-only strain of *Synechocystis* was constructed and confirmed in the same way as the LacI-enhanced strain.

Construction and characterization of the P*trc*-derived promoter library in *E. coli*

To investigate the limited repression of P*trc*, the limited induction of P*trc*2O that we previously encountered in *Synechocystis* [25], and the characteristics of LacI-mediated DNA-looping in cyanobacteria, we designed and constructed a series of P*trc*-derived promoters. This collection consists of control promoters (Figure 3A) and the P*trc*2O-library itself (Figure 3B). Among the control promoters BBa_J23101 serves as a constitutive control and reference. PA1*lacO*-1 [36] is a strongly-repressed yet highly inducible promoter in *E. coli* that was recently found to be both repressed and inducible in *Synechocystis* [28]. The original P*trc*2O promoter works as a reference to our previous experiment [25] and P*trc*1O-distal as a P*trc*-derived promoter with the symmetric *lac* operator (*lac*Osym) inserted immediately upstream of the −35 element. P*trc*1O-core with a slightly truncated *lac*Osym inserted in the core of the promoter serves as a control for PA1*lacO*-1 that also contains a core-located *lacO*. P*trc*1O-proximal is identical to the original P*trc* [24] that we previously characterized [25] except for that we switched the proximal *lacO1* with *lac*Osym. If the low repressibility of P*trc* that we previously observed in *Synechocystis* is due to lack of LacI binding *lacO1*, we should observe an increase in repression by switching *lacO1* for

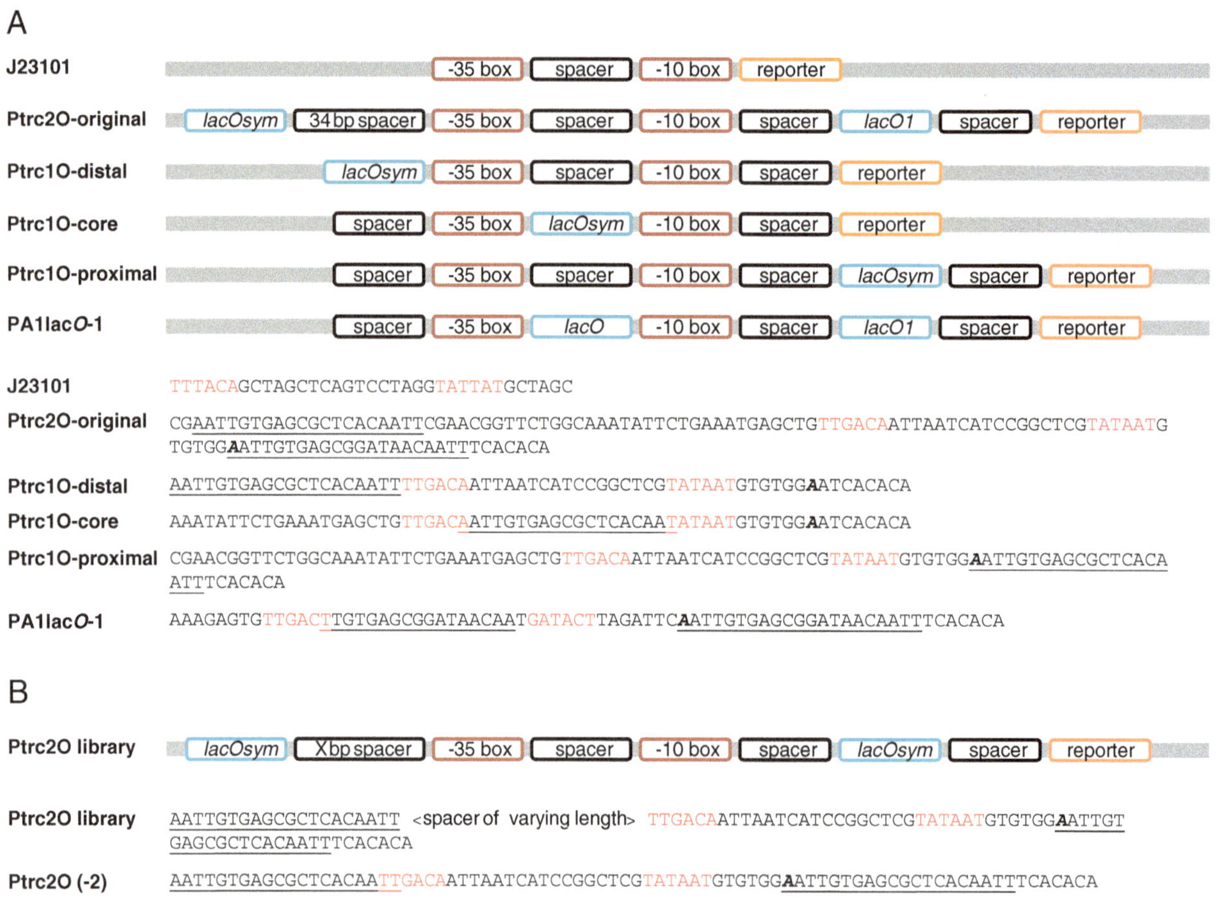

Figure 3 P*trc*-library promoter design overview and sequences. A. Control promoters. J23101 corresponds to BioBrick BBa_J23101 (iGEM Registry [34]) and is a constitutive control promoter without *lac* operator (*lacO*). P*trc*2O-original is the P*trc*2O version previously characterized [25] with *lacO1* in a proximal position and the symmetric *lacO* (*lacO*sym) in a distal position. P*trc*1O-distal, -core, and-proximal are P*trc*-derived promoters with only one *lacO*sym in the indicated positions. PA1*lacO*-1 is a previously used promoter with *lacO* in the core and proximal positions [36]. **B**. Overview of the P*trc*2O-library. The spacer sequence between the −35 box and the distal *lacO*sym varies in length from −2 bp (where 2 bp of *lacO*sym is part of the −35 box, as illustrated) to 21 bp. The variable-length spacer sequence is derived from the P*trc*2O-original spacer sequence. Transcriptional start sites are marked in bold italic.

the about 6-fold stronger *lac*Osym operator [13]. P*trc*1O-proximal also serves as a proximal-*lac*Osym-only control for the P*trc*2O-library. The P*trc*2O-library was built based on P*trc*2O-original but, identically to P*trc*1O-proximal, the proximal *lac*O1 has been exchanged with *lac*Osym to enhance repression, and the distance between the primary proximal and the secondary distal *lac*Osym was varied by varying the length of a spacer sequence between the −35 element and the distal *lac*Osym. This spacer sequence varies in length from −2 bp, where the distal *lac*Osym goes 2 bp into the −35 element (Figure 3B), and 21 bp, and consists of the sequence previously used in the 34 bp spacer of P*trc*2O-original. For characterization, the P*trc*-derived promoter library was cloned with the same EYFP reporter construct as the constitutive promoters.

The P*trc*-derived promoter library was characterized in the *E. coli* NEB5α F' *lacI*^{q} strain (New England Biolabs) using a setup with LB medium in 96-well plates supplemented with relevant antibiotics and 0 mM or 1 mM IPTG, for non-induced and induced conditions, respectively. For the control promoters (Figure 4A), we could observe that the constitutive control BBa_J23101 displays stable expression under both induced conditions and non-induced. P*trc*2O-orig and P*trc*1O-prox display the same limited repression as previously characterized [25] where P*trc*2O-orig is more strongly repressed than P*trc*1O-prox. The distal *lac*Osym control promoter P*trc*1O-dist displays no detectable repression whereas P*trc*1O-core displays repression similar to P*trc*1O-prox. PA1*lacO*-1, which contains a core and a proximal *lacO*, displays strong repression yet highly inducible expression as expected from its original characterization [36]. For the P*trc*2O-library (Figure 4B), repression is a periodic function of the distance between the two *lac*Osym with a period of 11 bp, approximately the expected bp per turn of the chromatin DNA in vivo [20], for both the non-

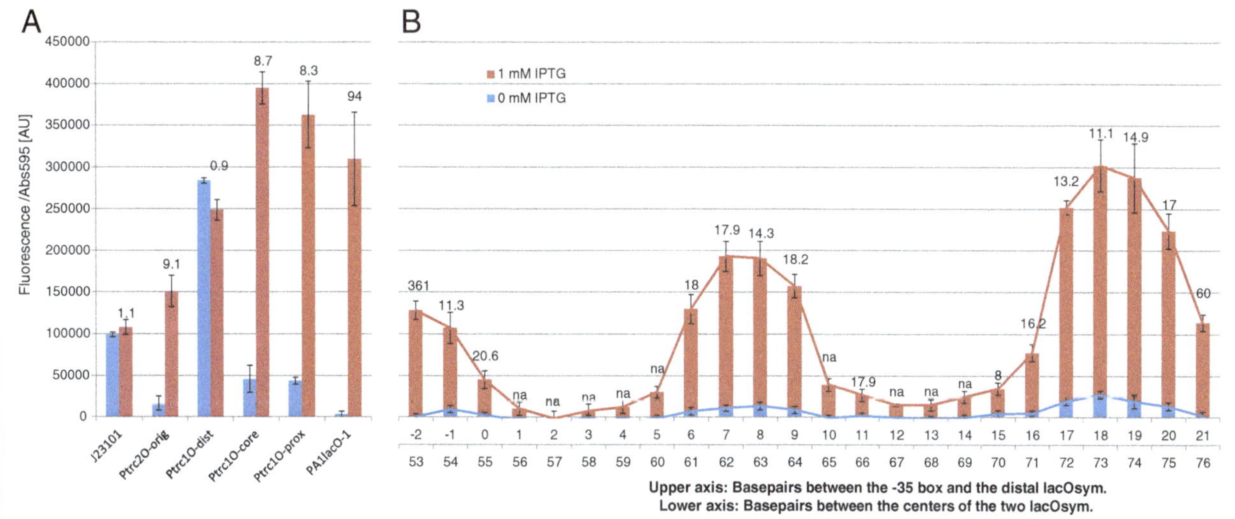

Figure 4 Promoter activities per cell under non-induced (0 mM IPTG) and induced (1 mM IPTG) conditions, characterized using an EYFP reporter construct carried on pPMQAK1 [25] in *Escherichia coli* NEB5α F′ I^q (New England Biolabs). A. Control promoters. J23101 corresponds to BioBrick BBa_J23101 (iGEM Registry [34]) and is a constitutive control promoter without *lac* operator (*lacO*). P*trc*2O-original is the P*trc*2O version previously characterized [25] with *lacO1* in the proximal position and the symmetric *lacO* (*lacO*sym) in the distal position. P*trc*1O-distal, -core, and-proximal are P*trc*-derived promoters with only one *lacO*sym at the indicated position. PA1*lacO*-1 is a previously used promoter with *lacO* in the core and proximal positions [36]. **B.** The P*trc*2O-library. The length of the spacer sequence between the −35 box and the distal *lacO*sym is indicated on the upper x-axis and the distance between the two *lacO*sym is indicated on the lower x-axis. Promoter activities per cell are given in fluorescence/absorbance at 595 nm in arbitrary units [AU]. Error bars indicate standard error of the mean (n = 6). The numbers on top of the columns are the induction ratios. na, not available (repressed fluorescence under detection limit). IPTG, Isopropyl β-D-1-thiogalactopyranoside.

induced and the induced conditions. Maximal repression is represented by the two troughs of the pattern, separated by one turn of DNA or 11 bp, and minimal repression is represented by the two peaks of the pattern, also separated by one turn of DNA. Maximal repression is not observed for inter-*lacO*sym distances that are even multiples of 11 bp, corresponding to 55 and 66 bp, which is expected to be optimal if the average number per turn over the promoter sequence was 11 bp. Instead, repression maxima appear at 57 and 67 or 68 bp, a deviation of 2 bp for the shorter loop and 1-2 bp for the longer one. Another observed trend for both conditions is that repression decreases with the distance of the secondary distal *lacO*sym from the promoter and the primary proximal *lacO*sym. The induction ratio values (Figure 4A and B) should be interpreted with caution for highly repressed promoters, as these values are close to the detection limit and even small variation due to noise will give large effects on the ratio.

Characterization of the P*trc*-derived promoter library in the LacI-expressing strain of *Synechocystis*

The complete P*trc*-derived promoter library (Figure 3A and B) was characterized in the LacI-expressing strain of *Synechocystis* using a setup with BG11 medium in 96-well plates supplemented with relevant antibiotics and 0 mM or 1 mM IPTG, for non-induced and induced conditions, respectively. As a control for promoter activity in the absence of LacI, selected promoters (BBa_J23101, P*trc* 1O-core, PA1*lacO*-1 and the P*trc*2O-library constructs P*trc*2O-2, 13 and 18) were characterized also in the Chloramphenicol resistance-only strain of *Synechocystis*. This was done similarly as the characterization in the LacI-expressing strain of *Synechocystis*, but a setup with BG11 in 13 mL growth tubes was used instead of 96-well plates. For the control promoters (Figure 5A) we could observe that the constitutive control BBa_J23101 displays stable expression under both non-induced and induced conditions. Further, when comparing BBa_J23101-driven expression between the constitutive promoter assay (Figure 1) and the LacI-repression assay (Figure 5A), we observe that the level of fluorescence per cell is very similar even though the conditions were different. Wild-type *Synechocystis* was used in the first case whereas the LacI-enhanced and the Chloramphenicol resistance-only strains were used in the second, the starter seed cultures were treated differently in the two cases, and 5 ml 6-well plate cultures were used in the first assay whereas 200 μl 96-well plate cultures or 1 mL cultures in 13 mL growth tubes were used in the second. Hence, under the conditions tested here, BBa_J23101 is stably constitutive in *Synechocystis*. For P*trc*2O-orig we largely reproduced our previous characterization result [25], even though the assays differed in the expression of LacI and the growth conditions. There is a small difference in that P*trc*2O-orig is not as well-repressed here, which could indicate that the

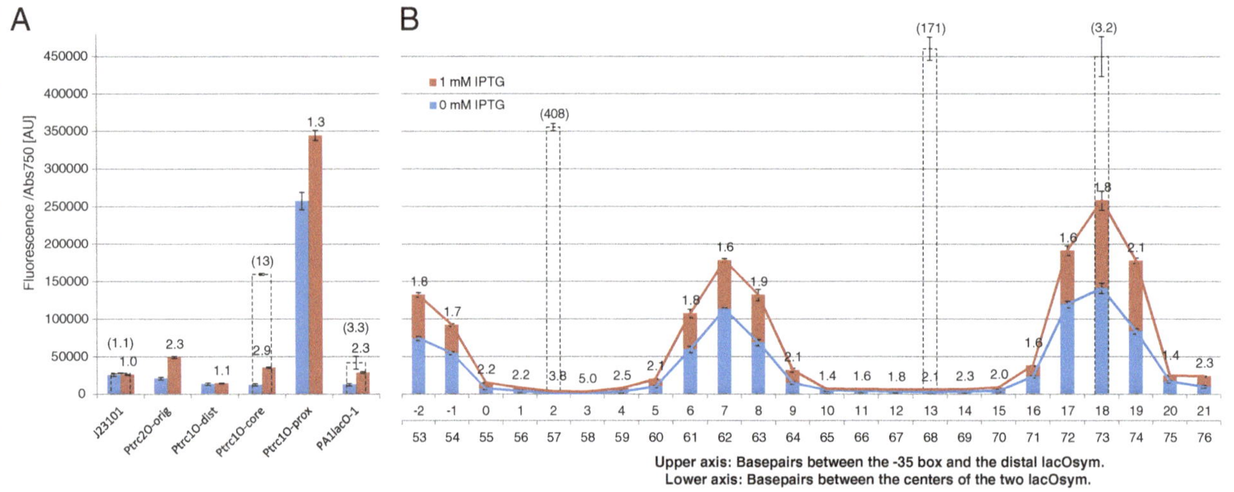

Figure 5 Promoter activities per cell under non-induced (0 mM IPTG) and induced (1 mM IPTG) conditions, characterized using an EYFP reporter construct carried on pPMQAK1 [25] in an engineered *Synechocystis* PCC 6803 strain expressing LacI (this work). A. Control promoters. J23101 corresponds to BioBrick BBa_J23101 (iGEM Registry [34]) and is a constitutive control promoter without *lac* operator (*lacO*). P*trc*2O-original is the P*trc*2O version previously characterized [25] with *lacO1* in the proximal position and the symmetric *lacO* (*lac*Osym) in the distal position. P*trc*1O-distal,-core, and-proximal are P*trc*-derived promoters with only one *lac*Osym at the indicated position. PA1*lacO*-1 is a previously used promoter with *lacO* in the core and proximal positions [36]. **B.** The P*trc*2O-library. The length of the spacer sequence between the–35 box and the distal *lac*Osym is indicated on the upper x-axis and the distance between the two *lac*Osym is indicated on the lower x-axis. Promoter activities per cell are given in fluorescence/absorbance at 750 nm in arbitrary units [AU]. Broken line columns represent promoter activity per cell for the corresponding promoter tested in the chloramphenicol resistance-only strain of *Synechocystis* (LacI negative, this work). Error bars indicate standard error of the mean for the *Synechocystis* PCC 6803 strain expressing LacI (n = 4) and the chloramphenicol resistance-only *Synechocystis* strain (n = 2). The numbers on top of the normal columns are the induction ratios, the numbers in parenthesis on top of the broken columns are the repression ratios. IPTG, Isopropyl β-D-1-thiogalactopyranoside.

levels of LacI are somewhat lower than in our previous study. The P*trc*1O-prox, which differs from the P*trc* we used previously [25] in that *lac*O1 of P*trc* has been switched for the more efficient *lac*Osym, largely behaves like P*trc* under non-induced and induced conditions, being very poorly repressed. This contrasts with the situation in *E. coli* where it is more efficiently repressed. The distal *lac*Osym promoter P*trc*1O-dist does not show strong activity, whereas it does in *E. coli*, and displays no difference between non-induced or induced conditions, there agreeing with the *E. coli* results. The core *lac*Osym promoter P*trc*1O-core behaves very similarly to PA1*lacO*-1, which carries one slightly truncated *lacO* in the core and *lacO1* in the proximal position, both displaying limited repression and induction. This contrasts strongly with the situation in *E. coli* where PA1*lacO*-1 is highly repressed yet displays strong expression under induction. When tested without LacI in the Chloramphenicol resistance-only strain of *Synechocystis*, P*trc*1O-core displays 13-times higher activity than under LacI-repressed conditions. This shows that the lack of induction in *Synechocystis* is not due to an inherently weak P*trc*1O-core promoter but due to the presence of induced LacI. Still, the activity of P*trc*1O-core is weaker than that of P*trc*1O-prox, even when comparing P*trc*1O-core without LacI and P*trc*1O-prox with non-induced LacI (Figure 5B). However in *E. coli*, P*trc*1O-core and P*trc*1O-prox are very similar under both repressed and induced conditions (Figure 4A). PA1*lacO*-1 however, displays a relatively low activity in *Synechocystis* even when LacI is absent (Figure 5A), and hence cannot be expected to be induced to higher levels. For the P*trc*2O-library (Figure 5B), the repression pattern is similar to the situation in *E. coli*, displaying periodic repression as a function of the distance between the two *lac*Osym with a period of 11 bp. Also as in *E. coli*, maximal repression can be observed for the troughs in the periodic repression pattern, minimal repression is represented as the two expression peaks separated by 11 bp at the same positions as in *E. coli*, and repression efficiency decreases with the distance of the secondary distal *lacO*-sym to the promoter and the primary proximal operator. However, even though repression of the P*trc*2O-library is very effective under both non-induced and induced conditions for optimal *lac*Osym positions, it is very ineffective in the least optimal positions represented by the expression peaks, even under non-induced conditions. This differs strongly from the more efficiently repressed non-optimal distal *lac*Osym positions in *E. coli*, which, even though they are clearly visible as expression peaks, are repressed even under non-induced conditions. Further, repression in the geometrically-favored configurations is represented in *Synechocystis* by much wider troughs than in *E. coli*, for both conditions. This contrasts with the most unfavorable looping-geometries, represented by the narrower expression peaks in *Synechocystis* than in *E. coli*. To test if the very efficient repression for favorable DNA-looping geometries in *Synechocystis* is due to non LacI-related promoter inhibition of the P*trc*2O-library, we characterized the activity of P*trc*2O-2 and 13 in the absence of LacI in the Chloramphenicol resistance-only strain. These two highly repressed promoters, which are located in the centers of the repression troughs, display very high promoter activity in the absence of LacI (Figure 5B). Hence, they work as very high dynamic range LacI-regulated promoters, if expression could only be efficiently induced. Conversely, to test if the very non-efficient repression for non-favorable DNA-looping geometries in *Synechocystis* is due to non LacI-related promoter activation effects on the P*trc*2O-library, we also characterized the activity of P*trc*2O-18 in the absence of LacI. This inefficiently repressed promoter displayed a close to identical activity to the P*trc*2O-13 promoter, illustrating that the lack of repression of the P*trc*2O-library for non favorable DNA-looping geometries in *Synechocystis* is not due to inherently stronger promoters. Taken together, these results show that the periodical repression pattern of the P*trc*2O-library in the LacI-expressing strain of *Synechocystis* is due to LacI-facilitated DNA-looping.

Characterization of the proximal *lacO*-library in *E. coli* DH5αZ1 and *Synechocystis*

Differences in how the *E. coli* and the *Synechocystis* RNAP interact with LacI bound to the proximal region of *trc*-derived promoters could be investigated by comparing IPTG-induction or LacI-repression patterns of a proximal *lacO*-library between *E. coli* and *Synechocystis*. Because induction or repression ratios of the same promoters are studied, expression differences between promoters caused by differences in the 5′-UTRs will be cancelled, and the ratios will reflect the efficiencies of LacI repression for specific promoters. Such induction or repression ratios of promoter libraries where the position of single *lac* operators relative to the promoter was varied by single base-pairs have been studied before to investigate variation in LacI repression [19,20]. We reason that a potential difference in the DNA-binding activity between the *E. coli* and the *Synechocystis* RNAP might be visible in differences in the pattern of this ratio. Therefore, we designed a promoter library in which we vary the distance between the center of a symmetric *lacO* and the TSS of a P*trc* core promoter from −2 to +18 bp by adjusting the length of a spacer in 1 bp steps and from +18 to +24 bp in 2 bp steps (Table 1). We inserted this proximal *lacO*-library into an *E. coli* strain that expresses high amounts of LacI repressor, DH5αZ1, to enable the calculation of IPTG-induction ratios, and into both the LacI-expressing *Synechocystis* strain and the chloramphenicol resistance-only strain, to enable the calculation of both IPTG-induction and LacI-repression ratios.

The characterization of the proximal *lacO*-library in *E. coli* DH5αZ1 was performed similarly to the characterization of the P*trc*-derived promoter library in NEB5α, but a key difference is that the proximal *lacO*-library was followed in time by measuring the same plates at different time-points and then continuing the experiment. The induction ratios were calculated from the fluorescence/absorbance values of the induced/non-induced cultures and the standard errors of the ratios were estimated (Figure 6A). An apparent steady-state of fluorescence/cell for most of the promoters (data not shown) was observed at ca 3.8 hours after the start of the experiment. At this time point, the induction ratios are also the highest; the J23101 constitutive control has reached its expected ratio of ca 1 and the ratios diminish progressively in value at later time-points. The induction ratio pattern established from ca 3.8 hours remains, however, also at later time points (Figure 6A). As shown in previous studies of induction and repression ratios of libraries of proximal *lacO*-promoters in *E. coli* [19], the *lacO* centered at +11 relative to the TSS (prox-11 in Figure 6A) provides the most efficient repression, closely followed by the +10 promoter. The +11 position corresponds to the position of *lacO1* in the wild-type *E. coli lac* promoter [10] and is one helix turn away from the TSS, putting LacI on the same side of the DNA as the RNAP.

The characterization of the proximal *lacO*-library in *Synechocystis* was performed similarly to the characterization of the P*trc*-derived promoter library, but 13 mL growth tubes were used instead of 96-well plates, and, as compared with the *E. coli* DH5αZ1 experiment, only one time-point was used. The induction ratios were calculated from the fluorescence/absorbance values of the induced/non-induced LacI-expressing *Synechocystis* cultures and the standard errors of the ratios were estimated (Figure 6B). The repression ratios were calculated from the fluorescence/absorbance values of the chloramphenicol resistance-only cultures/non-induced LacI-expressing *Synechocystis* cultures and the standard errors of the ratios were estimated (Figure 6C). First, we observe that the ratios are much smaller, with a smaller difference between the different promoters, than in *E. coli* DH5αZ1. However, both the induction and the repression ratios in *Synechocystis* are similar to the induction ratios in *E. coli* DH5αZ1. Also in *Synechocystis*, the centered +11 position of the *lacO* relative to the TSS produces the most efficiently repressed proximal *lacO* promoter. To summarize, no difference in the induction or repression patterns between *E. coli* and *Synechocystis* can be detected.

Table 1 Promoter DNA sequences

Promoter name	Sequence 5′ to 3′
BBa_J23100	TTGACGGCTAGCTCAGTCCTAGGTACAGTGCTAGC
BBa_J23101	TTTACAGCTAGCTCAGTCCTAGGTATTATGCTAGC
BBa_J23102	TTGACAGCTAGCTCAGTCCTAGGTACTGTGCTAGC
BBa_J23105	TTTACGGCTAGCTCAGTCCTAGGTACTATGCTAGC
BBa_J23106	TTTACGGCTAGCTCAGTCCTAGGTATAGTGCTAGC
BBa_J23109	TTTACAGCTAGCTCAGTCCTAGGGACTGTGCTAGC
BBa_J23113	CTGATGGCTAGCTCAGTCCTAGGGATTATGCTAGC
BBa_J23114	TTTATGGCTAGCTCAGTCCTAGGTACAATGCTAGC
P*nirA*	CAAGCTCAGAATGCTGCGGGGAGAAGGGCAACTCCC CACCAGCCCCAAATTTTTGCTGGCGATAAATATTTTTCGGTTTAATTGTTCACA AAGCTTTTTGAATTTGAGTTTATAGAAATTTATTGGCTGG TAATGCTTTTTTGCCCCCCTGCTGGACTTCATTGATCCTTGCCTATAC CATCAATATCATTGGTCAATAATGATGATGATTGACTAAAACATGTT TAACAAAATTTAACGCATATGCTAAATGCGTAAACTGCATATGCCTT GGCTGAGTGTAATTTACGTTACAAATTTTAACGAAACGGGAACCCT ATATTGATCTCTACTGTTATCTGGCT
P*petE*	TCATAGCGGTTGCCCAATCTAACTCAGGGAGCGACTTCAGCCCAC AAAAAACACCACTGGGCCTACTGGGCTATTCCCATTATCATCTACA TTGAAGGGATAGCAAGCTAATTTTTATGACGGCGATCGCCAAAAAC AAAGAAAATTCAGCAATTACCGTGGGTAGCAAAAAATCCCCATCTAA AGTTCAGTAAATATAGCTAGAACAACCAAGCATTTTCGGCAAAGTAC TATTCAGATAGAACGAGAAATGAGCTTGTTCTATCCGCCCGGGGCTG AGGCTGTATAATCTACGACGGGCTGTCAAACATTGTGATACCATGGG CAGAAGAAAGGAAAAACGTCCCTGATCGCCTTTTTGGGCACGGAGT AGGGCGTTACCCCGGCCCGTTCAACCACAAGTCCCTATAGATACAATCGCCAAGAAGT
P*rnpB*	TTCAATGCGGTCCAATACCTCCCCTGCCCAACTGGGTAAGCTCGCGGC TCCACTGAGTAATACAGACAAGGCTAAACAGGCAAATTTTTTCATTGG TCAACTCCTAGCACCAATTTCCCAAGACTACGGAGGGGGCAATGAAGTT TCAATTAATTGGGGTCACAAACCACAGCGGCCTATGGCTCTAATCAATGGCACACTAGAAAAA
P*trc*2O-(−2)	AATTGTGAGCGCTCACAATTGACAATTAATCATCCGGCTCGTATAATGTGTGGAATTGTGAG CGCTCACAATTTCACACA
P*trc*2O-(−1)	AATTGTGAGCGCTCACAATTTGACAATTAATCATCCGGCTCGTATAATGTGTGGAATTGTGA GCGCTCACAATTTCACACA
P*trc*2O-0	AATTGTGAGCGCTCACAATTTTGACAATTAATCATCCGGCTCGTATAATGTGTGGAATTGT GAGCGCTCACAATTTCACACA
P*trc*2O-1	AATTGTGAGCGCTCACAATTGTTGACAATTAATCATCCGGCTCGTATAATGTGTGGAATT GTGAGCGCTCACAATTTCACACA
P*trc*2O-2	AATTGTGAGCGCTCACAATTTGTTGACAATTAATCATCCGGCTCGTATAATGTGTGGAATT GTGAGCGCTCACAATTTCACACA
P*trc*2O-3	AATTGTGAGCGCTCACAATTCTGTTGACAATTAATCATCCGGCTCGTATAATGTGTGGAA TTGTGAGCGCTCACAATTTCACACA
P*trc*2O-4	AATTGTGAGCGCTCACAATTGCTGTTGACAATTAATCATCCGGCTCGTATAATGTGTGG AATTGTGAGCGCTCACAATTTCACACA
P*trc*2O-5	AATTGTGAGCGCTCACAATTAGCTGTTGACAATTAATCATCCGGCTCGTATAATGTGTGG AATTGTGAGCGCTCACAATTTCACACA
P*trc*2O-6	AATTGTGAGCGCTCACAATTGAGCTGTTGACAATTAATCATCCGGCTCGTATAATGTGTG GAATTGTGAGCGCTCACAATTTCACACA
P*trc*2O-7	AATTGTGAGCGCTCACAATTTGAGCTGTTGACAATTAATCATCCGGCTCGTATAATGTGT GGAATTGTGAGCGCTCACAATTTCACACA
P*trc*2O-8	AATTGTGAGCGCTCACAATTATGAGCTGTTGACAATTAATCATCCGGCTCGTATAATGTG TGGAATTGTGAGCGCTCACAATTTCACACA
P*trc*2O-9	AATTGTGAGCGCTCACAATTAATGAGCTGTTGACAATTAATCATCCGGCTCGTATAATGTG TGGAATTGTGAGCGCTCACAATTTCACACA
P*trc*2O-10	AATTGTGAGCGCTCACAATTAAATGAGCTGTTGACAATTAATCATCCGGCTCGTATAATGT GTGGAATTGTGAGCGCTCACAATTTCACACA

Table 1 Promoter DNA sequences *(Continued)*

P*trc*2O-11	AATTGTGAGCGCTCACAATTGAAATGAGCTGTTGACAATTAATCATCCGGCTCGTATAATGTGTGGAATTGTGAGCGCTCACAATTTCACACA
P*trc*2O-12	AATTGTGAGCGCTCACAATTTGAAATGAGCTGTTGACAATTAATCATCCGGCTCGTATAATGTGTGGAATTGTGAGCGCTCACAATTTCACACA
P*trc*2O-13	AATTGTGAGCGCTCACAATTCTGAAATGAGCTGTTGACAATTAATCATCCGGCTCGTATAATGTGTGGAATTGTGAGCGCTCACAATTTCACACA
P*trc*2O-14	AATTGTGAGCGCTCACAATTTCTGAAATGAGCTGTTGACAATTAATCATCCGGCTCGTATAATGTGTGGAATTGTGAGCGCTCACAATTTCACACA
P*trc*2O-15	AATTGTGAGCGCTCACAATTTTCTGAAATGAGCTGTTGACAATTAATCATCCGGCTCGTATAATGTGTGGAATTGTGAGCGCTCACAATTTCACACA
P*trc*2O-16	AATTGTGAGCGCTCACAATTATTCTGAAATGAGCTGTTGACAATTAATCATCCGGCTCGTATAATGTGTGGAATTGTGAGCGCTCACAATTTCACACA
P*trc*2O-17	AATTGTGAGCGCTCACAATTTATTCTGAAATGAGCTGTTGACAATTAATCATCCGGCTCGTATAATGTGTGGAATTGTGAGCGCTCACAATTTCACACA
P*trc*2O-18	AATTGTGAGCGCTCACAATTATATTCTGAAATGAGCTGTTGACAATTAATCATCCGGCTCGTATAATGTGTGGAATTGTGAGCGCTCACAATTTCACACA
P*trc*2O-19	AATTGTGAGCGCTCACAATTAATATTCTGAAATGAGCTGTTGACAATTAATCATCCGGCTCGTATAATGTGTGGAATTGTGAGCGCTCACAATTTCACACA
P*trc*2O-20	AATTGTGAGCGCTCACAATTAAATATTCTGAAATGAGCTGTTGACAATTAATCATCCGGCTCGTATAATGTGTGGAATTGTGAGCGCTCACAATTTCACACA
P*trc*2O-21	AATTGTGAGCGCTCACAATTCAAATATTCTGAAATGAGCTGTTGACAATTAATCATCCGGCTCGTATAATGTGTGGAATTGTGAGCGCTCACAATTTCACACA
P*trc*2O-orig	CGAATTGTGAGCGCTCACAATTCGAACGGTTCTGGCAAATATTCTGAAATGAGCTGTTGACAATTAATCATCCGGCTCGTATAATGTGTGGAATTGTGAGCGGATAACAATTTCACACA
P*trc*1O-dist	AATTGTGAGCGCTCACAATTTTGACAATTAATCATCCGGCTCGTATAATGTGTGGAATCACACA
P*trc*1O-core	AAATATTCTGAAATGAGCTGTTGACAATTGTGAGCGCTCACAATATAATGTGTGGAATCACACA
P*trc*1O-prox	CGAACGGTTCTGGCAAATATTCTGAAATGAGCTGTTGACAATTAATCATCCGGCTCGTATAATGTGTGGAATTGTGAGCGCTCACAATTTCACACA
PA1*lacO*-1	AAAGAGTGTTGACTTGTGAGCGGATAACAATGATACTTAGATTCAATTGTGAGCGGATAACAATTTCACACA
PtrcOs-prox-8	TTGACAATTAATCATCCGGCTCGTATAATGTGTAATTGTGAGCGCTCACAATT
PtrcOs-prox-9	TTGACAATTAATCATCCGGCTCGTATAATGTGTGAATTGTGAGCGCTCACAATT
PtrcOs-prox-10	TTGACAATTAATCATCCGGCTCGTATAATGTGTGGAATTGTGAGCGCTCACAATT
PtrcOs-prox-11	TTGACAATTAATCATCCGGCTCGTATAATGTGTGGAAATTGTGAGCGCTCACAATT
PtrcOs-prox-12	TTGACAATTAATCATCCGGCTCGTATAATGTGTGGAGAATTGTGAGCGCTCACAATT
PtrcOs-prox-13	TTGACAATTAATCATCCGGCTCGTATAATGTGTGGAGTAATTGTGAGCGCTCACAATT
PtrcOs-prox-14	TTGACAATTAATCATCCGGCTCGTATAATGTGTGGAGTCAATTGTGAGCGCTCACAATT
PtrcOs-prox-15	TTGACAATTAATCATCCGGCTCGTATAATGTGTGGAGTCGAATTGTGAGCGCTCACAATT
PtrcOs-prox-16	TTGACAATTAATCATCCGGCTCGTATAATGTGTGGAGTCGTAATTGTGAGCGCTCACAATT
PtrcOs-prox-17	TTGACAATTAATCATCCGGCTCGTATAATGTGTGGAGTCGTCAATTGTGAGCGCTCACAATT
PtrcOs-prox-18	TTGACAATTAATCATCCGGCTCGTATAATGTGTGGAGTCGTCCAATTGTGAGCGCTCACAATT
PtrcOs-prox-20	TTGACAATTAATCATCCGGCTCGTATAATGTGTGGAGTCGTCCAGAATTGTGAGCGCTCACAATT
PtrcOs-prox-22	TTGACAATTAATCATCCGGCTCGTATAATGTGTGGAGTCGTCCAGACAATTGTGAGCGCTCACAATT
PtrcOs-prox-24	TTGACAATTAATCATCCGGCTCGTATAATGTGTGGAGTCGTCCAGACTCAATTGTGAGCGCTCACAATT

Discussion

To elucidate the design principles of LacI-repressed promoters and DNA-looping in cyanobacteria, we constructed a P*trc*-derived promoter library consisting of single *lac* operator control promoters (Figure 3A) and dual *lacO* promoters with LacI-facilitated DNA-loop capability (Figure 3B). For characterizing the repression and induction of this library, we first characterized several artificial minimal promoters in *Synechocystis* and used this knowledge to engineer a constitutive LacI-expressing strain of *Synechocystis*. Finally, to investigate if differences between the *E. coli* and *Synechocystis* RNA polymerases would

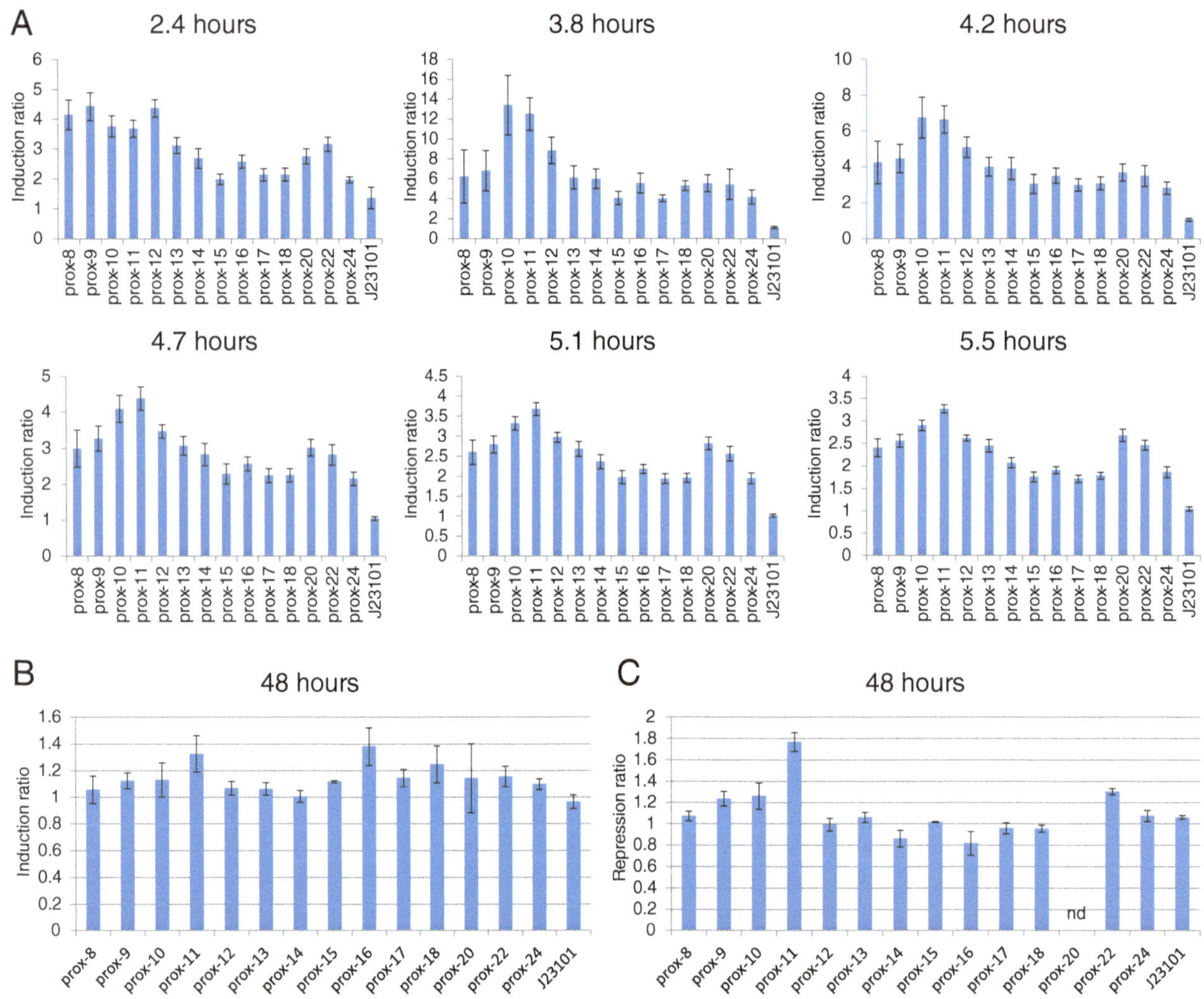

Figure 6 Patterns of induction and repression ratios for the proximal *lacO* promoter library characterized using a SGFP2 reporter construct carried on pPMQAK1 [25]. The number in prox-# indicates the distance between the center of the proximal symmetric *lac* operator and the transcriptional start site of a *Ptrc* core promoter. The time in hours represents time from start of experiment. J23101 corresponds to BioBrick BBa_J23101 (iGEM Registry [34]) and is a constitutive control promoter without *lac* operator. **A.** Time-course study of induction ratios (fluorescence per cell values of 2 mM IPTG induced cultures/the values of the non-induced cultures) in *E. coli* DH5αZ1. Error bars represent the estimation of the standard error of the mean for the induction ratios (n = 6). **B.** Induction ratios in the *Synechocystis* PCC 6803 strain expressing LacI (this work). Error bars represent the estimation of the standard error of the mean for the induction ratios (n = 2). **C.** Repression ratios (fluorescence per cell values of cultures from the chloramphenicol resistance-only *Synechocystis* strain (LacI negative, this work)/the values of cultures from the *Synechocystis* PCC 6803 strain expressing LacI). Error bars represent the estimation of the standard error of the mean for the repression ratios (n = 2). nd, not determined.

affect LacI-repression of promoters with proximally located *lacO*, we engineered and characterized a library of proximal *lacO* promoters in both *E. coli* and *Synechocystis*.

Minimal and artificial promoters as orthogonal alternatives to native cyanobacterial promoters

As orthogonal parts can help minimize or eliminate cross-talk and unwanted interactions between parts or between parts and the chassis [7], we first set out to characterize several minimal and artificial promoters that could potentially drive the constitutive expression of LacI in *Synechocystis*. The artificial, minimal and constitutive promoters BBa_J23100, BBa_J23101, BBa_J23102, BBa_J23105, BBa_J23106, BBa_J23109, BBa_J23113 and BBa_J23114 were found to span a greater range of expression levels, high and low, as compared to the three commonly used native promoters P*nirA*, P*petE* and P*rnpB* from *Synechocystis*. The artificial promoters are derived from a consensus *E. coli* promoter, which −35 and −10 elements are almost identical to the consensus −35 and −10 boxes of *Synechocystis* [37]. These types of promoters, belonging to the group 1 promoters in cyanobacteria, are served by the primary cyanobacterial sigma factor SigA during normal growth conditions [26]. Hence, all the artificial promoters tested here can be classified as cyanobacterial group 1 promoters, and as orthogonal and minimal promoters, minimizing the risk for cross-talk with native transcription factors or sRNAs, they are expected to remain constitutive under normal growth conditions. In favor of this, the BBa_J23101 promoter was found to display stable levels of expression of EYFP per cell under the different conditions of light-intensity, culture volumes and mixing, presence of the exogenous chemical IPTG, and three chassis (wild-type, LacI-enhanced and chloramphenicol resistance-only strains of *Synechocystis*) that were used in this study. Further, the weak BBa_J23114 promoter was used with RBS* to express relatively low levels of LacI/cell (about 190 tetramers of LacI/cell) in the LacI-enhanced strain of *Synechocystis*. Therefore, we suggest that promoters from the iGEM Registry's Anderson collection could be used as minimal orthogonal alternatives to native promoters for constitutive expression in cyanobacteria, or as core promoters for the engineering of regulated promoters. Further, as the BBa_J23101 promoter has been used as a standard reference promoter in *E. coli* [38], we propose that it could fulfill the same role in cyanobacteria. As all the P*trc*-derived promoters used in this study are close-to-consensus group 1 cyanobacterial promoters, orthogonal to *Synechocystis*, and minimal in the sense that they only contain the necessary elements for recognition by RNAP and regulation by LacI, they are also expected to be constitutive in the absence of LacI, which we showed herein for selected members of the library. The orthogonality also means that they should display no or minimal cross-talk with native transcription factors and sRNAs. However, to truly test the variation of expression, and find constitutive promoters with minimal variation of expression, a more rigorous study of diverse environmental conditions than the present work is warranted. For example, cyanobacteria naturally grow in day and night cycles and have a circadian rhythm that affects a large part of the transcriptome, seemingly mainly through modulating DNA topology [39]. How or if circadian rhythms would modulate orthogonal promoters in cyanobacteria would have to be tested. Further, a change in DNA topology could change the relative spatial positioning or phasing on the helix of a promoter and the binding site of a transcription factor, such as LacI, possibly rendering repression even by an orthogonal repressor circadian rhythm-dependent.

Other environmental conditions are also relevant to test for orthogonal promoters, such as higher variations in light intensity and duration of darkness, changes in salinity and temperature. It is well known that different environmental variations give rise to sigma-factor switching, a phenomenon that helps cyanobacteria adapt to diverse changes by switching between different sigma-factors that activate the expression of stress responses [26]. To further decouple synthetic circuits from cell regulation and phenomena such as sigma-switching, system orthogonality could be increased by using orthogonal RNAP and their corresponding promoters, which are decoupled from the native transcriptional machinery of the host. For example, the bacteriophage T7 RNAP can be engineered to produce orthogonal genetic circuits incorporating several mutually orthogonal T7 RNAP [40] or orthogonal split T7 RNAP [41].

LacI-repressed promoters in *E. coli* and *Synechocystis*

Since the following analysis builds on comparisons of promoter repression patterns, it is important to note that expression also depends on the 5′ untranslated region, the coding sequence of the particular gene of interest (GOI) and their combined interactions. The folding status of the 5′ UTR and in particular the ribosome binding site (RBS) strongly affects translation initiation [42]. Since the first part of the GOI sequence is located close to the 5′ UTR on the mRNA, its identity will affect potential mRNA secondary structures involving the RBS. Further, internal GOI sequence elements such as codon identities and secondary structures will affect translation elongation [43]. Consequently, when comparing promoter strengths indirectly through the expression of a reporter gene it only makes sense to compare constructs with the same 5′ UTR and coding sequence. In this study, the same reporter cassette featuring the same RBS and coding sequence is used for all promoters within a library. However, some of the promoters in the P*trc*-derived library differ in the sequences that come after the transcription start site (TSS), meaning that these promoters will have slightly different 5′ UTRs (Figures 3A and B). Among these promoters, P*trc*1O-proximal and the whole P*trc*2O-library, except P*trc*2O-original, share the same 5′ UTR. Further, P*trc*1O-distal and P*trc*1O-core share the same 5′ UTR whereas BBa_J23101, P*trc*2O-original and PA1*lacO*-1 do not share 5′ UTRs. However, the induction ratios can still be compared between promoters that do not share exact 5′ UTRs as 5′ UTR effects that affect both the induced state and the non-induced state will be cancelled. This is true also for the proximal *lacO* promoter library (Table 1), where all promoters differ slightly in their 5′ UTRs but only induction or repression ratios are compared. Finally, it should be noted that the fluorescence from promoters that are highly repressed are close to the detection limit, meaning that even small amounts of variation due to noise will affect the induction or repression ratio values strongly. Hence, in general caution should be exercised when comparing promoter activity ratios for strongly repressed promoters.

We engineered a strain of *Synechocystis* to express similar levels of LacI as in our previous study [25] for repression of the P*trc*-derived promoter library. Also, we previously observed that P*trc* and P*trc*2O behave as constitutive promoters of approximately the same strength in the absence of heterologously expressed LacI, displaying stable expression levels irrespective of the presence of IPTG. Therefore, and since we are herein interested in analyzing the patterns of repression and induction depending on *lacO* location(s), we did not analyze the majority of the P*trc*-derived promoter library in the absence of LacI. Further, by comparing repression and induction levels of P*trc*2O (here termed P*trc*2O-original to distinguish it from the other P*trc*2O promoters) in *Synechocystis* between the present (Figure 5A) and our previous study [25], we observe a similar pattern but that the induction/repression ratio is about two-fold lower in this study. As the two studies differ in both the characterization growth conditions and the LacI expression cassettes we cannot confidently conclude the reason for this difference, but the discrepancy leads us to estimate that the levels of LacI in the present study are similar to or lower than in our previous study. We estimated the number of LacI tetramers per cell to be about 190 for the engineered LacI-expressing strain developed here, so the number can be expected to be higher than 190 in our previous study [25].

To investigate if a lack of LacI binding the proximal *lacO1* of P*trc* was the reason for the limited repression in *Synechocystis* that we previously observed [25], we exchanged *lacO1* with the stronger *lacO*sym operator. However, no increase in the induction ratio was observed (Figure 5A). This however, could be due to the possibly lower levels of LacI in this study, or that the 1 bp shorter *lacO*sym places LacI in a spatially less optimal position relative to the promoter's TSS than *lacO1*. Nonetheless, when comparing the repression and induction results of P*trc*1O-proximal between *E. coli* (Figure 4A) and *Synechocystis* (Figure 5A) we managed to confirm our previous results [25] that the proximal *lacO* position downstream to and overlapping 1 bp with the TSS leads to more efficient repression in *E. coli* as compared to in *Synechocystis*. Another study observed the same lack of repression of P*trc* in *Synechocystis* when using another construct to express LacI [28]. This reproducibility, and the fact that the levels of LacI expressed in the LacI-enhanced strain of *Synechocystis* are sufficient for much more effective repression of other *lacO* promoters (Figures 5A and B) lead us to hypothesize that the lack of repression of P*trc* and its twin P*trc*1O-proximal in *Synechocystis* is due to the characteristics of RNAP, and not due to the lack of LacI. In bacteria, the RNAP holoenzyme consists of the α_2, β, β' and ω subunits plus a σ-factor that gives promoter specificity [44]. A time-resolved *E. coli* RNAP-promoter footprinting study showed that RNAP makes first contact with the upstream part of the promoter, and then extends downstream towards the −10 element, the TSS and all the way to about +20 bp from the TSS as it proceeds from a closed to an open complex [45]. It was recently shown that the RNAP clamp, consisting of pincer-like parts of the β' and the β subunits, progressively closes around the double stranded DNA downstream from the TSS as the complex approaches transcription initiation, and then remains closed during elongation [46]. In cyanobacteria, the β' subunit has been split up in two parts: γ, which corresponds to the N-terminal part of the bacterial β' subunit, and β', which corresponds to the C-terminal part [31,32]. It was observed in a later study that enteric and cyanobacterial RNAP transcribe promoters differently, and differences in RNAP architecture, including a large insertion in the β' subunit was suggested to be the cause [47]. Since then, these differences between the enteric and the cyanobacterial RNAP has been suggested to explain differences in promoter activity due to single nucleotide mutations [48] and more recently differences in elongation rate and fidelity [49].

Hence we hypothesize that the differences between the *E. coli* RNAP β' subunit and the cyanobacterial RNAP γ + β' subunits could explain the apparent decreased sensitivity of *Synechocystis* RNAP towards LacI binding the proximal *lac* operator of P*trc* and P*trc*1O-proximal. Since the β' and β clamp of enteric RNAP closes around the DNA downstream of the TSS during transcription initiation, and β' is the subunit that differs the most between cyanobacteria and *E. coli*, we speculate that the RNAP clamp of cyanobacteria may possess an increased stability or a different DNA footprint than enteric RNAP, blocking or removing a proximally bound LacI more effectively. If the DNA footprint differs, the proximal promoter binding site location for maximizing repression by a transcription factor is likely to be different for cyanobacteria as compared to enterobacteria. To test if the DNA footprint of *E. coli* and *Synechocystis* RNAP differ for the proximal promoter region close to and downstream the TSS, we designed a library of proximal *lacO* promoters in which we vary the distance between the center of a symmetric *lacO* and the TSS of a P*trc* core promoter by steps of 1 or 2 bp. We characterized the induction and/or the repression ratio patterns for this proximal promoter library in both *E. coli* and *Synechocystis*, but could not detect any differences. Hence, our results do not support the hypothesis that there is a difference in the DNA footprint between the *E. coli* and *Synechocystis* RNAP at the promoter proximal location. If the cyanobacterial RNAP possesses an increased DNA-binding stability in the promoter proximal region remains to be tested, and what the potential effect of the β' subunit split and insertion in cyanobacteria is remains to be shown.

Finally, the fact that P*trc* can be more efficiently repressed and induced in *Synechococcus elongatus* PCC 7942 [27] than in *Synechocystis* [25,28] may simply be explained by a higher number of LacI/cell in the *Synechococcus* study than in the *Synechocystis* studies. Indeed, the numbers of LacI/cell expressed in the LacI-enhanced strain of *Synechocystis* herein presented is about five times lower than the per cell value for *E. coli* NEB5α, and about sixteen times lower than in DH5αZ1.

Recently, the in *E. coli* well-repressed yet highly inducible promoter PA1*lacO*-1 [36] was characterized in *Synechocystis* using a plasmid-carried *lacI^q* gene for expressing LacI and an ethylene-forming enzyme (EFE) as a reporter [28]. Characterization was carried out by measuring the rate of ethylene formation per volume, time and cell density and comparing IPTG-inducing and non-inducing conditions for different cell densities. The repression of PA1*lacO*-1 was found to be higher at lower cell densities and lower at higher cell densities. Further, they tested P*trc* with the same setup and found that PA1*lacO*-1 could be induced to the same level as P*trc*. To enable the comparison of a regulated high dynamic range *E. coli* promoter with its activity in *Synechocystis*, we characterized PA1*lacO*-1 with the same EYFP reporter construct as the rest of the P*trc*-derived library (Figures 3A and B). As expected, we found that PA1*lacO*-1 was strongly repressed yet highly inducible in *E. coli* (Figure 4A). However, even though we found that PA1*lacO*-1 is repressed in *Synechocystis*, in line with the previous study [28], we found that it could not be induced to the activity of P*trc*1O-prox (Figure 5A), which is almost identical in its repression and induction behavior to P*trc* [25]. Still, it is problematic to directly compare P*trc*1O-prox/P*trc* with PA1*lacO*-1 since they have slightly different 5′ UTRs, as well as comparing the two studies that use different reporters, both factors potentially affecting expression. Further, for promoter activity quantification in general, it is problematic to infer promoter activities indirectly from the expression of protein reporters. This problem can be solved by directly measuring promoter mRNA output using quantitative reverse transcription PCR. However, this poses a problem of its own as it cannot be done in living cells. Hence, it may sometimes still be preferable to characterize promoters indirectly through the use of fluorescent reporters, since this can be done in vivo. For instance, absolute quantification of promoters can be done using fluorescent protein reporters together with quantitative models to derive promoter strengths [50] or the number of RNAP that clears the promoter per second [38]. Finally, using the activity of an enzymatic protein reporter for inferring promoter activity adds an additional level of uncertainty over using fluorescent proteins, as enzymes depend on many factors, for instance substrate availability. These considerations make it difficult to speculate about the reason(s) for the difference in induced expression of P*trc*1O-prox/P*trc* versus PA1*lacO*-1 when comparing this and the previous study [28]. Further, PA1*lacO*-1 consists of a proximal *lacO1* 1 bp downstream in position as compared to P*trc* and P*trc*1O-proximal and a core *lacO* at the same position as P*trc*1O-core (Figure 3A). As the proximal *lacO* position of P*trc* and P*trc*1O-proximal has proven ineffective for repression and because PA1*lacO*-1 looks similar to P*trc*1O-core in its repression and induction pattern (Figure 5A), we agree with the previous study [28] that the core-located *lacO* appears to be causing the drop in expression. To test this, we characterized both P*trc*1O-core and PA1*lacO*-1 in the LacI-negative chloramphenicol resistance-only strain of *Synechocystis.* Indeed, both promoters were weaker than P*trc1O*-proximal, even when their activity without LacI is compared with the repressed activity of P*trc1O*-proximal (Figure 5A). Further, the activity of PA1*lacO*-1 was much lower than that of P*trc*1O-core in the absence of LacI, leading to a repression ratio of only 3.3 for PA1*lacO*-1 and 13 for P*trc*1O-core. It is possible that the slightly lower induction ratio for PA1*lacO*-1 as compared to the previous study [28] is due to lower levels of LacI/cell in this study. Finally, the reduced inherent promoter activity of P*trc*1O-core and PA1*lacO*-1 in *Synechocystis* as compared with *E. coli* further stresses our previous conclusion that tools to be used in a specific bacterium have to be functionally characterized in that particular chassis [4,25] and illustrates the need for orthogonal and hence portable RNAP systems.

LacI-mediated DNA-looping in *E. coli* and *Synechocystis*

To investigate the effect of tetrameric LacI-cooperativity on repression and DNA-looping in cyanobacteria, we designed a P*trc*-based library with two *lacO* to enable LacI-mediated DNA looping in analogy with previous designs for *E. coli* [20,23,51]. We added an auxiliary *lac*Osym at different distances, ranging from −2 to 21 bp, upstream of the P*trc*1O-proximal −35 element, while keeping the core P*trc* promoter and proximal *lac*Osym of P*trc*1O-proximal intact. As the P*trc*2O-library and P*trc*1O-proximal share identical promoters from the −35 box to the proximal *lac* operator and further produce identical 5′ UTRs (Figures 3A and B), this enables us to directly compare them and detect effects from LacI-mediated DNA looping. Further, single *lac*Osym operators positioned upstream of a *lac*UV5 promoter −35 element have previously been shown to produce only minor effects in *E. coli* on expression in the presence of LacI and the absence of DNA-looping [51]. However, a subsequent study showed loop-independent effects of placing a single distal *lac*Osym too close to the promoter −35 element, especially for edge-to-edge distances between *lac*Osym and the −35 hexamer of 2 and 6 bp [20]. This effect was suggested to be due to interactions between LacI and the *E. coli* RNAP, or due to sequence-specific effects affecting the nearby promoter itself. Such sequence-specific effects are possible, and have been illustrated recently for cyanobacteria [5]. Therefore, the P*trc*2O-library can be compared with P*trc*1O-proximal to draw conclusions about the possible effects of DNA-looping, but caution should be taken in interpreting the data for the closest spacings of the −35 box to the distal *lac*Osym operator.

First of all, we can conclude that the repression and induction patterns of the P*trc*2O-library in *E. coli* (Figure 4B) and *Synechocystis* (Figure 5B) share many similarities. Both species produce a periodical repression and induction pattern with a period of 11 bp, illustrating that maximal repression can be achieved when both *lacO* are positioned on the same side of the DNA helix, and that minimal repression is produced when the operators are on opposite sides. To confirm that the maximal repression displayed in the troughs of the periodical pattern, and the minimal repression displayed in the peaks, are not due to promoter activity effects unrelated to LacI, we tested two highly repressed promoters and one weakly repressed promoter without LacI. Both the two highly repressed promoters were highly active in the absence of LacI, and the activity of the weakly repressed promoter was similar to the activity of the highly repressed promoters (Figure 5B), confirming that the periodical repression pattern is due to the presence of LacI. This confirms the numerous previous studies on LacI-mediated DNA looping in *E. coli* [14,16-18,20,23,51], and demonstrates that DNA-looping as a mechanism of transcriptional regulation works in cyanobacteria, displaying the same number of average bp per turn of the supercoiled DNA helix as *E. coli*. This was anticipated, as structural DNA-binding proteins like HU that have been demonstrated to facilitate LacI-mediated DNA looping in *E. coli* [52] are present also in cyanobacteria (the *Synechocystis* HU gene is *sll1712* in Cyanobase [53]). That a periodical repression pattern remains even under full IPTG induction in both *E. coli* and *Synechocystis* is not surprising. As stated previously, the binding of IPTG to LacI decreases the affinity of LacI to *lacO* approximately 1000-fold [11]. Under circumstances when LacI is in excess to its operators, or its binding to a high-affinity operator is stabilized through DNA-looping, LacI-mediated DNA-loops could remain even under induced conditions. In fact, it has been shown that saturating induction does not completely remove loops in vivo [51] or in vitro [54]. To decrease the effect of residual repression even under full IPTG-induction, concentrations of LacI should not be higher than what is required for satisfactory repression. Further, just as in *E. coli*, the repression both for the non-induced and the induced conditions decrease with longer inter-*lacO* distances in *Synechocystis*. This has been shown before in *E. coli* and illustrates the drop in contribution to the local concentration of LacI at the primary proximal repression site from the distal site for longer distances [23,55]. On the other hand, it was recently demonstrated that the LacI-mediated DNA-loop itself is enough to repress transcription from a T7 promoter positioned inside the loop [55]. As this was demonstrated using the T7 RNAP, it is not tested whether bacterial RNAP will share the same sensitivity to DNA-loops, but likely, as the polymerases have evolved to perform the same function in similar ways. Hence, the decrease in repression for longer inter-*lacO* distances in *Synechocystis* may not only be due to a decreased local concentration of LacI at the primary repression site, but also due to a larger DNA-loop, which could exhibit a smaller hindrance for RNAP than a tighter one. Another similarity with *E. coli* is that the inter-*lacO* distances that represent maximum repression are not even multiples of 11 bp, corresponding to repression maxima at 55 and 66 bp, but one to two bp longer. This has been previously observed in vivo [23] and may be due to distortion of the intervening DNA sequence of the promoter by RNAP binding or the promoter DNA-sequence itself.

In *Synechocystis*, repression in the least optimal geometrical *lacO* positions, as represented by the expression peaks (Figure 5B), is very low even under non-induced conditions. As the P*trc*2O-library is identical to P*trc*1O-proximal, which is even less efficiently repressed, from the −35 element down to the TSS and the 5′ UTR, P*trc*2O-library promoters are expected to be similar to P*trc*1O-proximal in repression without DNA-looping. In fact, in the absence of LacI, the two highly repressed promoters P*trc*2O-2 and 13 were as or even more active than the fully induced P*trc*1O-prox (Figure 5A and B). Thus, P*trc*2O-library promoters whose *lacO* are out of phase are much less efficiently repressed in *Synechocystis* because of the weakly functioning proximal *lac* operator of P*trc*1O-proximal (Figure 5A and B), whereas their repression is stronger in *E. coli* for which P*trc*1O-proximal is more efficiently repressed (Figure 4A and B). As stated before, this difference in repression between *E. coli* NEB5α and *Synechocystis* for P*trc* and P*trc*1O-prox could be due to the about five times lower numbers of LacI/cell in the LacI-expressing *Synechocystis* strain. Thus, it is remarkable that repression of the P*trc*2O-library when the *lacO* are in phase is very efficient in *Synechocystis*, despite the weak primary proximal operator site and the lower levels of LacI/cell. In fact, the repression for geometrically optimal *lacO* positions is so efficient that IPTG induction almost fails to raise expression in *Synechocystis*, whereas it is possible to induce expression more in *E. coli*. Additionally, the repression troughs, which represent more optimal DNA-loop geometries, are wider in *Synechocystis* as compared to *E. coli*, and the expression peaks are narrower. Taken together, the stronger repression of an extremely leaky proximal *lac*Osym in optimal LacI-mediated DNA-loop configurations, the wider repression troughs, narrower expression peaks and the lower amounts of LacI/cell in *Synechocystis* as compared to *E. coli* have at least two possible implications. It implies that the *Synechocystis* RNAP is either more sensitive to DNA-looping than *E. coli* RNAP, and/or that *Synechocystis* chromatin DNA has lower in vivo torsion-resistance than *E. coli* chromatin DNA, lowering the barrier for forming twisted-DNA loops is the cyanobacterium.

Conclusions

Artificial and minimal promoters orthogonal to cyanobacteria can be used for stable constitutive expression and pose an alternative to the use of native promoters. The proximal *lac* operator site of P*trc* and its sibling P*trc*1O-proximal is insufficient for efficient repression of transcription in *Synechocystis* under the amounts of LacI/cell tested here. We show that in vivo DNA-looping works as a mechanism of transcriptional regulation in cyanobacteria similarly as in *E. coli*. DNA-looping appears to be an efficient mechanism of repression in cyanobacteria, as promoters carrying the inefficient proximal *lacO* site could be more strongly repressed in combination with auxiliary distal *lac* operators in optimal phases in *Synechocystis* than in *E. coli*. The weak repression of promoters with only the proximal *lac* operator site in *Synechocystis* could simply be due to insufficient numbers of LacI/cell and target operator. Yet, as some P*trc*2O promoters are more efficiently repressed in *Synechocystis* than in *E. coli* despite the lower levels of Lac/cell in the cyanobacterium, we suggest that the repression differences could be due to previously observed differences in RNAP subunits. We tested for differences in the RNAP footprint using a proximal *lacO* promoter library but could not detect any discrepancy in pattern. The reasons for the more efficient repression of favorably-phased P*trc*2O promoters in *Synechocystis* than in *E. coli* are unclear. We speculate that it could be due to differences in the ease with which the chromatin DNA is looped or twisted, or differences in RNAP sensitivity to deformed DNA. As DNA-looping is very efficient for LacI-repression of promoters in *Synechocystis*, we suggest that an optimal design should incorporate this mechanism. Further, to avoid unpredictable expression levels of different genes due to interactions between coding sequences and 5' UTRs on the mRNA level [56], future optimally repressed promoters would preferably involve DNA looping between core and distally located *lacO*. This allows for the exclusion of proximal *lac* operators from the 5' UTR and any possible effects they may have on the mRNA. Finally, the Ptrc2O (2) and (13) promoters display very high repression ratios in *Synechocystis* (about 400 and 170 times, respectively, Figure 5B), and could be used for genetic switches or gene expression systems that do not require subsequent induction. Alternatively, lower levels of LacI or unstable versions of LacI could be used to improve their inducibility.

Methods

DNA parts and sequences

All BioBrick parts, denoted by "BBa_", and BioBrick plasmids were obtained from the iGEM Registry of Standard Biological Parts [34] (referred to as the iGEM Registry). Colony PCR on BioBrick plasmids was performed using the VF2 (BBa_G00100) and VR (BBa_G00101) flanking primers. Primer sequences (Table 2), promoter sequences (Table 1) and construct sequences (Additional file 1) are available.

Strains and culture conditions

Escherichia coli strain NEB5α F' I^q (*lacI*^{q} strain, New England Biolabs, referred to as *E. coli*) was used for cloning and promoter characterization of the constitutive, control and P*trc*2O-library promoters. *E. coli* strain DH5αZ1 (carries two *lacI*^{q} genes at the *attB* locus, Expressys) was used for cloning and promoter characterization of the proximal *lacO*-library. The *lacI*^{q} gene is a promoter-mutated version of the wild-type *lacI* gene that causes about ten times as much LacI to be expressed as compared with the wild-type strain [57]. The strains were cultured in LB medium at 37°C shaking at 225 rpm, or on LB-agar plates, and were supplemented with 25 μg/mL kanamycin for pPMQAK1 [25] [iGEM Registry: BBa_J15 3000, Genbank: GU933126, Addgene: 26052] constructs, 50 μg/mL kanamycin for pSB1AK3 [34] constructs or 35 μg/mL chloramphenicol for the *Synechocystis lacI* or Chloramphenicol resistance-only recombination vectors.

Synechocystis PCC 6803 (the glucose tolerant strain ATCC 27184, referred to as *Synechocystis*) was cultured in BG11 medium [58], or on BG11-agar plates, supplemented with 50 μg/mL kanamycin for pPMQAK1 constructs. The *Synechocystis* engineered strain expressing LacI and the strain carrying only Chloramphenicol resistance were cultured in BG11 supplemented with 25 μg/mL chloramphenicol, or both 25 μg/mL kanamycin and 25 μg/mL chloramphenicol when containing pPMQAK1 constructs. All strains were grown at 30°C under continuous white light of 20-50 μmol photons m^{-2} s^{-1} shaking at 120 rpm unless otherwise stated.

Construction and characterization of constitutive promoters in *Synechocystis*

Our selection criteria for promoters expressing the Lac repressor in *Synechocystis* were minimal constitutive promoters orthogonal to *Synechocystis*. For this purpose, we selected several members of an artificial σ-70 promoter

Table 2 Primer DNA sequences

Primer name	Sequence 5′ to 3′
VF2 (BBa_G00100)	TGCCACCTGACGTCTAAGAA
VR (BBa_G00101)	ATTACCGCCTTTGAGTGAGC
RBS*-EYFP-BB-f1	CTTTCTAGAGTAGTGGAGGTTACTAGATGGTGAGCAAGGGCG
B0015-BB-r1	GAACTGCAGCGGCCGCTACTAGTATATAAACGCAGAAAGGCCC
j5_00029	GCGGCCGCTGCAGCCCGTAGAAAAGATCAAAGGATCTTCTTGAGA
j5_00030	GCCACGTAGGGGTCTCTAGAAGCGGCCGCGAATTCATGTGAGCAAAAGGCCAGCA
j5_00031	GCCGCTTCTAGAGACCCCTACGTGGCCGGCAATGGTCC
j5_00032	GGTGAAACTGACCGAACATAGGAGACTTTGGTGGGCTGGCCG
j5_00033	CCACCAAAGTCTCCTATGTTCGGTCAGTTTCACCTGATTTACG
j5_00079	CTGGTTTCACATTCACCATGATCAAACCTCCACTACTCTAGTAGCTAGCATTGTACCTAGGACTG AGCTAGCCATAAAGACAGTCATTCATCTTTCTGCCCCTCC
j5_00042	GTAGTGGAGGTTTGATCATGGTGAATGTGAAACCAGTAACGTTATACGATG
j5_00072	GTTCGTTAAGGCTTGATCTCTATTATTACTGCCCGCTTTCCAGTCGG
j5_00073	GGAAAGCGGGCAGTAATAATAGAGATCAAGCCTTAACGAACTAAGACCCC
j5_00088	CCCGATCAACTCGTGTCTGCTCCTCGGTTATGTTTTTAAGGTC
j5_00090	CGAGGAGCAGACACGAGTTGATCGGGCACGTAAGAGG
j5_00092	GCCATCCACTTCCTTGGTCTGACAGCTCGAGGCTTGG
j5_00093	GCTGTCAGACCAAGGAAGTGGATGGCCCCGTATTGC
j5_00039	CTTTGATCTTTTCTACGGGCTGCAGCGGCCGCTACTAGTAGGTCCCAAGTTTGTGCTGTGGC
slr0168-UR-f2	AAAGGGGACGAAGCCGCAGTTC
slr0168-DR-r2	ATCGCCGTGGTTAAATCCGTGG
P2OLib-r	TGAGCTACTAGTATGTGTGAAATTGTGAGCGCTCACAATTCCAC
P2OLib-f1O	TCAGAATTCGCGGCCGCTTCTAGAGCGAACGGTTCTGGCAAATATTC
P2OLib-f-2	TCAGAATTCGCGGCCGCTTCTAGAGAATTGTGAGCGCTCACAATTGACAATTAATCATCCGGCTCG
P2OLib-f-1	TCAGAATTCGCGGCCGCTTCTAGAGAATTGTGAGCGCTCACAATTTGACAATTAATCATCCGGCTCG
P2OLib-f0	TCAGAATTCGCGGCCGCTTCTAGAGAATTGTGAGCGCTCACAATTTTGACAATTAATCATCCGGCTC
P2OLib-f1	TCAGAATTCGCGGCCGCTTCTAGAGAATTGTGAGCGCTCACAATTGTTGACAATTAATCATCCGGC
P2OLib-f2	TCAGAATTCGCGGCCGCTTCTAGAGAATTGTGAGCGCTCACAATTTGTTGACAATTAATCATCCGG
P2OLib-f3	TCAGAATTCGCGGCCGCTTCTAGAGAATTGTGAGCGCTCACAATTCTGTTGACAATTAATCATCCGG
P2OLib-f4	TCAGAATTCGCGGCCGCTTCTAGAGAATTGTGAGCGCTCACAATTGCTGTTGACAATTAATCATCCGG
P2OLib-f5	TCAGAATTCGCGGCCGCTTCTAGAGAATTGTGAGCGCTCACAATTAGCTGTTGACAATTAATCATCCG
P2OLib-f6	TCAGAATTCGCGGCCGCTTCTAGAGAATTGTGAGCGCTCACAATTGAGCTGTTGACAATTAATCATCCG
P2OLib-f7	TCAGAATTCGCGGCCGCTTCTAGAGAATTGTGAGCGCTCACAATTTGAGCTGTTGACAATTAATCATCC
P2OLib-f8	TCAGAATTCGCGGCCGCTTCTAGAGAATTGTGAGCGCTCACAATTATGAGCTGTTGACAATTAATCATCC
P2OLib-f9	TCAGAATTCGCGGCCGCTTCTAGAGAATTGTGAGCGCTCACAATTAATGAGCTGTTGACAATTAATCATCC
P2OLib-f10	TCAGAATTCGCGGCCGCTTCTAGAGAATTGTGAGCGCTCACAATTAAATGAGCTGTTGACAATTAATCATC
P2OLib-f11	TCAGAATTCGCGGCCGCTTCTAGAGAATTGTGAGCGCTCACAATTGAAATGAGCTGTTGACAATTAATCATC
P2OLib-f12	TCAGAATTCGCGGCCGCTTCTAGAGAATTGTGAGCGCTCACAATTTGAAATGAGCTGTTGACAATTAATC
P2OLib-f13	TCAGAATTCGCGGCCGCTTCTAGAGAATTGTGAGCGCTCACAATTCTGAAATGAGCTGTTGACAATTAATC
P2OLib-f14	TCAGAATTCGCGGCCGCTTCTAGAGAATTGTGAGCGCTCACAATTTCTGAAATGAGCTGTTGACAATTAATC
P2OLib-f15	TCAGAATTCGCGGCCGCTTCTAGAGAATTGTGAGCGCTCACAATTTTCTGAAATGAGCTGTTGACAATTA
P2OLib-f16	TCAGAATTCGCGGCCGCTTCTAGAGAATTGTGAGCGCTCACAATTATTCTGAAATGAGCTGTTGACAATT
P2OLib-f17	TCAGAATTCGCGGCCGCTTCTAGAGAATTGTGAGCGCTCACAATTTATTCTGAAATGAGCTGTTGACAATT
P2OLib-f18	TCAGAATTCGCGGCCGCTTCTAGAGAATTGTGAGCGCTCACAATTATATTCTGAAATGAGCTGTTGACAATT

Table 2 Primer DNA sequences *(Continued)*

P2OLib-f19	TCAGAATTCGCGGCCGCTTCTAGAGAATTGTGAGCGCTCACAATTAATATTCTGAAATGAGCTGTTGACAA
P2OLib-f20	TCAGAATTCGCGGCCGCTTCTAGAGAATTGTGAGCGCTCACAATTAAATATTCTGAAATGAGCTGTTGACA
P2OLib-f21	TCAGAATTCGCGGCCGCTTCTAGAGAATTGTGAGCGCTCACAATTCAAATATTCTGAAATGAGCTGTTGAC
P1Od-f1	AAGTCTAGAGAATTGTGAGCGCTCACAATTTTGACAATTAATCATCCGGCTCGTATAATGTGTGGAATCAC ACATACTAGAGTAGTGGAGGTTACTAGATGG
P1Oc-f1	AAGTCTAGAGAAATATTCTGAAATGAGCTGTTGACAATTGTGAGCGCTCACAATATAATGTGTGGAATCAC ACATACTAGAGTAGTGGAGGTTACTAGATGG
PA1lacO1-FP-f1	AAGTCTAGAGAAAGAGTGTTGACTTGTGAGCGGATAACAATGATACTTAGATTCAATTGTGAGCGGATAAC AATTTCACACATACTAGAGTAGTGGAGGTTACTAGATGG
PtrcOs-prox-8-f1	TTCTCTAGAGTTGACAATTAATCATCCGGCTCGTATAATGTGTAATTGTGAGCGCTCACAATTTACTAGAGTA GTGGAGGTTACTAGATGG
PtrcOs-prox-9-f1	TTCTCTAGAGTTGACAATTAATCATCCGGCTCGTATAATGTGTGAATTGTGAGCGCTCACAATTTACTAGAG TAGTGGAGGTTACTAGATGG
PtrcOs-prox-10-f1	TTCTCTAGAGTTGACAATTAATCATCCGGCTCGTATAATGTGTGGAATTGTGAGCGCTCACAATTTACTAGA GTAGTGGAGGTTACTAGATGG
PtrcOs-prox-11-f1	TTCTCTAGAGTTGACAATTAATCATCCGGCTCGTATAATGTGTGGAAATTGTGAGCGCTCACAATTTACTAG AGTAGTGGAGGTTACTAGATGG
PtrcOs-prox-12-f1	TTCTCTAGAGTTGACAATTAATCATCCGGCTCGTATAATGTGTGGAGAATTGTGAGCGCTCACAATTTACTA GAGTAGTGGAGGTTACTAGATGG
PtrcOs-prox-13-f1	TTCTCTAGAGTTGACAATTAATCATCCGGCTCGTATAATGTGTGGAGTAATTGTGAGCGCTCACAATTTACT AGAGTAGTGGAGGTTACTAGATGG
PtrcOs-prox-14-f1	TTCTCTAGAGTTGACAATTAATCATCCGGCTCGTATAATGTGTGGAGTCAATTGTGAGCGCTCACAATTTAC TAGAGTAGTGGAGGTTACTAGATGG
PtrcOs-prox-15-f1	TTCTCTAGAGTTGACAATTAATCATCCGGCTCGTATAATGTGTGGAGTCGAATTGTGAGCGCTCACAATTTA CTAGAGTAGTGGAGGTTACTAGATGG
PtrcOs-prox-16-f1	TTCTCTAGAGTTGACAATTAATCATCCGGCTCGTATAATGTGTGGAGTCGTAATTGTGAGCGCTCACAATTT ACTAGAGTAGTGGAGGTTACTAGATGG
PtrcOs-prox-17-f1	TTCTCTAGAGTTGACAATTAATCATCCGGCTCGTATAATGTGTGGAGTCGTCAATTGTGAGCGCTCACAATTT ACTAGAGTAGTGGAGGTTACTAGATGG
PtrcOs-prox-18-f1	TTCTCTAGAGTTGACAATTAATCATCCGGCTCGTATAATGTGTGGAGTCGTCCAATTGTGAGCGCTCACAAT TTACTAGAGTAGTGGAGGTTACTAGATGG
PtrcOs-prox-20-f1	TTCTCTAGAGTTGACAATTAATCATCCGGCTCGTATAATGTGTGGAGTCGTCCAGAATTGTGAGCGCTCACA ATTTACTAGAGTAGTGGAGGTTACTAGATGG
PtrcOs-prox-22-f1	TTCTCTAGAGTTGACAATTAATCATCCGGCTCGTATAATGTGTGGAGTCGTCCAGACAATTGTGAGCGCTCA CAATTTACTAGAGTAGTGGAGGTTACTAGATGG
PtrcOs-prox-24-f1	TTCTCTAGAGTTGACAATTAATCATCCGGCTCGTATAATGTGTGGAGTCGTCCAGACTCAATTGTGAGCGCT CACAATTTACTAGAGTAGTGGAGGTTACTAGATGG

library spanning a wide range of activities in *E. coli* for characterization in *Synechocystis* (BBa_J23100, BBa_J23101, BBa_J23102, BBa_J23105, BBa_J23106, BBa_J23109, BBa_J23113 and BBa_J23114, from the iGEM Registry's Anderson collection [34]). For comparison, we selected three promoters commonly used for engineered expression in *Synechocystis;* the nitrate reductase promoter P*nirA* (synthesized by Epoch Biolabs Inc.), the plastocyanin promoter P*petE* (BBa_K273019, from iGEM-team Uppsala-Sweden 2009) and the RNase P subunit B promoter P*rnpB* [25]. An EYFP reporter cassette containing RBS*, a ribosome binding site that is highly active in both *E. coli* and *Synechocystis* [4], EYFP itself (BBa_E0030) and a double transcriptional terminator (BBa_B0015) was constructed by PCR using the Phusion Hot-Start II High-Fidelity DNA polymerase (Thermo Scientific, referred to as Phusion) with primers RBS*-EYFP-BB-f1, B0015-BB-r1 and another EYFP reporter cassette as template (BBa_E0430) according to the manufacturer's instructions. The promoters were cloned together with the EYFP reporter cassette into pSB1AK3 using BioBrick 3A-assembly [59], subcloned into pPMQAK1 and confirmed by sequencing. The pPM QAK1-carried constructs plus an empty pPMQAK1 vector control plasmid [25] were transferred to *Synechocystis* through conjugation according to a previously published procedure [4] using *E. coli* strain HB101 with pRL443 [60] as the conjugal plasmid and BG11 plates with 50 μg/mL kanamycin for selection. Single *Synechocystis* colonies were grown in flat-bottomed tissue culture 6-well plates (Sarstedt) with lids taped shut with Micropore tape in 5 mL BG11 supplemented with 50 μg/mL kanamycin at ca 50 μmol photons m^{-2} s^{-1} (protecting cultures from light with white paper the first days) for about two weeks (corresponding to a final absorbance at 750 nm (Abs750) of 0.6-1.3 as measured in translucent 96-well tissue culture plates using a Chameleon V Microplate Reader (Hidex, referred to as plate reader)). The presence of the promoter-reporter plasmids was confirmed with colony PCR and the cultures were inspected for the absence of *E. coli* using microscopy. To produce promoter characterization seed cultures in the same growth phase, all cultures were diluted to Abs750 = 0.01 into E-flasks with 25 mL medium and grown for four days in the same conditions as before, reaching a final Abs750 of 0.3-0.45. The seed cultures were used to inoculate measurement cultures to Abs750 = 0.01, which were split into six replicate cultures each of 5 mL in 6-well plates with sealed lids. The measurement cultures were grown for 42 hours whereupon samples were taken for plate reader measurements of Abs750 and EYFP fluorescence using 485 nm emission and 535 nm excitation filters. Black 96-well plates with clear flat bottom (BD Falcon) were used to eliminate well-to-well fluorescence cross-talk. After subtracting medium and instrument background, fluorescence per cell values were calculated by normalizing fluorescence values with Abs750 values, averaging over the six replicates and estimating the standard error of the mean. Finally, specific fluorescence per cell values for each promoter construct was obtained by subtracting cellular background fluorescence from the empty pPMQAK1 control culture.

Construction of LacI-expressing and Chloramphenicol resistance-only strains of *Synechocystis*

As we have previously found that the P*rnpB* promoter in combination with the BBa_B0034 RBS express amounts of LacI strongly repressing P*trc*2O when both constructs are carried on the pPMQAK1 plasmid [25], we selected the weaker (as characterized in this study) and synthetic BBa_J23114 promoter in combination with RBS* [4] to express LacI from the chromosome. For insertion of the LacI-expressing cassette into the chromosome we selected *slr0168*, a previously used neutral site [61,62] that displays very low transcript levels [37]. The LacI-expression cassette, which is flanked by *slr0168* insertion sequences, consists of a forward terminator to insulate the cassette from potentially incoming chromosomal RNAP (*rnpB*_T1 from *E. coli*), BBa_J23114 in combination with RBS* to drive LacI expression, the *lacI* CDS, another forward terminator (*ilvGEDA*_T from *E. coli*) and a chloramphenicol resistance cassette (CmR) derived from pSB1AC3 (iGEM Registry). The LacI-expression cassette together with a pMB1 replicon for replication in *E. coli* were assembled using one-step isothermal assembly [63] using the j5 DNA assembly design automation software [64] for primer design. Specifically, primers j5_00029 and j5_00030 were used to PCR the pMB1 replicon using pBluescript II SK + (Stratagene) as template, j5_00031 and j5_00032 to PCR the *slr0168* upstream recombination recombination sequence using *Synechocystis* genomic DNA as template, j5_00033 and j5_00079 to PCR *rnpB*_T1 and the promoter-RBS region using *rnpB*_T1 synthetic DNA (Genscript) as template, j5_00042 and j5_00072 to PCR *lacI* using *E. coli* strain JM107 genomic DNA as template, j5_00073 and j5_00088 to PCR *ilvGEDA*_T using synthetic DNA (Genscript) as template, j5_00090 and j5_00092 to PCR CmR using pSB1AC3 as template and finally j5_00093 and j5_00039 to PCR the *slr0168* downstream recombination sequence using *Synechocystis* genomic DNA as template. The upstream recombination and the *rnpB*_T1-promoter-RBS fragments, and the *ilvGEDA*_T and CmR fragments, were spliced together by overlap extension (SOE) before one-step isothermal assembly. The chloramphenicol resistance-only construct was constructed analogously to the LacI-expressing construct, but all elements except the recombination sites, the second forward terminator (*ilvGEDA*_T) and the chloramphenicol resistance cassette were excluded. The sequences of both constructs are available in Additional file 1. All PCR and SOE reactions were made using Phusion. The resulting pMB1-replicated LacI-expression cassette and the chloramphenicol resistance-only construct were sequence-confirmed, transferred to *Synechocystis* by natural transformation and inserted into the genomic *slr0168* target site by homologous recombination according to previously published procedures [4]. Mutants were selected on BG11 plates supplemented with 10 μg/mL chloramphenicol and single colonies were streaked on BG11 25 μg/mL chloramphenicol plates twice to allow for full segregation. Finally, the fully segregated mutants were confirmed with colony PCR and sequencing of the insertion site using primers slr0168-UR-f2 and slr0168-DR-r2.

To confirm and quantify the expression of LacI in the LacI-expressing strain of *Synechocystis*, a quantitative western blot was performed with anti-LacI polyclonal antibodies (PAB10255, raised in rabbit, Abnova) and secondary HRP-conjugated anti-rabbit antibodies (ImmunStar GAR-HRP conjugate, #170-5046, BioRad). Briefly, a known number of cells from three biological replicates of DH5αZ1, two replicates of NEB5α, and three replicates of the LacI-expressing strain of *Synechocystis* carrying different promoter reporter constructs on pPMQAK1, were denatured in reducing SDS sample buffer for 5 mins at 95°C. The samples were run on a PAGE gel (any-kD precast 8-well gels, BioRad) and blotted to membranes (TransBlot turbo, midi transfer packs with PVDF membranes, BioRad). Western blotting was performed according to the manufacturer's instruction with the exception that the primary LacI antibody was hybridized over night. The blot was developed using HRP chemiluminescence (Immun-Star HRP substrate kit, BioRad) and the band intensities were quantified using QuantityOne software (BioRad). The average band intensities were normalized with the number of loaded cells to get relative values of LacI/cell for the different strains. For an absolute value of LacI tetramers/cell for all strains, a previous quantification of 3000 tetramers of LacI/cell for DH5αZ1 [36] was used as a standard.

Construction and characterization of the P*trc*-derived promoter library in *E. coli* and *Synechocystis*

The P*trc*2O-library and P*trc*1O-prox promoters were made with Phusion PCR using the reverse primer P2OLib-r to exchange *lac*O1 of the original P*trc* [24] and P*trc*2O [25] (here referred to as P*trc*2O-orig) to the symmetric *lac*Osym operator. Different forward primers adding another *lac*Osym upstream the −35 box with spacers of different length in between (primers P2OLib-f-n for the P*trc*2O-library, where n is a numeral representing the spacer length ranging from −2 to 21, and primer P2OLib-f1O for P*trc*1O-prox that does not include a second *lac*Osym) were used, together with P*trc*2O-orig as PCR template. The promoter PCR products were assembled with the same EYFP reporter cassette as the constitutive promoters into pPMQAK1 using 3A-assembly. The single *lac*Osym control promoters P*trc*1O-dist, with *lac*Osym immediately distal to the −35 box, and P*trc*1O-core, with a slightly truncated *lac*Osym in the core region between the −35 and the −10 boxes, and the PA1*lacO*-1 promoter [36], were made with Phusion PCR using the different forward primers P1Od-f1 (for P*trc*1O-dist), P1Oc-f1 (for P*trc*1O-core) and PA1lacO1-FP-f1 (for PA1*lacO*-1) and the same B0015-BB-r1 reverse primer using the EYFP reporter cassette as template. These PCR products were also cloned into pPMQAK1 using the BioBrick cloning site. After sequence confirmation using the VF2 and VR primers all promoters were first characterized in *E. coli* and then transferred to the LacI-expressing *Synechocystis* strain for subsequent characterization.

For characterization of the P*trc*-derived promoter library in *E. coli* the following protocol was repeated at three different days, producing a total of six biological measurement replicates. Two overnight cultures for each promoter construct were inoculated from frozen glycerol stocks in 2 mL LB supplemented with 25 μg/mL kanamycin in 13 mL growth tubes (Sarstedt) and grown at 37°C shaking at 225 rpm until all cultures had reached stationary phase (approximately 22 hours). The overnight cultures were put into ice-water to ensure that both the negative and positive Isopropyl β-D-1-thiogalactopyranoside (IPTG) cultures had the same starting conditions. For each culture, 1 μl was transferred to a 96-well tissue culture plate pre-loaded with 99 μl LB supplemented with 25 μg/mL kanamycin and either 0 or 1 mM IPTG for the non-induced plate and induced plate, respectively. Only the middle 60 wells were used as the evaporation rate was observed to be higher for the outermost wells. The plates were sealed with Breathe-Easy sealing membranes (Sigma-Aldrich) and incubated shaking at level 8 on a Delfia plate shaker (Wallac) at 37°C for 7 hours. Samples were taken and measured in black 96-well plates with clear flat bottom in the same way as for the constitutive promoters in *Synechocystis*, but Abs595 was used instead of Abs750.

For characterization in *Synechocystis* the P*trc*-derived promoter library was transferred to the LacI-expressing *Synechocystis* strain by conjugation as was done for the constitutive promoters to wild-type *Synechocystis*, but BG11 plates with 25 μg/mL kanamycin and 25 μg/mL chloramphenicol was used for selection. Two colonies for each promoter construct were inoculated into ca 2 mL BG11 with 25 μg/mL kanamycin and 25 μg/mL chloramphenicol in 13 mL growth tubes shaking in low light conditions (ca 5 μmol photons m^{-2} s^{-1}) for ca three weeks (reaching a final Abs750 ranging from 1 to 3.5). To ensure that the seed cultures are in the same growth phase all cultures were diluted to Abs750 = 0.05 and grown under the same conditions as the parent cultures for 44 hours. For measurements, all cultures were diluted once more in the 13 mL growth tubes to Abs750 = 0.05 and 200 μl of each was loaded into the middle 32 wells of 96-well tissue culture plates, where the induction plates were pre-loaded with 1 μl 200 mM IPTG, and mixed by pipetting. The plate measurement cultures were grown with the lids on shaking at level 4 on a Delfia plateshaker at 30°C in white light of ca 50 μmol photons m^{-2} s^{-1} for 46 hours, after which samples were taken for measurements. The cultures in the 13 mL growth tubes were put back in the same low-light conditions immediately after starting the first measurement cultures and after two days they were used to start new measurement cultures, leading to a total of four biological replicates for all promoter constructs. The second round of measurement cultures were produced in the same way as the first one, but the cultures were diluted to Abs750 = 0.05 directly in the 96-well plates, and grown under the same conditions for the same time. Samples were taken and measured in black 96-well plates with clear flat bottom in the same way as for the constitutive promoters in *Synechocystis*, but averages and standard errors of the mean were calculated for four replicates instead of six.

The control promoter constructs BBa_J23101, P*trc*1O-core, PA1*lacO*-1 and the P*trc*2O-library constructs P*trc*2O-

2, 13 and 18 were transferred into the chloramphenicol resistance-only strain of *Synechocystis* to enable characterization of these promoters without the presence of LacI. Conjugation and characterization was carried out in a similar way as for the LacI-expressing *Synechocystis* strain, but the measurement cultures were inoculated into 1 mL medium in 13 mL growth tubes instead of 96-well plates and grown at 10-15 μmol photons m^{-2} s^{-1}. Only one round of measurement cultures were produced (yielding two biological replicates). The cultures were measured in the same way as for the LacI-expressing *Synechocystis* strain but averages and standard errors of the mean were calculated for two replicates instead of four.

Construction and characterization of the proximal *lacO*-library in *E. coli* DH5αZ1 and *Synechocystis*

The proximal *lacO* promoter library was made with Phusion PCR using reverse primer B0015-BB-r1 together with the forward primers PtrcOs-prox-n-f1. The letter n is a numeral ranging from 8-18 in steps of one and from 18-24 in steps of two, representing the distance in bp between the center of a proximal *lacOsym* to the TSS of a P*trc* core promoter (Table 1). The PCR was performed using a strongly enhanced green fluorescent protein (SGFP2) [65] codon-optimized for *E. coli* (DNA 2.0, a gift from Erik Gullberg) reporter construct consisting of RBS*-SGFP2-BBa_B0015 (Additional file 1) as template. The PCR products were cloned into pPMQAK1 using the BioBrick cloning site. After sequence confirmation using the VF2 and VR primers all promoters were characterized in *E. coli* DH5αZ1 and transferred to both the LacI-expressing *Synechocystis* strain and the chloramphenicol resistance-only strain for subsequent characterization.

Characterization in *E. coli* DH5αZ1 was done similarly as for the P*trc*-derived promoter library in NEB5α but six biological replicates were characterized divided on two different days. The cultures were grown directly in black 96-well plates with clear flat bottom diluted 1:200 in 200 μl medium supplemented with water or 2 mM IPTG, shaking at level 7 on the plate shaker. The cultures were measured without lids at several time-points after the start of the experiment (approximately 0, 2.4, 3.8, 4.2, 4.7, 5.1 and 5.5 hours after dilution and induction) to observe for possible fluorescence/absorbance steady states and the temporal development of the induction ratios. The induction ratios and their estimated standard errors were calculated from the ratio of induced/non-induced fluorescence/absorbance data.

Conjugation to and characterization in the LacI-expressing *Synechocystis* strain and the chloramphenicol resistance-only strain was done similarly as for the P*trc*-derived promoter library in the LacI-expressing *Synechocystis* strain. The measurement cultures were inoculated into 1 mL medium supplemented with water or 2 mM IPTG in 13 mL growth tubes instead of 96-well plates, and grown at 10-15 μmol photons m^{-2} s^{-1}. Only one round of measurement cultures were produced (yielding two biological replicates). The cultures were measured in the same way as the P*trc*-derived promoter library in the LacI-expressing *Synechocystis* strain. The induction ratios and their estimated standard errors were calculated from the ratio of induced/non-induced fluorescence/absorbance data for the LacI-expressing *Synechocystis* strain and the repression ratios and their estimated standard errors were calculated from the ratio of the fluorescence/absorbance data of the chloramphenicol resistance-only strain/the data from the LacI-expressing strain.

Additional file

Additional file 1: DNA sequences of the EYFP reporter, the SGFP2 reporter, the LacI-expression cassette and the chloramphenicol resistance-only construct.

Competing interests
The authors declare that they have no competing interests.

Author's contributions
DC, TH and PL conceived of the research; DC designed the study and performed the experiments; DC and PL analyzed the data; DC and PL wrote the manuscript. All authors approved of the final manuscript.

Acknowledgements
We would like to acknowledge financial support from the Swedish Energy Agency; the Knut and Alice Wallenberg Foundation (project MoSE); and the European Union Seventh Framework Programme (FP7/2007-2013) under grant agreement number 308518 (CyanoFactory).

Author details
[1]Microbial Chemistry, Department of Chemistry–Ångström Laboratory, Science for Life Laboratory, Uppsala University, P.O. Box 523, SE-75120 Uppsala, Sweden. [2]Bioforsk, Frederik A Dahls vei 20, 1432 Ås Oslo, Norway.

References
1. Ducat DC, Way JC, Silver PA: **Engineering cyanobacteria to generate high-value products.** *Trends Biotechnol* 2011, **29:**95–103.
2. Lindblad P, Lindberg P, Oliveira P, Stensjö K, Heidorn T: **Design, engineering, and construction of photosynthetic microbial cell factories for renewable solar fuel production.** *Ambio* 2012, **41:**163–168.
3. Keasling JD: **Synthetic biology and the development of tools for metabolic engineering.** *Metab Eng* 2012, **14:**189–195.
4. Heidorn T, Camsund D, Huang HH, Lindberg P, Oliveira P, Stensjö K, Lindblad P: **Synthetic biology in cyanobacteria engineering and analyzing novel functions.** *Methods Enzymol* 2011, **497:**539–579.
5. Huang HH, Lindblad P: **Wide-dynamic-range promoters engineered for cyanobacteria.** *J Biol Eng* 2013, **7:**10.
6. Cardinale S, Arkin AP: **Contextualizing context for synthetic biology - identifying causes of failure of synthetic biological systems.** *Biotechnol J* 2012, **7:**856–866.
7. Rao CV: **Expanding the synthetic biology toolbox: engineering orthogonal regulators of gene expression.** *Curr Opin Biotechnol* 2012, **23:**689–694.
8. Wang B, Wang J, Zhang W, Meldrum DR: **Application of synthetic biology in cyanobacteria and algae.** *Front Microbiol* 2012, **3:**344.
9. Gilbert W, Müller-Hill B: **Isolation of the *lac* repressor.** *Proc Natl Acad Sci USA* 1966, **56:**1891–1898.

10. Lewis M: **The lac repressor.** *C R Biol* 2005, **328:**521–548.
11. Barkley MD, Riggs AD, Jobe A, Burgeois S: **Interaction of effecting ligands with *lac* repressor and repressor-operator complex.** *Biochemistry (Mosc)* 1975, **14:**1700–1712.
12. Sadler JR, Sasmor H, Betz JL: **A perfectly symmetric *lac* operator binds the *lac* repressor very tightly.** *Proc Natl Acad Sci U S A* 1983, **80:**6785–6789.
13. Oehler S, Amouyal M, Kolkhof P, Von Wilcken-Bergmann B, Muller-Hill B: **Quality and position of the three *lac* operators of *E. coli* define efficiency of repression.** *EMBO J* 1994, **13:**3348–3355.
14. Mossing MC, Record MT Jr: **Upstream operators enhance repression of the *lac* promoter.** *Science* 1986, **233:**889–892.
15. Krämer H, Niemoller M, Amouyal M, Revet B, Von Wilcken-Bergmann B, Muller-Hill B: ***lac* repressor forms loops with linear DNA carrying two suitably spaced *lac* operators.** *EMBO J* 1987, **6:**1481–1491.
16. Bellomy GR, Mossing MC, Record MT: **Physical properties of DNA in vivo as probed by the length pependence of the *lac* operator looping process.** *Biochemistry (Mosc)* 1988, **27:**3900–3906.
17. Oehler S, Eismann ER, Kramer H, Muller-Hill B: **The three operators of the *lac* operon cooperate in repression.** *EMBO J* 1990, **9:**973–979.
18. Law SM, Bellomy GR, Schlax PJ, Record MT Jr: **In vivo thermodynamic analysis of repression with and without looping in *lac* constructs: estimates of free and local *lac* repressor concentrations and of physical properties of a region of supercoiled plasmid DNA in vivo.** *J Mol Biol* 1993, **230:**161–173.
19. Garcia HG, Sanchez A, Boedicker JQ, Osborne M, Gelles J, Kondev J, Phillips R: **Operator sequence alters gene expression independently of transcription factor occupancy in bacteria.** *Cell Reports* 2012, **2:**150–161.
20. Bond LM, Peters JP, Becker NA, Kahn JD, Maher LJ: **Gene repression by minimal *lac* loops in vivo.** *Nucleic Acids Res* 2010, **38:**8072–8082.
21. Lee DH, Schleif RF: **In vivo DNA loops in araCBAD: size limits and helical repeat.** *Proc Natl Acad Sci USA* 1989, **86:**476–480.
22. Zhang Y, McEwen AE, Crothers DM, Levene SD: **Analysis of In-Vivo LacR-Mediated Gene Repression Based on the Mechanics of DNA Looping.** *PLoS One* 2006, **1:**e136.
23. Müller J, Oehler S, Müller-Hill B: **Repression of *lac* promoter as a function of distance, phase and quality of an auxiliary *lac* operator.** *J Mol Biol* 1996, **257:**21–29.
24. Brosius J, Erfle M, Storella J: **Spacing of the–10 and–35 regions in the *tac* promoter–effect on its in vivo activity.** *J Biol Chem* 1985, **260:**3539–3541.
25. Huang HH, Camsund D, Lindblad P, Heidorn T: **Design and characterization of molecular tools for a synthetic biology approach towards developing cyanobacterial biotechnology.** *Nucleic Acids Res* 2010, **38:**2577–2593.
26. Imamura S, Asayama M: **Sigma factors for cyanobacterial transcription.** *Gene Regul Syst Biol* 2009, **3:**65–87.
27. Geerts D, Bovy A, De Vrieze G, Borrias M, Weisbeek P: **Inducible expression of heterologous genes targeted to a chromosomal platform in the cyanobacterium *Synechococcus* sp. PCC 7942.** *Microbiology* 1995, **141**(Pt 4):831–841.
28. Guerrero F, Carbonell V, Cossu M, Correddu D, Jones PR: **Ethylene synthesis and regulated expression of recombinant protein in Synechocystis sp. PCC 6803.** *PLoS One* 2012, **7:**e50470.
29. de Boer HA, Comstock LJ, Vasser M: **The *tac* promoter: a functional hybrid derived from the *trp* and *lac* promoters.** *Proc Natl Acad Sci USA* 1983, **80:**21–25.
30. Elhai J: **Strong and regulated promoters in the cyanobacterium *Anabaena* PCC 7120.** *FEMS Microbiol Lett* 1993, **114:**179–184.
31. Schneider GJ, Haselkorn R: **RNA-polymerase subunit homology among cyanobacteria, other eubacteria, and archaebacteria.** *J Bacteriol* 1988, **170:**4136–4140.
32. Xie WQ, Jager K, Potts M: **Cyanobacterial RNA-Polymerase Genes *rpoc1* and *rpoc2* Correspond to *rpoc* of *Escherichia coli*.** *J Bacteriol* 1989, **171:**1967–1973.
33. Asayama M: **Regulatory system for light-responsive gene expression in photosynthesizing bacteria: cis-elements and trans-acting factors in transcription and post-transcription.** *Biosci Biotechnol Biochem* 2006, **70:**565–573.
34. The iGEM registry of standard biological parts. http://parts.igem.org/.
35. Griese M, Lange C, Soppa J: **Ploidy in cyanobacteria.** *FEMS Microbiol Lett* 2011, **323:**124–131.
36. Lutz R, Bujard H: **Independent and tight regulation of transcriptional units in Escherichia coli via the LacR/O, the TetR/O and AraC/I-1-I-2 regulatory elements.** *Nucleic Acids Res* 1997, **25:**1203–1210.
37. Mitschke J, Georg J, Scholz I, Sharma CM, Dienst D, Bantscheff J, Voss B, Steglich C, Wilde A, Vogel J, Hess WR: **An experimentally anchored map of transcriptional start sites in the model cyanobacterium *Synechocystis* sp PCC6803.** *Proc Natl Acad Sci U S A* 2011, **108:**2124–2129.
38. Kelly JR, Rubin AJ, Davis JH, Ajo-Franklin CM, Cumbers J, Czar MJ, De Mora K, Glieberman AL, Monie DD, Endy D: **Measuring the activity of BioBrick promoters using an in vivo reference standard.** *J Biol Eng* 2009, **3:**4.
39. Dong GG, Kim YI, Golden SS: **Simplicity and complexity in the cyanobacterial circadian clock mechanism.** *Curr Opin Genet Dev* 2010, **20:**619–625.
40. Temme K, Hill R, Segall-Shapiro TH, Moser F, Voigt CA: **Modular control of multiple pathways using engineered orthogonal T7 polymerases.** *Nucleic Acids Res* 2012, **40:**8773–8781.
41. Shis DL, Bennett MR: **Library of synthetic transcriptional AND gates built with split T7 RNA polymerase mutants.** *Proc Natl Acad Sci USA* 2013, **110:**5028–5033.
42. De Smit MH, Vanduin J: **Secondary structure of the ribosome binding-site determines translational efficiency–a quantitative-analysis.** *Proc Natl Acad Sci USA* 1990, **87:**7668–7672.
43. Welch M, Villalobos A, Gustafsson C, Minshull J: **You're one in a googol: optimizing genes for protein expression.** *J R Soc Interface* 2009, **6**(4):S467–S476.
44. Saecker RM, Record MT Jr, Dehaseth PL: **Mechanism of bacterial transcription initiation: RNA polymerase–promoter binding, isomerization to initiation-competent open complexes, and initiation of RNA synthesis.** *J Mol Biol* 2011, **412:**754–771.
45. Sclavi B, Zaychikov E, Rogozina A, Walther F, Buckle M, Heumann H: **Real-time characterization of intermediates in the pathway to open complex formation by *Escherichia coli* RNA polymerase at the T7A1 promoter.** *Proc Natl Acad Sci USA* 2005, **102:**4706–4711.
46. Chakraborty A, Wang D, Ebright YW, Korlann Y, Kortkhonjia E, Kim T, Chowdhury S, Wigneshweraraj S, Irschik H, Jansen R, *et al*: **Opening and closing of the bacterial RNA polymerase clamp.** *Science* 2012, **337:**591–595.
47. Schyns G, Jia L, Coursin T, Tandeau de Marsac N, Houmard J: **Promoter recognition by a cyanobacterial RNA polymerase: in vitro studies with the *Calothrix* sp. PCC 7601 transcriptional factors RcaA and RcaD.** *Plant Mol Biol* 1998, **36:**649–659.
48. Imashimizu M, Fujiwara S, Tanigawa R, Tanaka K, Hirokawa T, Nakajima Y, Higo J, Tsuzuki M: **Thymine at–5 is crucial for *cpc* promoter activity of *Synechocystis* sp. strain PCC 6714.** *J Bacteriol* 2003, **185:**6477–6480.
49. Imashimizu M, Tanaka K, Shimamoto N: **Comparative study of cyanobacterial and *E. coli* RNA Polymerases: misincorporation, abortive transcription, and dependence on divalent cations.** *Genet Res Int* 2011:572689.
50. Leveau JH, Lindow SE: **Predictive and interpretive simulation of green fluorescent protein expression in reporter bacteria.** *J Bacteriol* 2001, **183:**6752–6762.
51. Becker NA, Kahn JD, Maher LJ III: **Bacterial repression loops require enhanced DNA flexibility.** *J Mol Biol* 2005, **349:**716–730.
52. Becker NA, Kahn JD, Maher LJ III: **Eukaryotic HMGB proteins as replacements for HU in *E. coli* repression loop formation.** *Nucleic Acids Res* 2008, **36:**4009–4021.
53. *CyanoBase.* http://genome.microbedb.jp/CyanoBase.
54. Goodson KA, Wang Z, Haeusler AR, Kahn JD, English DS: **LacI-DNA-IPTG loops: equilibria among conformations by single-molecule FRET.** *J Phys Chem B* 2013, **117:**4713–4722.
55. Becker NA, Peters JP, Maher LJ III, Lionberger TA: **Mechanism of promoter repression by Lac repressor-DNA loops.** *Nucleic Acids Res* 2012, **41:**156–166.
56. Mutalik VK, Guimaraes JC, Cambray G, Lam C, Christoffersen MJ, Mai QA, Tran AB, Paull M, Keasling JD, Arkin AP, Endy D: **Precise and reliable gene expression via standard transcription and translation initiation elements.** *Nat Methods* 2013, **10:**353–354.
57. Müller-Hill B, Crapo L, Gilbert W: **Mutants that make more *lac* repressor.** *Proc Natl Acad Sci USA* 1968, **59:**1259–1264.

58. Stanier RY, Kunisawa R, Mandel M, Cohenbaz G: **Purification and properties of unicellular blue-green algae (Order Chroococcales).** *Bacteriol Rev* 1971, **35:**171.
59. Shetty RP, Endy D, Knight TF Jr: **Engineering BioBrick vectors from BioBrick parts.** *J Biol Eng* 2008, **2:**5.
60. Elhai J, Vepritskiy A, Muro-Pastor AM, Flores E, Wolk CP: **Reduction of conjugal transfer efficiency by three restriction activities of *Anabaena* sp. strain PCC 7120.** *J Bacteriol* 1997, **179:**1998–2005.
61. Milkowski C, Quinones A, Hagemann M: **A DNA fragment from the cyanobacterium *Synechocystis* sp. PCC 6803 mediates gene expression inducible by osmotic stress in *E. coli*.** *Curr Microbiol* 1998, **37:**108–116.
62. Kunert A, Hagemann M, Erdmann N: **Construction of promoter probe vectors for *Synechocystis* sp. PCC 6803 using the light-emitting reporter systems Gfp and LuxAB.** *J Microbiol Methods* 2000, **41:**185–194.
63. Gibson DG, Young L, Chuang RY, Venter JC, Hutchison CA, Smith HO: **Enzymatic assembly of DNA molecules up to several hundred kilobases.** *Nat Methods* 2009, **6:**U341–U343.
64. Hillson NJ, Rosengarten RD, Keasling JD: **j5 DNA assembly design automation software.** *ACS Synthetic Biol* 2012, **1:**14–21.
65. Kremers GJ, Goedhart J, van den Heuvel DJ, Gerritsen HC, Gadella TWJ: **Improved green and blue fluorescent proteins for expression in bacteria and mammalian cells.** *Biochemistry (Mosc)* 2007, **46:**3775–3783.

9

Electrospun nanofibers as versatile interfaces for efficient gene delivery

Slgirim Lee[†], Gyuhyung Jin[†] and Jae-Hyung Jang[*]

Abstract

The integration of gene delivery technologies with electrospun nanofibers is a versatile strategy to increase the potential of gene therapy as a key platform technology that can be readily utilized for numerous biomedical applications, including cancer therapy, stem cell therapy, and tissue engineering. As a spatial template for gene delivery, electrospun nanofibers possess highly advantageous characteristics, such as their ease of production, their ECM-analogue nature, the broad range of choices for materials, the feasibility of producing structures with varied physical and chemical properties, and their large surface-to-volume ratios. Thus, electrospun fiber-mediated gene delivery exhibits a great capacity to modulate the spatial and temporal release kinetics of gene vectors and enhance gene delivery efficiency. This review discusses the powerful characteristics of electrospun nanofibers, which can function as spatial interfaces capable of promoting controlled and efficient gene delivery.

Keywords: Gene delivery, Electrospun nanofibers, Controlled gene delivery, Tissue engineering, Sustained release

Introduction

Gene delivery has emerged as a powerful platform technology for a variety of biomedical applications, including tissue engineering, cancer therapy, and stem cell therapy. Altering genetic information through exogenous stimulation of target cells can accomplish numerous intended goals, such as differentiation of the targets cells into specialized cell types [1-3], activation of apoptosis signals in cancer cells [4,5], secretion of factors that cause autocrine or paracrine effects in tissues [6-8], or the production of cellular therapeutics [4,9]. Each of these functions of gene delivery technologies can be essential in a particular case, potentially offering promising strategies for improving the targeted function. Additionally, the identification of new genetic targets or sequences involved in human diseases through the completion of the Human Genome Project has enormously accelerated the progress of gene delivery technologies in numerous applications [10].

Using the full potential of gene delivery in numerous biomedical applications requires a series of toolkits that can help to overcome the limitations associated with gene delivery technologies, the first necessary step prior to employing gene delivery in applications is the selection of suitable gene delivery vehicles, which can be categorized into viral or non-viral vectors, depending on the target applications or cell types. The engineering of versatile gene delivery carriers, which can specifically target clinically valuable cell types (i.e., cancer or stem cells), can avoid immune system effects or toxicities, and can safely pass through complicated intracellular steps to reach the nucleus, is one of the most critical tasks in gene delivery [11-13]. However, the direct administration of gene vectors in liquid formulations to humans, a representative delivery mode, can lead to systemic spread in the body, presumably resulting in risks arising from gene expression in off-target regions [14]. Importantly, direct exposure to viral vectors, which have typically been known to increase gene transfer efficiencies compared with non-viral vectors, might cause severe immune responses against the vectors or even its gene products [15]. Direct injection of gene vectors typically boosts the vector or gene expression dosages in the blood stream or target regions within a short time, possibly leading to cellular toxicities or a short duration of gene expression [7]. Repeated and periodic administration of gene vectors, which can cause pain in patients, may be the only method that can extend the duration of gene expression. These aforementioned concerns about gene delivery are

* Correspondence: j-jang@yonsei.ac.kr
[†]Equal contributors
Department of Chemical and Biomolecular Engineering, Yonsei University, Seoul 120-749, Korea

mostly associated with extracellular delivery mechanisms and can raise safety issues, possibly delaying the immediate translation of gene delivery methods into clinical trials. Thus, novel technologies that can modulate gene delivery routes or profiles within extracellular environments must be used to facilitate the successful translation of gene delivery for human clinical use. Combining gene delivery with biomaterial systems has been commonly discussed as a powerful strategy that can provide opportunities to more effectively apply gene delivery for many biomedical applications [6,7,13,16].

In this review, we primarily focus on the strategy that combines gene delivery with electrospun nanofibers as one strategy among all of the interdisciplinary approaches of gene delivery with biomaterials, which have been typically employed as tissue engineering scaffolds [17-19], microparticles [20,21] and nano or micro devices [22,23]. This strategy addresses the concerns about both gene delivery and potential applications, and these nanofibers can be easily fabricated using a cost-effective method. Electrospun nanofibers have been extensively explored as spatial templates that can effectively mimic the structure or functions of extracellular matrices (ECMs), thereby working as highly effective interfaces that can retain cellular morphologies and efficiently deliver biomolecules to target cell types [24,25]. Thus, electrospun nanofibers have great potential as a physicochemical guide that can be used for numerous biomedical applications, including tissue engineering and drug or gene delivery [24,26]. This review will primarily discuss the powerful characteristics of electrospun nanofibers as spatial templates for gene delivery. Finally, successful employment of the combinatorial approaches of gene delivery with electrospun nanofibers will be classified depending upon the application, including tissue engineering, cancer therapy, and stem cell studies.

Figure 1 ECM-analogue morphology of electrospun PCL nanofibers imaged using scanning electron microscopy (SEM). The scale bar indicates 1 μm. The image was reproduced from [48], Copyright 2011, with permission from Elsevier.

Electrospun nanofibers as versatile spatial templates

Electrospinning is a versatile method for fabricating ultrafine polymeric nanofibrous structures through electrostatic interactions (Figure 1). Deposition of the resultant fibers on grounded collectors can produce non-woven fibrous matrices with high surface-to-volume ratios and diameters ranging from nanometers to micrometers [27-29]. One of the highly advantageous aspects of using electrospun matrices as building blocks for numerous biomedical applications is the feasibility of manipulating the physical and chemical characteristics of the resultant fibrous structures. Specifically, the surface morphology [30,31], mechanical strength [32,33], fiber orientation [34-36], and inner structure of the fibers [37-39] can be diversified by simply adjusting the various fabrication parameters, such as collector designs and nozzle alignments. Additionally, a variety of materials, including metals [40,41], ceramics [42-44], synthetic polymers [33,45-47], peptides [48,49], and viral solutions [50], can be applied to produce electrospun nanofibrous structures. Conventional electrospinning techniques typically generate two-dimensional sheet-like shapes. In contrast, advanced electrospinning tools, such as layer-by-layer deposition [45,46], E-beam etching [51,52], selective leaching [33], and plasma treatment [53,54], have recently been developed to produce well-defined or sophisticated three-dimensional fibrous structures, ultimately further expanding the scope for numerous biomedical applications.

Electrospun nanofibers for gene delivery

Owing to these multiple merits, highly porous fibrous structures have recently been employed as powerful templates for delivering genes of interest for various purposes (e.g., tissue engineering, cancer treatment or stem cell studies) in a spatially or temporally regulated manner [24-26]. Non-viral gene vectors, such as naked plasmid DNA or DNA/polyplexes, have primarily been incorporated within or onto electrospun nanofibers, presumably because of their ease of production or their capacity to maintain the intact properties compared with viral vectors [55-57]. The use of viral vectors, which are created by stripping the viral genomic sequences and subsequently packaging the genes of interest into the viral capsid, has recently been integrated with electrospun nanofibers, primarily to further increase the gene delivery efficiency or extend the duration of gene expression [30,33,48].

To ensure that electrospun fibers can be used as gene delivery templates, the lack of harmful effects of two major processing features on the bioactivities of gene vectors must be confirmed; the processing typically includes direct contact between gene vectors and organic solvents and exposure to a high electric voltage. No systematic studies have been performed to thoroughly investigate the detrimental effects of the electrospinning processes on the activities of gene vectors. However, no substantial loss of the capabilities of gene vectors as a result of the relatively harsh electrospinning processes has been observed in many previous studies [47,48,58]. The methods for loading gene vectors in electrospun nanofibers can be classified into two categories, which are summarized in Table 1: encapsulating gene vectors within the fiber interiors during the fiber production process, and immobilizing gene vectors onto the fiber exterior after the process. The former methods, which incorporate gene vectors within structures by simple mixing with the polymer solution, emulsification, and co-axial electrospinning, primarily focus on the controlled release of gene vectors by modulating the physical or chemical properties of fibrous constructs. In the latter methods, gene vectors are typically associated with electrospun nanofibers after the completion of all the fabrication procedures, and the main goal is to prevent the possible harmful effects of organic solvents or a high electric field on the activities of gene vectors and thus further increase gene delivery efficiencies. Additionally, post-adsorption approaches can be adjusted to allow the spatial control of gene delivery by using additional technological tools, such as surface chemistry [33,46,59] and vector modifications [60].

Advantageous aspects of electrospun nanofibers for gene delivery

Combining gene delivery with engineered polymeric biomaterials has been regarded as an indispensable strategy to increase delivery efficiencies and modulate gene delivery kinetics in a spatial and temporal manner. Genes delivered via a classic method (i.e., direct administration of a liquid formulation) diffuse freely within the body and thus suffer from the regulation of gene delivery rates or the localization of gene expression within a designated region [13,14]. In contrast, manipulating the physical or chemical properties of polymer matrices or modulating the molecular interactions of gene vectors with polymer materials can readily tune the release profiles of gene vectors, which can range from a few hours to more than months [6,8]. The sustained release of gene vectors from polymeric templates can increase their residence time within the cellular microenvironment, potentially enhancing the gene transfer efficiency and extending the duration of gene expression [16]. Importantly, the extremely large surface-to-volume ratios and the ECM-analogue nature, which are unique properties of electrospun nanofibrous structures, make nanofibers powerful alternatives for maximizing the capabilities and efficiency of gene delivery in a variety of biomedical fields [25,61].

The high porosity of electrospun nanofibers can facilitate increased cellular contacts with well-distributed gene vectors within or over large surface areas, thereby allowing more opportunities to internalize gene vectors across the cellular membrane. Furthermore, the pores of electrospun nanofibers are typically produced with interconnected open structures, which can improve the cellular penetration into fibrous interiors and increase the gene vector loading capacities, thereby increasing in the gene delivery efficiencies [62]. For example, the adsorption of adeno-associated viral (AAV) vectors on the surface of three-dimensional fluffy fibrous structures, whose specific pore volume was 4-times greater than that of two-dimensional electrospun mats, allowed a significant improvement in cellular transduction efficiencies compared with vectors associated with flat non-porous polymeric systems [33]. Consistently, immobilizing plasmid DNA on highly porous poly(lactic) acid (PLA) fibrous matrices resulted in approximately 1.5-fold increased gene delivery efficiencies compared with those for nonporous PLA films [46], confirming the superior contribution of nanofibrous pore structures to gene delivery.

Strategies for controlling electrospun nanofiber-mediated gene delivery

Owing to the ease of tuning structural variations of electrospun nanofibers, the capability of these fibers to mediate controlled and sustained gene delivery has been recognized as the most representative feature of electrospun nanofibers compared with other existing polymeric gene delivery templates. It is well known that variations in polymer degradation through hydrolysis and alterations of diffusion routes through porous structures have been key design parameters that can vary the release kinetics of biomolecules encapsulated within or adsorbed on polymeric templates [63]. In addition to these conventional ways to control release profiles, additional design variations in electrospun nanofibers have been explored to diversify the kinetics of the release of gene vectors from the fibrous structures. For example, the structural characteristics of individual fibers can be altered by changing the formulation of the core-sheath structures or by modifying the surface properties as an alternative strategy to vary the release rates of incorporated or adsorbed gene vectors from the electrospun fibers or to further increase delivery efficiencies [64,65]. The following section describes versatile methods capable of mediating controlled gene delivery from electrospun nanofibers; these methods were classified by the vector-loading approaches, in which the vector

Table 1 Classification of electrospun nanofiber-mediated gene delivery approaches

Vector loading methods	Vectors	Genes	Target cells	Results	Ref
Vector encapsulation	Virus (Ad)	GFP, RFP	HEK293T	• Sustained and controlled viral release for 30 days. • Localized gene expression from electrospun scaffolds.	[30]
	Plasmid DNA	β-Gal, GFP	MC3T3-E1	• Gene expression by the released DNA 48 h after seeding • Burst release of majority of encapsulated plasmid DNA within 30 minutes.	[47]
		Cdk2i, EGFPi	MCF-7 cell	• Sustained release over 21 days • ~40% decrease in proliferation of breast cancer cells compared with control scaffold	[115]
		EGFP	Rat fibroblasts	• Extended release of pDNA and transgene expression over 60 days.	[55]
	Virus (AAV)	GFP	NIH3T3	• Sustained viral release for 7 days. • Maintained transgene expression (>90%) on the scaffolds for 7 days.	[48]
	Plasmid DNA/chitosan nanoparticle	BMP-2	hMSC	• Sustained release for 45 ~ 55 days • DNA/chitosan nanoparticles encapsulated electrospun scaffolds as a favorable DNA delivery device with regard to cell transfection efficiency and cell viability	[56]
	Plasmid DNA/LEL polyplex	β-Gal, GFP	MC3T3	• Transgene expression on DNA-incorporating electrospun scaffolds 24 h after seeding. • Sustained release for 7 days.	[57,80]
	siRNA/CPP polyplex	Col1A1 silencing	Human dermal fibroblasts, *in vivo*	• Prolonged *in vitro* gene silencing duration by 2 ~ 3-fold. • *In vivo* gene silencing for 4 weeks.	[58]
	Plasmid DNA/PEI	EGFP	NIH3T3	• Controlled release time from 6 days to 25 days by internal structures and porogens. • 10-fold increased gene expression on the scaffolds compared to simple pDNA/PELA blends.	[64]
		VEGF/eGFP & bFGF/eGFP	HUVEC	• Sustained release for 4 weeks • Significantly higher vessel densities	[65]
		bFGF/GFP	BEF, *in vivo*	• Sustained release for 26 days • 4 ~ 6-fold increased bFGF expression compared with post-electrospinning delivery after 7 day incubation • Significantly higher wound recovery rate compared with post-electrospinning delivery	[84]
	siRNA/chitosan polyplex	EGFP silencing	EGFP expressing human lung carcinoma cell lines	• Sustained and controlled delivery for 30 days. • Prolonged *in vitro* gene silencing duration by 3 ~ 4-fold compared to the bolus delivery.	[67]
	siRNA	GAPDH silencing	HEK293, NIH3T3	• Sustained release of siRNA for 28 days. • Gene silencing on scaffolds in presence of additional transfection agents. • Enhanced gene silencing capability with additional transfection agents in the media.	[68]
	siRNA/transfect-ion reagent complex	GAPDH silencing	NIH3T3	• Sustained release of siRNA and gene silencing on the scaffolds for at least 28 days. • Improved gene silencing capability with transfection agents supplemented in the media.	[69]
	Solid-in-oil dispersion of plasmid DNA	Luciferase	N/A	• Release profile controlled by degrading rates of fibers. • 10-fold increases in functional integrity of released pDNA compared to mixed mesh.	[81]

Table 1 Classification of electrospun nanofiber-mediated gene delivery approaches *(Continued)*

	Plasmid DNA/calcium phosphate nanoparticle	VEGF/eGFP & bFGF/eGFP	HUVEC, hAoSMC	• Sustained release for 4 weeks • Significantly higher densities of blood vessels and mature vessels	[83]
Vector immobilization	Virus (AAV)	GFP, Luciferase	HEK293T	• Three-dimensional and uniaxially aligned transgene expression • 4-fold enhanced transgene expression levels compared to 2D electrospun scaffolds.	[33]
	Plasmid DNA	EGFP	Glioblastoma cells	• Transgene expression by the released DNA from the fibers (maximum transfection efficiency > 90%).	[45]
		Luciferase	COS-7	• Retained gene expression on the fibers for 5 days after seeding. • 2-fold increased gene delivery efficiency of electrospun fibers over that of flat films.	[46]
		GFP, Dsred	HEK293, MSC, *in vivo*	• 10-fold increase in gene expression intensity compared to PCL fibers *in vivo*.	[59]
		EGFP-N1	NIH3T3, *in vivo*	• MMP-2 responsive release of DNA • Significantly enhanced gene expression in wound tissue compared to naked DNA delivery	[66]
		Luciferase, KGF	NIH3T3, *in vivo*	• Sustained expression for 7 days • 65% smaller epithelial gap in KGF scaffold treated wounds than in untreated wounds	[76]
		hEGF	HDF, *in vivo*	• MMP-2 responsive release of DNA • Approximately 2-fold increased wound closure compared with non-treated wounds	[85]
		EGFP	MC3T3-E1	• Controlled gradients of pDNA concentration and gene expression level by spatially regulating rates of chemical reactions.	[98]
	Virus (AAV)	GFP	HEK293T	• Patterned and localized gene vectors and gene expression on the scaffolds. • 2-fold increase in transfection efficiency compared with unmodified virus delivery.	[60]
	Plasmid DNA/liposome	RUNX2/eGFP	hBMSC	• Long-term gene expression for 21 days • Improved osteogenic differentiation of stem cells	[77]
	siRNA/PEI polyplexes & siRNA/ transfection reagent complex	TSP-2 silencing	hAoSMC	• Down-regulated TSP-2 mRNA expression	[78]
	Plasmid DNA/chitosan nanoparticle	BMP-2	*in vivo*	• Different bone healing performance depending on the loading methods • Improved bone healing for DNA/chitosan nanoparticles adsorbed electrospun scaffolds at 4 weeks of treatment	[82]
	siRNA	MMP-2 silencing	HDF, *in vivo*	• MMP-2 responsive release of DNA • Faster wound recovery rate compared with siRNA solution delivery	[86]
	Plasmid DNA/ssPEI	Luciferase, RFP, VEGF	H9C2 myoblastic cell	• Enhanced transfection efficiency compared to bolus delivery • Successful expression of the VEGF gene in the cells	[87]
	siRNA/ transfection reagent complex	REST silencing	NPC	• Enhanced neural marker expression and neuronal differentiation	[88]

is encapsulated into the interior (Controlled release from nanofibers by encapsulating gene vectors and Core-sheath formulations for controlled release) or immobilized onto the exterior of nanofibers (Substrate-mediated gene delivery using electrospun nanofibers).

Controlled release from nanofibers by encapsulating gene vectors

The encapsulation of gene vectors within electrospun nanofibers for subsequent diffusion through porous routes can result in the sustained release of gene vectors, as well as controlled delivery via manipulations of the physical or chemical properties of the fibrous structures. Gene vectors can be incorporated within the inner space of fibrous structures, as illustrated in Figure 2, by simply mixing aqueous DNA buffers with polymer solutions in organic solvents, followed by electrospinning the nanofibers. The first trial involving loading gene vectors into electrospun nanofibers was performed by blending Tris-EDTA buffer solution containing plasmid DNA, which encoded β-galactosidase driven by the cytomegalovirus (CMV) promoter, with the block-copolymers of PLA and poly(ethylene glycol) (PEG) dissolved in N,N-dimethyl formamide [47]. In this study, the activity of the plasmid DNA was stably maintained during the encapsulation and electrospinning process. To reduce the steric repulsion between the hydrophobic PLA solution and plasmid DNA in the polar aqueous buffer, the hydrophilic polymer PEG was included in the mixture prior to electrospinning. The resultant DNA-blending approach within the fiber interior spaces containing the hydrophilic PEG resulted in a rapid DNA release within 2 hours, followed by a high level of gene expression at 48 hours post-transfection. The sustained release of the plasmid DNA, whose release kinetics was determined by both the pore morphologies and the contents of the copolymers used to produce the nanofibers, was maintained for at least 20 days, demonstrating the efficacy of DNA-blending approaches for efficient gene delivery as well as for an extended duration of gene delivery.

The release profiles of gene vectors within the inner space of nanofibers were readily regulated using environmental factors that can degrade polymer matrices (Figure 2B). It is well known that polymer degradation patterns, which can be categorized into either surface erosion or bulk degradation, can determine whether the release kinetics of incorporated gene vectors follow a sustained mode or burst release mode [63]. The addition of proteinase K into the release buffer accelerated the enzymatic degradation of polyester-based biodegradable polymers, such as poly(L-lactide-co-ε-caprolactone) (PLCL) or poly (caprolactone) (PCL), resulting in a rapid burst within a short time period [66]. Approximately the total quantity of the plasmid DNA incorporated in the PCL matrices could be released within 10 hours when the releasing buffer contained proteinase K, whereas quite slow release rates of plasmid DNA were observed without the enzyme in the buffer solution. Similarly, varying the acidity of the releasing buffers fixed the acidic or alkaline hydrolysis patterns of poly(D,L-lactide-co-glycolide) (PLGA), which typically results in the bulk or surface erosion of PLGA, respectively [67]. Consequently, the alteration of the polymer erosion modes via manipulations of the environmental acidity acted as a crucial parameter to regulate the release mechanisms of small interfering RNA (siRNA) encapsulated within PGLA nanofibers. The resultant sustained delivery of siRNA from the PLGA fibrous matrices, which were slowly degraded in acidic environments, prolonged the gene silencing effects for up to 30 days.

Another key factor that can modulate the release modes of gene vectors encapsulated within nanofibers is the inclusion of additives within fibers (Figure 2C and 2D). Water-soluble additives that are included within the hydrophobic polymer fibers can function as porogens that allow the incorporated gene vectors to readily pass through the routes, whose dimensions are newly expanded via the dissolution of the additives in aqueous solution (Figure 2C). For example, the incorporation of hydrophilic PEG content within hydrophobic fibers provided favorable hydrophilic traces that increased the affinity for gene vectors or siRNA complexes, possibly facilitating the diffusion of the incorporated vectors from the fibrous matrices [68]. Additionally, the addition of hydroxylapatite (HAp) nanoparticles to PLGA nanofibers regulated the DNA release rates and promoted cellular adherence on the fibers as well (Figure 2D) [56]. The incorporation of HAp nanoparticles triggered the association of DNA/chitosan complexes with PLGA nanofibers during the fiber fabrication process, substantially improving the DNA loading efficiencies. The inclusion of the hydrophilic inorganic nanoparticles, which

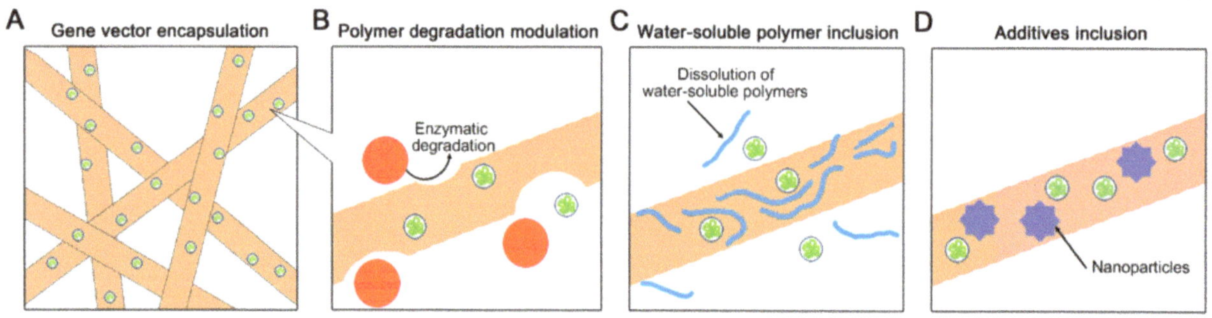

Figure 2 Blending approaches for controlled gene delivery. (A) Schematic illustration demonstrating gene vector encapsulations within electrospun fibers. The right three panels display strategies for controlled delivery using the blending-approaches via **(B)** modulating the environmental conditions, **(C)** including water-soluble polymers and **(D)** providing additives **(D)**.

altered the mechanical properties of the composite fibers, including their strain–stress behaviors and glass transition temperatures, accelerated the release of DNA/chitosan complexes and enhanced the gene delivery efficiencies as well as the cellular viabilities.

Altering the compositions of polymeric materials comprising nanofibers can be an additional option to tailor the release profiles of gene vectors, as well as the delivery efficiencies. The copolymerization of ethyl ethylene phosphate (EEP) with caprolactone facilitated the incorporation of gene vector elements into the inner space of nanofibers and resulted in a sustained release of siRNA complexes that exceeded 40 days [69]. The co-encapsulation of small-interfering RNA with either a transfection reagent or cell-penetrating peptides within poly(caprolactone-co-ethylethylene phosphate) (PCLEEP) nanofibers resulted in increased gene silencing efficiencies and extended the duration of gene silencing to over 14 days, thus prolonging the expression by approximately 2-3-fold compared with that for bolus delivery [58]. Consequently, the sustained delivery of siRNA suppressing the production of collagen type I substantially reduced the fibrous capsule thickness adjacent to nanofibrous scaffolds that were implanted subcutaneously. Additionally, blending the PCL solution with elastin-like polypeptides (ELP) in hexafluoro-2-propanol (HFP) triggered the release of AAV vectors and prolonged the viral delivery to more than 14 days [48]. In this study, the phase transition properties of ELPs at different temperature enabled versatile AAV release profiles at various temperatures. Taken together, these findings show that the presence of hydrophilic portions within hydrophobic building blocks can facilitate the porogen-assisted release of incorporated agents. Furthermore, in addition to these factors that affect the release profiles, the inclusion of hydrophilic materials into fibrous matrices can offer additional promising features, such as improved biocompatibility and cellular attachment.

Core-sheath formulations for controlled release

A notable structural feature of the electrospinning process is its capability to produce a core-sheath structure within individual fibers, where multiple biomolecules at each layer can be designed to diffuse out sequentially (Figure 3). The electrospun nanofibers with the core-sheath structures can be fabricated using co-axial electrospinning (Figure 3A) [33,39] or the emulsion electrospinning technique (Figure 3B) [64]. The core-sheath structures, whose representative morphology is demonstrated in Figure 3C, have been typically produced for the following: i) the protection of gene vectors from direct exposure to organic

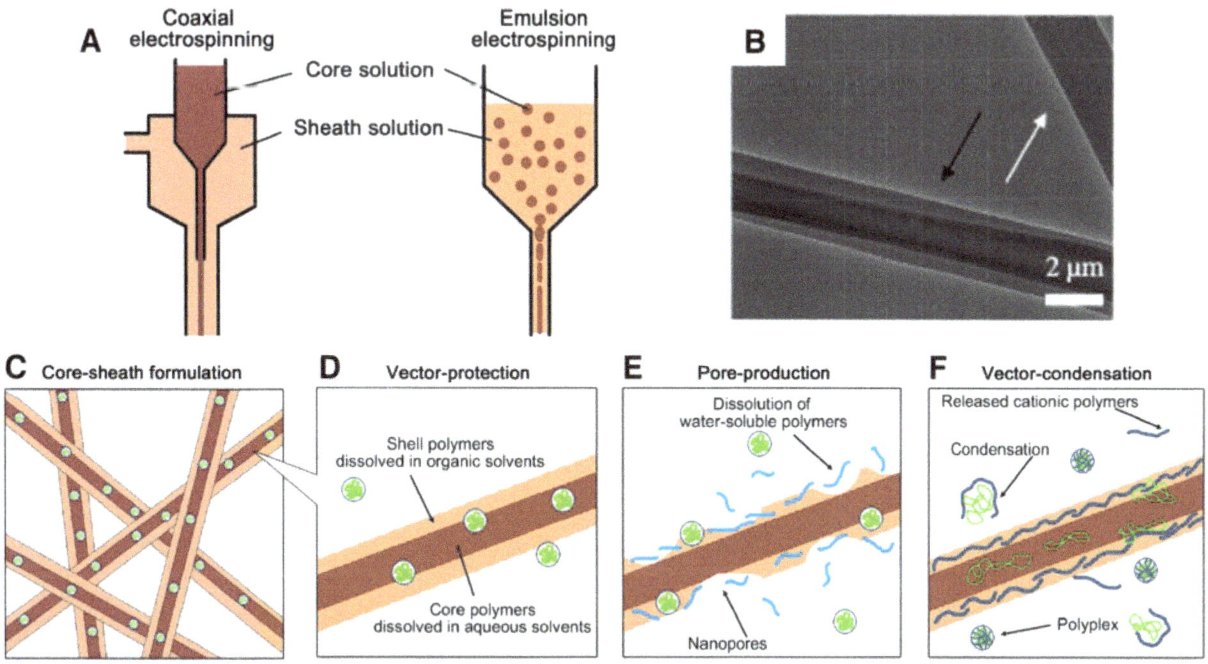

Figure 3 Core-sheath electrospun nanofibrous systems for controlled gene delivery. (A) Two representative methods to form core-sheath electrospun nanofibers: coaxial (left) and emulsion (right) electrospinning. **(B)** Transmission electron microscopy (TEM) image of an individual core-sheath nanofiber fabricated using coaxial electrospinning. Core and sheath are composed of viruses dispersed in Minimal Essential Medium and PCL, respectively. Scale bar is 2 μm. Reprinted from [30], Copyright 2009, with permission from Elsevier. **(C)** A scheme depicting gene vector encapsulation within the core layer for controlled release. The core-sheath fibrous formulations contribute **(D)** to preventing the direct contact of gene vectors in the core layer with organic solvents in the sheath layer, **(E)** to regulating delivery modes by producing porous sheath layers, and **(F)** to enhancing delivery efficiencies by modifying the sheath layers with polycationic polymers.

solvents and ii) the controlled release of gene vectors residing in core layers through modifying the shell structures. The inclusion of gene vectors in hydrophilic core-layers followed by encapsulation with hydrophobic shell-layers in organic solvents can prevent the direct contact of gene vectors with organic solvents (Figure 3D). Differences in the diffusion pathways of gene vectors through two layers composed of different materials can alter the release rates of the incorporated gene vectors in each layer, which have already been observed in many drug delivery studies using core-sheath structures [70,71]. Unfortunately, the sequentially controlled release of multiple gene vectors from each core-sheath layer has not been explored yet. Taken together, these possibilities for the integration of gene delivery technologies into the core-sheath fibrous matrices can provide an efficient means to control the sequential release of multiple vectors and can simultaneously protect gene vectors in the core-layer against the relatively harsh processes.

The presence of porogens in the shell layers can facilitate the sustained and controlled release of gene vectors incorporated in the core layers and can also extend the duration of gene delivery (Figure 3E). The production of shell layers with higher levels of PEG, which was deposited in the shell layer by varying the molecular weights and concentrations of the PEG, led to the rapid release of the plasmid DNA that was complexed with poly(ethylenimine) (PEI) [64]. The dissolution of PEG in aqueous environments (i.e., cell culture medium) might provide more space for the diffusional passage of gene vectors in the core layers to the outer spaces. He et al. used dual gene vectors that encoded different inducible factors (i.e., vascular endothelial growth factor (VEGF) and basic fibroblast growth factor (bFGF)) and loaded them together into the core poly(D,L-lactide)-poly(ethylene glycol) (PELA) layers that were encapsulated by a sheath layer containing PEG [65]. The sustained release of these multiple vectors from the core layer through the PEG-assisted routes in the sheath layer synergistically promoted mature blood vessel formation. In this particular case, the pores exist separately in the exterior layer, and the burst of gene vectors at the initial time points may be avoided, which is not possible in cases in which the pores coexist with vectors in the same layer (i.e., single fibers). This difference can be demonstrated by comparisons of the parallel data from several studies [56,65]. Importantly, the enlarged surface areas in the shell layers might create room for cell migration toward the inner spaces, where cells can encounter the gene vectors released from the core layer.

The variations in the PEG contents in the shell layers demonstrated the distinctive controlled release profiles of viral vectors that were encapsulated within the core layers. The release rates of adenoviral (Ad) vectors encapsulated within co-axially produced PCL fibers were dependent on the PEG concentrations in the shell layers [30]. As the PEG contents increased in the shell layers, highly rapid viral elution from the fibers in a short time period (~ a week) was observed, but the inclusion of intermediate quantities of PEG resulted in a gradual release with almost constant rates over 30 days. As a result, the continuous supply of Ad vectors to HEK293T cells that were cultured on highly porous PCL core-sheath fibers persistently induced high levels of cellular transduction. However, non-porous Ad-encapsulated PCL fibers exhibited almost no viral release over a month, resulting in extremely low levels of transduction throughout the time points. Ad capsid PEGylation, which was primarily performed to protect Ad vectors from immune rejection, has been shown to inherently reduce transduction efficiencies [72]. However, the extended duration of the substantially increased gene expression by PEG-assisted Ad delivery from PCL fibers demonstrates the superior capabilities of polymeric gene delivery compared with direct delivery approaches.

The core-sheath fibrous structures can provide multiple physical spaces for separately delivering dual factors, which possess individual roles (Figure 3F). Co-axially electrospun nanofibrous matrices were produced by incorporating plasmid DNA in the core PEG layer, along with a derivative of PEI conjugated with hyaluronic acids (HA-PEI) in the shell layer that contained PCL polymers dissolved in a chloroform and methanol mixture [55]. The inclusion of plasmid DNA in the core PEG region was likely attempted to minimize the exposure of the plasmid to organic solvents in the shell layer and to prevent additional processes (i.e., lyophilization) that could ultimately reduce the activity of the plasmid DNA [73,74]. The lyophilization process was typically required in blending approaches to uniformly spread DNA powders in hydrophobic polymer solutions prior to electrospinning. Interestingly, the cationic polymer HA-PEI in the outer sheath layer self-assembled with the negative plasmid DNA vectors that were released from the core layer, thereby enhancing the efficiencies of cellular internalization in a rat fibroblast cell line [55]. The coordination of the release profiles of both factors (i.e., plasmid DNA and HA-PEI), whose release rates were further tuned by additional parameters (e.g., concentrations or molecular weights of the polymers in each layer), resulted in persistent gene expression over 60 days. The increases in the delivery efficiencies were significantly greater than those for PCL matrices containing pDNA alone. Owing to the capacity to modulate the release profiles of multiple biomolecules in each layer, electrospun nanofibers with core-sheath structures have great potential as a platform template for tailoring the delivery kinetics of multiple gene vectors, potentially leading to the broad application of these fibers in numerous biomedical fields.

Substrate-mediated gene delivery using electrospun nanofibers

An alternative approach capable of delivering gene vectors from electrospun nanofibers is to immobilize gene vectors onto the surfaces of completed nanofibrous structures; this technique is termed substrate-mediated delivery. This approach can block any attempts to expose gene vectors to both organic solvents and a high electric field because the gene vectors are adsorbed on the surface of fibrous constructs in the last step after the manipulation of the fibrous constructs is completely finished. Thus, the relatively harsh processing steps could have no harmful effects on the bioactivities of gene vectors. Importantly, this delivery method can place gene vectors in close proximity to the desired location within the cellular microenvironment and can extend the residence time of the gene vectors within the boundary layers; thus, this method can overcome mass transfer limitations to deliver the genes of interest to target cells [7,75]. Increasing the extent of physical contact of gene vectors with target cells that are seeded on fibrous structures is thought to significantly increase the delivery efficiencies of gene vectors, which is a primary goal of the substrate-mediated delivery systems. Gene vectors can be immobilized on the fibrous interfaces primarily by either simple random adsorption [45,46,59,76] or specific adherence [77,78]. Importantly, tuning the interactions of gene vectors with fibrous surfaces by altering the surface chemistry or by modifying the vectors themselves can result in the spatially and temporally controlled delivery of gene vectors, which can be a unique feature of substrate-mediated gene delivery. This section classifies substrate-mediated gene delivery using electrospun nanofibers according to the method for immobilizing the gene vectors on the fiber surfaces.

Non-specific adsorption of gene vectors on the fibrous surfaces The most representative method of immobilizing gene vectors on electrospun fibers is non-specific random adsorption, which is typically accomplished using van der Waals, hydrophobic, and electrostatic interactions between gene vectors and the fiber surfaces (Figure 4A) [7,13,79]. Zhang et al. adsorbed plasmid DNA on PCL nanofibrous matrices blended with a cationic PEI through the electric attraction between the negatively charged DNA and the cationic PEI (Figure 4B) [59]. While the non-specific adsorption onto the PCL fibers without the inclusion of PEI resulted in substantially lower quantities of plasmid DNA, charging the PCL fibers with cations by blending with PEI substantially increased the DNA adsorption and yielded high transfection efficiencies in both human embryonic kidney cells and mesenchymal stem cells. Similar to the substrate-mediated delivery of non-viral vectors, the non-specific random adsorption of AAV capsid particles on the serum-coated three-dimensional moldable PCL fluffy matrices resulted in highly potent gene expression throughout the entire volumetric matrices; this process led to approximately 5-fold increased luciferase expression compared with that on two-dimensional PCL fibrous sheets [33]. This observation confirms the crucial contribution of large-surface, porous structures to higher gene delivery efficiencies.

Producing multiple DNA layers on fibrous surfaces was proposed as a means to increase gene delivery (Figure 4C) [45,46,76]. Layer-by-layer (LBL) DNA films were constructed by iterative accumulations of plasmid DNA on fibrous matrices, which were designed to possess cationic properties by electrospinning polycationic poly(β-amino ester) (PBAE) poly(1,4-butanediol diacrylate-co-4-amino-1-butanol) end-capped with 1-(3-aminopropyl)-4methylpiperazine (447) [45]. The concentration of the polymer 447 was one of the key parameters that could influence the total quantity of the multiple DNA layers and ultimately worked as a crucial factor to tune the DNA release profiles and improve the gene delivery to primary human glioblastoma cells. Interestingly, the number of DNA layers that accumulated on the fibrous surfaces was directly related to the level of

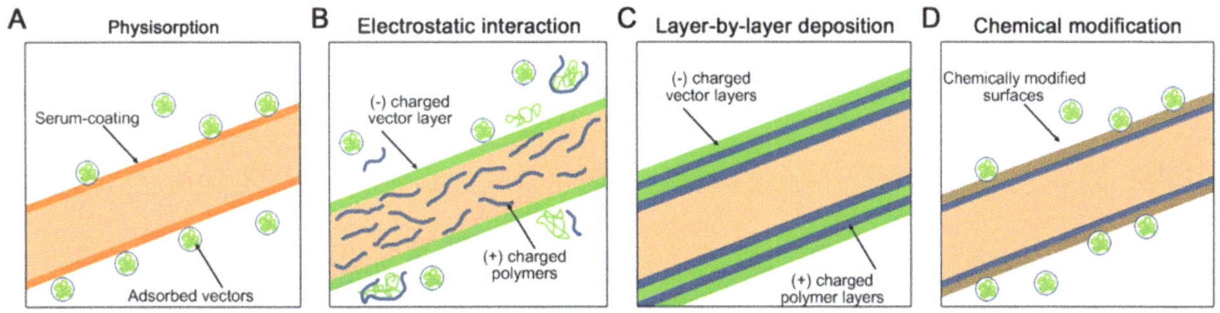

Figure 4 Substrate-mediated gene delivery from electrospun nanofibers. Gene vectors can be simply adsorbed on the fibrous surfaces via **(A)** physisorption, **(B)** electrostatic interaction, and **(C)** layer-by-layer deposition techniques. **(D)** Gene vectors can be specifically immobilized on the chemically modified fibrous surfaces to further enhance the mutual interactions between vectors and fibers.

transgene expression, whose levels improved as the number of DNA layers increased [76]. Kobsa et al. formed multi-layered DNA/PEI films on electrospun fibrous matrices composed of PLA or PCL; this process resulted in persistent luciferase expression for at least 7 days and consequently accelerated the wound re-epithelialization, keratinocyte proliferation, and granulation response [76]. These combinatorial approaches with the LBL technique will expand the scope of substrate-mediated gene delivery systems for applications that typically require long-term gene expression.

Electrospun nanofiber-mediated gene delivery approaches using naked plasmid DNA typically require additional process, such as condensing the plasmid DNA to further increase the delivery efficiencies; the delivery capability for naked plasmid DNA is substantially lower than that for viral vectors. Different methods were employed to decrease the size of plasmid DNA upon electrospinning: i) solvent-induced condensation [57,80], ii) emulsion-induced condensation [81], and iii) polycation-induced condensation [56,82,83]. Exposing DNA to a co-solvent system composed of 94% DMF and 6% TE buffer resulted in a transition from a random coil structure to a structure containing both coil and globule features [57,80], and the resultant DNA complexes that were released from the PLGA fibrous matrices showed significantly increased cellular transfection compared with that for non-complexed DNA. Additionally, a solid-in-oil (S/O) emulsion-induced condensation, which formed a solid phase with lyophilized DNA in cyclohexane, maintained the activity of DNA upon electrospinning with dichloromethane containing PLCL [81]. For the resultant S/O nanodispersion, the bioactivity of gene vectors improved approximately 12-fold improved compared with that for the system produced by simple mixing; this improvement potentially provides potent gene vectors that can increase delivery efficiencies [81]. Polycation-induced condensation, which is performed prior to electrospinning, is the most widely used method for condensing non-viral vectors to further increase transfection efficiencies [56,82,83]. The complexation of plasmid DNA with PEI followed by electrospun fiber-mediated delivery was superior to bolus delivery of DNA complexes at increasing transfection efficiencies for extended time periods [84]. In addition to acting as a DNA-condensing agent, PEI has been combined with electrospun fibers as surface-coating [46,66,85,86] or blending agent [59,64,65,76,78,84,87]; all of these functions were utilized to develop the most efficient fibrous systems to release non-viral vectors by balancing the release profiles of DNA complexes, the cytotoxicity of PEI itself, and the interaction with cellular membranes [11,59].

Surface immobilization of gene vectors for substrate-mediated gene delivery The chemical modification of fibrous surfaces is an efficient way to adhere gene vectors onto fibrous interfaces, through increasing the specificity of the fibers for the gene vectors (Figure 4D) [77,78,88]. Monteiro et al. used the affinity of liposomes with thiol groups exposed on PCL nanofibrous matrices to specifically attach DNA/liposome complexes onto SH-functionalized surfaces [77]. The specific immobilization of the DNA/liposome complexes resulted in the prolonged expression of runt-related transcription factor 2 (RUNX2) from human bone-marrow-derived mesenchymal stem cells (hBMSCs) and increased the viability of the hBMSCs on the modified PCL fibers. In another study, PCL fibers that were functionalized by coating with adhesive mussel-inspired polydopamine (PD) facilitated the adsorption of RE-1 silencing transcription factor (REST)-directed siRNAs and subsequently resulted in knocked down REST for 5 days in neural progenitor cells (NPCs) [88]. Compared with the simple non-specific random adsorption approaches, the functionalization of fibrous interfaces to coordinate the interactions between gene vectors and fiber surfaces readily inhibited the aggregation of vectors deposited on the surfaces and thus allowed for homogeneous spreading of the vectors, thereby possibly promoting efficient transfer of the vectors across the cellular membrane [17].

Spatially patterned or localized gene delivery on fibrous surfaces

One of the key unique characteristics of electrospinning compared with other fabrication tools is that it can readily produce uni-axially aligned or patterned fibrous structures (Figure 5). The uniaxial alignment of polymeric nanofibers has been widely used in numerous tissue engineering applications that require specialized tissue orientations or patterns, such as nerve or muscle regeneration [89-92]. Numerous fibrous devices demonstrating morphological orientations with specialized patterns have already been developed mainly through the use of approaches such as collector modification [35,93] or electromagnetic methods [94], but a few patterned matrices have been employed for gene delivery templates. Recently, Lee et al. developed highly fluffy three-dimensional, uni-axially patterned PCL nanofibrous matrices using co-axial electrospinning on a rotating mandrel followed by a selective leaching process [33]. Human embryonic kidney cells were cultured on the patterned fibers where AAV vectors encoding for green fluorescent protein (GFP) were pre-immobilized by non-specific adsorption. Consequently, the patterned-physical guide arranged the GFP-expressing cells uni-axially, demonstrating the potential of this platform tool to induce spatially patterned gene expression.

The modification of either fibrous surfaces or gene vectors can allow spatially patterned or localized gene

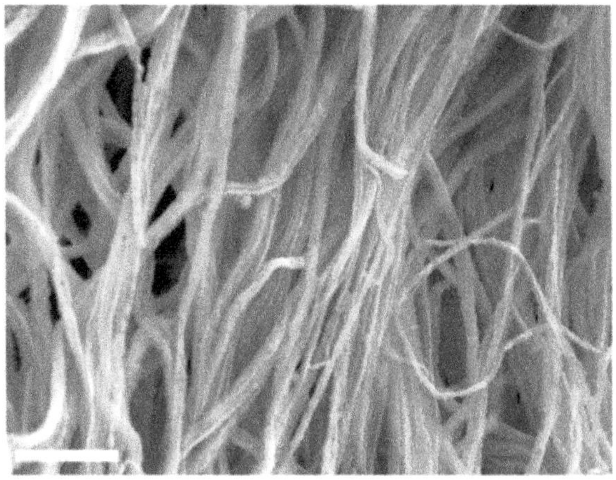

Figure 5 SEM image of uniaxially patterned PCL nanofibers fabricated by utilizing rotating mandrels. The scale bar indicates 10 μm. The image was adapted with permission from [33]. Copyright 2014 American Chemical Society.

delivery on the fibrous matrices, and these matrices can possibly be used for patterned tissue regeneration [95-97]. Covalently coupling gelatin onto poly(D,L-lactide) (PDLLA) via sequentially involving an aminolysis reaction and Schiff base formation created fibrous matrices with gradients in HAp contents, which ultimately led to gradients in the plasmid DNA content through the mineralization process [98]. The gradient deposition of plasmid DNA on the modified surfaces induced spatially patterned gene transfection, which corresponded to the graded patterns of cell distribution as well as osteoblastic differentiation [98]. Additionally, the chemical alterations of the gene vector itself played a key role in inducing spatially patterned gene expression on fibrous PCL matrices [60]. The complexation of negatively charged AAV vectors with positively charged catechol-conjugated PEI generated adhesive viral vectors, facilitating the immobilization of AAV vectors onto the fibrous PCL matrices due to their stickiness [60]. Adjusting the stickiness properties of AAV vectors regulated the gene delivery efficiencies and optimized the increase in gene transfer compared with that for non-modified AAV vectors. The resultant sticky viral system aided in the spatially patterned deposition of viral vectors via a simple pipette drawing technique and soft lithography. In principle, creating the concentration gradients of gene vectors in a localized region can induce patterned gene expression or oriented growth factor generation that corresponds to the gradients of the gene vectors; thus, this strategy can offer basic tools to promote patterned tissue regeneration. Similarly, a recent study demonstrated that manipulating the adhesive properties of surfaces can work as a key design parameter to control the release profiles of immobilized gene vectors for inducing sustained gene expression; this strategy can be further translated into electrospun nanofibers [99].

Potential applications of electrospun nanofibers that release gene vectors

Tissue engineering

Owing to both their versatile capabilities that can be used to coordinate the release profiles of gene vectors and their ECM-analogue nature (Figure 1), gene vector-eluting electrospun nanofibrous structures have primarily been used for tissue regeneration approaches. In this combinatorial gene delivery approach using electrospun fibers as tissue engineering scaffolds, the spatially or temporally regulated secretion of tissue inductive growth factors from cells can initiate either autocrine or paracrine effects to stimulate cellular processes for tissue formation; the cells are transfected or transduced by gene vectors carrying the specific genes [7,8,18,95]. Compared with the direct protein delivery approach, this approach allows for the sustained secretion of fresh growth factor proteins from the cells, and this secretion can play a key role in maintaining the stability of the proteins, potentially maximizing the efficacy of tissue formation [7,8,100]. Additionally, the capabilities of inducing localized gene expression adjacent to polymeric systems can further increase the efficacy to promote tissue formation in a defined area [18]. Thus, electrospun nanofibers that release gene vectors have been employed as a platform scaffold for mediating the regeneration of tissue, such as bone [56,77,82], skin [66,76,84-86], blood vessels [65,78,83,87], and nervous system tissues [88].

Bone tissue engineering

In addition to the ability of electrospun fibers to precisely mimic bone ECM, these fibers have large surface-to-volume ratios, allowing vascularization across the newly produced tissues within fibrous structures; these features provide strong rationales for the use of nanofibers as a guide to regenerate bone tissues [101]. Bone morphogenetic protein 2 (BMP-2) is a representative osteoinductive protein that plays an important role in directing the cellular processes that regenerate bone or cartilage [82]. Wang et al. fabricated electrospun scaffolds comprised of a PLGA/HAp composite; these scaffolds released plasmid DNA encoding for BMP-2 to promote bone tissue formation *in vitro* [56] and *in vivo* [82]. The sustained release of chitosan/DNA-BMP-2 complexes (Figure 6A) localized the BMP-2 expression at the region adjacent to the PLGA-HAp fibrous matrices. Consequently, the coordination of the release modes of the chitosan/DNA-BMP-2 complexes regulated both the transfection efficiencies and the cellular viabilities [56], ultimately resulting in improved healing of segmental bone defects in mouse tibias (Figure 6B) [82]. Additionally, the delivery of

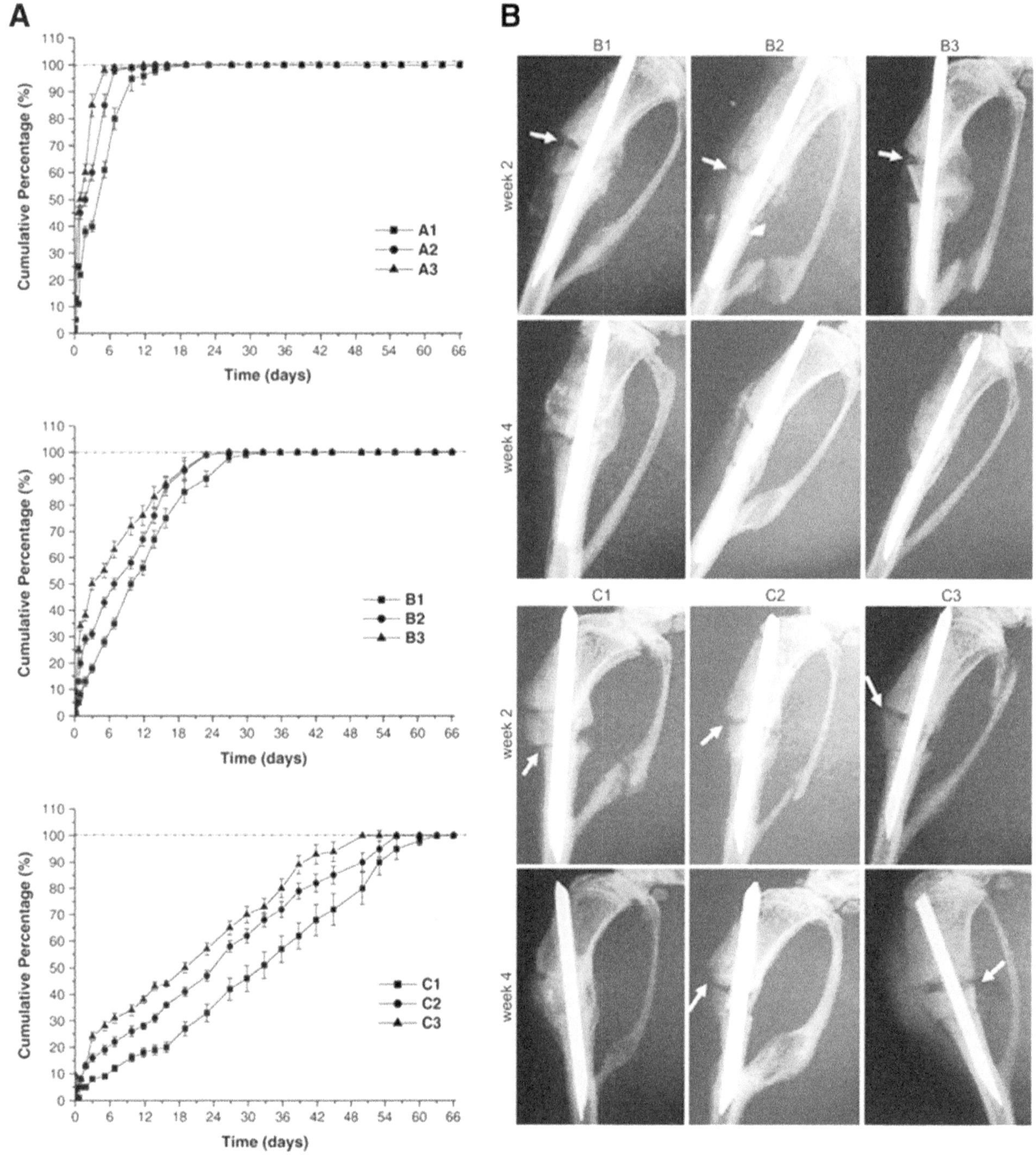

Figure 6 BMP-2 plasmid loaded electrospun scaffolds for bone tissue engineering. (A) *In vitro* release curve of three groups of scaffolds [56], Copyright 2007. Reproduced with permission from Elsevier. **(B)** Radiographs of nude mice tibias after 2 and 4 weeks of implantation of scaffolds. Bone fragment without implantation of any scaffold is denoted as control and white arrows identify bone defects [82], Copyright 2009. Reproduced with permission from Elsevier. (Group A: PLGA/HAp composite fiber with naked DNA coated outside, Group B: PLGA/HAp composite fiber with DNA-loaded chitosan nanoparticles coated outside, Group C: PLGA/HAp composite fiber with DNA-loaded chitosan nanoparticles encapsulated inside. The number indicates HAp contents in composite. X1: 0/100, X2: 5/95, X3: 10/90 (HAp/PLGA w/w%)).

plasmid DNA encoding a transcription factor, which regulates the cascades for the expression of multiple endogenous genes or for intracellular signals, can act as a key tool to promote bone tissue formation. The PCL nanofiber-mediated delivery of liposomes programmed to up-regulate RUNX2, a factor that induces cellular differentiation into the osteoblast phenotype, increased the osteogenic differentiation of hBMSCs [77]. As previously mentioned, the electrospun fibers aided in the reduction of the aggregation of liposome-RUNX2 and the cellular toxicity, leading to improved delivery efficiencies and cellular differentiation. Eventually, orchestrating the osteogenesis, angiogenesis, and inflammation at injured sites will be a crucial factor to repair or form new bone tissues functionally, which is currently a critical challenge [102]. Thus, creating synergistic effects from multiple factors, including osteogenic factors (e.g., transforming growth factor-β (TGF-β or growth differentiation factor (GDF)), angiogenic factors (e.g., VEGF or platelet-derived growth factor (PDGF)), and inflammatory inhibitory factors, through coordination of the delivery modes of these factors from electrospun fibers would be the next challenge in bone tissue engineering. Additionally, many advanced electrospinning technologies capable of readily manipulating pore sizes, mechanical properties, and three-dimensional morphologies would be required to further improve the efficiency of bone tissue engineering [103].

Skin tissue engineering

Skin tissue engineering has emerged as a powerful means to promote wound healing, a sophisticated cellular process that can be achieved by coordinating hemostasis, inflammation, epithelialization, angiogenesis, and collagen synthesis [104-106]. Importantly, highly porous electrospun nanofibrous sheets, which provide suitable environments for precisely mimicking the morphology of skin ECM components [105], have been considered as robust templates for promoting skin tissue regeneration. Kim et al. developed matrix metalloproteinase (MMP)-responsive electrospun nanofibrous matrices that release plasmid DNA encoding that encodes human epidermal growth factor (hEGF) to generate skin tissue in diabetic ulcer animal models [66,85,86]. The MMP-cleavable linker was conjugated to the amine group on the linear PEI, which was present on the surface of the fibrous matrices to anchor the plasmid DNA-hEGF through electrostatic interactions. Subsequently, the exposure to the MMPs, which are inherently overexpressed in diabetic ulcers, cleaved the DNA-hEGF for subsequent localization of the gene expression in a controlled manner [66]. The resultant electrospun fiber systems were used to increase the expression levels of hEGF in primary human dermal fibroblasts (HDFs) [85], ultimately accelerating the wound healing rates in animal models [66,85]. In a different study, the suppression of MMP in diabetic animal models through treatment with siRNA-decorated nanofibrous sheets for 7 days dramatically increased neo-collagen accumulation at dorsal wound sites, which subsequently triggered improvements in the wound recovery rates [86]. Additionally, the sustained delivery of PEI/DNA complexes encoding basic fibroblast growth factor (bFGF) from core-sheath fibers that were generated using PELA/PEG blends significantly increased the efficiency of transfecting mouse embryonic fibroblasts, resulting in improved skin regeneration in dorsal wound diabetic rat models (Figure 7) [84]. The accumulation of multiple layers of keratinocyte growth factor (KGF)-encoding plasmid DNA on the electrospun PLA/PCL fibers achieved robust re-epithelialization, keratinocyte proliferation, and granulation responses, thereby inducing full-thickness wound recovery in mouse cutaneous wound models [76]. Taken together, these findings show that the spatiotemporally regulated delivery of multiple inducible factors [106] and the use of *ex vivo* stem cell transplantation [107] with electrospun fibrous structures would be a successful strategy for constructing dermal or epidermal tissue layers.

Vascular tissue engineering

Both endothelial cell lining and smooth muscle cell layers on the luminal surface of vessel architectures have been regarded as crucial design factors to explore in vascular tissue engineering [51,108]. Additionally, providing both the secretion of an angiogenic growth factor and a physical guide are of great importance in vascular regeneration. Thus, gene vector-releasing electrospun fibers with an ECM-analogue, which acts as a support for the endothelial cells to proliferate, can provide efficient physical cues for vascular tissue engineering [109]. Subcutaneous implantation of PELA fibrous scaffolds eluting dual plasmids encoding VEGF and bFGF accelerated the maturation of blood vessels compared with single plasmid DNA delivery approaches [65,83]. Nonspecific adsorption of PEI/DNA complexes encoding for VEGF for the subsequent substrate-mediated delivery of the vectors resulted in the robust expression of VEGF in H9C2 myoblast cells and triggered local angiogenesis adjacent to the electrospun fibers, demonstrating the future potential of this method for the treatment of myocardial infarction [87]. Importantly, an abnormal tissue response in conjunction with excessive vascularization may cause the failure of tissue grafts. Thus, suppressing the upregulation of angiogenic factors to reduce the intimal hyperplasia may sometimes be required to produce functional tissue grafts with suitable vascularized structures. The amine-functionalized poly(ethylene terephthalate) (PET) electrospun fiber-mediated delivery of PEI/siRNA-thrombospondin-2 (TSP-2), an anti-angiogenic matricellular protein, was designed to lead to TSP-2 knockdown in primary human aortic smooth muscle cells (AoSMCs), eventually improving the biocompatibility between implanted materials and host tissues [78]. Importantly, the sequential delivery of multiple gene vectors in a spatiotemporal manner may be appropriate to regenerate blood vessels composed of double layers, including vascular

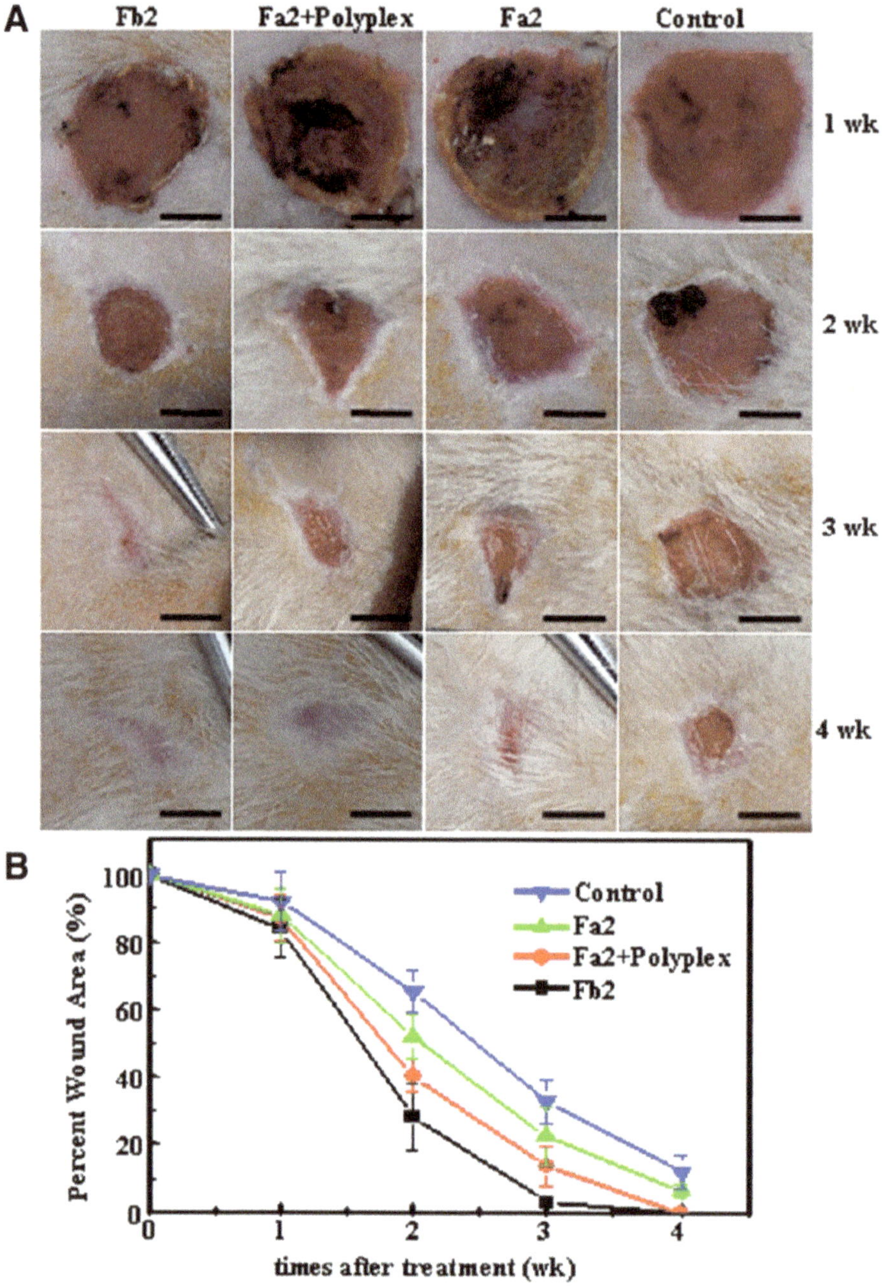

Figure 7 Improved skin regeneration by electrospun fibers with plasmid bFGF polyplex. (A) The representative images of skin wounds after treatment with Fb2 (pbFGF polyplex, blend), Fa2 (without pbFGF) and Fa2 + Polyplex (pbFGF polyplex infiltrated Fa2) fibrous mats for 1, 2, 3, and 4 weeks, using untreated wound as control. Bars represent 10 mm. **(B)** Wound areas at different time points after treatment (n = 10). Reprinted with permission from [84]. Copyright 2012 American Chemical Society.

endothelial cells in the interior layer and vascular smooth muscle cells in the outer layer [51]. Thus, mimicking the double layered structures with highly porous fibrous structures that can release multiple gene vectors capable of stimulating cells individually in each layer will be useful to maximize the efficacy of vascular grafts, as well as for vascular tissue engineering.

Neural tissue engineering

Employing neural stem cells, which have the capabilities of self-renewal as well as neuronal differentiation, and providing patterned physical guidance along with biochemical cues (e.g., growth factors or genes) have been regarded as pivotal factors to induce neural regeneration [110,111]. Thus, the feasibility of adjusting the orientation or topographical changes of electrospun fibers during the electrospinning process facilitated the use of these fibers as versatile scaffolds to guide neuronal cell growth or direct neuronal differentiation on their surfaces [112,113]. The topological effect of nanofibers along with the knockdown of REST in NPCs synergistically promoted the neuronal differentiation of NPCs on the adhesive mussel-inspired PD-coated PCL fibrous surfaces while also reducing astrocytic and oligodendrocytic differentiation (Figure 8) [88]. The physical alignment of the electrospun fibers that release neuronal inducible factors at nerve lesion sites can guide the neurite outgrowth along with the orientation; these capabilities will be required for the functional recovery of injured neurons [95,96]. Therefore, the ability to easily produce patterned structures using electrospun fibers will lead to extensive applications of these fibers for numerous incurable chronic neurodegenerative diseases.

Other applications

Electrospun fibers have been recently employed in a stem cell study or a cancer therapy approach in addition to their use in tissue engineering applications. Fei et al. fabricated a nanofiber-based sandwich electroporation (NSE) device capable of *in situ* gene transfection in mouse embryonic stem cells [114]. Unlike a conventional bulk electroporation, this NSE system did not require the removal of adherent cells from the substrate, thereby improving cellular viability and ultimately enhancing transfection efficiency. This study demonstrated that gene vector-releasing electrospun nanofibers can be potentially employed as powerful templates to elucidate stem cell behavior or biology. Meanwhile, Achille et al. utilized electrospun PCL fibers as a gene carrier for the treatment of breast cancer [115]. A plasmid encoding short hairpin RNA (shRNA) suppressing the level of the cell cycle specific protein, cyclin-dependent kinase 2 (Cdk2), was blended with PCL solutions to construct highly porous electrospun fibers. The interruption of the cell cycle by the silencing effect of the shRNA, which was gradually released from the electrospun PCL fibers, robustly led to the suppression of the proliferation of breast cancer cell lines. This study also indicates that the combination of gene delivery with electrospun nanofibers, which can create ECM-analogous environments and tunable gene delivery in a spatial and temporal manner, can provide an alternative powerful means to specifically target a variety of cancer cells.

Conclusions and challenges

Gene therapy has demonstrated increasing promise for treating a variety of human diseases, including inherited or acquired disorders, infectious diseases, tissue loss, and organ failure. The development of highly efficient gene delivery systems that can deliver a gene of interest

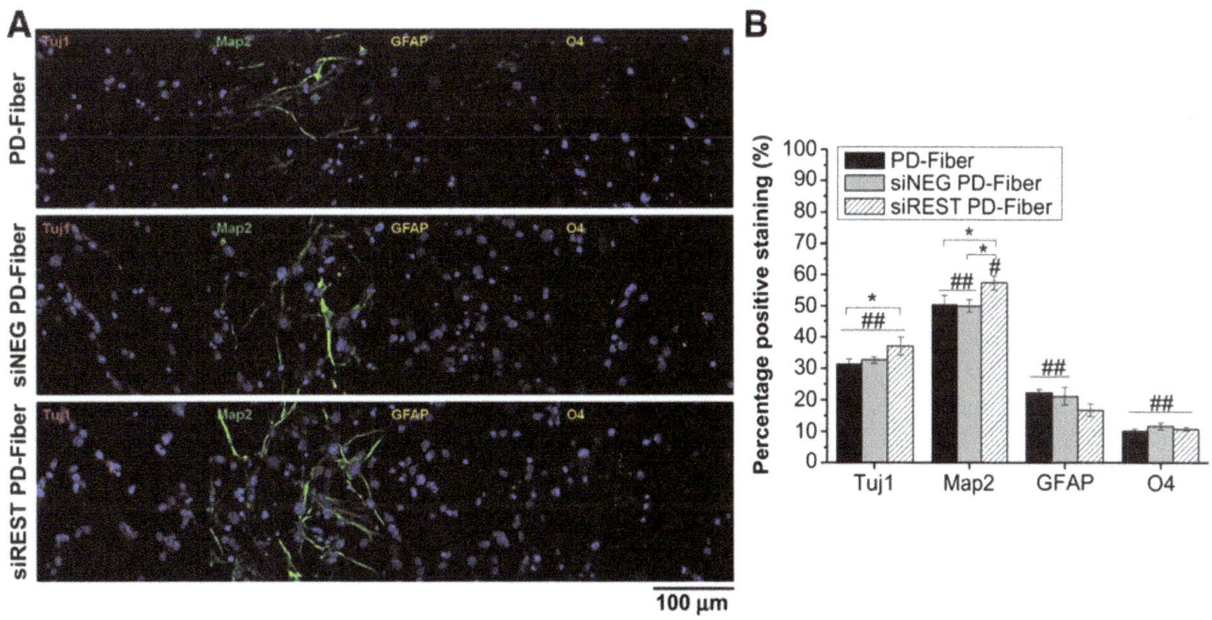

Figure 8 Immunofluorescence analysis of NPCs differentiated on plain, siNEG PD-fiber and siREST PD-fiber for 7 days. (A) Immunostaining for Tuj1 (early neuronal marker, red), Map2 (late neuronal marker, green), GFAP (glial marker, yellow) and O4 (glial marker, yellow). Nuclei were counterstained with DAPI (blue). **(B)** Quantification of immunostaining results showing percentage positive staining of various neural and glial cell markers for NPCs cultured on all PD-fiber samples. *indicates $p < 0.05$ (ANOVA). # and ## indicate $p < 0.05$ and $p < 0.01$ (ANOVA) when PD-fiber samples were compared to respective PD-film samples. Mean ± SE, n = 3 [88], Copyright 2013. Reproduced with permission from Elsevier.

safely to particular target cells has been always regarded as a large hurdle that must be cleared for the further advancement of gene delivery technologies. Electrospun fibers have highly advantageous characteristics, including ease of production, an ECM-analogue nature, a broad range of choices for materials, the feasibility of producing structures with varied physical and chemical properties, and large surface-to-volume ratios. Because of these characteristics, electrospun nanofibers have recently been highlighted as versatile and powerful templates that can be applied to numerous biomedical fields. Therefore, the integration of gene delivery with electrospun nanofibers is a highly promising strategy to improve gene delivery for a broad range of applications.

For further innovation of approaches that combine gene delivery and electrospun nanofibers, there are several critical challenges, especially improving the structural aspects of electrospun nanofibers. Electrospinning typically produces flat two-dimensional sheet-like meshes, which may not fully represent the three-dimensional extracellular environments in the body. These structural characteristics may limit the ease of application of electrospun nanofibers to many biomedical fields. Numerous studies have created three-dimensional electrospun nanofibers, typically by modifying collectors or by utilizing an extra apparatus [116]. Additionally, engaging the structural flexibility of electrospun nanofibers, which is required to precisely mimic the various shapes of tissues or organs, may be another challenge that must be addressed. Recently, clay-like moldable electrospun nanofibers were created by adjusting the electric repulsion between different materials used to produce the core-sheath layers and then selectively removing sacrificial fibers in the sheath layer [33]. Owing to their moldable clay-like properties, any desired forms, such as the human nose, a ball, or tubes, could be manually shaped. The combination of the clay-like electrospun fibers with viral gene delivery resulted in a high level of gene expression throughout the entire fibrous structure [33]. Adapting micro- and nanofabrication technologies to generate electrospun nanofibrous structures to mimic the spatial and temporal control of the expression of multiple genes in tissues or organs is another challenge that must be addressed. Electrospun nanofibers are powerful candidates for mediators of the spatio-temporal delivery modes of multiple gene vectors because their structure makes it possible to generate patterned fibers and core-sheath structures. However, the mechanical strength of electrospun fibers, which is inherently lower than that of existing polymeric scaffolds, must be reinforced to resist cellular contractile forces upon implantation and to maintain the structural integrity of the scaffolds. Finally, the development of non-invasive electrospun nanofibers, such as injectable formulations, would contribute to the increasing promise of fibrous systems as spatial templates. Overall, further innovation in improving the structures of electrospun nanofibers will be pivotal to extend their use to a variety of biomedical applications, especially gene delivery applications.

Abbreviations

AAV: Adeno-associated virus; Ad: Adenovirus; AoSMC: Aortic smooth muscle cell; bFGF: Basic fibroblast growth factor; bFGF: Basic fibroblast growth factor; BMP-2: Bone morphogenetic protein-2; Cdk2: Cyclin-dependent kinase 2; CMV: Cytomegalovirus; ECM: Extracellular matrix; EEP: Ethyl ethylene phosphate; ELP: Elastin-like polypeptides; GDF: Growth differentiation factor; GFP: Green fluorescent protein; HAp: Hydroxyapatite; HA-PEI: PEI conjugated with hyaluronic acids; hBMSC: Human bone-marrow-derived mesenchymal stem cell; HDF: Human dermal fibroblasts; hEGF: Human epidermal growth factor; HFP: Hexafluoro-2-propanol; KGF: Keratinocyte growth factor; LBL: Layer-by-layer; MMP: Matrix metalloproteinase; NPC: Neural progenitor cell; NSE: Nanofibers-based sandwich electroporation; PBAE: Polycationic poly(β-amino ester); PCL: Poly(caprolactone); PCLEEP: Poly(caprolactone-co-ethylethylene phosphate); PD: Polydopamine; PDGF: Platelet-derived growth factor; PDLLA: Poly(D,L-lactide); PEG: Poly(ethylene glycol); PEI: Poly(ethylenimine); PELA: Poly(D,L-lactide)-poly(ethylene glycol); PET: Poly(ethylene terephthalate); PLA: Poly(lactic) acid; PLCL: Poly(L-lactide-co-ε-caprolactone); PLGA: Poly(D,L-lactide-co-glycolide); REST: RE-1 silencing transcription factor; siNEG: Negative siRNA; RUNX2: Runt-related transcription factor 2; shRNA: Short hairpin RNA; siRNA: Small interfering RNA; TGF-β: Transforming growth factor-β; TSP-2: Thrombospondin-2; VEGF: Vascular endothelial growth factor.

Competing interests

The authors declare that they have no competing interests.

Authors' contributions

SL, GJ, and JHJ drafted the manuscript. SL participated in the review of advantageous features of electrospun nanofibers for controlled and efficient gene delivery. GJ participated in the overview of current biomedical applications of gene delivery approaches combined with electrospun nanofiber systems. JHJ coordinated all sections, revised the manuscript critically, and finalized the manuscript. All authors read and approved the final manuscript.

Acknowledgments

This work was supported by the National Research Foundation of Korea (NRF) grant through the Active Polymer Center for Pattern Integration (NRF-2007-0056558), Basic Science Research Program (NRF-2012R1A1A1003397), and Bio & Medical Technology Development Program (NRF-2013M3A9D3046431) funded by the Ministry of Science, ICT & Future Planning (MSIP). This research was also supported by a grant of the Korea Health Technology R&D Project through the Korea Health Industry Development Institute (KHIDI), funded by the Ministry of Health & Welfare, Republic of Korea (HI14C1564).

References

1. Maitra A, Arking DE, Shivapurkar N, Ikeda M, Stastny V, Kassauei K, Sui G, Cutler DJ, Liu Y, Brimble SN, Noaksson K, Hyllner J, Schulz TC, Zeng X, Freed WJ, Crook J, Abraham S, Colman A, Sartipy P, Matsui S, Carpenter M, Gazdar AF, Rao M, Chakravarti A: **Genomic alterations in cultured human embryonic stem cells.** *Nat Genet* 2005, **37:**1099–1103.
2. Lai K, Kaspar BK, Gage FH, Schaffer DV: **Sonic hedgehog regulates adult neural progenitor proliferation in vitro and in vivo.** *Nat Neurosci* 2003, **6:**21–27.
3. Gropp M, Itsykson P, Singer O, Ben-Hur T, Reinhartz E, Galun E, Reubinoff BE: **Stable genetic modification of human embryonic stem cells by lentiviral vectors.** *Mol Ther* 2003, **7:**281–287.
4. Walther W, Schlag PM: **Current status of gene therapy for cancer.** *Curr Opin Oncol* 2013, **25:**659–664.

5. Rejiba S, Bigand C, Parmentier C, Masmoudi A, Hajri A: **Oncosuppressive suicide gene virotherapy "PVH1-yCD/5-FC" for pancreatic peritoneal carcinomatosis treatment: NFkappaB and Akt/PI3K involvement.** *PLoS ONE* 2013, **8:**e70594.
6. Ochiya T, Nagahara S, Sano A, Itoh H, Terada M: **Biomaterials for gene delivery: atelocollagen-mediated controlled release of molecular medicines.** *Curr Gene Ther* 2001, **1:**31–52.
7. Jang JH, Houchin TL, Shea LD: **Gene delivery from polymer scaffolds for tissue engineering.** *Expert Rev Med Devices* 2004, **1:**127–138.
8. Shea LD, Smiley E, Bonadio J, Mooney DJ: **DNA delivery from polymer matrices for tissue engineering.** *Nat Biotechnol* 1999, **17:**551–554.
9. Yates F, Daley GQ: **Progress and prospects: gene transfer into embryonic stem cells.** *Gene Ther* 2006, **13:**1431–1439.
10. Marshall E: **Genome sequencing. Claim and counterclaim on the human genome.** *Science* 2000, **288:**242–243.
11. Sung SJ, Min SH, Cho KY, Lee S, Min YJ, Yeom YI, Park JK: **Effect of polyethylene glycol on gene delivery of polyethylenimine.** *Biol Pharm Bull* 2003, **26:**492–500.
12. Dey D, Inayathullah M, Lee AS, LeMieux MC, Zhang X, Wu Y, Nag D, De Almeida PE, Han L, Rajadas J, Wu JC: **Efficient gene delivery of primary human cells using peptide linked polyethylenimine polymer hybrid.** *Biomaterials* 2011, **32:**4647–4658.
13. Jang JH, Schaffer DV, Shea LD: **Engineering biomaterial systems to enhance viral vector gene delivery.** *Mol Ther* 2011, **19:**1407–1415.
14. Selkirk SM: **Gene therapy in clinical medicine.** *Postgrad Med J* 2004, **80:**560–570.
15. Jang JH, Lim KI, Schaffer DV: **Library selection and directed evolution approaches to engineering targeted viral vectors.** *Biotechnol Bioeng* 2007, **98:**515–524.
16. Pannier AK, Shea LD: **Controlled release systems for DNA delivery.** *Mol Ther* 2004, **10:**19–26.
17. Jang JH, Bengali Z, Houchin TL, Shea LD: **Surface adsorption of DNA to tissue engineering scaffolds for efficient gene delivery.** *J Biomed Mater Res A* 2006, **77:**50–58.
18. Jang JH, Rives CB, Shea LD: **Plasmid delivery in vivo from porous tissue-engineering scaffolds: transgene expression and cellular transfection.** *Mol Ther* 2005, **12:**475–483.
19. Keeney M, van den Beucken JJ, van der Kraan PM, Jansen JA, Pandit A: **The ability of a collagen/calcium phosphate scaffold to act as its own vector for gene delivery and to promote bone formation via transfection with VEGF(165).** *Biomaterials* 2010, **31:**2893–2902.
20. Cohen JA, Beaudette TT, Cohen JL, Broaders KE, Bachelder EM, Frechet JM: **Acetal-modified dextran microparticles with controlled degradation kinetics and surface functionality for gene delivery in phagocytic and non-phagocytic cells.** *Adv Mater* 2010, **22:**3593–3597.
21. Zhang L, Sinclair A, Cao Z, Ella-Menye JR, Xu X, Carr LR, Pun SH, Jiang S: **Hydrolytic cationic ester microparticles for highly efficient DNA vaccine delivery.** *Small* 2013, **9:**3439–3444.
22. Chong RH, Gonzalez-Gonzalez E, Lara MF, Speaker TJ, Contag CH, Kaspar RL, Coulman SA, Hargest R, Birchall JC: **Gene silencing following siRNA delivery to skin via coated steel microneedles: In vitro and in vivo proof-of-concept.** *J Control Release* 2013, **166:**211–219.
23. DeMuth PC, Su X, Samuel RE, Hammond PT, Irvine DJ: **Nano-layered microneedles for transcutaneous delivery of polymer nanoparticles and plasmid DNA.** *Adv Mater* 2010, **22:**4851–4856.
24. Sill TJ, von Recum HA: **Electrospinning: applications in drug delivery and tissue engineering.** *Biomaterials* 2008, **29:**1989–2006.
25. Agarwal S, Wendorff JH, Greiner A: **Progress in the field of electrospinning for tissue engineering applications.** *Adv Mater* 2009, **21:**3343–3351.
26. Williams GR, Chatterton NP, Nazir T, Yu DG, Zhu LM, Branford-White CJ: **Electrospun nanofibers in drug delivery: recent developments and perspectives.** *Ther Deliv* 2012, **3:**515–533.
27. Papkov D, Zou Y, Andalib MN, Goponenko A, Cheng SZD, Dzenis YA: **Simultaneously strong and tough ultrafine continuous nanofibers.** *ACS Nano* 2013, **7:**3324–3331.
28. Pham QP, Sharma U, Mikos AG: **Electrospun Poly(ε-caprolactone) microfiber and multilayer Nanofiber/Microfiber scaffolds: characterization of scaffolds and measurement of cellular infiltration.** *Biomacromolecules* 2006, **7:**2796–2805.
29. Kwon IK, Kidoaki S, Matsuda T: **Electrospun nano- to microfiber fabrics made of biodegradable copolyesters: structural characteristics, mechanical properties and cell adhesion potential.** *Biomaterials* 2005, **26:**3929–3939.
30. Liao IC, Chen S, Liu JB, Leong KW: **Sustained viral gene delivery through core-shell fibers.** *J Control Release* 2009, **139:**48–55.
31. Zander N: **Hierarchically structured electrospun fibers.** *Polymers* 2013, **5:**19–44.
32. Yao J, Bastiaansen C, Peijs T: **High strength and high modulus electrospun nanofibers.** *Fibers* 2014, **2:**158–186.
33. Lee S, Cho S, Kim M, Jin G, Jeong U, Jang JH: **Highly moldable electrospun clay-like fluffy nanofibers for three-dimensional scaffolds.** *ACS Appl Mater Interfaces* 2014, **6:**1082–1091.
34. Li D, Wang Y, Xia Y: **Electrospinning nanofibers as uniaxially aligned arrays and layer-by-layer stacked films.** *Adv Mater* 2004, **16:**361–366.
35. Lu LX, Wang YY, Mao X, Xiao ZD, Huang NP: **The effects of PHBV electrospun fibers with different diameters and orientations on growth behavior of bone-marrow-derived mesenchymal stem cells.** *Biomed Mater* 2012, **7:**015002.
36. Xie J, MacEwan MR, Schwartz AG, Xia Y: **Electrospun nanofibers for neural tissue engineering.** *Nanoscale* 2010, **2:**35–44.
37. Jin G, Lee S, Kim SH, Kim M, Jang JH: **Bicomponent electrospinning to fabricate three-dimensional hydrogel-hybrid nanofibrous scaffolds with spatial fiber tortuosity.** *Biomed Microdevices* 2014, **16:**793–804.
38. Xiong Y, Mayers BT, Xia Y: **Some recent developments in the chemical synthesis of inorganic nanotubes.** *Chem Commun (Camb)* 2005, **40:**5013–5022.
39. Qu H, Wei S, Guo Z: **Coaxial electrospun nanostructures and their applications.** *J Mater Chem A Mater Engergy Sustain* 2013, **1:**11513.
40. Wu H, Hu L, Rowell MW, Kong D, Cha JJ, McDonough JR, Zhu J, Yang Y, McGehee MD, Cui Y: **Electrospun metal nanofiber webs as high-performance transparent electrode.** *Nano Lett* 2010, **10:**4242–4248.
41. Hansen NS, Cho D, Joo YL: **Metal nanofibers with highly tunable electrical and magnetic properties via highly loaded water-based electrospinning.** *Small* 2012, **8:**1510–1514.
42. Li D, Xia Y: **Direct fabrication of composite and ceramic hollow nanofibers by electrospinning.** *Nano Lett* 2004, **4:**933–938.
43. Li D, Wang Y, Xia Y: **Electrospinning of polymeric and ceramic nanofibers as uniaxially aligned arrays.** *Nano Lett* 2003, **3:**1167–1171.
44. Ganesh VA, Dinachali SS, Raut HK, Walsh TM, Nair AS, Ramakrishna S: **Electrospun SiO2 nanofibers as a template to fabricate a robust and transparent superamphiphobic coating.** *RSC Advances* 2013, **3:**3819.
45. Li C, Tzeng SY, Tellier LE, Green JJ: **(3-aminopropyl)-4-methylpiperazine end-capped poly(1,4-butanediol diacrylate-co-4-amino-1-butanol)-based multilayer films for gene delivery.** *ACS Appl Mater Interfaces* 2013, **5:**5947–5953.
46. Sakai S, Yamada Y, Yamaguchi T, Ciach T, Kawakami K: **Surface immobilization of poly(ethyleneimine) and plasmid DNA on electrospun poly(L-lactic acid) fibrous mats using a layer-by-layer approach for gene delivery.** *J Biomed Mater Res A* 2009, **88:**281–287.
47. Luu YK, Kim K, Hsiao BS, Chu B, Hadjiargyrou M: **Development of a nanostructured DNA delivery scaffold via electrospinning of PLGA and PLA–PEG block copolymers.** *J Control Release* 2003, **89:**341–353.
48. Lee S, Kim JS, Chu HS, Kim GW, Won JI, Jang JH: **Electrospun nanofibrous scaffolds for controlled release of adeno-associated viral vectors.** *Acta Biomater* 2011, **7:**3868–3876.
49. Kim JS, Chu HS, Park KI, Won JI, Jang JH: **Elastin-like polypeptide matrices for enhancing adeno-associated virus-mediated gene delivery to human neural stem cells.** *Gene Ther* 2012, **19:**329–337.
50. Lee S-W, Belcher AM: **Virus-based fabrication of micro- and nanofibers using electrospinning.** *Nano Lett* 2004, **4:**387–390.
51. Zhang H, Jia X, Han F, Zhao J, Zhao Y, Fan Y, Yuan X: **Dual-delivery of VEGF and PDGF by double-layered electrospun membranes for blood vessel regeneration.** *Biomaterials* 2013, **34:**2202–2212.
52. Seo DK, Jeun JP, Kim HB, Kang PH: **Preparation and characterization of the carbon nanofiber mat produced from electrospun PAN/lignin precursors by electron beam irradiation.** *Rev Adv Mater Sci* 2011, **28:**31–34.
53. Chen JP, Su CH: **Surface modification of electrospun PLLA nanofibers by plasma treatment and cationized gelatin immobilization for cartilage tissue engineering.** *Acta Biomater* 2011, **7:**234–243.
54. Martins A, Pinho ED, Faria S, Pashkuleva I, Marques AP, Reis RL, Neves NM: **Surface modification of electrospun polycaprolactone nanofiber meshes**

by plasma treatment to enhance biological performance. *Small* 2009, **5**:1195–1206.
55. Saraf A, Baggett LS, Raphael RM, Kasper FK, Mikos AG: **Regulated non-viral gene delivery from coaxial electrospun fiber mesh scaffolds.** *J Control Release* 2010, **143**:95–103.
56. Nie H, Wang CH: **Fabrication and characterization of PLGA/HAp composite scaffolds for delivery of BMP-2 plasmid DNA.** *J Control Release* 2007, **120**:111–121.
57. Liang D, Luu YK, Kim K, Hsiao BS, Hadjiargyrou M, Chu B: **In vitro non-viral gene delivery with nanofibrous scaffolds.** *Nucleic Acids Res* 2005, **33**:e170.
58. Rujitanaroj PO, Jao B, Yang J, Wang F, Anderson JM, Wang J, Chew SY: **Controlling fibrous capsule formation through long-term down-regulation of collagen type I (COL1A1) expression by nanofiber-mediated siRNA gene silencing.** *Acta Biomater* 2013, **9**:4513–4524.
59. Zhang J, Duan Y, Wei D, Wang L, Wang H, Gu Z, Kong D: **Co-electrospun fibrous scaffold-adsorbed DNA for substrate-mediated gene delivery.** *J Biomed Mater Res A* 2011, **96**:212–220.
60. Kim E, Song IT, Lee S, Kim JS, Lee H, Jang JH: **Drawing sticky adeno-associated viruses on surfaces for spatially patterned gene expression.** *Angew Chem Int Ed* 2012, **51**:5598–5601.
61. Burger C, Chu B: **Functional nanofibrous scaffolds for bone reconstruction.** *Colloids Surf B: Biointerfaces* 2007, **56**:134–141.
62. Ji W, Sun Y, Yang F, van den Beucken JJ, Fan M, Chen Z, Jansen JA: **Bioactive electrospun scaffolds delivering growth factors and genes for tissue engineering applications.** *Pharm Res* 2011, **28**:1259–1272.
63. Langer R: **New methods of drug delivery.** *Science* 1990, **249**:1527–1533.
64. Yang Y, Li X, Cheng L, He S, Zou J, Chen F, Zhang Z: **Core-sheath structured fibers with pDNA polyplex loadings for the optimal release profile and transfection efficiency as potential tissue engineering scaffolds.** *Acta Biomater* 2011, **7**:2533–2543.
65. He S, Xia T, Wang H, Wei L, Luo X, Li X: **Multiple release of polyplexes of plasmids VEGF and bFGF from electrospun fibrous scaffolds towards regeneration of mature blood vessels.** *Acta Biomater* 2012, **8**:2659–2669.
66. Kim HS, Yoo HS: **MMPs-responsive release of DNA from electrospun nanofibrous matrix for local gene therapy: in vitro and in vivo evaluation.** *J Control Release* 2010, **145**:264–271.
67. Chen M, Gao S, Dong M, Song J, Yang C, Howard KA, Kjems J, Besenbacher F: **Chitosan/siRNA nanoparticles encapsulated in PLGA nanofibers for siRNA delivery.** *ACS Nano* 2012, **6**:4835–4844.
68. Cao H, Jiang X, Chai C, Chew SY: **RNA interference by nanofiber-based siRNA delivery system.** *J Control Release* 2010, **144**:203–212.
69. Rujitanaroj PO, Wang YC, Wang J, Chew SY: **Nanofiber-mediated controlled release of siRNA complexes for long term gene-silencing applications.** *Biomaterials* 2011, **32**:5915–5923.
70. Yan S, Xiaoqiang L, Shuiping L, Xiumei M, Ramakrishna S: **Controlled release of dual drugs from emulsion electrospun nanofibrous mats.** *Colloids Surf B: Biointerfaces* 2009, **73**:376–381.
71. Su Y, Su Q, Liu W, Lim M, Venugopal JR, Mo X, Ramakrishna S, Al-Deyab SS, El-Newehy M: **Controlled release of bone morphogenetic protein 2 and dexamethasone loaded in core–shell PLLACL–collagen fibers for use in bone tissue engineering.** *Acta Biomater* 2012, **8**:763–771.
72. Eto Y, Yoshioka Y, Mukai Y, Okada N, Nakagawa S: **Development of PEGylated adenovirus vector with targeting ligand.** *Int J Pharm* 2008, **354**:3–8.
73. Allison SD, Molina MC, Anchordoquy TJ: **Stabilization of lipid/DNA complexes during the freezing step of the lyophilization process: the particle isolation hypothesis.** *Biochim Biophys Acta* 2000, **1468**:127–138.
74. Kasper JC, Pikal MJ, Friess W: **Investigations on polyplex stability during the freezing step of lyophilization using controlled ice nucleation–the importance of residence time in the low-viscosity fluid state.** *J Pharm Sci* 2013, **102**:929–946.
75. Segura T, Volk MJ, Shea LD: **Substrate-mediated DNA delivery: role of the cationic polymer structure and extent of modification.** *J Control Release* 2003, **93**:69–84.
76. Kobsa S, Kristofik NJ, Sawyer AJ, Bothwell AL, Kyriakides TR, Saltzman WM: **An electrospun scaffold integrating nucleic acid delivery for treatment of full-thickness wounds.** *Biomaterials* 2013, **34**:3891–3901.
77. Monteiro N, Ribeiro D, Martins A, Faria S, Fonseca NA, Moreira JN, Reis RL, Neves NM: **Instructive nanofibrous scaffold comprising runt-related transcription factor 2 gene delivery for bone tissue engineering.** *ACS Nano* 2014, **8**:8082–8094.
78. Nabzdyk CS, Chun MC, Oliver-Allen HS, Pathan SG, Phaneuf MD, You JO, Pradhan-Nabzdyk LK, LoGerfo FW: **Gene silencing in human aortic smooth muscle cells induced by PEI-siRNA complexes released from dip-coated electrospun poly(ethylene terephthalate) grafts.** *Biomaterials* 2014, **35**:3071–3079.
79. Bengali Z, Pannier AK, Segura T, Anderson BC, Jang JH, Mustoe TA, Shea LD: **Gene delivery through cell culture substrate adsorbed DNA complexes.** *Biotechnol Bioeng* 2005, **90**:290–302.
80. Liang D, Hsiao BS, Hadjiargyrou M, Chu B: **Scaffolds with encapsulated DNA for non-viral gene delivery.** *J Non Cryst Solids* 2006, **352**:4394–4399.
81. Okuda T, Tahara Y, Kamiya N, Goto M, Kidoaki S: **S/O-nanodispersion electrospun fiber mesh effective for sustained release of healthy plasmid DNA with the structural and functional integrity.** *J Biomater Sci Polym Ed* 2013, **24**:1277–1290.
82. Nie H, Ho ML, Wang CK, Wang CH, Fu YC: **BMP-2 plasmid loaded PLGA/HAp composite scaffolds for treatment of bone defects in nude mice.** *Biomaterials* 2009, **30**:892–901.
83. Chen F, Wan H, Xia T, Guo X, Wang H, Liu Y, Li X: **Promoted regeneration of mature blood vessels by electrospun fibers with loaded multiple pDNA-calcium phosphate nanoparticles.** *Eur J Pharm Biopharm* 2013, **85**:699–710.
84. Yang Y, Xia T, Chen F, Wei W, Liu C, He S, Li X: **Electrospun fibers with plasmid bFGF polyplex loadings promote skin wound healing in diabetic rats.** *Mol Pharm* 2012, **9**:48–58.
85. Kim HS, Yoo HS: **In vitro and in vivo epidermal growth factor gene therapy for diabetic ulcers with electrospun fibrous meshes.** *Acta Biomater* 2013, **9**:7371–7380.
86. Kim HS, Yoo HS: **Matrix metalloproteinase-inspired suicidal treatments of diabetic ulcers with siRNA-decorated nanofibrous meshes.** *Gene Ther* 2013, **20**:378–385.
87. Che H-L, Muthiah M, Ahn Y, Son S, Kim WJ, Seonwoo H, Chung JH, Cho C-S, Park I-K: **Biodegradable particulate delivery of vascular endothelial growth factor plasmid from Polycaprolactone/Polyethylenimine electrospun nanofibers for the treatment of myocardial infarction.** *J Nanosci Nanotechnol* 2011, **11**:7073–7077.
88. Low WC, Rujitanaroj PO, Lee DK, Messersmith PB, Stanton LW, Goh E, Chew SY: **Nanofibrous scaffold-mediated REST knockdown to enhance neuronal differentiation of stem cells.** *Biomaterials* 2013, **34**:3581–3590.
89. Ren Y-J, Zhang S, Mi R, Liu Q, Zeng X, Rao M, Hoke A, Mao H-Q: **Enhanced differentiation of human neural crest stem cells towards the Schwann cell lineage by aligned electrospun fiber matrix.** *Acta Biomater* 2013, **9**:7727–7736.
90. Yang F, Murugan R, Wang S, Ramakrishna S: **Electrospinning of nano/micro scale poly(l-lactic acid) aligned fibers and their potential in neural tissue engineering.** *Biomaterials* 2005, **26**:2603–2610.
91. Zhao W, Ju YM, Christ G, Atala A, Yoo JJ, Lee SJ: **Diaphragmatic muscle reconstruction with an aligned electrospun poly(ε-caprolactone)/collagen hybrid scaffold.** *Biomaterials* 2013, **34**:8235–8240.
92. Wang HB, Mullins ME, Cregg JM, Hurtado A, Oudega M, Trombley MT, Gilbert RJ: **Creation of highly aligned electrospun poly-L-lactic acid fibers for nerve regeneration applications.** *J Neural Eng* 2009, **6**:016001.
93. Vaquette C, Cooper-White JJ: **Increasing electrospun scaffold pore size with tailored collectors for improved cell penetration.** *Acta Biomater* 2011, **7**:2544–2557.
94. Liu Y, Zhang X, Xia Y, Yang H: **Magnetic-field-assisted electrospinning of aligned straight and wavy polymeric nanofibers.** *Adv Mater* 2010, **22**:2454–2457.
95. De Laporte L, Huang A, Ducommun MM, Zelivyanska ML, Aviles MO, Adler AF, Shea LD: **Patterned transgene expression in multiple-channel bridges after spinal cord injury.** *Acta Biomater* 2010, **6**:2889–2897.
96. Houchin-Ray T, Huang A, West ER, Zelivyanskaya M, Shea LD: **Spatially patterned gene expression for guided neurite extension.** *J Neurosci Res* 2009, **87**:844–856.
97. Houchin-Ray T, Whittlesey KJ, Shea LD: **Spatially patterned gene delivery for localized neuron survival and neurite extension.** *Mol Ther* 2007, **15**:705–712.
98. Zou B, Liu Y, Luo X, Chen F, Guo X, Li X: **Electrospun fibrous scaffolds with continuous gradations in mineral contents and biological cues for manipulating cellular behaviors.** *Acta Biomater* 2012, **8**:1576–1585.
99. Kim E, Lee S, Hong S, Jin G, Kim M, Park KI, Lee H, Jang JH: **Sticky "delivering-from" strategies using viral vectors for efficient human neural**

stem cell infection by bioinspired catecholamines. *ACS Appl Mater Interfaces* 2014, **6**:8288–8294.

100. Bonadio J, Smiley E, Patil P, Goldstein S: Localized, direct plasmid gene delivery in vivo: prolonged therapy results in reproducible tissue regeneration. *Nat Med* 1999, **5**:753–759.
101. Shin M, Yoshimoto H, Vacanti JP: In vivo bone tissue engineering using mesenchymal stem cells on a novel electrospun nanofibrous scaffold. *Tissue Eng* 2004, **10**:33–41.
102. Vo TN, Kasper FK, Mikos AG: Strategies for controlled delivery of growth factors and cells for bone regeneration. *Adv Drug Deliv Rev* 2012, **64**:1292–1309.
103. Yoshimoto H, Shin YM, Terai H, Vacanti JP: A biodegradable nanofiber scaffold by electrospinning and its potential for bone tissue engineering. *Biomaterials* 2003, **24**:2077–2082.
104. Ma L: Collagen/chitosan porous scaffolds with improved biostability for skin tissue engineering. *Biomaterials* 2003, **24**:4833–4841.
105. Kumbar SG, Nukavarapu SP, James R, Nair LS, Laurencin CT: Electrospun poly(lactic acid-co-glycolic acid) scaffolds for skin tissue engineering. *Biomaterials* 2008, **29**:4100–4107.
106. Peng L-H, Tsang S-Y, Tabata Y, Gao J-Q: Genetically-manipulated adult stem cells as therapeutic agents and gene delivery vehicle for wound repair and regeneration. *J Control Release* 2012, **157**:321–330.
107. Bi H, Jin Y: Current progress of skin tissue engineering: Seed cells, bioscaffolds, and construction strategies. *Burns Trauma* 2013, **1**:63–72.
108. Cleary MA, Geiger E, Grady C, Best C, Naito Y, Breuer C: Vascular tissue engineering: the next generation. *Trends Mol Med* 2012, **18**:394–404.
109. Ku SH, Park CB: Human endothelial cell growth on mussel-inspired nanofiber scaffold for vascular tissue engineering. *Biomaterials* 2010, **31**:9431–9437.
110. Horne MK, Nisbet DR, Forsythe JS, Parish CL: Three-dimensional nanofibrous scaffolds incorporating immobilized BDNF promote proliferation and differentiation of cortical neural stem cells. *Stem Cells Dev* 2010, **19**:843–852.
111. Mahairaki V, Lim SH, Christopherson GT, Xu L, Nasonkin I, Yu C, Mao HQ, Koliatsos VE: Nanofiber matrices promote the neuronal differentiation of human embryonic stem cell-derived neural precursors in vitro. *Tissue Eng Part A* 2011, **17**:855–863.
112. Cao H, Liu T, Chew SY: The application of nanofibrous scaffolds in neural tissue engineering. *Adv Drug Deliv Rev* 2009, **61**:1055–1064.
113. Ghasemi-Mobarakeh L, Prabhakaran MP, Morshed M, Nasr-Esfahani MH, Ramakrishna S: Electrospun poly(epsilon-caprolactone)/gelatin nanofibrous scaffolds for nerve tissue engineering. *Biomaterials* 2008, **29**:4532–4539.
114. Fei Z, Wu Y, Sharma S, Gallego-Perez D, Higuita-Castro N, Hansford D, Lannutti JJ, Lee LJ: Gene delivery to cultured embryonic stem cells using nanofiber-based sandwich electroporation. *Anal Chem* 2013, **85**:1401–1407.
115. Achille C, Sundaresh S, Chu B, Hadjiargyrou M: Cdk2 silencing via a DNA/PCL electrospun scaffold suppresses proliferation and increases death of breast cancer cells. *PLoS ONE* 2012, **7**:e52356.
116. Sun B, Long YZ, Zhang HD, Li MM, Duvail JL, Jiang XY, Yin HL: Advances in three-dimensional nanofibrous macrostructures via electrospinning. *Prog Polym Sci* 2014, **39**:862–890.

10

Endometrial stem cells in regenerative medicine

Javad Verdi[1,2], Aaron Tan[1,3], Alireza Shoae-Hassani[2] and Alexander M Seifalian[1,4*]

Abstract

First described in 2004, endometrial stem cells (EnSCs) are adult stem cells isolated from the endometrial tissue. EnSCs comprise of a population of epithelial stem cells, mesenchymal stem cells, and side population stem cells. When secreted in the menstrual blood, they are termed menstrual stem cells or endometrial regenerative cells. Mounting evidence suggests that EnSCs can be utilized in regenerative medicine. EnSCs can be used as immuno-modulatory agents to attenuate inflammation, are implicated in angiogenesis and vascularization during tissue regeneration, and can also be reprogrammed into induced pluripotent stem cells. Furthermore, EnSCs can be used in tissue engineering applications and there are several clinical trials currently in place to ascertain the therapeutic potential of EnSCs. This review highlights the progress made in EnSC research, describing their mesodermal, ectodermal, and endodermal potentials both *in vitro* and *in vivo*.

Keywords: Endometrial stem cells, Menstrual blood stem cells, Endometrial regenerative cells, Endometrium, Regenerative medicine

Introduction

Cells in the earliest developmental stages in the embryo can generate embryonic and extra-embryonic tissues [1]. The ability to generate other cell types is known as potency. In the inner cell mass of the blastocyst in the embryo, the cells pluripotent, meaning they can give rise to the ectoderm, mesoderm, and endoderm lineages (entire organism). Pluripotent stem cells include embryonic stem cells (ESCs) from the inner cell mass of blastocysts, epiblast derived stem cells from embryos after implantation, embryonic germ cells (EGCs) from primordial germ cells, embryonic carcinoma cells (ECCs) derived from germ cells tumors, and germ line stem cells from testicular tissue [2-4]. As the reproductive cycle progresses, cell potency decreases and only a small fraction of cells retain their potency, namely adult stem and progenitor cells. These are limited to tissue generation within specific lineages. Adult stem cells (ASCs) are found in numerous human tissues including: intestines [5], muscles [6], skin [7], blood [8], nervous system [9-11], endometrium [12], heart, liver [13,14], dental pulp, adipose tissue, synovial membrane, umbilical cord blood, amniotic fluid [15,16], and the endometrium [17]. In comparison to pluripotent stem cells, ASCs are considered safer for therapeutic purposes and several are currently used in clinical trials. The concept of using ASCs over embryonic stem (ES) cells in regenerative medicine has recently gained traction. The rationale for this trend include: (1) The complicated control of the culture conditions of ESCs; (2) the existence of several intermediate stages before reaching terminal differentiation of ESCs; (3) Teratoma formation is a major obstacle for clinical development of ESCs; (4) immunological rejection of cells derived from ES cells to the recipient and (5) ethical scrutiny involved in ESCs application.

* Correspondence: a.seifalian@ucl.ac.uk
[1]Centre for Nanotechnology and Regenerative Medicine, UCL Division of Surgery & Interventional Science, University College London (UCL), London NW3 2QG, UK
[4]Royal Free London NHS Foundation Trust Hospital, London, UK
Full list of author information is available at the end of the article

Endometrium and endometrial stem cells (EnSCs)

The human endometrium is an extraordinary model of controlled tissue remodeling, unparalleled in other organs, which grows about 7 mm within one week in every menstrual cycle [18]. At this rate, there is a very rapid rate of angiogenesis for approximately 500 cycles within a tightly controlled manner in a woman's lifetime. The human endometrium, derived from the mucosal lining of the fused paramesonephric tubes during embryogenesis, is a dynamic tissue (Figure 1). It is comprised of two major zones: (1) the functionalis, a transient layer containing glands extending from the surface epithelium as well as the supportive stroma, and (2) the basalis, comprised of the basal region of the glands, stroma,

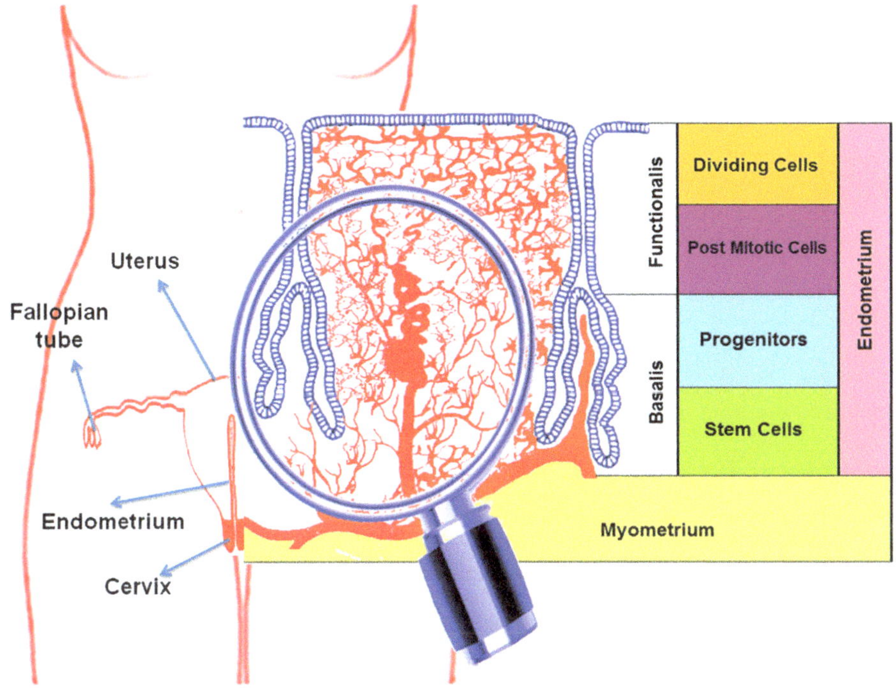

Figure 1 The endometrium. Location of endometrial stem cells in human endometrium. In human endometrium, epithelial progenitor cells are located in the base of the glands in the basalis, while endometrial stem cells are located near blood vessels even in functionalis.

supporting vasculature, and lymphoid aggregates. Progenitor cells are located in the basal layer (Figure 1) of human endometrium [19]. These rapidly proliferating cells (transient cells) then move to the functional layer [20] and actively participate in the regeneration and remodeling of the endometrium. Endometrial mesenchymal stem cells (MSCs) have been identified in the endometrium (lining of the uterus). Indeed, cloning of human endometrial cells confirmed the existence of endometrial epithelial progenitor cells for the first time in 2004 [21]. There are three distinct endometrial stem cells including: epithelial progenitor cells, MSCs and endothelial progenitor cells [22]. Recent studies reported the isolation of multipotent EnSCs from menstrual blood [23,24] or endometrial biopsies [25]. These cells are called endometrial regenerative cells (ERCs), and are heterogenic and have morphological differences among their different isolation sites. Stromal cells have been cultured from menstrual blood [24,26], suggesting that endometrial MSCs are shed during menstruation [22] but epithelial cells have not been detected [27], suggesting that epithelial progenitors may not be shed during menstruation and more likely reside in the basalis. During the proliferative menstrual phase, the endometrium prepares for recruiting new EnSCs to cover the new blood vessels. The variation in the numbers of endothelial progenitor cells and stem cell derived factor-1 is a pivotal factor for release and homing of stem cells [28]. These EnSCs have properties similar to bone marrow or adipose tissue stem cells. They proliferate and differentiate throughout the estrous cycle and during pregnancy under the control of the ovarian hormones. Clonogenic MSCs are responsible for regenerating the endometrium, and they exist in peri-menopausal women, postmenopausal women, and women on oral contraceptives [29]. Irregular function of these SCs may contribute to the pathogenesis of endometriosis and the growth of tissue outside the uterine cavity, causing dysmenorrhoea, subfertility and endometrial carcinoma [30]. Stem cells that are ectopically distributed through lymphovascular metastasis [31] may also contribute to the pathogenetic process, because their high proliferation promotes rapid clonal expansion [32].

The progenitor cells in the endometrium has a high proliferative potential that can generate 6×10^{11} cells from a single cell, with the ability to differentiate into large cytokeratin-expressing structures when cultured in Matrigel [17], comprising laminin, collagen IV and heparan sulfate proteoglycan. It was used instead of mouse embryonic fibroblast feeder layers for ES cells proliferation. Our data shows that mesenchymal stem cells from endometrium can be grown extensively both *in vivo* and *in vitro*. These cells are not tumorigenic in nude mice and could be passaged approximately 40 times *in vitro*.

The fact that about 600,000 hysterectomies are performed yearly in the United States creates a potential source of endometrial cells [33]. Studies have indicated that EnSCs reside in the superficial layers accessible by endometrial biopsy [34,35]. Indeed, endometrial biopsies may represent a viable technique in harvesting EnSCs, as they are routinely performed for gynecologic purposes, and does not impair endometrial function. Thus, endometrial tissue specimens from women who are not receiving hormonal therapy are suitable for isolation of stem cells (Figure 2).

Endometrial stem cells may be derived from fetal stem cells or bone marrow stem cells, including haemopoietic SCs, MSCs and endothelial progenitor cells [36-38]. The human endometrium shows expression of the pluripotency factors Sox-2, Oct-4 and Nanog. Sox-2 co-localized with telomerase, contribute to the immortality of embryonic stem cells [39]. Reprogramming of differentiated cells to an induced pluripotent stem (iPS) cell is through induction or increasing in the expression of Oct-4, Sox-2, Klf-4, and c-Myc [40] that easily could be achieved from EnSCs that already express these factors. As a result, iPS cells from EnSCs have been established as early as 12 days after transduction rather than the usually reported 4 weeks for other cell types [41]. As patient- or disease-specific biomedical research using iPS cells becomes more widespread, the endometrium may play a key role as reprogrammable cells into donor- or disease-specific pluripotent cell lines for the female population.

Markers of human EnSCs

Flow cytometric analysis of important endometrial stem cell markers are indicated in Figure 3. Cultured EnSCs express typical markers used for bone marrow stem cells; CD9, CD13, CD14, CD29, CD31, CD44, CD73, CD90, CD105, CD117, CD133, CD146 [42], but not STRO-1, CD31 (endothelial) and CD34 (haemopoietic stem cell and endothelial) [43]. The EnSCs could be purified on the basis of their co-expression of two perivascular markers CD140b and CD146 [44]. Cells with a hematopoietic stem cell phenotype (CD34 + CD45+) co-expressing CD7 and CD56 have been identified in human endometrial cell suspensions and may be lymphoid progenitors [45]. Musashi-1, an RNA-binding protein in neural stem cells and an epithelial progenitor cell marker that regulates self-renewal signaling pathways, has been recognized in human endometrium. NAC1 is a transcription repressor that is involved in self-renewal and the maintenance of pluripotency in embryonic stem cells [46]. Ishikawa et al. demonstrated that NAC1 was over-expressed in the normal cyclic endometrium in the early and mid proliferative phases [47]. MSI1 and NOTCH1, which maintain SCs in an undifferentiated state, are expressed in EnSCs. Tissue non-specific alkaline phosphatase, also immune-localizes to a perivascular location in human endometrium and may be useful for the prospective isolation of EnSCs [48]. Leucine rich repeat containing G protein-coupled receptor-5 (Lgr-5) is also expressed in human endometrial epithelial progenitor cells and endometrial [49] stromal cells [50]. W5C5 is a single marker for purifying EnSC, which differentiates into adipogenic, osteogenic, chondrogenic, and myogenic cell lineages [51]. One-third of these EnSCs expressed α-smooth muscle actin (αSMA) but not

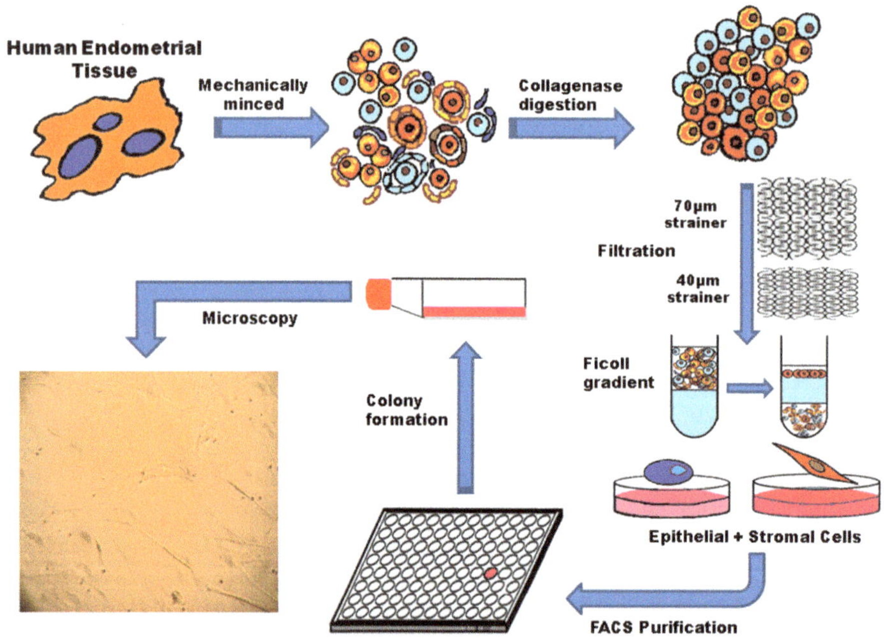

Figure 2 The endometrial stem cell isolation. A diagram that shows the stepwise isolation of endometrial stem cells by collagenase digestion, then serial filtration through 70 µm and 40 µm meshes, ficoll separation and selection the colony forming cells.

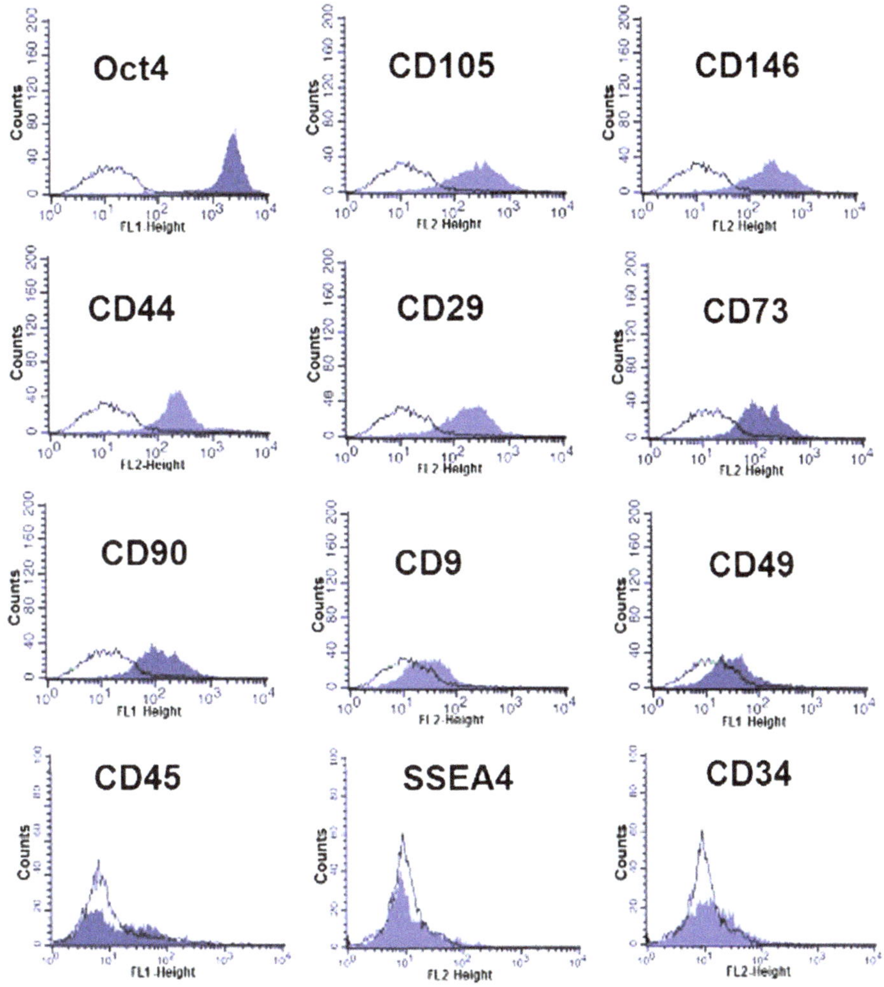

Figure 3 Flow-cytometric analysis of endometrial stem cell markers. Further phenotypic analysis of EnSCs after enrichment in culture. Expression of Oct4, CD90, CD105, CD146 and CD44 have more percentage of EnSC markers.

CD31, suggesting that they are vascular smooth muscle cells and that they occupy a peri-vascular niche.

Immunogenicity of EnSCs

Utilizing ES cells in regenerative therapy is associated with many hazards. Without adequate ES cells screening procedures, patients would face tissue rejection similar as rejection of organ transplants. A solution to this problem is to create patient specific stem cells. Unlike ES cells, adult stem cells are not rejected by the immune system [52]. The human endometrium has unique immunological requirements. In pregnancy, the endometrium must tolerate the invading embryo, which expresses both paternal and maternal antigens. In 2012, Peron et al. demonstrated that endometrial MSCs were able to suppress neuro-inflammation in a mouse model of multiple sclerosis. This suppression is dependent on the secretion of anti-inflammatory cytokines such as IL-10 and IL-27, and also the expression of indoleamine-2,3-dioxigenase [53]. Menstrual blood SCs were tested for allogeneic response in culture with peripheral blood mononuclear cells (PBMN). The SCs were analyzed at a 1:2 stimulator (mitomycin C treated PBMN) to responder cell (menstrual blood SCs) ratio for 6 days. The menstrual blood SCs demonstrated a moderately weak stimulatory response in a mixed lymphocyte reaction (MLR). An emerging bank of data allows for the classification of the mesenchymal stem cell from bone marrow as immunosuppressive cell source derived from studies with human cells [54,55]. Some studies support that SC administration is a potential way to suppress tumor growth. It has been demonstrated that human skin-derived progenitor cells have selective tropism for malignant tissues. Also more interestingly, these cells have the ability to inhibit tumor growth [56]. MSCs could selectively integrate into gliomas after intravascular or local delivery [57-59]. ERC administration inhibits C6 tumor growth and its administration associated with reduced neovascularization [60]. Despite the angiogenic potential of ERC in the hindlimb ischemia model, these data support a paradoxical tumor inhibitory activity of ERC. Further studies are needed to determine the qualitative differences between physiological angiogenesis, which seems to be supported by ERC and tumor angiogenesis which appeared to be inhibited. Other studies have demonstrated that MSCs directly secrete tumor inhibitory factors [61,62].

Potency and differentiation of EnSCs

The main features of mesenchymal stem cells are their ability for self-renewal, colony-forming ability, and differentiation into mesoderm-derived lineages including adipogenic, chondrogenic and osteogenic differentiation [63-65]. Endometrial SCs that are persistent in the uterine endometrium from the fetal to the postmenopausal period [66] were found to differentiate into adipocytes, osteoblasts and chondrocytes as well [67,68] (Figure 4). The modulation of p38 and c-jun might play an important role for the differentiation and proliferation of normal human endometrial cells [69]. These differentiations can identify by positive staining, with Oil Red O (for lipid droplets), alizarin red (for osteogenesis), von Kossa (for calcified extracellular matrix), and Alcian blue (for sulfated proteoglycans), respectively. Clonogenicity of EnSCs does not vary from the proliferative to secretory stage of the menstrual cycle, or between active, cycling and inactive endometrium for both epithelial and stromal cells [29].

Identification and isolation of differentiated cells

Research pertaining to stem cell biology in the female reproductive tract is still in its infancy, and although surface markers of prospective isolation of human endometrial stromal colony-forming cells (putative endometrial stromal/progenitor cells) have recently identified [17,44], there remains a need for definitive markers of both myometrial and endometrial stem cells for more selective isolation and enrichment. Complete characterization of uterine stem/ progenitor cells will improve our understanding of the mechanisms supporting physiological regeneration of the female reproductive tract. In addition, such studies will enhance our understanding of uterine cancer, hyperplasia, endometriosis, leiomyomas and adenomyosis. Indeed, our data and the published observations from other laboratories have enabled us to propose a novel

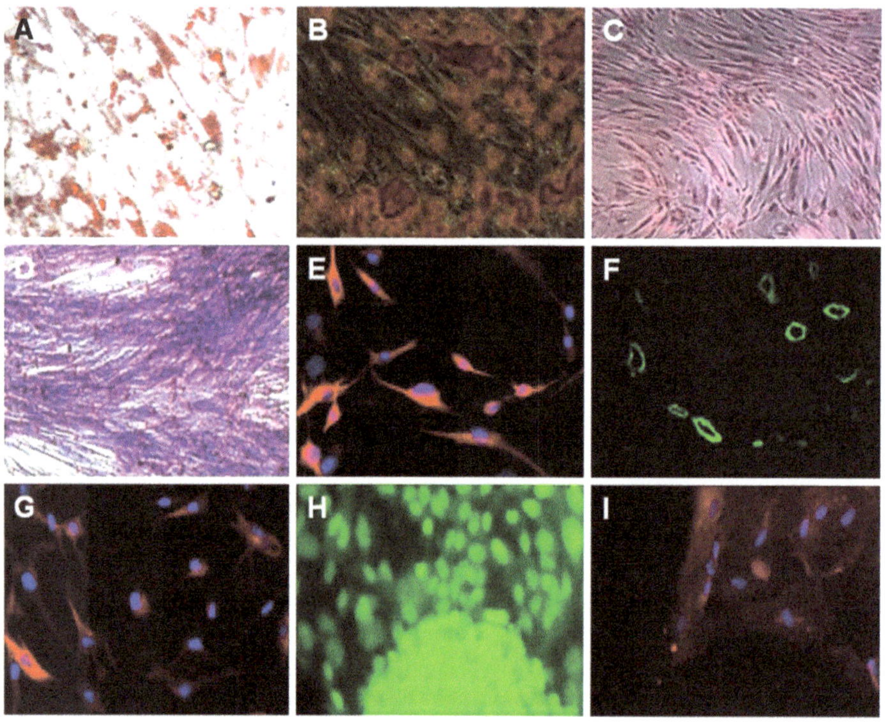

Figure 4 Multipotent differentiation of endometrial stem cells. Endometrial stem cells were cultured under appropriate differentiation media as described in box1-15 and assessed for differentiation using the indicated staining methods. **A)** Adipocytic differentiation, the red color indicates lipid vacuoles stained by Oil-Red. **B)** Osteocytic differentiation, red indicates calcium stained by Alizarin Red and **C)** Von-Kossa staining of osteoclasts differentiated from EnSCs. **D)** Chondrocytic differentiation, the blue color resulted from Alcian blue staining. **E)** Neuronal differentiation, the red conjugated mAb indicates pre-neural marker Nestin stain. **F)** Angiogenesis of EnSCs, the green indicated CD31 stain. **G)** Oligodendrocytic differentiation, the red conjugated mAb indicates receptor interacting protein (RIP). **H)** Hepatocytic differentiation, green indicates albumin stain. **I)** Myocytic differentiation, red indicates α-actinin stain.

hypothetical model for eutopic and ectopic endometrial regeneration. Finally, availability of these stem cells suggests new approaches to reconstruction of the human uterus and perhaps other organs as well.

The final step for processing stem cells before transplantation would be the identification and isolation of the differentiated cells. Recognition and selection of desired cells can be done via reverse transcriptase polymerase chain reaction, flow cytometry or fluorescence-activated cell sorting (FACS). Unfortunately, these methods would destroy many of the differentiated cells, thereby rendering them unusable. Also, flow cytometry and FACS is dependent on the presence of specific markers on the cell surface, which are not always displayed on every type of differentiated cells [70]. Thus, a more viable technique for identification of differentiated cells can be made with the transient transfection of a reporter gene into a stem cell source. The reporter gene can encode a fluorescent (green fluorescent protein) or a drug selectable protein. The transient transfection would be favorable for human cell therapy because the reporter is only meant for selection purposes.

Comparison between EnSCs, MSCs, and umbilical cord blood stem cells

Genetic profiling of EnSCs and bone marrow MSCs revealed that they had similar (although not identical) cytokine production, miRNAs, and gene expression [71]. It was reported that EnSCs express higher levels of ICAM-1 and IL-8, while bone marrow MSCs express higher levels of TGFβ1, TGFβ2, IL-6, and vascular endothelial growth factor (VEGF). This observation suggests that EnSCs may play a role in acute inflammation, and may be more suitable for tissue engineering applications. It was also reported that, compared to MSCs, EnSCs had a 27-fold higher level of PDGF-BB, and 14-fold higher level of angiopoietin. This indicates that EnSCs may activate alternative angiogenesis pathways. However, more studies are required to ascertain if EnSCs stimulate higher levels of angiogenesis than bone marrow MSCs *in vivo*. It was shown that EnSCs produced collagen type II and proteoglycans when cultured on nanofibrous polycaprolactone material, suggesting their cartilaginous tissue differentiation potential [72]. It was also observed that proliferation of EnSCs was faster than bone marrow MSCs, and EnSCs express the pluripotency marker OCT-4, but lack the MSC marker STRO1 [72]. Umbilical cord blood was first used in bone marrow transplant in a patient with Fanconi's anaemia in 1988 [73]. Considering that a large proportion of haematopoietic stem cells are found in the fetal circulation, there is also a correspondingly large proportion of haematopoietic stem cells in the umbilical cord. Bone marrow MSCs are distinct from haematopoietic stem cells, as they are $CD45^-$. Mature bone marrow MSCs are a heterogeneous population of cells that are able to support haematopoiesis, and capable of differentiating into different lineages such as neural cells [74,75], muscle [76-78], liver cells [79,80], among many others. Various studies have reported that a single mesenchymal progenitor cell is responsible for all lineages are produced from bone marrow MSC *in vitro*. It has been shown that bone marrow MSCs contain more mesenchymal progenitor cells when compared to umbilical cord blood stem cells. Indeed, the proportion of MSCs within umbilical cord blood is extremely low and difficult to detect [81]. Nevertheless, more studies comparing EnSCs, MSCs and umbilical cord blood stem cells are required to accurately ascertain their similarities and differences.

EnSCs in regenerative therapeutics

To date, EnSCs have been used in several pre-clinical and small animal studies. The regenerative potential of EnSCs was first seen in an experimental model of Duchenne muscular dystrophy in immunodeficient mdx mice, whereby EnSCs that were transplanted into atrophied skeletal muscle fibres contributed to muscle repair [26]. Although the exact mechanism of repair has not been elucidated, it was postulated that cell fusion and *in situ* differentiation might have been involved, in part due to the observation that the transplanted EnSCs homed to peri-muscle fibre regions and promoted angiogenesis. The concept of angiogenesis was also supported by another study, where EnSCs were improved critical limb ischaemia induced by femoral artery ligation [82].

The use of EnSCs to treat myocardial infarction in a murine model was also seen. In this study, EGFP-labelled EnSCs were grafted into the infarct area of nude rat hearts, which subsequently differentiated into α-actinin$^+$, troponin$^+$ striated cardiac muscle cells [83]. Furthermore, it was observed that a significantly larger reduction in infarct area was seen in animals treated with EnSCs, compared to control bone marrow MSCs.

Gargett et al., the first group that reported the existence of EnSCs in 2004 [21], are currently developing an autologous tissue engineered scaffold using artificial meshes and EnSCs for the treatment of pelvic organ prolapse, and was tested *in vivo*, on an animal skin wound repair model [84,85]. Results indicated that EnSCs promoted neovascularization, increased tissue integration, reduced chronic inflammation, increased deposition of collagen fibres and distensibility of the mesh at 3 months post-implantation. They also found that EnSCs downregulated foreign body reactions and enhanced mesh integration, indicating they have a role to play in modulating tissue response with regards to implanted foreign materials.

Our group has also advocated the tissue engineering approach to the application of EnSCs. To this end, we have fabricated nanofibrous silk-collagen fibres that were seeded with EnSCs for the reconstruction of urinary bladder wall for women [86]. We had also shown that EnSCs were capable of differentiating into smooth muscle cells [86], potentially functioning as an autologous cell source for bladder tissue engineering.

Intraperitoneal delivery of EnSCs were also studied in a murine model of encephalomyelitis [53]. It was founds that EnSCs exerted a an anti-inflammatory effect, as evidenced by a lower number of infiltrating mononuclear cells in the lesions, upregulation of IL-10 and IL-27 in the spleen, and reduced recruitment of Th1 and Th17 cells in the central nervous system.

EnSCs were also used to demonstrate neural regenerative capabilities, whereby reduced neuronal cell death was seen when oxygen-deprived primary neuronal cell cultures were exposed to EnSCs [87]. It was also observed that EnSCs released neuroprotective trophic factors such as neurotrophin-3 (NT-3), brain-derived neurotrophic factor (BDNF), and VEGF. The *in vivo* part of the study was done in a murine model of ischaemic stroke, whereby injection of EnSCs resulted in significantly lower histological and behavioural impairments. It was reported that EnSCs exerted a trophic effect, releasing factors that promoted survival of neural cells.

The use of EnSCs to treat glioma was observed in a murine model. In this study, EnSCs were administered intravenously in a murine model of intracranial glioma. Results revealed a reduction of tumour size of almost 50%, possibly due to its anti-angiogenic effects [60].

The applications of EnSCs have also been reported in several human studies. The first reported use of EnSCs was demonstrated by Zhong et al. [88]. Clinical-grade menstrual blood-derived EnSCs have been used in a small Phase I clinical trial of 4 patients suffering from multiple sclerosis, whereby EnSCs were delivered via intravenous and intrathecal routes. Results showed no immunological reactions or adverse side effects after 1 year [88].

Another human study involved a patient suffering from Duchenne muscular dystrophy that was given intramuscular injections of EnSCs. Follow-up observations reported no adverse effects even after 3 years, and increased muscle strength and decreased respiratory infections was also reported [89].

The third reported use of EnSCs in human was a patient with congestive heart failure, who was given intravenous administration of EnSCs. Results revealved that ejection fraction of the patient increased from 30% to 40%, decreased basic natriuretic peptide values (Pro-BNP), and decreased Minnesota Living with Heart Failure Questionnaire score at 1-year follow up [90].

The promise and limitations of EnSCs

EnSCs are an attractive source of stem cells for regenerative therapeutics as they are easily obtainable and easily expandable in culture, as has been demonstrated to be safe for clinical use. Protocols and methods for extraction and isolation of EnSCs are well established, as purified EnSCs can be obtained using magnetic bead sorting using the W5C5/SUSD2 marker. In addition, clinical-grade good manufacturing practice (cGMP) are currently being developed for culture expansion of EnSCs, and have been tested in animals. However, there is a lack of published information on the exact cGMP protocols in place for the production of EnSCs. This is compounded by the fact that there is no general scientific consensus regarding specific MSC markers to detect EnSCs; rather, researchers rely on the ability of MSCs to adhere to plastic. Hence, the purity of EnSCs is not guaranteed as the cultures could potentially contain fibroblasts.

EnSCs can be obtained from menstrual blood; hence no invasive procedures are needed to harvest these cells. A menstrual cup is used to collect menstrual blood over several hours on days 2 to 3 of the menstrual period. Although there is a potential risk of infection via vaginal contact, there have been no reports of any complications after antibiotic use.

Although the ability of EnSCs to re-integrate into tissue *in vivo* has been demonstrated, there is a theoretical risk that endometriosis could develop from using EnSCs. However, none of the animal model studies have reported this. Nevertheless, it is an aspect of EnSC application that warrants attention. Indeed, transdifferentiation (sometimes referred to as adult stem cell plasticity) is a controversial topic in the field of stem cells. It is highly probable that nuclear reprogramming and altered transcriptional activity of important developmental genes are in part responsible for transdifferentiation. Transdifferentiation can be seen as a form of metaplasia due to alterations in the extracellular environment, and often happens during tissue damage [91]. Hence, one probable cause of endometriosis is thought to be due to metaplasia of the peritoneal lining. However, there is also evidence to suggest that circulating stem cells can be a source of metaplastic transdifferentiation when it gains access into the pelvic cavity. Genetically-tagged bone marrow MSCs were tracked in a murine transplant model, and it was established that a small population of it integrated into endometriosis lesions and transdifferentiated into stromal cells (0.1%) and epithelial cells (<0.04%) [92]. It was found that these bone marrow MSCs contributed to the progression of endometriosis rather than initiating it [93]. Changes in cellular phenotype may involve processes that are associated with both embryogenesis and carcinogenesis, such as mesenchymal epithelial transition (MET) and epithelial mesenchymal transition (EMT) [94]. Correspondingly, the invasiveness of endometriotic cells may be attributed to the changes in cellular phenotype in endometriotic lesions.

For instance, well differentiated CK$^+$E$^+$-Cadherin cells, CK$^-$ E-Cadherin- stromal cells, and an invasive CK$^+$E-Cadherin$^-$N-Cadherin$^+$ epithelial cells (similar to carcinoma micrometastasis) are found in endometriotic lesions [95]. In line with carcinoma characteristics, a regression of endometriotic lesion is seen in estrogen depletion therapy, but recur when therapy stops. This indicates that a quiescent stem cell population is present within the lesion, which reactivates on exposure to estrogen. Thus it may also follow that EnSCs within lesions may also contribute to subsequent lesions.

In terms of regenerative medicine applications, it is envisioned that bone marrow MSCs are more suited for osteogenic differentiation and therefore used for bone tissue engineering, while EnSCs are more suited for soft tissue engineering, such as bladder reconstruction. Due to the intense interest generated with using EnSCs for regenerative medicine, we believe that EnSCs were introduced too quickly into human tests without adequately assessing its impact on large animal models. Furthermore, the exact mechanism of how EnSCs exert is regenerative potential is not clearly understood. Hence more robust large animal studies are needed to fill this information gap that currently exists.

Concluding remarks and future perspectives

Although EnSCs would form a non-invasive and steady supply of autologous stem cells for women, we must not forget that this means that 50% of the population (i.e. the male population) is excluded. Furthermore, non-invasive harvesting of EnSCs would not be possible from post-menopausal women (in this case, endometrial biopsy might be possible, but data regarding the potency of EnSCs in post-menopausal women has not been published). In addition, the issue of EnSCs storage could also develop into the same debates we are facing regarding the storage of cord blood; is there enough scientific evidence to justify storage of EnSCs for future use?

Nevertheless, given the significant progress made in the application of EnSCs in regenerative medicine, it is envisioned that human clinical trials would be conducted in the near future. Indeed, the RECOVER-ERC trial was recently launched by Medistem Inc (a start-up company based in San Diego, California), which aims to evaluate the potency of EnSCs in 60 heart failure patients in a double-blind placebo controlled Phase II clinical trial [96]. Furthermore, a search in the *clinicaltrials.gov* database reveals that several clinical trials regarding EnSCs are underway, with applications ranging from treating critical limb ischaemia (NCT 01558908), liver cirrhosis (NCT01483248), type I diabetes (NCT01496339), and *in vitro* fertilization (NCT01649752). Hence, it can be envisioned that data from human patients would be available in due course to evaluate the clinical impact of EnSCs.

In conclusion, the discovery of EnSCs represents a paradigm shift for the use of adult stem cells by offering an "off the shelf" therapeutic application regenerative medicine. The EnSCs niche and *in situ* role highlight these cells as being important for cell and tissue regeneration. This work highlights crucial features of an interesting population of mesenchymal progenitors isolated form endometrial tissue. Cells grown *in vitro* were characterized by a high clonogenic potential and a long-term survival. Advantages in comparison with MSC from other sources include a greater ease of supply and the protracted availability during a woman's lifetime with the additional benefit of deriving stem cells from a waste tissue, thus avoiding critical ethical issues.

In the light of their capacity to differentiate into mesenchymal tissues and their propensity to undergo genetic manipulation, EnSCs hold considerable promise for novel therapeutic approaches for diseases in the wider field of regenerative medicine.

Competing interests
The authors declare that no competing interests exist.

Authors' contributions
JV, AT, AS, AMS conceived and planned the structure of the manuscript. JV, AT, AS, AMS collated and analyzed the data. JV, AT, AS, AMS wrote the manuscript. All authors read and approved the final manuscript.

Author details
[1]Centre for Nanotechnology and Regenerative Medicine, UCL Division of Surgery & Interventional Science, University College London (UCL), London NW3 2QG, UK. [2]Applied Cell Sciences Department, School of Advanced Technologies in Medicine, Tehran University of Medical Sciences, Tehran, Iran. [3]UCL Medical School, University College London (UCL), London, UK. [4]Royal Free London NHS Foundation Trust Hospital, London, UK.

References

1. Kelly SJ: **Studies of the developmental potential of 4-and 8-cell stage mouse blastomeres.** *J Exp Zool* 1977, **200**(3):365–376.
2. Conrad S, Renninger M, Hennenlotter J, Wiesner T, Just L, Bonin M, Aicher W, Bühring H-J, Mattheus U, Mack A: **Generation of pluripotent stem cells from adult human testis.** *Nature* 2008, **456**(7220):344–349.
3. Ko K, Araúzo-Bravo MJ, Tapia N, Kim J, Lin Q, Bernemann C, Han DW, Gentile L, Reinhardt P, Greber B: **Human adult germline stem cells in question.** *Nature* 2010, **465**(7301):E1–E1.
4. Stadtfeld M, Hochedlinger K: **Induced pluripotency: history, mechanisms, and applications.** *Genes Dev* 2010, **24**(20):2239–2263.
5. Bjerknes M, Cheng H: **Clonal analysis of mouse intestinal epithelial progenitors.** *Gastroenterology* 1999, **116**(1):7–14.
6. Jankowski R, Deasy B, Huard J: **Muscle-derived stem cells.** *Gene Ther* 2002, **9**(10):642–647.
7. Alonso L, Fuchs E: **Stem cells of the skin epithelium.** *Proc Natl Acad Sci U S A* 2003, **100**(Suppl 1):11830–11835.
8. Spangrude GJ, Smith L, Uchida N, Ikuta K, Heimfeld S, Friedman J, Weissman I: **Mouse hematopoietic stem cells.** *Blood* 1991, **78**(6):1395–1402.
9. Morrison SJ, White PM, Zock C, Anderson DJ: **Prospective identification, isolation by flow cytometry, and in vivo self-renewal of multipotent mammalian neural crest stem cells.** *Cell* 1999, **96**(5):737–749.
10. Uchida N, Buck DW, He D, Reitsma MJ, Masek M, Phan TV, Tsukamoto AS, Gage FH, Weissman IL: **Direct isolation of human central nervous system stem cells.** *Proc Natl Acad Sci U S A* 2000, **97**(26):14720–14725.

11. Rietze RL, Valcanis H, Brooker GF, Thomas T, Voss AK, Bartlett PF: **Purification of a pluripotent neural stem cell from the adult mouse brain.** *Nature* 2001, **412**(6848):736–739.
12. Chan RW, Gargett CE: **Identification of Label-Retaining Cells in Mouse Endometrium.** *Stem Cells* 2006, **24**(6):1529–1538.
13. Beltrami AP, Cesselli D, Bergamin N, Marcon P, Rigo S, Puppato E, D'Aurizio F, Verardo R, Piazza S, Pignatelli A: **Multipotent cells can be generated in vitro from several adult human organs (heart, liver, and bone marrow).** *Blood* 2007, **110**(9):3438–3446.
14. Crisan M, Yap S, Casteilla L, Chen C-W, Corselli M, Park TS, Andriolo G, Sun B, Zheng B, Zhang L: **A perivascular origin for mesenchymal stem cells in multiple human organs.** *Cell Stem Cell* 2008, **3**(3):301–313.
15. da Silva ML, Chagastelles PC, Nardi NB: **Mesenchymal stem cells reside in virtually all post-natal organs and tissues.** *J Cell Sci* 2006, **119**(11):2204–2213.
16. Kalervo Väänänen H: **Mesenchymal stem cells.** *Ann Med* 2005, **37**(7):469–479.
17. Gargett CE, Schwab KE, Zillwood RM, Nguyen HP, Wu D: **Isolation and culture of epithelial progenitors and mesenchymal stem cells from human endometrium.** *Biol Reprod* 2009, **80**(6):1136–1145.
18. McLennan CE, Rydell AH: **Extent of endometrial shedding during normal menstruation.** *Obstet Gynecol* 1965, **26**(5):605–621.
19. Padykula HA: **Regeneration in the primate uterus: the role of stem cells.** *Ann NY Acad Sci* 1991, **622:**47–56.
20. Okulicz WC, Ace CI, Scarrell R: **Zonal changes in proliferation in the rhesus endometrium during the late secretory phase and menses.** *Exp Biol Med* 1997, **214**(2):132–138.
21. Chan RW, Schwab KE, Gargett CE: **Clonogenicity of human endometrial epithelial and stromal cells.** *Biol Reprod* 2004, **70**(6):1738–1750.
22. Gargett CE, Masuda H: **Adult stem cells in the endometrium.** *Mol Hum Reprod* 2010, **16**(11):818–834.
23. Meng X, Ichim TE, Zhong J, Rogers A, Yin Z, Jackson J, Wang H, Ge W, Bogin V, Chan KW: **Endometrial regenerative cells: a novel stem cell population.** *J Transl Med* 2007, **5**(1):57.
24. Patel AN, Park E, Kuzman M, Benetti F, Silva FJ, Allickson JG: **Multipotent menstrual blood stromal stem cells: isolation, characterization, and differentiation.** *Cell Transplant* 2008, **17**(3):303–311.
25. Schüring AN, Schulte N, Kelsch R, Röpke A, Kiesel L, Götte M: **Characterization of endometrial mesenchymal stem-like cells obtained by endometrial biopsy during routine diagnostics.** *Fertil Steril* 2011, **95**(1):423–426.
26. Cui CH, Uyama T, Miyado K, Terai M, Kyo S, Kiyono T, Umezawa A: **Menstrual blood-derived cells confer human dystrophin expression in the murine model of Duchenne muscular dystrophy via cell fusion and myogenic transdifferentiation.** *Mol Biol Cell* 2007, **18**(5):1586–1594.
27. Musina R, Belyavski A, Tarusova O, Solovyova E, Sukhikh G: **Endometrial mesenchymal stem cells isolated from the menstrual blood.** *Bull Exp Biol Med* 2008, **145**(4):539–543.
28. Elsheikh E, Sylven C, Ericzon BG, Palmblad J, Mints M: **Cyclic variability of stromal cell-derived factor-1 and endothelial progenitor cells during the menstrual cycle.** *Int J Mol Med* 2011, **27**(2):221–226.
29. Schwab KE, Chan RWS, Gargett CE: **Putative stem cell activity of human endometrial epithelial and stromal cells during the menstrual cycle.** *Fertil Steril* 2005, **84:**1124–1130.
30. Gargett C: **Uterine stem cells: what is the evidence?** *Hum Reprod Update* 2007, **13**(1):87–101.
31. Sasson IE, Taylor HS: **Stem cells and the pathogenesis of endometriosis.** *Ann NY Acad Sci* 2008, **1127**(1):106–115.
32. Kim C, Oh Y, Cho S, Chung D, Hwang J, Park K, Cho D, Choi Y, Lee B: **Increased telomerase activity and human telomerase reverse transcriptase mRNA expression in the endometrium of patients with endometriosis.** *Hum Reprod* 2007, **22**(3):843–849.
33. Wu JM, Wechter ME, Geller EJ, Nguyen TV, Visco AG: **Hysterectomy rates in the United States, 2003.** *Obstet Gynecol* 2007, **110**(5):1091–1095.
34. Götte M, Wolf M, Staebler A, Buchweitz O, Kelsch R, Schüring A, Kiesel L: **Increased expression of the adult stem cell marker Musashi-1 in endometriosis and endometrial carcinoma.** *J Pathol* 2008, **215**(3):317–329.
35. Shoae Hassani A, Mortazavi Tabatabaei SA, Sharif S, Seifalian AM, Azimi A, Samadikuchaksaraei A, Verdi J: **Differentiation of human endometrial stem cells into urothelial cells on a three-dimensional nanofibrous silk-collagen scaffold: an autologous cell resource for reconstruction of the urinary bladder wall.** *J Tissue Eng Regen Med* 2013, [In press].
36. Mayani H, Alvarado-Moreno JA, Flores-Guzmán P: **Biology of human hematopoietic stem and progenitor cells present in circulation.** *Arch Med Res* 2003, **34**(6):476–488.
37. He Q, Wan C, Li G: **Concise review: multipotent mesenchymal stromal cells in blood.** *Stem Cells* 2007, **25**(1):69–77.
38. Robb A, Mills N, Smith I, Short A, Tura-Ceide O, Barclay G, Blomberg A, Critchley H, Newby D, Denison F: **Influence of menstrual cycle on circulating endothelial progenitor cells.** *Hum Reprod* 2009, **24**(3):619–625.
39. Kyo S, Masutomi K, Maida Y, Kanaya T, Yatabe N, Nakamura M, Tanaka M, Takarada M, Sugawara I, Murakami S: **Significance of immunological detection of human telomerase reverse transcriptase: re-evaluation of expression and localization of human telomerase reverse transcriptase.** *Am J Pathol* 2003, **163**(3):859–867.
40. Takahashi K, Tanabe K, Ohnuki M, Narita M, Ichisaka T, Tomoda K, Yamanaka S: **Induction of pluripotent stem cells from adult human fibroblasts by defined factors.** *Cell* 2007, **131**(5):861–872.
41. Park JH, Daheron L, Kantarci S, Lee BS, Teixeira JM: **Human endometrial cells express elevated levels of pluripotent factors and are more amenable to reprogramming into induced pluripotent stem cells.** *Endocrinology* 2011, **152**(3):1080–1089.
42. Dominici M, Le Blanc K, Mueller I, Slaper-Cortenbach I, Marini F, Krause D, Deans R, Keating A, Prockop D, Horwitz E: **Minimal criteria for defining multipotent mesenchymal stromal cells. The International Society for Cellular Therapy position statement.** *Cytotherapy* 2006, **8**(4):315–317.
43. Dimitrov R, Timeva T, Kyurkchiev D, Stamenova M, Shterev A, Kostova P, Zlatkov V, Kehayov I, Kyurkchiev S: **Characterization of clonogenic stromal cells isolated from human endometrium.** *Reproduction* 2008, **135**(4):551–558.
44. Schwab K, Gargett C: **Co-expression of two perivascular cell markers isolates mesenchymal stem-like cells from human endometrium.** *Hum Reprod* 2007, **22**(11):2903–2911.
45. Lynch L, Golden-Mason L, Eogan M, O'Herlihy C, O'Farrelly C: **Cells with haematopoietic stem cell phenotype in adult human endometrium: relevance to infertility?** *Human Reprod* 2007, **22**(4):919–926.
46. Wang J, Rao S, Chu J, Shen X, Levasseur DN, Theunissen TW, Orkin SH: **A protein interaction network for pluripotency of embryonic stem cells.** *Nature* 2006, **444**(7117):364–368.
47. Ishikawa M, Nakayama K, Yeasmin S, Katagiri A, Iida K, Nakayama N, Miyazaki K: **NAC1, a potential stem cell pluripotency factor expression in normal endometrium, endometrial hyperplasia and endometrial carcinoma.** *Int J Oncol* 2010, **36**(5):1097–1103.
48. Sobiesiak M, Sivasubramaniyan K, Hermann C, Tan C, Örgel M, Treml S, Cerabona F, de Zwart P, Ochs U, Müller CA: **The mesenchymal stem cell antigen MSCA-1 is identical to tissue non-specific alkaline phosphatase.** *Stem Cells Dev* 2009, **19**(5):669–677.
49. Krusche CA, Kroll T, Beier HM, Classen-Linke I: **Expression of leucine-rich repeat-containing G-protein-coupled receptors in the human cyclic endometrium.** *Fertil Steril* 2007, **87**(6):1428–1437.
50. Gil Sanchis C, Cervelló I, Mas A, Faus A, Pellicer A, Simón C: **Leucine-rich repeat-containing G-protein-coupled receptor 5 (Lgr5) as a putative human endometrial stem cell marker.** *Mol Hum Reprod* 2013, **19**(7):407–414.
51. Masuda H, Anwar SS, Bühring H-J, Rao JR, Gargett CE: **A novel marker of human endometrial mesenchymal stem-like cells.** *Cell Transplant* 2012, **21**(10):2201–2214.
52. Awong G, Herer E, Surh CD, Dick JE, La Motte-Mohs RN, Zúñiga-Pflücker JC: **Characterization in vitro and engraftment potential in vivo of human progenitor T cells generated from hematopoietic stem cells.** *Blood* 2009, **114**(5):972–982.
53. Peron J, Jazedje T, Brandao W, Perin P, Maluf M, Evangelista L, Halpern S, Nisenbaum M, Czeresnia C, Zatz M: **Human endometrial-derived mesenchymal stem cells suppress inflammation in the central nervous system of EAE mice.** *Stem Cell Rev* 2012, **8**(3):940–952.
54. Tse WT, Pendleton JD, Beyer WM, Egalka MC, Guinan EC: **Suppression of allogeneic T-cell proliferation by human marrow stromal cells: implications in transplantation.** *Transplantation* 2003, **75**(3):389–397.
55. Kiss J, Urban V, Dudics V, Vas V: **Uher F: [Mesenchymal stem cells and the immune system–immunosuppression without drugs?].** *Orv Hetil* 2008, **149**(8):339–346.
56. Pisati F, Belicchi M, Acerbi F, Marchesi C, Giussani C, Gavina M, Javerzat S, Hagedorn M, Carrabba G, Lucini V: **Effect of human skin-derived stem cells on vessel architecture, tumor growth, and tumor invasion in brain tumor animal models.** *Cancer Res* 2007, **67**(7):3054–3063.

57. Ehtesham M, Kabos P, Kabosova A, Neuman T, Black KL, John SY: The use of interleukin 12-secreting neural stem cells for the treatment of intracranial glioma. Cancer Res 2002, 62(20):5657–5663.
58. Ehtesham M, Kabos P, Gutierrez MA, Chung NH, Griffith TS, Black KL, John SY: Induction of glioblastoma apoptosis using neural stem cell-mediated delivery of tumor necrosis factor-related apoptosis-inducing ligand. Cancer Res 2002, 62(24):7170–7174.
59. Nakamizo A, Marini F, Amano T, Khan A, Studeny M, Gumin J, Chen J, Hentschel S, Vecil G, Dembinski J: Human bone marrow–derived mesenchymal stem cells in the treatment of gliomas. Cancer Res 2005, 65(8):3307–3318.
60. Han X, Meng X, Yin Z, Rogers A, Zhong J, Rillema P, Jackson JA, Ichim TE, Minev B, Carrier E: Inhibition of intracranial glioma growth by endometrial regenerative cells. Cell Cycle 2009, 8(4):606–610.
61. Qiao L, Xu ZL, Zhao TJ, Ye LH, Zhang XD: Dkk-1 secreted by mesenchymal stem cells inhibits growth of breast cancer cells < i > via</i > depression of Wnt signalling. Cancer Lett 2008, 269(1):67–77.
62. Qiao L, Zhao TJ, Wang FZ, Shan CL, Ye LH, Zhang XD: NF-kappaB downregulation may be involved the depression of tumor cell proliferation mediated by human mesenchymal stem cells. Acta Pharmacol Sin 2008, 29(3):333–340.
63. Pittenger MF, Mackay AM, Beck SC, Jaiswal RK, Douglas R, Mosca JD, Moorman MA, Simonetti DW, Craig S, Marshak DR: Multilineage potential of adult human mesenchymal stem cells. Science 1999, 284(5411):143–147.
64. Colter DC, Class R, DiGirolamo CM, Prockop DJ: Rapid expansion of recycling stem cells in cultures of plastic-adherent cells from human bone marrow. Proc Natl Acad Sci U S A 2000, 97(7):3213–3218.
65. Zuk PA, Zhu M, Mizuno H, Huang J, Futrell JW, Katz AJ, Benhaim P, Lorenz HP, Hedrick MH: Multilineage cells from human adipose tissue: implications for cell-based therapies. Tissue Eng 2001, 7(2):211–228.
66. Cho NH, Park YK, Kim YT, Yang H, Kim SK: Lifetime expression of stem cell markers in the uterine endometrium. Fertil and Steril 2004, 81(2):403–407.
67. Gargett CE, Chan RW, Schwab KE: Endometrial stem cells. Curr Opin Obstet Gynecol 2007, 19(4):377–383.
68. Kao AP, Wang KH, Chang CC, Lee JN, Long CY, Chen HS, Tsai CF, Hsieh TH, Tsai EM: Comparative study of human eutopic and ectopic endometrial mesenchymal stem cells and the development of an in vivo endometriotic invasion model. Fertil Steril 2011, 95(4):1308–1315. e1301.
69. Hong I-S, Kim S-H, Koong MK, Jun JH, Kim S-H, Lee Y-S, Kang K-S: Roles of p38 and c-jun in the differentiation, proliferation and immortalization of normal human endometrial cells. Hum Reprod 2004, 19(10):2192–2199.
70. Giudice A, Trounson A: Genetic modification of human embryonic stem cells for derivation of target cells. Cell Stem Cell 2008, 2(5):422–433.
71. Wang H, Jin P, Sabatino M, Ren J, Civini S, Bogin V, Ichim TE, Stroncek DF: Comparison of endometrial regenerative cells and bone marrow stromal cells. J Transl Med 2012, 10(1):207.
72. Kazemnejad S, Akhondi M-M, Soleimani M, Zarnani AH, Khanmohammadi M, Darzi S, Alimoghadam K: Characterization and chondrogenic differentiation of menstrual blood-derived stem cells on a nanofibrous scaffold. Int J Artif Organs 2012, 35(1):55–66.
73. Gluckman E, Broxmeyer H, Auerbach AD, Friedman HS, Douglas GW, Devergie A, Esperou H, Thierry D, Socie G, Lehn P: Hematopoietic reconstitution in a patient with Fanconi's anemia by means of umbilical-cord blood from an HLA-identical sibling. N Engl J Med 1989, 321(17):1174–1178.
74. Zhao LR, Duan WM, Reyes M, Keene CD, Verfaillie CM, Low WC: Human bone marrow stem cells exhibit neural phenotypes and ameliorate neurological deficits after grafting into the ischemic brain of rats. Exp Neurol 2002, 174(1):11–20.
75. Meletis K, Frisen J: Have the bloody cells gone to our heads? J Cell Biol 2001, 155(5):699–702.
76. Hakuno D, Fukuda K, Makino S, Konishi F, Tomita Y, Manabe T, Suzuki Y, Umezawa A, Ogawa S: Bone marrow–derived regenerated cardiomyocytes (CMG cells) express functional adrenergic and muscarinic receptors. Circulation 2002, 105(3):380–386.
77. Wakitani S, Saito T, Caplan AI: Myogenic cells derived from rat bone marrow mesenchymal stem cells exposed to 5-azacytidine. Muscle Nerve 1995, 18(12):1417–1426.
78. Prockop DJ: Marrow stromal cells as stem cells for nonhematopoietic tissues. Science 1997, 276(5309):71–74.
79. Theise ND, Badve S, Saxena R, Henegariu O, Sell S, Crawford JM, Krause DS: Derivation of hepatocytes from bone marrow cells in mice after radiation-induced myeloablation. Hepatology 2000, 31(1):235–240.
80. Theise ND, Nimmakayalu M, Gardner R, Illei PB, Morgan G, Teperman L, Henegariu O, Krause DS: Liver from bone marrow in humans. Hepatol 2000, 32(1):11–16.
81. Wexler SA, Donaldson C, Denning Kendall P, Rice C, Bradley B, Hows JM: Adult bone marrow is a rich source of human mesenchymal 'stem'cells but umbilical cord and mobilized adult blood are not. Br J Haematol 2003, 121(2):368–374.
82. Murphy MP, Wang H, Patel AN, Kambhampati S, Angle N, Chan K, Marleau AM, Pyszniak A, Carrier E, Ichim TE, Riordan NH: Allogeneic endometrial regenerative cells: an "Off the shelf solution" for critical limb ischemia? J Transl Med 2008, 6:45.
83. Hida N, Nishiyama N, Miyoshi S, Kira S, Segawa K, Uyama T, Mori T, Miyado K, Ikegami Y, Cui C: Novel Cardiac Precursor-Like Cells from Human Menstrual Blood-Derived Mesenchymal Cells. Stem Cells 2008, 26(7):1695–1704.
84. Ulrich D, Edwards SL, White JF, Supit T, Ramshaw JA, Lo C, Rosamilia A, Werkmeister JA, Gargett CE: A preclinical evaluation of alternative synthetic biomaterials for fascial defect repair using a rat abdominal hernia model. PLoS One 2012, 7(11):e50044.
85. Edwards SL, Werkmeister JA, Rosamilia A, Ramshaw JA, White JF, Gargett CE: Characterisation of clinical and newly fabricated meshes for pelvic organ prolapse repair. J Mech Behav Biomed Mater 2013, 23:53–61.
86. Shoae-Hassani A, Sharif S, Seifalian AM, Mortazavi-Tabatabaei SA, Rezaie S, Verdi J: Endometrial stem cell differentiation into smooth muscle cell: a novel approach for bladder tissue engineering in women. BJU Int 2013, 112(6):854–863.
87. Borlongan CV, Kaneko Y, Maki M, Yu S-J, Ali M, Allickson JG, Sanberg CD, Kuzmin-Nichols N, Sanberg PR: Menstrual blood cells display stem cell-like phenotypic markers and exert neuroprotection following transplantation in experimental stroke. Stem Cells Dev 2010, 19(4):439–452.
88. Zhong Z, Patel AN, Ichim TE, Riordan NH, Wang H, Min W-P, Woods EJ, Reid M, Mansilla E, Marin GH: Feasibility investigation of allogeneic endometrial regenerative cells. J Transl Med 2009, 7(15):29–37.
89. Ichim TE, Alexandrescu DT, Solano F, Lara F, Campion RDN, Paris E, Woods EJ, Murphy MP, Dasanu CA, Patel AN: Mesenchymal stem cells as anti-inflammatories: implications for treatment of Duchenne muscular dystrophy. Cell Immunol 2010, 260(2):75–82.
90. Ichim TE, Solano F, Lara F, Rodriguez JP, Cristea O, Minev B, Ramos F, Woods EJ, Murphy MP, Alexandrescu DT, Patel AN, Riordan NH: Combination stem cell therapy for heart failure. Int Arch Med 2010, 3(1):5.
91. Tosh D, Slack JM: How cells change their phenotype. Nat Rev Mol Cell Biol 2002, 3(3):187–194.
92. Du H, Taylor HS: Contribution of Bone Marrow-Derived Stem Cells to Endometrium and Endometriosis. Stem Cells 2007, 25(8):2082–2086.
93. Du H, Taylor HS: Reviews: stem cells and female reproduction. Reprod Sci 2009, 16(2):126–139.
94. Polyak K, Weinberg RA: Transitions between epithelial and mesenchymal states: acquisition of malignant and stem cell traits. Nature Rev Cancer 2009, 9(4):265–273.
95. Starzinski-Powitz A, Zeitvogel A, Schreiner A, Baumann R: In search of pathogenic mechanisms in endometriosis: the challenge for molecular cell biology. Curr Mol Med 2001, 1(6):655–664.
96. Bockeria L, Bogin V, Bockeria O, Le T, Alekyan B, Woods EJ, Brown AA, Ichim TE, Patel AN: Endometrial regenerative cells for treatment of heart failure: a new stem cell enters the clinic. J Transl Med 2013, 11:56.

The *Bacillus* BioBrick Box: generation and evaluation of essential genetic building blocks for standardized work with *Bacillus subtilis*

Jara Radeck[1†], Korinna Kraft[1†], Julia Bartels[1], Tamara Cikovic[1], Franziska Dürr[1], Jennifer Emenegger[1], Simon Kelterborn[1], Christopher Sauer[1,3], Georg Fritz[1,2], Susanne Gebhard[1] and Thorsten Mascher[1*]

Abstract

Background: Standardized and well-characterized genetic building blocks are a prerequisite for the convenient and reproducible assembly of novel genetic modules and devices. While numerous standardized parts exist for *Escherichia coli*, such tools are still missing for the Gram-positive model organism *Bacillus subtilis*. The goal of this study was to develop and thoroughly evaluate such a genetic toolbox.

Results: We developed five BioBrick-compatible integrative *B. subtilis* vectors by deleting unnecessary parts and removing forbidden restriction sites to allow cloning in BioBrick (RFC10) standard. Three empty backbone vectors with compatible resistance markers and integration sites were generated, allowing the stable chromosomal integration and combination of up to three different devices in one strain. In addition, two integrative reporter vectors, based on the *lacZ* and *luxABCDE* cassettes, were BioBrick-adjusted, to enable β-galactosidase and luciferase reporter assays, respectively. Four constitutive and two inducible promoters were thoroughly characterized by quantitative, time-resolved measurements. Together, these promoters cover a range of more than three orders of magnitude in promoter strength, thereby allowing a fine-tuned adjustment of cellular protein amounts. Finally, the *Bacillus* BioBrick Box also provides five widely used epitope tags (FLAG, His_{10}, cMyc, HA, StrepII), which can be translationally fused N- or C-terminally to any protein of choice.

Conclusion: Our genetic toolbox contains three compatible empty integration vectors, two reporter vectors and a set of six promoters, two of them inducible. Furthermore, five different epitope tags offer convenient protein handling and detection. All parts adhere to the BioBrick standard and hence enable standardized work with *B. subtilis*. We believe that our well-documented and carefully evaluated *Bacillus* BioBrick Box represents a very useful genetic tool kit, not only for the iGEM competition but any other BioBrick-based project in *B. subtilis*.

Keywords: *Bacillus subtilis*, BioBrick standard, iGEM, *lux*, Luminescence, Epitope tag, Integrative vector, Plasmid, Inducible promoter

Introduction

One core aspect of synthetic biology projects that sets them apart from "classical" genetic work is the application of engineering principles such as abstraction, modularity and standardization to assembly strategies and procedures. The characterization and standardization of reusable genetic building blocks is one of the prerequisites for the engineering approach of building complex synthetic biological systems [1]. Towards that end, the Registry of Standard Biological Parts (partsregistry, [2]) was founded by the Massachusetts Institute of Technology in 2003 as a repository for the *i*nternational *G*enetically *E*ngineered *M*achine competition (iGEM) and now maintains and distributes over ten thousand standardized biological parts that adhere to the BioBrick standard as described in the "request for comments 10" (RFC 10) [3]. Such standardized genetic parts – which have for example

* Correspondence: mascher@bio.lmu.de
†Equal contributors
1Department Biology I, AG Synthetic Microbiology, Ludwig-Maximilians-Universität München, Grosshaderner Str. 2-4, D-82152 Planegg-Martinsried, Germany
Full list of author information is available at the end of the article

been successfully used for the construction of novel genetic circuitries, such as a bacterial camera or a push-on-push-off-switch [4,5] – not only significantly simplify devices assembly, but also increase the reproducibility of the resulting constructs [6].

While there are a number of other assembly techniques like Gibson assembly [7], Golden Gate shuffling [8,9] or MoClo [10], the BioBrick standard still plays a key role in the framework of the annual iGEM competition. Moreover, it is also very useful for any other lab, since it is based on standard type II restriction endonucleases used for routine cloning. While the use of such standardized parts and assembly strategies is organism-independent, there is nevertheless a need for specific parts that accommodate organism-specific requirements, such as G+C content, codon preference and different expression and/or regulatory signals.

The classical BioBrick standard allows the free combination of most parts, but does not work for translational fusions, e.g. for addition of *gfp* or epitope-tags to protein-coding sequences. For this purpose, a number of BioBrick-compatible adaptations were developed, as described in RFC 23 and RFC 25 [11,12]. In each case, the combination of parts is performed via standard restriction digests and subsequent ligations, preferably with vector backbones of different antibiotic resistances to allow the so-called 3A-assembly [13].

Currently, the vast majority of available parts in the Registry of Standard Biological Parts were designed for the Gram-negative model organism *Escherichia coli*, due to its central role in iGEM and other synthetic biology projects. For other organisms, such as the Gram-positive model organism *Bacillus subtilis*, the range of available BioBricks is still very limited, especially when looking for well-evaluated and reliable parts. This limitation is unfortunate, given the unique features, powerful genetics and biotechnological relevance of this bacterium.

B. subtilis, together with other *Bacillus* species, is one of the main producers of industrially relevant enzymes, such as proteases, amylases and lipases. Its excellent fermentation properties, the ability to efficiently secrete proteins and the lack of toxic by-products render it indispensable for the biotechnological industry [14]. *B. subtilis* is the by far best-characterized Gram-positive bacterium [15], due to its powerful genetics and advantages for industrial use. In addition, *B. subtilis* has also become a model organism for studying bacterial (multicellular) differentiation, because of its capability to form highly resistant endospores upon nutrient limitation [16-18]. Another transient differentiation strategy is to become naturally competent for genetic transformation by synthesizing the machinery necessary for DNA uptake [17,19,20]. The high efficiency of this process and its tight association with homologous recombination not only enables easy genetic manipulations of the chromosome, but has also led to the development of mostly integrative vectors for use in *B. subtilis*, even though replicative vectors can also be used [21]. The advantages of integrative vectors are their stable maintenance and hence also defined copy number inside the chromosome.

Based on all those features and differences, the 2012 LMU-Munich iGEM-team decided to develop and provide a set of essential genetic tools for the work with *B. subtilis*. First of all, empty compatible integrative vectors are needed that can carry any construct of choice. Second, well-evaluated promoters of different strength, both constitutive and inducible, are required to control gene expression levels. For their evaluation and as a tool for measuring the activity of other promoters, reporter vectors are also required. For that purpose, we chose the well-established *lacZ* reporter, encoding the β-galactosidase, as well as a luciferase reporter (*luxABCDE*), which we carefully evaluated for the first time. Moreover, a set of epitope tags is provided to enable the convenient detection and purification of proteins. The resulting toolbox presented here, called the *Bacillus* BioBrick Box, (Table 1) is now completely evaluated and freely available in two DNA repositories, the Registry of Standard Biological Parts [2] and the Bacillus Genetics Stock Center [22]. The vector sources and formatted GeneBank files are given in Table 2 and Additional file 1, respectively.

Results and discussion

The vectors of the *Bacillus* BioBrick Box

To allow the integration of new devices and modules into the chromosome of *B. subtilis*, we first constructed five vectors (see Table 1 and Figure 1) by modifying established *B. subtilis* vectors to comply with the BioBrick standard. As described in detail below, we chose the original *B. subtilis* vectors such that they harbor three different resistance cassettes and also compatible homology regions for integration into the *B. subtilis* chromosome. This way, all three vectors and derived plasmids can be combined in a single strain. For convenient cloning, they all contain the *bla* cassette for ampicillin resistance and an *E. coli* origin of replication (ori).

The original *B. subtilis* vectors pDG1662, pAX01, pDG1731 [23,24] were modified to generate "empty" vectors that lack promoters and reporter genes (Figure 1A). Their integrative part contains the flanking homology regions, a resistance cassette for selection in *B. subtilis* and the multiple cloning site (MCS), containing an *rfp*-cassette flanked by the restriction sites EcoRI, NotI, XbaI (upstream) and SpeI, NotI and PstI (downstream) (Figure 1C). They allow cloning in BioBrick standard [3] with selection for white colonies as a result of the removal of the *rfp*-insert, which – if still present – leads to formation of red colonies in *E. coli*. All unnecessary parts, such as promoters or

Table 1 Overview of the *Bacillus* BioBrick Box including a summary of the features

BioBrick	Description and comments[a]	Reference of source	BGSC #	Registry #
Vectors				
pBS1C	Empty vector, integration at *amyE*, amp^r, cm^r	pDG1662-derivative [23]	ECE257	BBa_K823023
pBS2E	Empty vector, integration at *lacA*, amp^r, mls^r	pAX01-derivative [24]	ECE258	BBa_K823027
pBS4S	Empty vector, integration at *thrC*, amp^r, spec^r	pDG1731-derivative [23]	ECE259	BBa_K823022
pBS1C*lacZ*	*lacZ*-reporter vector, integration at *amyE*, amp^r, cm^r	pAC6-derivative [25]	ECE260	BBa_K823021
pBS3C*lux*	*lux*-reporter vector, integration at *sacA*, amp^r, cm^r	pAH328-derivative [26]	ECE261	BBa_K823025
Promoters				
P_{veg}	Very strong constitutive promoter	[27]	ECE262	BBa_K823003
P_{liaG}	Constitutive promoter	[28]	ECE263	BBa_K823000
P_{lepA}	Strong constitutive promoter	[29]	ECE264	BBa_K823002
J23101	Very weak constitutive promoter	BBa_J23101	ECE266	BBa_K823005
P_{liaI}	Bacitracin-inducible promoter	[30,31]	ECE267	BBa_K823001
P_{xylA}	Xylose-inducible promoter	[32]	ECE268	BBa_K823015
Epitope-tags				
His_{10}	10xHis-tag	[33]	ECE269	BBa_K823037
FLAG	FLAG-tag	[34]	ECE270	BBa_K823034
StrepII	Streptactin-tag	[35,36]	ECE271	BBa_K823038
HA	HA-tag	[34,37]	ECE272	BBa_K823035
cMyc	cMyc-tag	[34,38]	ECE273	BBa_K823036

[a]amp^r, ampicillin resistance; cm^r, chloramphenicol resistance; kan^r, kanamycin resistance; spc^r, spectinomycin resistance; mls^r, erythromycin-induced resistance to *m*acrolide, *l*incosamide and *s*treptogramin B-antibiotics (MLS).

additional resistance genes were deleted (see methods for details on construction). In all vectors the restriction sites interfering with the RFC 10 standard were removed. Moreover, for the empty vectors we also removed all AgeI and NgoMIV sites, thus allowing direct cloning in "Freiburg" standard [11] for translational fusions. Alignments with original vectors are provided in Additional file 2.

The three empty vectors, designated pBS1C, pBS2E and pBS4S, integrate into the *amyE*, *lacA* and *thrC* locus and harbor *cat*, *erm* and *spc* resistance cassettes, respectively (Table 1). The reporter vector pBS1C*lacZ* was derived from pAC6 [25] and contains the β-galactosidase reporter gene *lacZ* from *E. coli* with the *sacB* ribosome binding site (RBS) from *B. subtilis* downstream of the MCS. The luciferase-reporter vector pBS3C*lux*, which was derived from pAH328 [26], contains the *luxABCDE*-operon from *Photorabdus luminescence* with all RBSs adjusted for use in *B. subtilis* (Figure 1B). Both reporter vectors allow the measurement of promoter activities based on transcriptional fusions and mediate chloramphenicol resistance. They integrate into the *amyE* (pBS1C*lacZ*) and *sacA* (pBS3C*lux*) loci, respectively.

For transformation of *B. subtilis*, the vectors and derived plasmids need to be linearized at a unique ScaI site inside

Table 2 Vector sources for the *Bacillus* BioBrick Box

Vector	Description[a]	Used for	Ref.
pAC6	Vector for transcriptional promoter fusions to *lacZ*; integrates at *amyE*; cm^r, amp^r	pBS1C*lacZ*	[25]
pAH328	Vector for transcriptional promoter fusions to *luxABCDE* (luciferase); integrates at *sacA*; cm^r, amp^r	pBS3C*lux*	[26]
pDG1662	Empty vector, integrates at *amyE*, cm^r, spc^r, amp^r	pBS1C	[23]
pDG1731	Empty vector; integrates at *thrC*, spc^r, mls^r, amp^r	pBS4S	[23]
pAX01	Vector for xylose-dependent gene expression; integrates at *lacA*, mls^r, amp^r	pBS2E	[24]
pXT	Vector for xylose-inducible gene expression; integrates in *thrC*; spc^r, amp^r	P_{xylA}	[39]
pSB1C3	Replicative *E. coli* vector, MCS features *rfp*-cassette; cm^r	MCS	[40]
pGFPamy	Vector for transcriptional promoter fusions to *gfpmut1*; integrates at *amyE*; cm^r, amp^r	*gfpmut1*	[41]

[a]cm^r, chloramphenicol resistance; kan^r, kanamycin resistance; spc^r, spectinomycin resistance; mls^r, erythromycin-induced resistance to *m*acrolid-, *l*incosamid- and *s*treptogramin B-antibiotics (MLS); MCS, multiple cloning site.

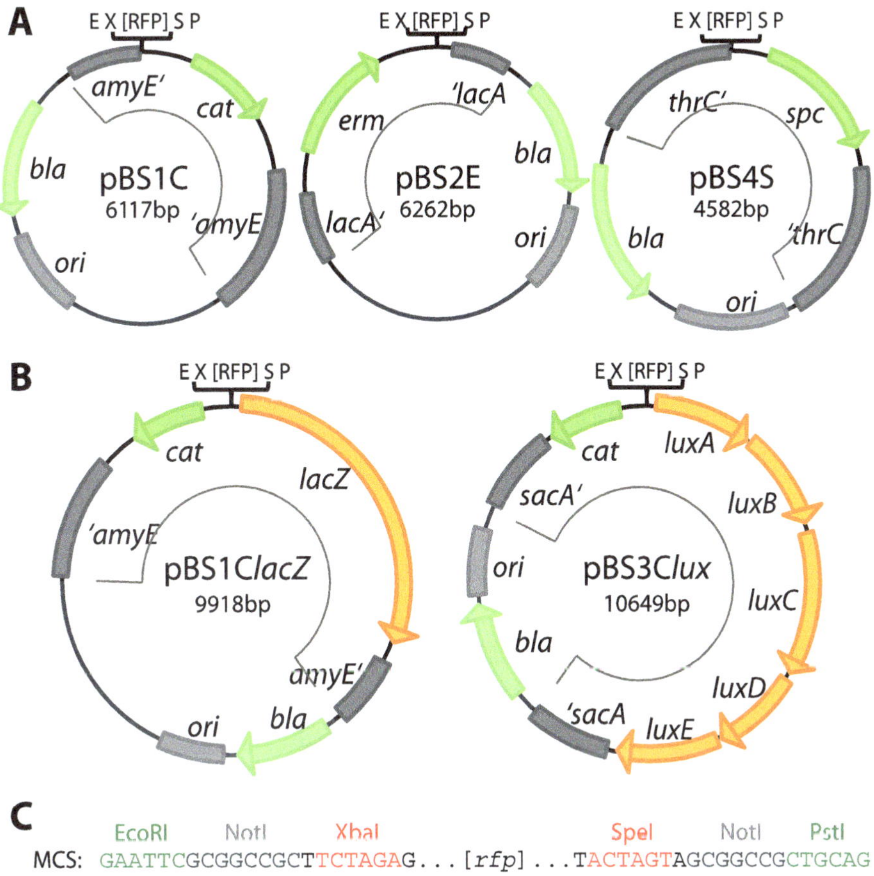

Figure 1 Maps of the modified *Bacillus* BioBrick Box-vectors in BioBrick (RFC10) standard. Resistance genes are indicated in green, reporter genes in orange and the origin of replication for *E. coli* as well as the integration loci are shown as gray boxes. The integrative part of the vectors is marked with a thin black line inside each vector map. The vector names are based on the following nomenclature: p = plasmid, BS = *B. subtilis*; the number refers to the integration locus: 1 = *amyE*, 2 = *lacA*, 3 = *sacA*, 4 = *thrC*; and the last letter codes for the resistance in *B. subtilis*: C = chloramphenicol (mediated by *cat*), E = MLS (mediated by *erm:* specifies resistance to macrolide, lincosamide and streptogramin B antibiotics if induced by erythromycin), S = spectinomycin (mediated by *spc*)). The addition in italics refers to the functional part of the reporter vectors: *lacZ* for β-galactosidase and *lux* for the *luxABCDE*-cassette encoding luciferase. **(A)** The three empty integrative vectors pBS1C, pBS2E and pBS4S contain different resistance genes (*cat*, *erm*, *spec*) and integration loci (*amyE*, *lacA*, *thrC*). This compatibility allows a combined usage in one single strain. They are derived from pDG1662, pAX01 and pDG1731, respectively. **(B)** Reporter vectors encoding β-galactosidase (*lacZ*) or luciferase (*luxABDCE*) as reporter genes. They integrate into *amyE* or *sacA*, respectively, and both mediate chloramphenicol resistance. All vectors harbor the RFC10-compatible multiple cloning site with an *rfp*-cassette, as shown in **(C)**. See text and material and methods for details on construction.

the *bla* gene. The only exception is pBS2E, which contains a second ScaI site within the integrative part. For this vector, BsaI or PciI can be used for linearization. The functionality of all empty vectors has been verified in a number of different projects (own unpublished results). The reporter vectors were subsequently used to evaluate the different promoters of the *Bacillus* BioBrick Box (see further below).

Characterization of the *thrC* integration locus

The four integration sites chosen for the vectors of the *Bacillus subtilis* BioBrick Box are commonly used for insertion of genetic constructs into the *B. subtilis* chromosome. The *sacA*, *lacA* and *amyE* genes encode metabolic enzymes required for the degradation of different alternative carbohydrates, i.e. sucrose, galactans and starch, respectively [42-44]. Accordingly, their expression is regulated by sugar availability [42,43,45], which is easy to control by experimental design. The fourth locus, *thrC*, is required for synthesis of threonine and amino acids derived from it. Expression of the *hom-thrC-thrB* operon was found to be high under many of the experimental conditions tested in a recent expression profiling study of *B. subtilis* [46]. While this is not surprising for an amino acid biosynthesis operon, it has major implications for use as an integration site. Regulation of the operon's promoter, P_{hom}, is not well understood in *B. subtilis*, and we therefore constructed a transcriptional P_{hom}-*luxABCDE* fusion to study the promoter activity under different growth conditions.

During growth in LB medium, representing amino acid-replete conditions, expression of P_{hom}-*luxABCDE* was at an intermediate level, reaching a maximum of 10^5 RLU/OD (Additional file 3: Figure S1). However, in the defined CSE medium, promoter activities were strongly increased and reached twelve times higher values than in LB medium. Addition of 0.1% casamino acids reduced these activities to two-fold above LB, and in the presence of 1% casamino acids expression was similar to growth in LB (Additional file 3: Figure S1). These data show that some considerations are required when using *thrC* as an integration site. That is, for applications in minimal or defined media, any construct to be integrated into *thrC* should contain a strong terminator upstream in order to prevent undesired read-through from P_{hom}.

The promoters of the *Bacillus* BioBrick Box

Promoters are crucial control elements for developing complex *in vivo* expression programs. For this purpose, they do not only have to be standardized, but also have to show a predictable and reliable activity under a variety of different conditions. For the optimization of metabolic pathways, the fine-tuning of gene expression is indispensable. Those features can sometimes be provided by inducible promoters or by developing promoter libraries [47-49], as described for *E. coli* [50,51]. To provide a first set of characterized promoter-parts for *B. subtilis*, we constructed and evaluated four constitutive promoters of different strength, as well as two inducible promoters that allow inducer-controlled gene expression levels.

Constitutive promoters

For our BioBrick Box, we chose three native σ^A-dependent *B. subtilis* promoters and eleven representatives of the Anderson promoter collection [51], which are often used as a gold standard for constitutive gene expression in *E. coli*. The native *B. subtilis* promoters comprise P_{liaG}, expressing regulatory components of the cell envelope stress response system [28], P_{lepA}, expressing highly conserved genes in bacteria and mitochondria necessary for exact protein translation [29], and P_{veg}, which is the strongest constitutive promoter in *B. subtilis* known to date [27]. The latter controls transcription of the *veg* gene, which seems to have an important function during the outgrowth of spores [52].

All relevant sequence features of the promoters evaluated in this study are shown in Figure 2; the complete sequences of the cloned fragments can be found in Additional file 1. To determine their activity in *B. subtilis*, the promoter fragments were cloned into the luciferase reporter vector pBS3C*lux* and assayed in LB medium as well as in chemically defined MCSE medium (MOPS-based CSE, see Methods for details).

Figure 3 shows both cell growth (Figure 3A/D) and the resulting luciferase activities (Figure 3B/E) over an extended growth period, ranging from exponential into the stationary growth phase in the two growth media. The activities, expressed as relative luminescence units (RLU) per OD_{600}, covered about three orders of magnitude (Figure 3B/E), ranging from ~10^3 RLU/OD for the Anderson promoter J23101 to ~10^6 RLU/OD for the strong P_{veg} promoter. For growth in LB, promoter activities varied within a factor of two to three during exponential phase, and declined during stationary phase by more than one order of magnitude (Figure 3B). These variations likely arise from metabolic changes that the cells undergo during growth in this complex medium, as also reflected in the ever-changing growth rate in LB medium (Figure 3A). In contrast, the growth rate in the chemically defined MCSE medium was more constant in exponential phase (Figure 3D) and luciferase activities displayed less dynamic variation than in LB medium (Figure 3E).

While the Anderson promoter J23101 is a strong promoter in *E. coli* [51], deviating from the *E. coli* consensus promoter sequence by only a single nucleotide exchange in the -35 promoter region (TTGACA to TTTACA), it was the only of eleven different Anderson promoters tested in *B. subtilis* that generated a measurable luminescence output at all (data not shown). Still, compared to the values measured in *E. coli*, its activity was surprisingly low and only about ten-fold above the background of the empty vector (Figure 3B/E). This result clearly demonstrates the context-dependence of expression signals and hence stresses the importance of establishing organism-specific parts.

The absolute luciferase activities as documented in Figure 3B/E do not only depend on the promoter strength, but are also dependent on experimental conditions and setups, which can vary significantly between labs and experiments. However, while absolute values might vary, the relative strength of the promoters has proven to be a rather robust and comparable unit between several experiments and even between different research groups [54]. Towards that end, a suitable standard promoter, measured under the same experimental conditions as the promoter under consideration, serves as reference value, and promoter activities are expressed relative to this standard promoter as *relative promoter units* (RPUs) [54]. The stable expression behavior and intermediate strength of P_{liaG} makes it ideally suited as a reference promoter, and hence we propose to use P_{liaG} for calculating RPUs in *B. subtilis*. The RPUs for all constitutive promoters, calculated by dividing their activity at an OD_{600} of 0.1 by the activity of P_{liaG} are shown in Figure 3C/F.

Promoter	Sequence (5'-3')
J23101	**TT**t**ACA**gctagctc-agtcctagg**TAT**t**AT**-N_6
P_{liaG}	N_{80}-g**T**at**CA**attcgcaagcttttcctt**TATAAT**-N_{11}
P_{lepA}	N_{99}-**TTGA**atctttacaa-tcctat**TG**a**TATAAT**-N_{29}
P_{veg}	N_{198}-**TTGACA**aaaatggg-ctcgtg**T**tg**TA**c**AAT**-N_9
P_{liaI}	N_{124}-*TACGACTCCGGTCTTATATAA*-N_{15}-**T**c**G**ttttgcatatcttccaac**T**tg**TATAA**g-N_{19}
P_{xylA}	N_{143}-**T**attg**A**aaatactga-cgagg**T**ta**TATAA**g-N_{14}-*GTTTGTTTAAACAACAAACT*-N_{60}

Figure 2 Nucleotide sequences of the *Bacillus* BioBrick Box promoter collection. Sequences are aligned according to their −35 and −10 region. All promoter fragments were designed with 3'-end exactly upstream of the RBS. The −10 and −35 regions (if conserved) are shown in bold and underlined. Regulator binding sites for XylR in P_{xylA} and LiaR in P_{liaI} are shown in italics and underlined. The native P_{xylA} also contains a catabolite responsive element (CRE) that suppresses the transcription in the presence of glucose [53]. This binding site is downstream of the promoter and is not present in the P_{xylA} –fragment used here, which is derived from the pXT vector [39].

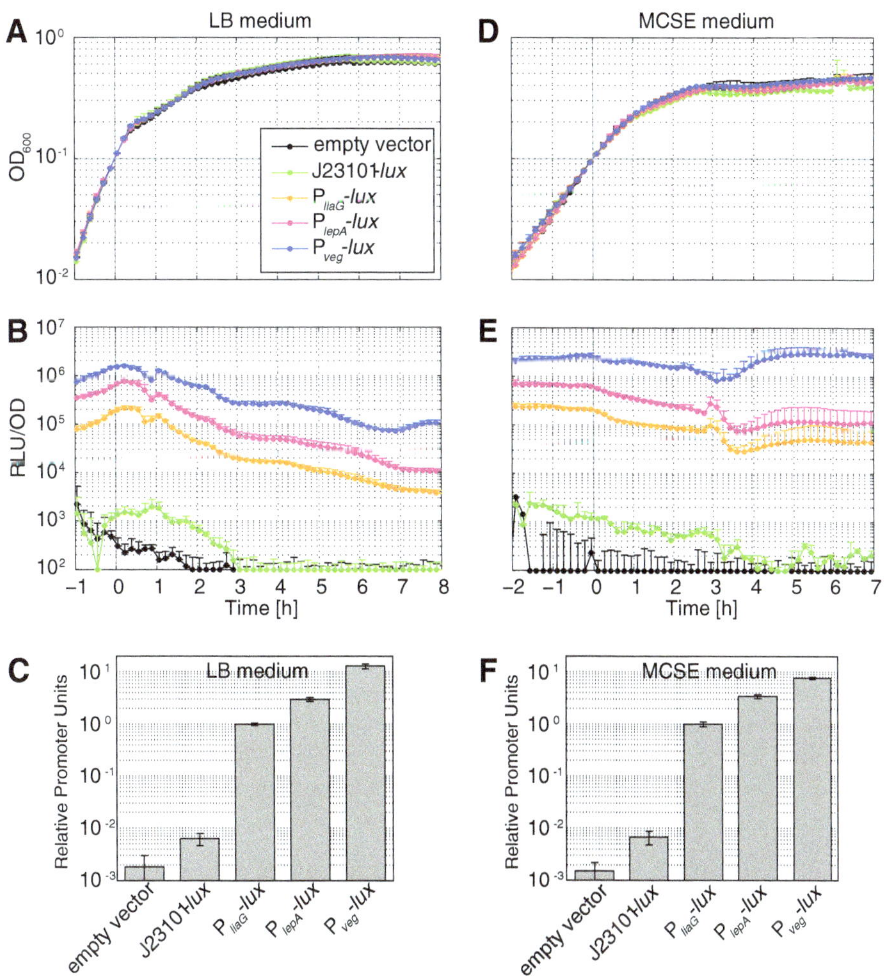

Figure 3 Luciferase activities of the constitutive *Bacillus* BioBrick Box promoters. Measurement of promoter activity was done based on pBS3C*lux*-derivatives via luminescence of 100 µl cultures growing in a plate reader (Biotek®, Synergy™2; 96-well plate, 37°C, shaking) over time. The graphs are aligned to an OD_{600} = 0.1 at time point 0 (*grey line*). **(A)**-**(C)** show data of growth in complex LB medium; **(D)**-**(F)** show data of growth in the defined medium MCSE. **(A)** and **(D)** Growth curves (OD_{600}) from cultures of the reporter strains TMB1872 (*lux* without promoter), TMB1862 (J23101-*lux*), TMB1856 (P_{liaG}-*lux*), TMB1860 (P_{lepA}-*lux*) and TMB1930 (P_{veg}-*lux*). **(B)** and **(E)** Promoter activities of the same strains expressed as luminescence in relative luminescence units (RLU) per OD_{600} and using the same color code as in panels **(A)** and **(D)**. **(C)** and **(F)** Bar graphs of relative promoter units (RPUs, normalized to P_{liaG}) at time point 0, corresponding to OD_{600} = 0.1. All graphs show mean values and standard deviations (error bars) of at least four biological replicates.

The bacitracin-inducible promoter P_{liaI}

Next, we chose the bacitracin-inducible promoter P_{liaI} for our BioBrick Box, due to its high dynamic range and low basal activity [30,31]. P_{liaI} controls the *liaIH* operon of *B. subtilis*, which is the main target of the envelope stress-inducible two-component system LiaRS. Upon induction with the cell wall antibiotic bacitracin, LiaS activates the cognate response regulator LiaR, which in turn strongly induces P_{liaI} [28,55].

For quantitative analysis, P_{liaI} was cloned into the luciferase reporter vector pBS3C*lux*, grown to exponential phase and induced with different concentrations of bacitracin (see Methods for details). Both in LB as well as in MCSE medium we observed a transient, concentration-dependent increase of luciferase activities for about two to four hours post-induction, which decreased about 100-fold towards stationary phase (Figure 4B/E). The maximal activity of about 300-fold induction was reached 1 hour after bacitracin treatment in LB (30 minutes in MCSE) at a concentration of 30 μg ml^{-1} bacitracin (Figure 4C/F), as shown previously [31]. Furthermore, induction with 30 μg ml^{-1} bacitracin leads to minor growth defects, while higher bacitracin concentrations resulted in the lysis of the culture (Figure 4A/D). The respective dose-response curves expressed in RPUs (normalized to P_{liaG}) indicate that P_{liaI} responded more sensitively to low bacitracin concentrations in LB compared to MCSE medium (Figure 4C/F). This might reflect higher stress levels experienced due to faster growth in LB compared to MCSE medium (Figure 4A/D). Notably, the basal activity of P_{liaI} was approximately 0.03 RPU and hence – in the absence of a suitable inducer – perfectly fits into the "activity gap" between the constitutive promoters J23101 (0.006 RPU) and P_{liaG} (1 RPU).

The xylose-inducible promoter P_{xylA}

Natively, P_{xylA} drives the expression of the *xylAB* operon in *B. subtilis*, which is required for using xylose as carbon and energy source. This promoter is regulated by XylR and CcpA and hence only induced by xylose if glucose is simultaneously absent, due to carbon catabolite repression [53]. The P_{xylA} derivative used in this study is amplified from the expression vector pXT [39] and lacks the catabolite-responsive element (CRE). As a result, the xylose-inducible promoter of the BioBrick Box is not subject to repression. Nevertheless, the choice of carbon source can still affect the xylose-dependent induction of P_{xylA} as described in more detail below.

The P_{xylA} reporter responded to xylose in a concentration-dependent manner in both LB and MCSE medium and reached its maximum activity about 1-1.5 hours post-induction (Figure 5B/E). Due to a significantly increased basal promoter activity in LB, there was only a ten-fold difference between minimal and maximal activity (Figure 5C), compared to a 100-fold difference in MCSE (Figure 5F). While the maximal activity of P_{xylA} was similar to P_{liaI}, the basal activity of the non-induced promoter, especially in LB medium, was significantly higher for P_{xylA}, resulting in a lower dynamic range. On the other hand, the activity of P_{xylA} was sustained, instead of transient for P_{liaI}, and the sensitivities towards its inducer xylose were comparable in LB and MCSE medium (Figure 5C/F).

As mentioned above, removal of the CRE sequence from P_{xylA} removed the repressive effects of glucose via CcpA, but we still observed significantly reduced promoter activities during growth on glucose as carbon source. It was previously shown that glucose-6-phosphate competes with xylose for binding to the xylose repressor protein XylR and prevents its de-activation [56]. We therefore tested xylose-dependent induction of P_{xylA} during growth in CSE medium containing different carbon sources. During growth on glucose, addition of 0.2% xylose led to a 20-fold promoter induction, while up to 70-fold changes were observed during growth on fructose, mannitol or arabinose (Additional file 3: Figure S4). Metabolism of the latter three sugars does not produce glucose-6-phosphate as an intermediate and thus bypasses its repressive effect. The reporter strain showed similar growth rates during growth on glucose, fructose or mannitol (doubling time ca. 50 min), while growth on arabinose was markedly slower at ca. 65 min doubling times (Additional file 3: Figure S4 and data not shown). Based on these results we suggest using fructose as the carbon source for routine experiments in minimal or defined media when working with P_{xylA}.

The reporters of the *Bacillus* BioBrick Box

Reporter genes are important genetic tools for the measurement of promoter activities. The BioBrick Box therefore provides the established *lacZ* [57] and the *luxABCDE* reporter in the aforementioned reporter vectors pBS1C*lacZ* and pBS3C*lux*. To evaluate both vectors, we used the bacitracin-inducible *liaI* promoter to compare the characteristics of both reporters. The Miller units obtained for the *lacZ* reporter were compared to the RLU/OD_{600} values of the *luxABCDE* reporter one hour post-induction at identical inducer concentrations.

Both reporters showed a high dynamic range over about four orders of magnitude between the empty vector as background and the maximal P_{liaI} activity (Additional file 3: Figure S2). Importantly, we observed a linear correlation between P_{liaI}-*lux* and P_{liaI}-*lacZ* activity, demonstrating that each of the reporters provides an accurate measurement of promoter activity. Nevertheless, the *lux*-reporter has several advantages over *lacZ*: luminescence can be monitored online in growing cultures and neither requires the addition of a substrate nor performing a separate

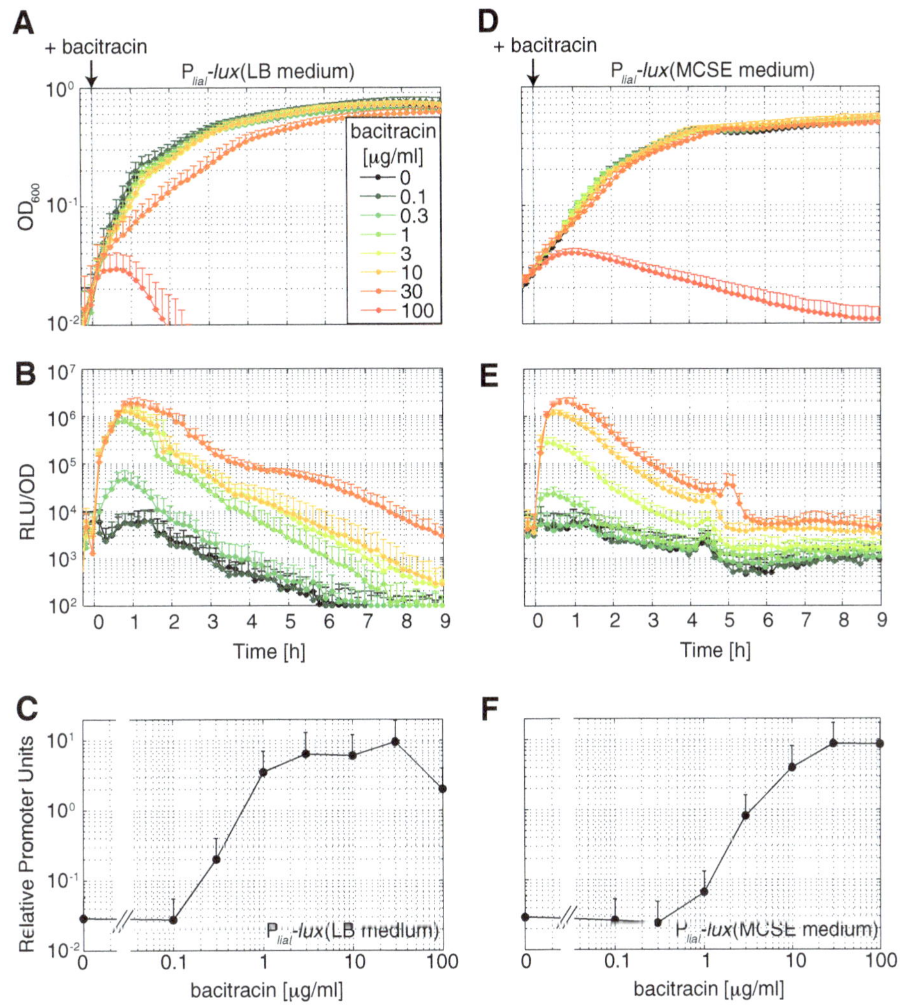

Figure 4 Luciferase activity of the bacitracin-inducible P_{liaI} promoter. Measurement of P_{liaI} activity in pBS3C*lux* upon induction with various bacitracin concentrations indicated in the legend. Addition of water was used as non-induced reference. For details of growth and measurement conditions, see legend of Figure 3 and Methods. **(A)**-**(C)** Growth in LB medium. **(D)**-**(F)** Growth in MCSE medium. **(A)** and **(D)** Growth curves (OD_{600}) from cultures of the reporter strain TMB1858 (P_{liaI}-*lux*). **(B)** and **(E)** Promoter activities upon bacitracin induction expressed as luminescence in RLU/OD_{600}. **(C)** and **(F)** Dose–response curves one hour post-induction in RPUs (normalized to P_{liaG}). The data on induction with 100 µg/ml bacitracin was omitted in **(B)** and **(E)** due to lysis of the culture. All graphs show mean values and standard deviations (error bars) of four biological replicates.

assay. Moreover, we have determined an average half-life of only 4.2 ± 0.3 min for the luciferase signal (Additional file 3: Figure S3), compared to a lifetime in the order of hours for β-galactosidase (data not shown). Hence, the *lux* reporter is particularly very well suited to resolve transcriptional responses occurring at timescales shorter than the cell-cycle. In fact, given this short lifetime one might expect that the luminescence activity follows the promoter activity almost instantly.

In order to test how closely the luminescence may follow the underlying promoter activity, we devised a simplified mathematical model that includes the processes of *lux* mRNA transcription and degradation as well as Lux protein translation and decay (Figure 6A); see Methods for all assumptions and details. This model then enabled us to relate the time-dependent luminescence signal back to the time-dependent changes in the apparent promoter activity, by making the simplifying assumption that all changes in the luminescence signal are caused by changes in the transcription rate alone. As a proof-of-principle, we fitted the experimental luminescence activity of the P_{liaI} promoter induced with 30 µg/ml bacitracin in MCSE (Figure 6B; *symbols*) by our model (Figure 6B; *colored lines*), and extracted the underlying time-dependent apparent promoter activity (Figure 6C). Here it is clearly visible that the fit of the experimental data is best for an mRNA half-life of 5 min or shorter, since otherwise rapid changes in the luminescence signal cannot be reproduced

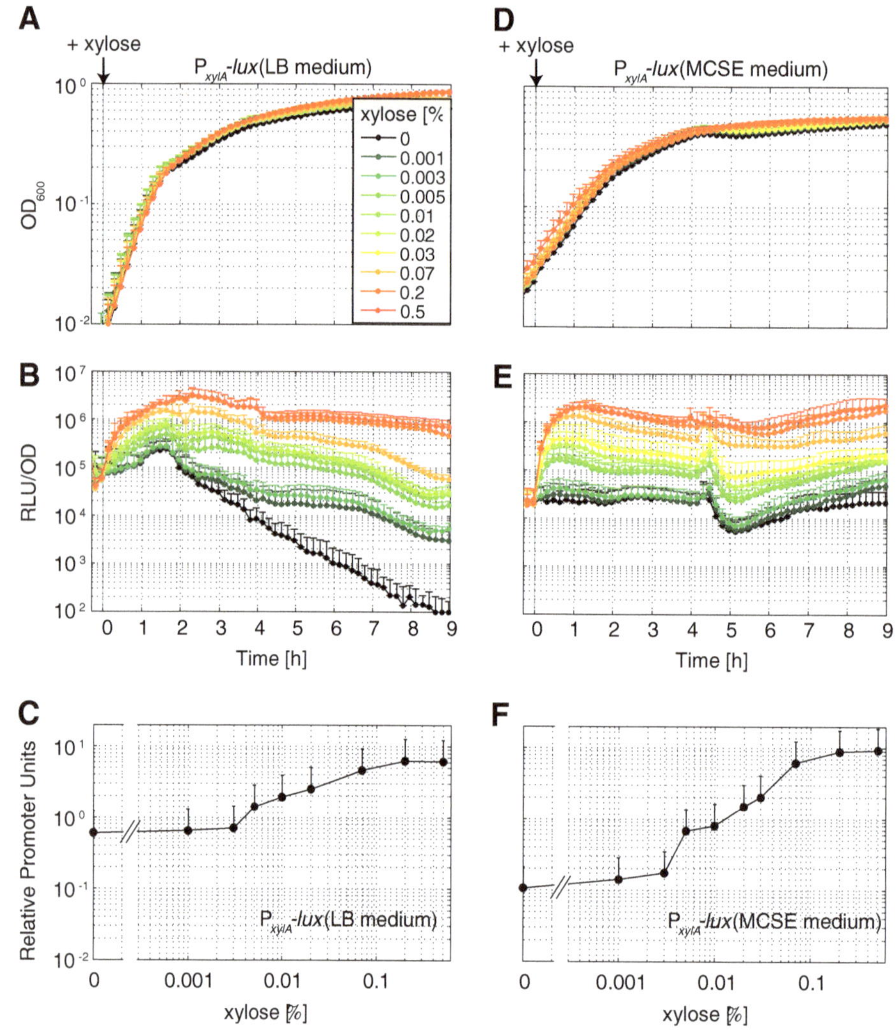

Figure 5 Luciferase activity of the xylose-inducible P_{xylA} promoter. Measurement of P_{xyl}-activity in pBS3C*lux* upon induction with various concentrations (see legend) of xylose. Addition of water was used as non-induced reference. For details of growth and measurement conditions, see legend of Figure 3 and methods. **(A)-(C)** Growth in LB **(D)-(F)** Growth in MCSE. **(A)** and **(D)** Growth curves (OD_{600}) from cultures of the reporter strain TMB1931 (P_{xylA}-*lux*) in semi-logarithmic scale. **(B)** and **(E)** Promoter activities upon xylose-induction expressed as luminescence in RLU/OD_{600}. **(C)** and **(F)** Dose–response curves one hour post-induction in RPUs (normalized to P_{liaG}). All graphs show mean values and standard deviations (error bars) of at least four biological replicates.

within this model (Figure 6B; *arrow*). Strikingly, the model shows that the time course of the apparent promoter activity is almost identical to the luminescence activity, with only a ~10 min delay between promoter activity and the luminescence signal (Figure 6C), further underlining the power of the *lux* reporter in resolving highly dynamic transcriptional responses. It should be noted that the large fluctuations in apparent promoter activities at values below 10^{-2} arbitrary units are likely due to experimental noise and do not necessarily reflect true changes in transcription rate.

The epitope tags of the *Bacillus* BioBrick Box

Epitope tags are useful tools for both protein detection by Western blot analysis using tag-specific antibodies and for purification purposes or protein interaction studies. In our *Bacillus* BioBrick Box, we provide five widely used tags – FLAG-, His_{10}-, StrepII-, HA-, and cMyc-tag – in a modified Freiburg standard that was developed to allow the generation of both N- and C-terminal translational fusions (Figure 7A). The DNA and protein sequences of all tags are shown in Figure 7B. To evaluate our tags, fusions to *gfpmut1* under control of a constitutive promoter were generated as described in Methods. Western blot analyses with GFP-specific antibodies showed that all proteins could be detected in the soluble protein fraction and were therefore correctly produced (Figure 7C). Analyses with tag-specific antibodies showed that addition of the different tags led to a slight shifting of apparent sizes between

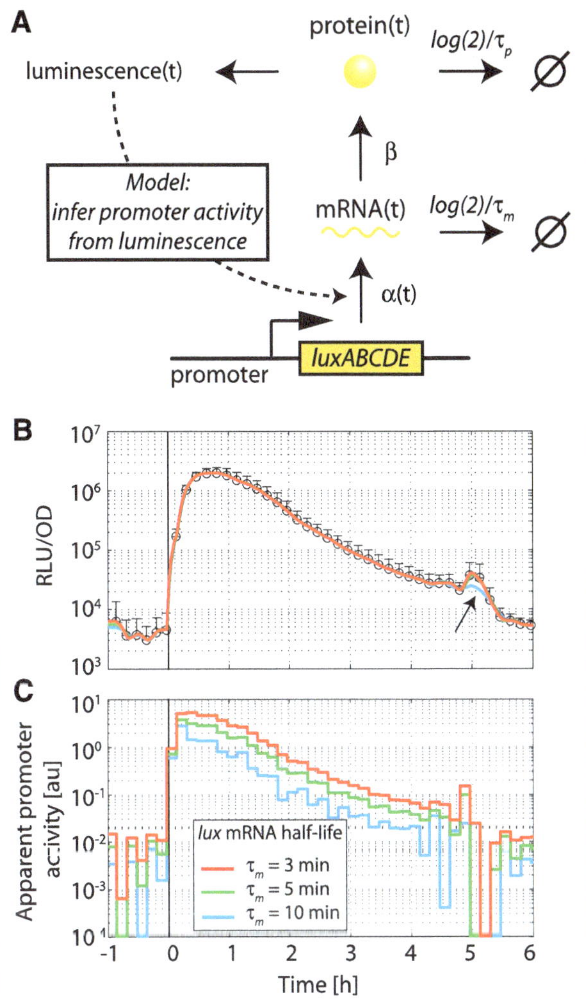

Figure 6 Inference of promoter activity from luminescence measurements. (A) Quantitative mathematical model used to infer the underlying promoter activity from luminescence time-series. Here, the promoter activity is reflected in a time-dependent transcription rate, α*(t)*. *luxABCDE* mRNA is degraded at rate $\log(2)/\tau_m$, translated at rate β into Lux proteins, which are in turn degraded at rate $\log(2)/\tau_p$. For simplicity, the model assumes a single Lux protein species, which is rate-limiting for luciferase activity and considered to be directly proportional to luminescence output. **(B)** As a test case for our method, the luciferase activity of pBS3C*lux*-P_{lial} (TMB1858) induced with 30 µg/ml bacitracin (*symbols*) was fitted by our quantitative mathematical model (*solid lines*). Since the *luxABCDE* mRNA half-life is unknown in *B. subtilis*, we fitted the data using a physiological spectrum of half-lives (τ_m = 3, 5 and 10 min) and found that the model can reproduce the highly dynamic luciferase activity (*arrow*) only for mRNA half-lives of 5 min and shorter. **(C)** Underlying apparent promoter activity in arbitrary units (au) for the fits in **(B)**. For all details see main text.

the N- and C-terminal fusions in the range of 25 to 30 kDa on the SDS-PAGE (Figure 7C).

The original FLAG-tag was developed as hydrophilic hexapeptide marker sequence (Lys-Asp-Asp-Asp-Asp-Lys) that is located at the outside of protein surfaces. Adaptation to *B. subtilis* was performed by Kaltwasser *et al.* [34]. In Western blots, the C-terminal fusion to GFP can be detected very well, while the signal for the N-terminal fusion was weak. The His_{10}-tag was developed and is used for affinity purification on Ni^{2+}-NTA-columns and Western blotting [33]. Both the N- and C-terminal fusions to GFP were easily detectable in Western blots (Figure 7C). The Strep-tag II, which was optimized from the original Strep-tag to allow both N- and C-terminal fusion, can be used together with the streptavidin-derivative Strep-Tactin as its specific binding partner [35,36]. The expression of the N-terminal fusion seemed to be rather low (anti-GFP blot) and detection with anti-strep antibodies was weak. Nevertheless, Strep-tagII is the epitope of choice for co-immunoprecipitation in protein-protein interaction studies in *B. subtilis* due to the highly specific interaction with Strep-Tactin-covered magnetic beads or resins [58]. The HA-tag was named after the human influenza *h*em*a*gglutinin, a viral surface protein that is important for host cell infection [37]. The corresponding HA-tag was codon-optimized for *B. subtilis* [34] and worked very well for detection of C-terminally tagged GFP. The cMyc-tag is derived from the human *c-myc* proto-oncogene [38], which was also optimized for *B. subtilis* [34]. C-terminal fusion to *gfp* seemed to negatively influence the GFP-expression levels. However, the detection with anti-cMyc-antibodies worked well for both N- and C-terminal fusions.

In summary, all five epitope tags were shown to be functional. The ideal terminus for creating protein-tag fusions as well as the most suitable tag will have to be determined experimentally for each target protein and application. For such studies, the parts from the BioBrick Box offer a convenient and fast way for optimization of conditions and constructs.

Conclusion

In this study, we described the construction and evaluation of a toolkit containing essential vectors, promoters and epitope-tags for standardized genetic assembly in *B. subtilis*, a biotechnological workhorse and Gram-positive model organism.

We modified and adapted five integrative *B. subtilis* vectors to the BioBrick standard. Three are empty vectors that are fully compatible to each other and allow their combination in one single strain due to different integration loci and resistance cassettes. Two additional reporter vectors harbor *lacZ* and *luxABCDE* cassettes for evaluating transcriptional promoter fusions. While β-galactosidase is already long established as a reporter and needs no further introduction, we conducted a detailed study of the *lux*-reporter system. We determined the half-life of the luciferase signal as 4.2 ± 0.3 min, making the reporter instable enough to directly monitor both induction and repression of promoters almost in real time, as revealed by our mathematical model for the *lux*-reporter system. This

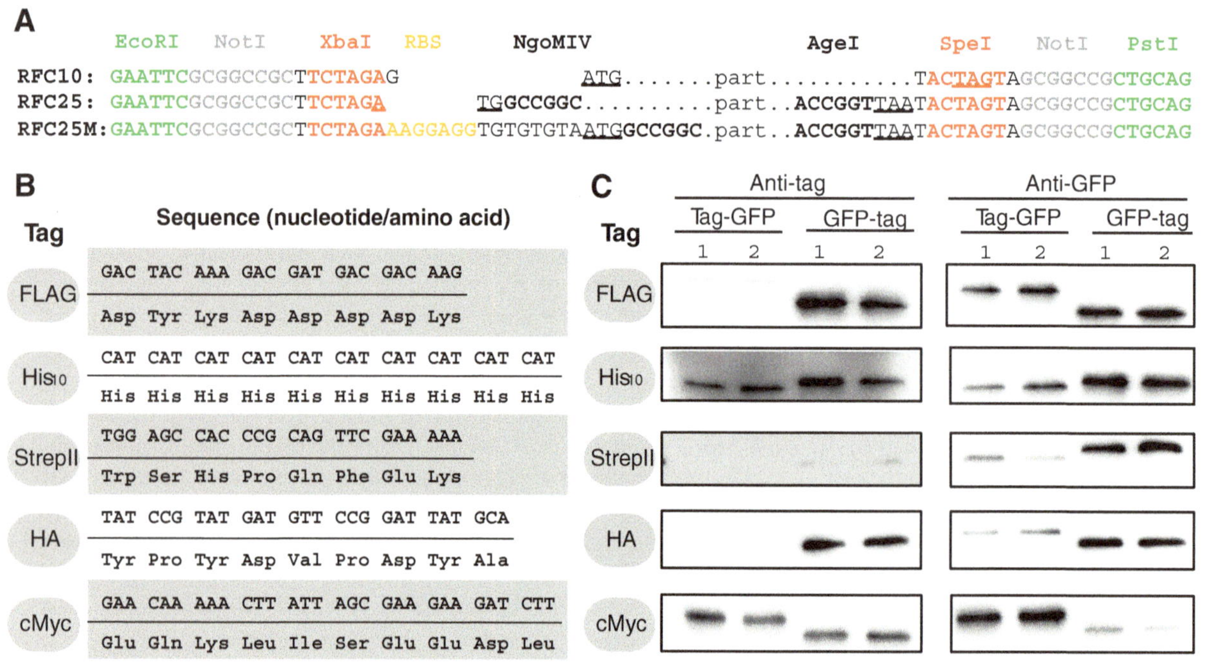

Figure 7 Epitope tags of the *Bacillus* BioBrick Box. (A) Pre- and suffixes of different cloning standards. The start- and the first stop-codon are underlined. RFC10 [3] describes the commonly used BioBrick standard. To combine two parts, the upstream part is cut with SpeI, the downstream one with XbaI (*red sites*). The compatible overhangs of the two isocaudomers can be ligated, and result in an eight basepair scar that cannot be recut but is optimal for RBS spacing. This causes a frameshift that is not suitable for translational fusions. RFC25 [11] describes the "Freiburg standard" which was developed by the 2007 iGEM-Team from Freiburg to allow translational fusions. Two parts can be combined similar to RFC10, but using NgoMIV and AgeI as restriction enzymes. The resulting scar ACCGGC codes for Threonine and Glycine. RFC25M refers to our modification of RFC25 by adding an optimized RBS (*yellow*). While this convenient modification simplifies the cloning procedure, it should be noted that as a consequence it also prevents downward compatibility, since the resulting BioBrick directly adds two functional DNA parts in a single step. The tag-BioBricks and *gfpmut1* are designed in that standard. **(B)** Nucleotide and amino acid sequences of the epitope tags, without the RFC25M pre- and suffixes shown in **(A)**. **(C)** Western blots of N- and C-terminal fusions of each tag to GFP, using the strains TMB1920 (Flag-*gfp*), TMB1921 (*gfp*-Flag), TMB1922 (HA-*gfp*), TMB1923 (*gfp*-HA), TMB1924 (cMyc-*gfp*), TMB1925 (*gfp*-cMyc), TMB1926 (His-*gfp*), TMB1927 (*gfp*-His), TMB1928 (StrepII-*gfp*) and TMB1929 (*gfp*-StrepII). For each construct, two independent clones were tested with epitope tag- and GFP-specific antibodies as a positive control.

feature, together with the very low background activity and the possibility to determine promoter activities directly in growing cultures, makes this an ideal reporter system, especially since the *lux* cassette from *P. luminescence* present in pBS3C*lux* does not require the addition of any luciferase substrate.

Four constitutive promoters and the uninduced P_{liaI} offer an activity range of three orders of magnitude with stable expression levels in defined MCSE medium and during exponential growth phase in LB medium. We adopted the concept of RPUs for *B. subtilis* and propose the P_{liaG} promoter to be used as a reference standard for RPU determination in this organism. The *Bacillus* BioBick Box also contains two inducible promoters, P_{liaI} and P_{xylA}, which were thoroughly evaluated with the inducers bacitracin and xylose, respectively. Using inducer concentrations over about three orders of magnitude, the two promoters revealed a dynamic range of 300- or 100-fold activity over background, respectively. P_{liaI} has a transient activity in complex and defined medium, whereas P_{xylA} displays sustained induction.

Finally, we evaluated five widely used epitope tags as N- and C-terminal fusions to GFP via Western blotting. All are functional, but results for the test case GFP indicate a context-dependence in immuno-detection: The FLAG- and HA-tag were only suitable for C-terminal fusions to GFP, whereas His_{10} and cMyc worked both for N- and C-terminal fusions.

Taken together, we provide a set of well-evaluated standardized tools for the work in *B. subtilis*. We hope that this toolkit will inspire not only future iGEM-teams to choose *B. subtilis* as their chassis organism, but also greatly facilitate the engineering of this versatile model organism for synthetic biology applications in the future.

Methods

Bacterial strains and growth conditions

B. subtilis and *E. coli* were routinely grown in Luria-Bertani (LB) medium (1% (w/v) tryptone, 0.5% (w/v) yeast extract, 1% (w/v) NaCl) at 37°C with agitation (200 rpm). All strains used in this study are listed in Additional file 3: Table S2.

Selective media for *B. subtilis* contained kanamycin (10 µg ml^{-1}), spectinomycin (100 µg ml^{-1}), chloramphenicol (5 µg ml^{-1}), or erythromycin in combination with lincomycin (1 µg ml^{-1}; 25 µg ml^{-1} for mls^r). Selective media for *E. coli* contained ampicillin (100 µg ml^{-1}) or chloramphenicol (35 µg ml^{-1}). Solid media additionally contained 1.5% (w/v) agar.

Transformation

E. coli (XL1 blue, Stratagene) competent cells were prepared and transformed according to OpenWetWare [59]. Transformations of *B. subtilis* were carried out as described previously [21]. The integration of plasmids into the *B. subtilis* genome was checked on starch plates (*amyE*), with minimal medium (*thrC*) or colony PCR (*sacA*, *lacA*). Detailed protocols can be found in the Additional file 3 and [60].

MOPS-based chemically defined medium with succinate and glutamate (MCSE)

The chemically defined CSE medium (3.3 g l^{-1} $(NH_4)_2SO_4$, 29 mM KH_2PO_4, 70 mM K_2HPO_4, 1 × III'-salts (100 × III'-salts: 0.232 g l^{-1} $MnSO_4 \times 4\ H_2O$, 12.3 g l^{-1} $MgSO_4 \times 7\ H_2O$), 50 mg l^{-1} Tryptophan, 22 mg l^{-1}ammonium ferric citrate, 0.8% (w/v) K-glutamate, 0.6% (w/v) Na-succinate, up to 2.5% (w/v) C-source) is used for experiments with stable growth conditions and a defined C-source [61]. In long-term experiments (18 hours), we noticed a white precipitate disturbing the OD_{600}-measurements in the microplate reader. We tested the use of a phosphate-reduced 3-(N-Morpholino)-Propanesulfonic acid (MOPS) buffer (10 × MOPS buffer: 83.72 g l^{-1} MOPS, 33 g l^{-1} $(NH_4)_2SO_4$, 3.85 mM KH_2PO_4, 6.15 mM K_2HPO_4; adjusted to pH 7 with KOH), which did not cause precipitation. This is the composition of the newly developed MOPS-based MCSE medium: 1 × MOPS, 50 mg l^{-1} Tryptophan, 22 mg l^{-1}ammonium ferric citrate, 1 × III'-salts, 0.8% (w/v) K-glutamate, 0.6% (w/v) Na-succinate, 0.2% (w/v) fructose. The C-source for MCSE can be varied according to the experimental requirements.

DNA manipulation and plasmid construction

General cloning procedure, such as endonuclease restriction digest, ligation and PCR, was performed with enzymes and buffers from New England Biolabs (NEB; Ipswich, MA, USA) according to the respective protocols. For subsequent cloning, Phusion® polymerase was used for PCR amplifications, otherwise OneTaq® was the polymerase of choice. PCR-purification was performed with the *HiYield PCR Gel Extraction/PCR Clean-up Kit* (Süd-Laborbedarf GmbH (SLG), Gauting, Germany). Plasmid preparation for mutagenesis-PCR templates was performed with the *HiYield Plasmid Mini-Kit* (SLG) according to the manufacturer's protocol, otherwise by alkaline lysis plasmid preparation. All plasmids created during this study are listed in Additional file 3: Table S1, all primer sequences are given in Additional file 3: Table S3.

Vectors of the *Bacillus* BioBrick Box

In all vectors, the PstI site in the *bla* gene was removed by site-directed mutagenesis using the oligonucleotides TM2206 and TM2207. The resulting vector is marked with the auxiliary "*bla*mut". Primer design and mutagenesis were performed according to the manufacturer's instructions for the QuikChange® II Site-Directed Mutagenesis Kit (Agilent Technologies). All plasmids were test-digested to confirm removal of the restriction site.

To create the empty vector **pBS1C**, pDG1662*bla*mut was cut with PstI to remove one XbaI site and the spc^r outside of the integrative part. The 6 kb fragment was religated. The remaining PstI site was mutated via site-directed mutagenesis with the primers TM2845 and TM2846. To insert the MCS, the vector was cut with EcoRI. The MCS was amplified by PCR from pSB1C3 with the Primers TM2843 and TM2844, cut with EcoRI and BsaI (EcoRI-compatible overhang) and ligated into the vector. The remaining NgoMIV sites were removed by subsequent site-directed mutagenesis with TM3005 + TM3006, TM3011 + TM3012 and TM3013 + TM3014, respectively, resulting in pBS1C.

To create the empty vector **pBS2E**, pAX01 was cut with SacI. The 6.3 kb fragment was religated to remove the *xylR*-P_{xylA}-fragment. The PstI site in *bla* was removed. The vector was cut with XbaI and the 6 kb fragment religated to reduce the amount of forbidden restriction sites. The lost terminator of *erm* was replaced by the PCR-amplified terminator with BsaI-overhangs (primers: TM2975 and TM2976). The vector was opened with XbaI and ligated with the PCR product cut with BsaI (XbaI-compatible overhang). The correct direction was checked by sequencing and removal of the XbaI-site was confirmed. Finally, the MCS was amplified from pSB1C3 (TM2608 and TM2609) with PstI and NsiI overhangs, cut with PstI and NsiI and ligated into the PstI-cut vector. Correct orientation of the insert was confirmed by restriction digest and sequencing. The remaining NgoMIV sites were removed by subsequent site-directed mutagenesis with TM3011 + TM3012 and TM3028 + TM3029, respectively, resulting in pBS2E.

To create the empty vector **pBS4S**, the *erm*-resistance outside of the integrative part of pDG1731 was removed via cut-ligation with MluI and BssHI, resulting in a 4.7 kb vector. The PstI site in *bla* was removed as described above. The PstI site in *thrB* was removed performing site-directed mutagenesis with the primers TM2835 and TM2836. The *spc*-promoter with upstream PstI-overhang was amplified with the primers TM2837 and TM2838 and cut with PstI and PciI. The MCS was cut from pSB1C3 with EcoRI and

PstI. The vector was cut with EcoRI and PciI and the 4.3 kb-fragment ligated with both cut DNA-fragments. The remaining NgoMIV sites were removed by subsequent site-directed mutagenesis with TM3005 + TM3006 and TM3011 + TM3012, respectively, resulting in pBS4S.

To create the reporter vector **pBS1C*lacZ***, pAC6*bla*mut was cut with EcoRI and PstI. The primers TM2301 and TM2302 (1 pM) were mixed, heated (95°C, 10 min), re-annealed (50°C, 10 min) to become double stranded with 4 bp overhangs and ligated into the vector. The vector was cut with EcoRI and PstI and ligated with the MCS from pSB1C3 cut with EcoRI and PstI. A PstI site in the *E. coli* ori was removed by cutting with BglII and religation of the 11 kb fragment.

To create the reporter vector **pBS3C*lux***, the PstI site in the *sacA* locus of pAH328*bla*mut was removed by site-directed mutagenesis with the primers TM2885 and TM2886. Then, the XbaI site of *luxD* was removed similarly with the Primers TM2887 and TM2888. The MCS was amplified by PCR from pSB1C3 with the primers TM2843 and TM2884, cut with EcoRI and BsaI (SpeI-overhang) and ligated into the EcoRI/SpeI-cut vector.

Promoters of the *Bacillus* BioBrick Box

The promoter J23101 was used from the parts registry [2]. The promoters P_{liaG}, P_{lepA}, P_{veg} and P_{liaI} were amplified from *B. subtilis* W168 genomic DNA using the primer pairs TM2891/TM2892, TM2899/TM2900, TM2903/TM2904 and TM2895/TM2896, respectively. P_{xylA} was amplified from pXT [39] with the primers TM2968 + TM2969. The PCR products were cut with EcoRI and SpeI, ligated into pSB1C3, pBS3C*lux* and pBS1C*lacZ* (P_{liaI} only) and verified by sequencing.

Epitope-tags of the *Bacillus* BioBrick Box

All eptitope tags were synthesized with pre- and suffixes (Figure 7A) together as a single part in a plasmid from GeneArt (Life Technologies Corporation, USA). The plasmid was cut with EcoRI and PstI and, without purification steps, ligated into pSB1C3 cut with EcoRI and PstI. The products were checked by sequencing for their correct insert. *gfpmut1* was adapted to modified Freiburg standard by PCR with the primers TM2934/TM2935 and pGFPamy as template and cloned into pSB1C3 (both cut with EcoRI and SpeI). For translational fusion, the pSB1C3-tag-constructs were cut open with AgeI + PstI (tag is N-terminal) or EcoRI + NgoMIV (tag is C-terminal) and purified. *gfp* was cut with the corresponding restriction enzymes of the Freiburg standard, purified from an agarose gel and ligated into the vector. The tag-*gfp* or *gfp*-tag fusions as well as pBS0K*Pspac** were cut with EcoRI and PstI, purified from agarose gels, ligated and verified by sequencing.

Luciferase assay

Luciferase activities of strains harbouring pBS3C*lux*-derivates were assayed using a Synergy™2 multi-mode microplate reader from BioTek® (Winooski, VT, USA). The reader was controlled using the software Gen5™. Culture volumes were 100 µl per well in 96-well plates (black walls, clear bottom; Greiner Bio-One, Frickenhausen, Germany), and incubation occurred at 37°C with agitation (intensity: medium). Cell growth was monitored by optical density at 600 nm wavelength (OD_{600}). Raw luminescence output (relative luminescence units, RLU) was normalized to cell density by dividing each data-point by its corresponding corrected OD_{600} value (RLU/OD).

For constitutive promoters, LB or MCSE medium was inoculated 1:500 from overnight cultures of each strain. Cultures were incubated at 37°C with agitation and OD_{600} as well as luminescence were monitored every 10 min for at least 13 hours. For inducible promoters, 10 ml of LB or MCSE medium were inoculated 1:500 from overnight cultures and grown to $OD_{600} = 0.2\text{-}0.5$. Those pre-cultures were diluted to $OD_{600} = 0.05$ (MCSE) or 0.01 (LB), respectively, and transferred to eight (P_{liaI}) or ten (P_{xylA}) wells of a 96-well plate. The OD_{600} as well as luminescence were monitored every 10 min for one hour. At an $OD_{600} \sim 0.1$ (corresponding to $OD_{600} \sim 0.4$ in cuvettes of 1 cm light path length) 5 µl of the inducer was added to final concentrations of 0.1, 0.3, 1, 3, 10, 30 or 100 µg ml^{-1} Zn^{2+}-bacitracin or 0.001, 0.003, 0.005, 0.01, 0.02, 0.03, 0.07, 0.2, 0.5% (w/v) xylose. To one culture-well, 5 µl of water was added as uninduced control. Cultures were incubated at 37°C with agitation and the OD_{600} as well as luminescence were monitored every 10 min for at least 13 hours.

β-Galactosidase assay

LB medium (30 ml) was inoculated 1:100 from fresh overnight culture of strains harboring pBS1C*lacZ*-P_{liaI} or pBS1C*lacZ* without promoter and grown at 37°C, shaking at 200 rpm. At $OD_{600} = 0.4\text{-}0.6$, the culture was split into 3 ml samples and induced with 30 µl of Zn^{2+}-bacitracin to the final concentrations of 0.1, 0.3, 1, 3, 10, 30 or 100 µg ml^{-1}. To one sample, 30 µl of water was added as a negative control. After incubation for another 30 min, 2 ml of the samples were harvested by centrifugation (13000 × *g*, 1-3 min) and the pellets were stored at –20°C. For the assay, the pellets were resuspended in 1 ml of working buffer and assayed for β-galactosidase activity as described elsewhere, with normalization to cell density [62].

Western blot analysis

To verify the functionality of the epitope tags, Western blot analyses of the strains TMB1920-TMB1929 were performed. LB medium (15 ml) was inoculated 1:100

from overnight culture and grown at 37°C and 200 rpm to $OD_{600} \sim 0.5$. Of this, 10 ml were harvested by centrifugation (8000 × *g*, 5 min) and the pellets stored at −20°C. Pellets were resuspended in 1 ml disruption buffer (50 mM Tris–HCl pH 7.5, 100 mM NaCl) and lysed by sonication. Samples (12 µl of lysate) were loaded per lane on two 12.5% SDS-polyacrylamide gels and SDS-PAGE was performed according standard procedure [60]. One gel was stained with colloidal coomassie, the other one was used for protein transfer to a PVDF membrane (Merck Millipore, Billerica, MA, USA) by submerged blotting procedure (Mini Trans-Blot Electrophoretic Transfer Cell (Bio-Rad, Hercules, CA, USA)). After protein transfer, the membranes were treated with the following antibodies and conditions. Detailed protocols can be found in the Additional file 3.

GFP

Probing with primary antibodies takes place with rabbit anti-GFP antibodies (1:3000, Epitomics, No. 1533). Horseradish-peroxidase (HRP)-conjugated anti-rabbit antibodies (1:2000, Promega, W401B) were used as secondary antibody. Hybridization of both antibodies was carried out in Blotto-buffer (2.5% (w/v) skim milk powder, 1 × TBS (50 mM Tris–HCl pH 7.6, 0.15 M NaCl)).

FLAG

Rabbit anti-FLAG (1:2000, Sigma, Anti-Flag polyclonal, F7425) and anti-rabbit-HRP (1:2000, Promega, W401B) were used in Blotto-buffer.

His_{10}

Mouse-anti-Penta-His (1:2000, Qiagen, Penta-His, No. 34660) in 1 × TBS, 5% bovine serum albumin (BSA, Fraction V; Carl ROTH) and anti-mouse-HRP (1:2000, Promega, W402B1) in 1 × TBS, 10 % (w/v) skim milk powder were used.

StrepII

Strep-Tactin-HRP conjugate (IBA, Strep-Tactin-HRP conjugate, No. 2-1502-001) 1:100 in 1 × PBS (4 mM KH_2PO_4; 16 mM Na_2HPO_4; 115 mM NaCl) with 0.1% (w/v) Tween20 was used.

HA

Rabbit anti-HA (1:500, Sigma, H6908) in TBS, 0.05% (w/v) Tween20, 5% (w/v) skim milk powder and anti-rabbit-HRP (1:2000, Promega, W401B) in Blotto-buffer were used.

cMyc

Rabbit anti-Myc (1:2000, Abcan, ab9106) in TBS, 0.05% (w/v) Tween20, 5% (w/v) skim milk powder and anti-rabbit-HRP (1:2000, Promega, W401B) in Blotto-buffer were used.

Chemiluminescence signals were detected after addition of the HRP-substrate Ace Glow (Peqlab, Erlangen, Germany) using a Fusion™ imaging system (Peqlab).

Determination of luciferase half-life

To determine the half-life of the output of the luciferase reporter system, *B. subtilis* harboring the P_{xyl}-*luxABCDE* reporter construct was grown in CSE medium in the presence of 0.15% (w/v) xylose under the conditions described for luciferase assays with constitutive promoters. When luciferase activities reached approximately 10^5 RLU/OD_{600} (early exponential phase), further protein synthesis was stopped by the addition of 500 µg ml^{-1} tetracycline, and luminescence and OD_{600} were monitored every 5 min for 1 h. The half-life of the luminescence output was determined from a fit of the data from eight replicate assays with an exponential decay function (Additional file 3: Figure S3).

Data analysis of luciferase assays

For each individual sample, the optical density (OD_{600}) and luminescence values (RLU) were background-corrected by subtracting the respective values measured for wells containing 100 µl of the corresponding medium. Only for the measurements of the constitutive promoters, in which wells were inoculated with cells right from the start of the experiment, the OD_{600} of pure medium was initially higher than the OD_{600} of culture-filled wells, such that the average values of culture-filled wells at t = 0 h were used as blank for OD_{600}. In order to compensate for sample-to-sample variations in the growth lag of individual cultures, we aligned all individual measurements of the constitutive promoters on the time-axis, such that $OD_{600} = 0.1$ was shifted to t = 0 h. Subsequently RLU/OD_{600} values were calculated for individual measurements and for each condition both mean and standard deviation of RLU/OD_{600} values were determined from at least three biological replicates.

Mathematical model for promoter activity

To get an estimate for the time-dependence of the promoter activity that results in a given luciferase activity, we set up a simple mathematical model for the expression of the Lux-system (Figure 6). For simplicity, the model assumes that one of the Lux proteins is rate-limiting for light production, such that the luminescence is proportional to this protein species. The corresponding differential equations for the concentration of *luxABCDE* mRNA, *m(t)*, and the rate-limiting Lux protein, *p(t)*, read

$$\frac{d}{dt}m(t) = \alpha(t) - \frac{log(2)}{\tau_m}m(t) \qquad (1)$$

and

$$\frac{d}{dt}p(t) = \beta m(t) - \frac{log(2)}{\tau_p}p(t), \quad (2)$$

where $\alpha(t)$ is the time-dependent apparent transcription rate, β the translation rate and τ_m and τ_p the mRNA and protein half-lives, respectively. Although mRNA and protein stability might also depend on cell physiology, time depended changes in these parameters were ignored to avoid further unknown parameters. Likewise, we did not model the kinetics of inducer uptake/signal transduction explicitly, but subsumed these processes in the kinetics of the time-dependent apparent transcription rate. For simplicity, the translation rate was directly expressed in units of RLU/mRNA/min, such that *p(t)* has units of RLU. The model in Eqs. (1) and (2) was solved analytically for *p(t)*, and then fitted to the experimental luminescence trajectories. To this end, all parameters except $\alpha(t)$ were fixed to the following values: $\tau_p = 4.2$ min (experimentally determined in Additional file 3: Figure S3), $\beta = 10^4$ RLU/mRNA/min (arbitrary, but adjusted to meet typical transcription rates), $\tau_m = 3$, 5 and 10 min (typical values for bacterial mRNA half-lives [63]). Then $\alpha(t)$ was discretized into N intervals of constant transcription rate α_i $(i = 1,...N)$ (the length of the intervals was adjusted to $\Delta t = 10$ min as in the luciferase assay). Finally, the α_i were estimated by fitting *p(t)* to the luminescence trajectories using a trust-region reflective Newton method for least-squares minimization (MATLAB, The MathWorks, Inc.).

Additional files

Additional file 1: GeneBank files of *Bacillus* BioBricks.
Additional file 2: Alignment of original and final vectors.
Additional file 3: Supplemental Figures, Tables and Text.

Competing interests
The authors declare that they have no competing interests.

Authors' contributions
TM, JR, KK, JB, TC, FD, JE and SK conceptualized the contents of the *Bacillus* BioBrick Box and designed strategies for the modification of the vectors. JR constructed all vectors. JR and KK cloned and evaluated the promoters. KK cloned the tag-*gfp*-fusions and performed all Western blots. SG and CS performed the analyses of P_{hom} activities and analyzed the influence of C-sources on P_{xyl} activation. GF performed the quantitative data analysis and mathematical modeling. JR, GF, SG and TM wrote the manuscript. TM supervised the 2012 iGEM team LMU-Munich and coordinated the experimental work. All authors read and approved the final manuscript.

Authors' information
Jara Radeck, Korinna Kraft, Julia Bartels, Tamara Cikovic, Franziska Dürr, Jennifer Emenegger and Simon Kelterborn were members of the 2012 iGEM-team LMU-Munich.

Acknowledgements
The authors would like to thank Tina Wecke for supervising the iGEM-Team 2012 and all initiators, organizers and volunteers of iGEM, because this work would never have been initiated without this competition. We thank Richard Losick for providing the pAH328, Angelika Böttger for providing antibodies against the cMyc- and HA-tag, and Ainhoa Revilla-Guarinos, Tobias Bauer and Nicole Treichelt for initial cloning and preliminary data.
This work was supported by grants from the Deutsche Forschungsgemeinschaft (DFG MA2837/3-1 to TM and GE2164/3-1 to SG) and from the Fonds der Chemischen Industrie (to SG). Finally, the LMU-Munich 2012 iGEM team would like to acknowledge the generous support from all of their sponsors: BioTek, DSM, ERA Synbio, GeneArt (Life Technologies), LMU Faculty of Biology, the LMU genomics service unit, Lehre@LMU, Münchner Universitätsgesellschaft, Sarstedt, and Semadeni (gold sponsors); AppliChem and Bio-Rad (silver sponsors); the Bacillus Genetic Stock Center BGSC and Süd-Laborbedarf GmbHSLG (bronze sponsors).

Author details
[1]Department Biology I, AG Synthetic Microbiology, Ludwig-Maximilians-Universität München, Grosshaderner Str. 2-4, D-82152 Planegg-Martinsried, Germany. [2]Ludwig-Maximilians-University Munich, Arnold Sommerfeld Center for Theoretical Physics, Theresienstr. 37, D-80333 München, Germany. [3]Present affiliation: Institute of Cell and Molecular Biosciences, Newcastle University, Centre for Bacterial Cell Biology, Richardson Road, NE2 4AX Newcastle upon Tyne, UK.

References
1. Endy D: **Foundations for engineering biology.** *Nature* 2005, **438:**449–453.
2. **Registry of standard biological parts.** http://partsregistry.org.
3. **Draft standard for BioBrick biological parts.** http://hdl.handle.net/1721.1/45138.
4. Levskaya A, Chevalier AA, Tabor JJ, Simpson ZB, Lavery LA, Levy M, Davidson EA, Scouras A, Ellington AD, Marcotte EM, *et al*: **Synthetic biology: engineering *Escherichia coli* to see light.** *Nature* 2005, **438:**441–442.
5. Lou C, Liu X, Ni M, Huang Y, Huang Q, Huang L, Jiang L, Lu D, Wang M, Liu C, *et al*: **Synthesizing a novel genetic sequential logic circuit: a push-on push-off switch.** *Mol Sys Biol* 2010, **6:**350.
6. Müller KM, Arndt KM: **Standardization in synthetic biology.** *Methods Mol Biol* 2012, **813:**23–43.
7. Gibson DG: **Synthesis of DNA fragments in yeast by one-step assembly of overlapping oligonucleotides.** *Nucleic Acids Res* 2009, **37:**6984–6990.
8. Engler C, Gruetzner R, Kandzia R, Marillonnet S: **Golden gate shuffling: a one-pot DNA shuffling method based on type IIs restriction enzymes.** *PLoS ONE* 2009, **4:**e5553.
9. Engler C, Kandzia R, Marillonnet S: **A one pot, one step, precision cloning method with high throughput capability.** *PLoS ONE* 2008, **3:**e3647.
10. Werner S, Engler C, Weber E, Gruetzner R, Marillonnet S: **Fast track assembly of multigene constructs using golden gate cloning and the MoClo system.** *Bioeng Bugs* 2012, **3:**38–43.
11. **Fusion Protein (Freiburg) BioBrick assembly standard.** [http://hdl.handle.net/1721.1/45140].
12. **A New BioBrick Assembly Strategy Designed for Facile Protein Engineering.** [http://hdl.handle.net/1721.1/32535].
13. Shetty R, Lizarazo M, Rettberg R, Knight TF: **Assembly of BioBrick standard biological parts using three antibiotic assembly.** *Methods Enzymol* 2011, **498:**311–326.
14. van Dijl JM, Hecker M: ***Bacillus subtilis*: from soil bacterium to super-secreting cell factory.** *Microb Cell Fact* 2013, **12:**3.
15. Earl AM, Losick R, Kolter R: **Ecology and genomics of *Bacillus subtilis*.** *Trends Microbiol* 2008, **16:**269–275.
16. Errington J: **Regulation of endospore formation in *Bacillus subtilis*.** *Nat Rev Microbiol* 2003, **1:**117–126.
17. Lopez D, Vlamakis H, Kolter R: **Generation of multiple cell types in *Bacillus subtilis*.** *FEMS Microbiol Rev* 2009, **33:**152–163.
18. McKenney PT, Driks A, Eichenberger P: **The *Bacillus subtilis* endospore: assembly and functions of the multilayered coat.** *Nat Rev Microbiol* 2013, **11:**33–44.
19. Chen I, Christie PJ, Dubnau D: **The ins and outs of DNA transfer in bacteria.** *Science* 2005, **310:**1456–1460.
20. Chen I, Dubnau D: **DNA uptake during bacterial transformation.** *Nat Rev Microbiol* 2004, **2:**241–249.
21. Harwood CR, Cutting SM: *Molecular Biological Methods for Bacillus.* Chichester: John Wiley & Sons; 1990.

22. **Bacillus Genetic Stock Center.** http://www.bgsc.org/.
23. Guerout-Fleury AM, Frandsen N, Stragier P: **Plasmids for ectopic integration in *Bacillus subtilis*.** *Gene* 1996, **180**:57–61.
24. Hartl B, Wehrl W, Wiegert T, Homuth G, Schumann W: **Development of a new integration site within the *Bacillus subtilis* chromosome and construction of compatible expression cassettes.** *J Bacteriol* 2001, **183**:2696–2699.
25. Stülke J, Martin-Verstraete I, Zagorec M, Rose M, Klier A, Rapoport G: **Induction of the *Bacillus subtilis ptsGHI* operon by glucose is controlled by a novel antiterminator, GlcT.** *Mol Microbiol* 1997, **25**:65–78.
26. Schmalisch M, Maiques E, Nikolov L, Camp AH, Chevreux B, Muffler A, Rodriguez S, Perkins J, Losick R: **Small genes under sporulation control in the *Bacillus subtilis* genome.** *J Bacteriol* 2010, **192**:5402–5412.
27. Lam KH, Chow KC, Wong WK: **Construction of an efficient *Bacillus subtilis* system for extracellular production of heterologous proteins.** *J Biotechnol* 1998, **63**:167–177.
28. Jordan S, Junker A, Helmann JD, Mascher T: **Regulation of LiaRS-dependent gene expression in *Bacillus subtilis*: Identification of inhibitor proteins, regulator binding sites and target genes of a conserved cell envelope stress-sensing two-component system.** *J Bacteriol* 2006, **188**:5153–5166.
29. Qin Y, Polacek N, Vesper O, Staub E, Einfeldt E, Wilson DN, Nierhaus KH: **The highly conserved LepA is a ribosomal elongation factor that back-translocates the ribosome.** *Cell* 2006, **127**:721–733.
30. Mascher T, Zimmer SL, Smith TA, Helmann JD: **Antibiotic-inducible promoter regulated by the cell envelope stress-sensing two-component system LiaRS of *Bacillus subtilis*.** *Antimicrob Agents Chemother* 2004, **48**:2888–2896.
31. Toymentseva AA, Schrecke K, Sharipova MR, Mascher T: **The LIKE system, a novel protein expression toolbox for *Bacillus subtilis* based on the *liaI* promoter.** *Microb Cell Fact* 2012, **11**:143.
32. Kreuzer P, Gartner D, Allmansberger R, Hillen W: **Identification and sequence analysis of the *Bacillus subtilis* W23 *xylR* gene and *xyl* operator.** *J Bacteriol* 1989, **171**:3840–3845.
33. Hochuli E, Bannwarth W, Dobeli H, Gentz R, Stuber D: **Genetic approach to facilitate purification of recombinant proteins with a novel metal chelate adsorbent.** *Nat Biotech* 1988, **6**:1321–1325.
34. Kaltwasser M, Wiegert T, Schumann W: **Construction and application of epitope and green fluorescent protein-tagging integration vectors for *Bacillus subtilis*.** *Appl Env Microbiol* 2002, **68**:2624–2628.
35. Skerra A, Schmidt TG: **Applications of a peptide ligand for streptavidin: the strep-tag.** *Biomol Eng* 1999, **16**:79–86.
36. Voss S, Skerra A: **Mutagenesis of a flexible loop in streptavidin leads to higher affinity for the strep-tag II peptide and improved performance in recombinant protein purification.** *Protein Eng* 1997, **10**:975–982.
37. Wilson IA, Niman HL, Houghten RA, Cherenson AR, Connolly ML, Lerner RA: **The structure of an antigenic determinant in a protein.** *Cell* 1984, **37**:767–778.
38. Evan GI, Lewis GK, Ramsay G, Bishop JM: **Isolation of monoclonal antibodies specific for human *c-myc* proto-oncogene product.** *Mol Cell Biol* 1985, **5**(12):3610–3616.
39. Derre I, Rapoport G, Msadek T: **The CtsR regulator of stress response is active as a dimer and specifically degraded in vivo at 37°C.** *Mol Microbiol* 2000, **38**:335–347.
40. **Part: pSB1C3.** [http://parts.igem.org/wiki/index.php/Part:pSB1C3].
41. Bisicchia P, Botella E, Devine KM: **Suite of novel vectors for ectopic insertion of GFP, CFP and IYFP transcriptional fusions in single copy at the *amyE* and *bglS* loci in *Bacillus subtilis*.** *Plasmid* 2010, **64**(3):143–149.
42. Arnaud M, Vary P, Zagorec M, Klier A, Debarbouille M, Postma P, Rapoport G: **Regulation of the *sacPA* operon of *Bacillus subtilis*: identification of phosphotransferase system components involved in SacT activity.** *J Bacteriol* 1992, **174**:3161–3170.
43. Daniel RA, Haiech J, Denizot F, Errington J: **Isolation and characterization of the *lacA* gene encoding beta-galactosidase in *Bacillus subtilis* and a regulator gene, *lacR*.** *J Bacteriol* 1997, **179**:5636–5638.
44. Yamaguchi K, Nagata Y, Maruo B: **Genetic control of the rate of alpha-amylase synthesis in *Bacillus subtilis*.** *J Bacteriol* 1974, **119**:410–415.
45. Kim JH, Guvener ZT, Cho JY, Chung KC, Chambliss GH: **Specificity of DNA binding activity of the *Bacillus subtilis* catabolite control protein CcpA.** *J Bacteriol* 1995, **177**:5129–5134.
46. Nicolas P, Mäder U, Dervyn E, Rochat T, Leduc A, Pigeonneau N, Bidnenko E, Marchadier E, Hoebeke M, Aymerich S, *et al*: **Condition-dependent transcriptome reveals high-level regulatory architecture in *Bacillus subtilis*.** *Science* 2012, **335**:1103–1106.
47. Dehli T, Solem C, Jensen PR: **Tunable promoters in synthetic and systems biology.** *Subcell Biochem* 2012, **64**:181–201.
48. Hammer K, Mijakovic I, Jensen PR: **Synthetic promoter libraries–tuning of gene expression.** *Trends Biotechnol* 2006, **24**:53–55.
49. Mutalik VK, Guimaraes JC, Cambray G, Lam C, Christoffersen MJ, Mai QA, Tran AB, Paull M, Keasling JD, Arkin AP, *et al*: **Precise and reliable gene expression via standard transcription and translation initiation elements.** *Nat Methods* 2013, **10**:354–360.
50. Braatsch S, Helmark S, Kranz H, Koebmann B, Jensen PR: ***Escherichia coli* strains with promoter libraries constructed by Red/ET recombination pave the way for transcriptional fine-tuning.** *Biotechniques* 2008, **45**:335–337.
51. **Anderson promoter collection.** [http://parts.igem.org/Promoters/Catalog/Anderson].
52. Fukushima T, Ishikawa S, Yamamoto H, Ogasawara N, Sekiguchi J: **Transcriptional, functional and cytochemical analyses of the *veg* gene in *Bacillus subtilis*.** *J Biochem* 2003, **133**:475–483.
53. Jacob S, Allmansberger R, Gartner D, Hillen W: **Catabolite repression of the operon for xylose utilization from *Bacillus subtilis* W23 is mediated at the level of transcription and depends on a cis site in the *xylA* reading frame.** *Mol Gen Genet* 1991, **229**:189–196.
54. Kelly JR, Rubin AJ, Davis JH, Ajo-Franklin CM, Cumbers J, Czar MJ, de Mora K, Glieberman AL, Monie DD, Endy D: **Measuring the activity of BioBrick promoters using an in vivo reference standard.** *J Biol Eng* 2009, **3**:4.
55. Schrecke K, Jordan S, Mascher T: **Stoichiometry and perturbation studies of the LiaFSR system of *Bacillus subtilis*.** *Mol Microbiol* 2013, **87**:769–788.
56. Dahl MK, Schmiedel D, Hillen W: **Glucose and glucose-6-phosphate interaction with Xyl repressor proteins from *Bacillus* spp. may contribute to regulation of xylose utilization.** *J Bacteriol* 1995, **177**:5467–5472.
57. Miller JH: *Experiments in molecular genetics.* Cold Spring Harbor, NY: Cold Spring Harbor Laboratory Press; 1972.
58. Herzberg C, Weidinger LA, Dorrbecker B, Hübner S, Stülke J, Commichau FM: **SPINE: a method for the rapid detection and analysis of protein-protein interactions in vivo.** *Proteomics* 2007, **7**:4032–4035.
59. **Competent *E. coli* cells.** http://openwetware.org/wiki/TOP10_chemically_competent_cells.
60. Sambrook J, Russell DW: *Molecular Cloning - A Laboratory Manual.* Cold Spring Harbor, N.Y.: Cold Spring Harbor Laboratory Press; 2001.
61. Stülke J, Hanschke R, Hecker M: **Temporal activation of β-glucanase synthesis in *Bacillus subtilis* is mediated by the GTP pool.** *J Gen Microbiol* 1993, **139**:2041–2045.
62. Miller JH: *Experiments in molecular genetics.* Cold Spring Harbor Laboratory: Cold Spring Harbor, New York; 1972.
63. Bernstein JA, Khodursky AB, Lin P-H, Lin-Chao S, Cohen SN: **Global analysis of mRNA decay and abundance in *Escherichia coli* at single-gene resolution using two-color fluorescent DNA microarrays.** *Proc Natl Acad Sci U S A* 2002, **99**:9697–9702.

12

Employing aromatic tuning to modulate output from two-component signaling circuits

Rahmi Yusuf[1] and Roger R Draheim[2*]

Abstract

Two-component signaling circuits (TCSs) govern the majority of environmental, pathogenic and industrial processes undertaken by bacteria. Therefore, controlling signal output from these circuits in a stimulus-independent manner is of central importance to synthetic microbiologists. Aromatic tuning, or repositioning the aromatic residues commonly found at the cytoplasmic end of the final TM helix has been shown to modulate signal output from the aspartate chemoreceptor (Tar) and the major osmosensor (EnvZ) of *Escherichia coli*. Aromatic residues are found in a similar location within other bacterial membrane-spanning receptors, suggesting that aromatic tuning could be harnessed for a wide-range of applications. Here, a brief synopsis of the data underpinning aromatic tuning, the initial successes with the method and the inherent advantages over those previously employed for modulating TCS signal output are presented.

Keywords: Synthetic microbiology, Aromatic tuning, Receptor engineering, Signal modulation, Synthetic circuits

Introduction

Two-component signaling circuits (TCSs) are a ubiquitous mechanism by which bacteria sense, respond and adapt to external stimuli. TCSs facilitate responses to a wide range of environmental parameters, such as ambient temperature, availability of nutrients or external osmolarity [1]. Medically relevant multiorganism phenomena including quorum-sensing [2] and host-pathogen interaction are also governed by TCSs [3]. Furthermore, these systems control essential agricultural and environmental processes such as chloroplast synthesis [4] and nitrogen fixation [5], which is involved in root nodule formation. More than 400,000 open reading frames (ORFs) believed to encode TCSs have been sequenced and annotated suggesting that these systems control an almost unlimited number of processes that could be harnessed by synthetic microbiologists [6, 7]. Based on this premise, engineering TCSs with novel functionality is a very active field [8], especially because it has been recently shown that these modified systems can be systematically transferred to mammalian cells [9]. However, rationally designing TCSs, either by moving intact components from one microorganism to another or by forming chimeric components from the domains of two different proteins, can result in aberrant signal output or loss of stimulus-perception, or both [10]. Here, an aromatic tuning methodology that has been established to circumvent these issues and is generally applicable to both native and rationally designed TCSs is described.

* Correspondence: rogerdraheim@gmail.com
[2]School of Pharmacy and Biomedical Sciences, University of Portsmouth, St. Michael's Building, White Swan Road, Portsmouth PO1 2DT, England, UK
Full list of author information is available at the end of the article

Two-component signaling circuits

Traditionally, a TCS consists of a membrane-spanning sensor histidine kinase (SHK) and a cytoplasmic response regulator (RR) [1], however, various higher-order protein architectures have been identified [11]. The largest group of membrane-spanning SHKs possesses a periplasmic or extracellular domain responsible for stimulus perception with signal transduction to the cell interior occurring via the adjacent transmembrane (TM) domain [12]. The extent of input stimulus usually controls the ratio of phosphorylation (kinase activity) to dephosphorylation (phosphatase activity) of the cognate RR, thereby regulating the intracellular level of phosphorylated RR [13]. Phosphorylation modulates the activity of the output domain, which normally interacts with DNA to control transcription of genes appropriate for mediating a response to the perceived stimulus (Fig. 1) [1]. Aromatic tuning, depicted as a red box at the cytoplasmic end of the TM domain in Fig. 1,

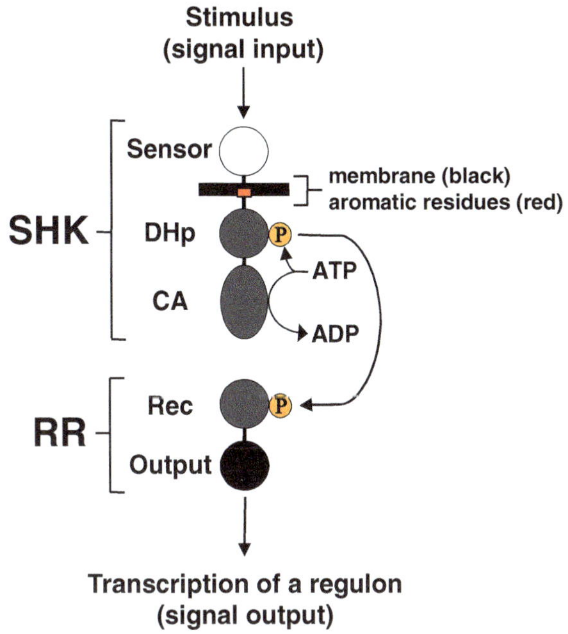

Fig. 1 Modularity of two-component signaling circuits (TCSs). When the sensor domain of a canonical SHK perceives stimulus, communication occurs across the membrane (black line) resulting in increased kinase activity of the catalytic ATPase (CA) domain. This enhances phosphorylation of the conserved histidyl residue within the domain responsible for dimerization and histidylphosphotransfer (DHp). These nascent phosphoryl groups are subsequently transferred to an aspartyl residue within the receiver domain of the RR, which usually increases the DNA-binding activity of the output domain leading to transcription of a group of genes, known as a regulon, related to the cognate stimulus [1]. Aromatic tuning, or moving aromatic residues (red box) at the cytoplasmic end of the transmembrane (TM) domain, facilitates stimulus-independent modulation of signaling circuits by mimicking the presence of cognate stimulus and thus altering SHK output, and in turn, transcription of the associated regulon

facilitates stimulus-independent modulation of signaling circuits by mimicking the presence of cognate stimulus and thus altering SHK output.

Biophysical and biochemical underpinnings of aromatic tuning

Aromatic tuning was conceived based on studies with peptides that possess an aliphatic core of Ala-Leu repeats flanked by Trp (WALP) or Tyr (YALP) residues (Fig. 2a). When these WALP [14] and YALP [15] peptides were mixed with various synthetic bilayers of an appropriate thickness they were shown to adopt a transmembrane α-helical confirmation based on circular dichroism (CD), ^{2}H NMR and ^{31}P NMR spectroscopy. Initially, hydrophobic mismatch, or differences between the length of the aliphatic core of these peptides and the thickness of the hydrophobic bilayer core was considered to be the crucial determinant in how α-helical peptides interact with the surrounding lipidic environment [14]. Subsequently, it was shown that the contribution of interfacial anchoring, i.e., interactions between Trp residues flanking the aliphatic core and the polar/hydrophobic interfaces near the boundaries of a lipid bilayer, can dominate over the effects introduced by hydrophobic mismatch alone [16]. Although this was only explicitly demonstrated for Trp residues, it was proposed that Tyr residues would also facilitate interfacial anchoring due to possessing similar physiochemical properties [16]. This elegant series of biophysical experiments is summarized in Fig. 2a. Addition of WALP or YALP peptides possessing a shorter distance between their flanking aromatic residues than the thickness of the hydrophobic core between the interfacial regions of a synthetic bilayer to which they are added results in induction of inverted hexagonal (H_{II}) phases that can be detected by ^{31}P NMR and sucrose density centrifugation [14, 16]. Conversely, when the distance between the aromatic residues was larger than the thickness of the hydrophobic core, the acyl chains within the bilayer became more ordered as determined by ^{2}H NMR, demonstrating that the membrane is slightly expanding to accommodate these "longer" peptides [15, 17]. In essence, these experiments demonstrate that amphipathic aromatic residues, namely Trp and Tyr, possess affinity for the interfacial regions where the polar phospholipid headgroups attach to the hydrophobic acyl chains.

From a biochemical perspective, another methodology, known as glycosylation mapping, also demonstrated the contribution of aromatic residues to positioning α-helices within a lipid bilayer [18]. Glycosylation mapping utilizes the lumenally positioned endoplasmic reticulum enzyme oligosaccharyltransferase (OST) to add a glycan to the Asn residue in Asn-Xaa-(Ser/Thr) glycosylation acceptor sites. In this manner, OST acts as a molecular ruler because each acceptor site will be glycosylated to an extent that correlates with the distance between the active site of OST and the acceptor site. In Fig. 2b, the red acceptor sites are not far enough from the lumenal membrane to become glycosylated, whereas the green sites are distal enough to become glycosylated. From this information, minimum glycosylation distance (MGD), or the distance required to attain half-maximal glycosylation, can be determined as schematically shown in Fig. 2c. These MGDs can then be compared between segments to determine whether any helix repositioning occurred. Upon comparison, a decrease in MGD suggests that lumenal boundary of TM2 has been displaced into the lumen, while an increase in MGD demonstrates that the TM2 boundary has been repositioned into the membrane (Fig. 2c). A series of TM segments that are identical except for the position of the flanking aromatic residues at one end can be assessed for changes in MGD that can be assigned to the repositioning ability of those aromatic residues.

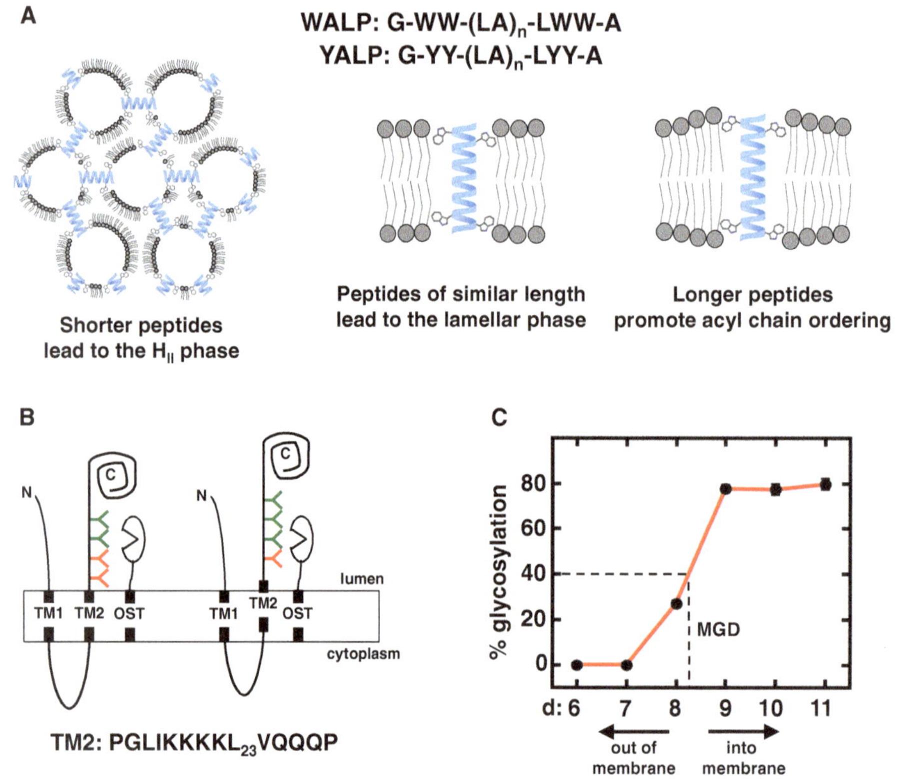

Fig. 2 Schematic summary of the evidence supporting the affinity of amphipathic aromatic residues for polar/hydrophobic interfaces. **a** Mixing peptides consisting of a poly-Ala-Leu core of different lengths flanked by Trp residues (WALP peptides) with synthetic bilayers of different thicknesses demonstrates the affinity of the Trp residues for the polar/hydrophobic interfaces [14, 15]. When the distance between the Trp residues was sufficiently shorter than the distance between the polar/hydrophobic interfaces, the lipids adopted an inverted hexagonal (H_{II}) phase to accommodate interactions between the Trp residues and the interfacial regions (*left panel*). When the distance between the flanking aromatic residues was increased so that it matched the distance between the interfacial regions, lamellar phases were observed (*center panel*). As the distance between Trp residues was increased, the acyl chains became more ordered suggesting a slight expansion of the bilayer to accommodate these "larger" peptides (*right panel*). **b** Glycosylation-mapping analysis employs a Lep model protein with two TM helices [18]. Segments to be analyzed are inserted at TM2 and a glycosylation-accepting site is positioned between 6 and 11 residues away from the lumenal boundary of TM2. Glycosylation-accepting sites shown in green are distal enough to become glycosylated while those in red are not. Proximity of an accepting site to the active site of oligosaccharyltransferase (OST) correlated with the extent of glycosylation. Therefore, repositioning of TM2 will change these relative positions and hence the extent of glycosylation. **c** Calculation of minimum glycosylation distance (MGD). MGD simply indicates the number of residues required for half-maximal glycosylation. Repositioning of TM2 into the membrane will increase MGD, while outward displacements of TM2 result in a reduction of MGD. Previous changes in MGD have demonstrated that moving Trp residues about the lumenal end of TM2 resulted in bidirectional, i.e., into and out of the membrane, displacement of the TM helix [19]

Changes in MGD have been reported when Trp residues are moved throughout the C-terminal half of a poly-Leu TM segment [19], thereby demonstrating that bidirectional repositioning of the TM segment is due to the affinity of Trp residues for the polar/hydrophobic interfaces (Fig. 2b). In a slightly modified version of this assay that considers all possible positions within the TM helix, Tyr residues where shown to possess character similar to Trp residues [20]. Thus, before aromatic tuning was conceived, previously existing biophysical and biochemical evidence suggested that Trp and Tyr residues possess affinity for the polar/hydrophobic regions located within membranes and suggested that moving them within an intact membrane-spanning protein could reposition individual TM helices.

Aromatic residues at the ends of Tar TM2 govern its position in the membrane

With the aforementioned biophysical and biochemical evidence suggesting that moving flanking aromatic residues could reposition a TM helix within an intact protein, it was necessary to select an initial bacterial membrane-spanning receptor in which to examine the role of these residues in governing signal output. The aspartate chemoreceptor of *E. coli* (Tar) was chosen because it governs a well-characterized downstream signaling pathway. It is important to note that Tar is not a canonical sensor histidine kinase (SHK), requires CheW and CheA to form functional intracellular signaling complexes, and controls flagellar rotation rather than gene transcription (Fig. 3a) [21–25]. When aromatic tuning of Tar was originally

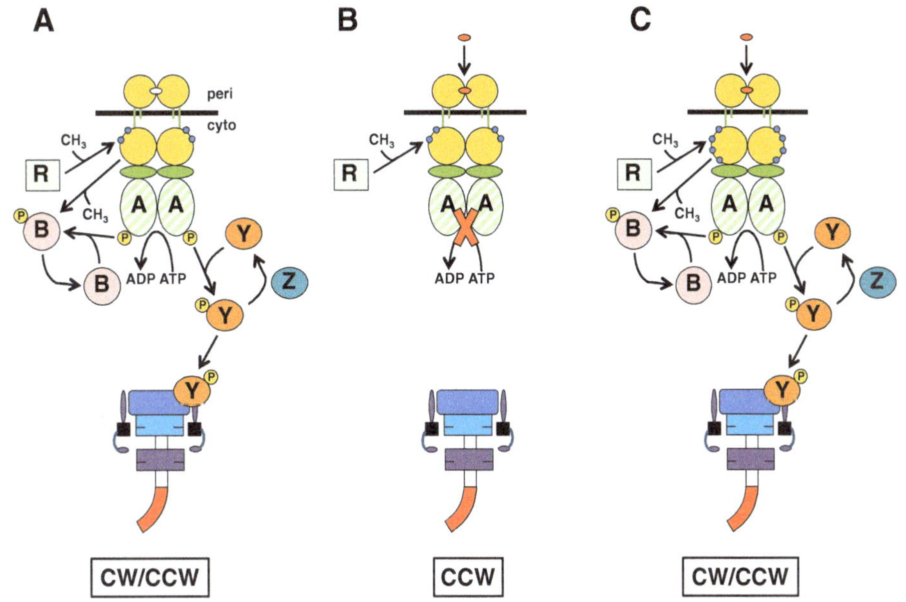

Fig. 3 The chemotactic circuit underlying control of flagellar rotation. **a** In the absence of chemoeffectors, baseline CheA activity maintains phsopho-CheY levels that produce the three-dimensional random walk underlying canonical bacterial chemotaxis. **b** Binding of attractant (red oval in the periplasm; peri) to the chemoreceptor, i.e., aspartate to Tar, abolishes CheA activity, thereby decreasing intracellular phospho-CheY levels. This also results in reduced methylesterase activity due to reduced CheB-P levels. Transmembrane communication (across the black line) is believed to occur via a piston-type displacement of TM2 toward the cytoplasm (cyto; center panel). **c** Adaptive methylation (blue dots in the cytoplasm) due to reduced CheB-P levels, restores the ability of the chemoreceptor to stimulate CheA activity when it is occupied by an attractant ligand [23]. In summary, this circuit was an excellent initial target for aromatic tuning because binding of attractant leads to displacement of TM2 toward the cytoplasm, reduced CheA kinase activity and increased levels of covalent modification. Conversely, displacements of TM2 toward the periplasm are consistent with increased CheA activity and reduced levels of covalent modification. In addition, signal output from Tar that is biased beyond the compensatory extent of methylation can be detected by monitoring rotation of individual flagella

undertaken, the mechanistic models for TM signaling were based upon piston-type displacements of the second transmembrane helix (TM2) [26–31]. This model proposes that displacement of TM2 toward the cytoplasm occurs upon binding of cognate ligand, i.e., aspartate, to the periplasmic domain (Fig. 3b). Binding of aspartate results in squelched CheA kinase activity, which, in turn, reduces the intracellular level of phospho-CheY that promotes clockwise (CW) flagellar rotation (Fig. 3b). Phospho-CheB, also phosphorylated by CheA, is the active form of the methylesterase and is also found in reduced intracellular levels upon binding of aspartate to Tar. This results in an increase in the extent of methylation within the cytoplasmic domain of Tar, which allosterically counteracts the effect of aspartate binding and leads to restoration a normal flagellar rotational bias (Fig. 3c). One advantage to initially employing this signaling circuit was that each of these activities could be independently assessed to determine whether changes in Tar signal output have occurred. For example, it is possible to monitor CheA kinase activity *in vitro* without the complicating effects of covalent methylation [22, 26, 27]. Likewise, the extent of methylation can be examined *in vitro* in the absence of CheA kinase activity [30]. Finally, flagellar rotation can be examined *in vivo*, which will allow for determination of when signal output from Tar is biased beyond the compensatory range of methylation [26, 27].

Armed with these outputs to monitor changes in Tar signaling, initial arginyl- and cysteinyl-scanning mutagenesis of Tar TM2 from *Salmonella enteritica* serovar Typhimurium was performed and resulted in several mutations that highlighted the importance of aromatic residues in governing helix position within the cytoplasmic membrane [30]. Large changes in signal output consistent with a displacement of TM2 toward the periplasm, similar to those observed in the *apo* conformation of Tar, namely increased CheA kinase activity and decreased methylation, were observed when either Phe-189 or Trp-192 was replaced with an Arg residue at the periplasmic end of TM2. This was also seen when Phe-189 was replaced with a Cys residue. Conversely, reduced kinase activity and increased levels of covalent methylation, both consistent with the aspartate-bound confirmation of Tar, were observed when Trp-209 was replaced with Arg [30]. Independently, the importance of Trp-192 and Trp-209 in maintaining baseline signal output from *E. coli* Tar was demonstrated when they were substituted for alanyl residues [27]. In the case of the Tar W192A variant, the receptor was undermethylated and exhibited higher signal output, consistent with *apo* state and displacement of TM2 toward the periplasm. In contrast, the W209A mutant was overmethylated and possessed reduced signal output, suggesting a displacement of TM2 toward the cytoplasm as proposed to occur within the aspartate-bound confirmation. In summary, these results demonstrate that aromatic residues within TM2 of Tar are critical for maintaining normal signal output and suggest that they govern the baseline position of the helix within the membrane to an extent that could be employed to manipulate Tar signal output.

Incremental tuning of Tar signal output

Based on the contributions of the individual aromatic residues to maintenance of signal output and the biophysical and biochemical data described above, it was originally hypothesized that aromatic tuning would displace TM2 of Tar within the membrane in a step-wise manner [26]. To examine this hypothesis, a series of Tar receptors was created in which the Trp-Tyr tandem found at the cytoplasmic end of TM2 was moved up to three residue steps in either direction (Fig. 4a).

Previous results show that the intracellular concentration of CheY-P governs the clockwise flagellar (CW) rotational bias (Fig. 4b). Therefore, Tar signal output could be estimated from the aromatically tuned variants by monitoring flagellar rotation. To accomplish this, individual flagella from single *E. coli* cells were monitored for 20 s and classified into one of five categories that represent increasing levels of intracellular CheY-P: those rotating exclusively CCW, flagella rotating mostly CCW with occasional switching to CW rotation, flagella rapidly switching between CCW and CW with no inherent bias, flagella rotating mostly CW with occasional switching to CCW rotation, and finally those that rotate exclusively CW [26]. When flagella from cells expressing the aromatically tuned variants described in Fig. 4a were analyzed, a clear shift from CCW to CW rotational bias was observed as the aromatic residues were moved from WY-3 toward WY + 3 (Fig. 4c). This indicates that the signal output from the minus-series of tuned Tar variants (i.e., WY-3 to WY-1) was so far biased toward reduced signal output that the compensatory action of covalent methylation failed to completely overcome this signaling bias. Conversely, the WY + 2 and WY + 3 variants were biased toward increased signal output to the extent that methylation could also not compensate. As

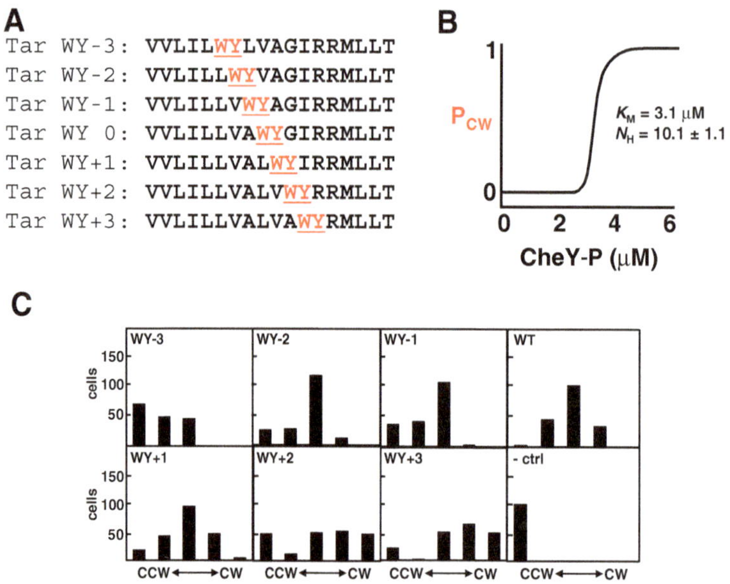

Fig. 4 Incremental tuning of signal output from the aspartate chemoreceptor of *E. coli* (Tar). **a** At the C-terminal end of Tar TM2, a Trp-Tyr (*red*) was moved about its original position at the cytoplasmic polar/hydrophobic interface [26]. **b** The intracellular level of CheY-P governs the probability of clockwise flagellar rotation (P_{CW}). Previous results demonstrate that increased intracellular CheY-P levels lead to an enhanced probability of CW rotation. In these experiments, a sharp transition was observed, i.e., a Hill coefficient of more than 10 [65]. **c** Rotation of a single flagellum from roughly 200 independent *E. coli* cells expressing one of the aromatically tuned variants were analyzed and classified into one of five categories (from left to right): rotating exclusively CCW, rotating primarily CCW with occasional reversals, rapidly switching between both rotational directions (CW/CCW), rotating primarily CW with occasional reversals and those rotating exclusively CW. As Tar signal output increases, the number of cells in each category shifts from CCW toward CW rotational bias. In summary, the lowest overall signal output was observed from cells expressing the WY-3 variant, while the greatest was observed from cells expressing the WY + 2 or WY + 3 variants. Therefore, in the case of Tar, the absolute vertical position of the aromatic residues correlates with signal output [26]

expected, in the absence of receptor (– ctrl), only CCW rotation was observed (Fig. 4c). It is also important to note that, as predicted, overmethylation of Tar was observed for the minus-series (WY-3 to WY-1) of receptors, while undermethylation was seen for the plus-series (WY + 1 to WY + 3) of aromatically tuned variants [26]. When taken together, these results demonstrate that signal output from Tar is modulated in an incremental manner when the aromatic residues were moved at the cytoplasmic end of TM2.

While these results served as an interesting foray into aromatic tuning, they do not empirically demonstrate that moving the aromatic residues at the cytoplasmic end of TM2 results in a physical repositioning of the helix. Therefore, two subsequent studies, one computational and one biochemical, were performed to assess repositioning of Tar TM2 upon moving the Trp-Tyr tandem. Coarse-grained molecular dynamics (CG-MD) simulations that were employed to examine the ability of aromatic tuning to displace Tar TM2 in the presence of an explicit membrane and solvent demonstrated that moving the Trp-Tyr residue was sufficient to induce small TM2 displacements of up to 1.5 Å [32]. In addition, the glycosylation-mapping assay described above (Fig. 2b and c) was performed in order to determine whether moving the Trp-Tyr pair at the end of TM2 would yield analogous results in a biological membrane. Assuming that the region in Lep that contains the glycosylation-accepting site is in an extended conformation, a shift in MGD of 0.5 residues as seen for the Tar constructs corresponds to a shift in the positioning of the TM2 helix of 1.6-1.7 Å [33], in close agreement with the CG-MD results [32]. Furthermore, within both the CG-MD simulations and MGD analysis, similar patterns of displacement were observed. A grouping of the minus-series of receptors with similar displacements toward the cytoplasm (WY-3 through WY-1), a baseline position for the wild-type (WY 0), two receptors that are slightly displaced toward the periplasm (WY + 1 and WY + 2) and a larger shift toward the periplasm for the WY + 3 variant was observed with both techniques. Therefore, this combination of *in vivo*, *in vitro* and *in silico* results demonstrate that repositioning the Trp-Tyr tandem at the cytoplasmic end is sufficient to generate a physical displacement of TM2 and the resulting modulation of Tar signal output in an incremental manner.

Non-incremental tuning of EnvZ signal output

In order to determine whether aromatic tuning would work within a canonical SHK, its effectiveness was examined within the major *E. coli* osmosensor (EnvZ), where a rotation of TM2 was initially proposed as the mechanism of transmembrane communication [34–38]. More recently, a regulated unfolding [39] model has been suggested, in which SHKs are modular proteins composed of individually folding domains that each contribute a distinct functionality. Regulated unfolding suggests that the effector domain within an SHK is maintained in an inactive conformation by a rigid connection between the stimulus perception and effector domains. Upon perception of stimulus, this connection disengages allowing the effector domain to adopt an active conformation [39]. This region connecting the TM to the cytoplasmic domain in bacterial receptors is colloquially referred to as a "control cable" because its residue composition governs coupling of signal transduction between adjacent domains [26, 27, 40–47].

EnvZ is a canonical SHK that responds to changes in the extracellular osmolarity of inner-membrane impermeable compounds by modulating the intracellular level of phosphorylated OmpR (Fig. 5a) [48–51]. In a laboratory environment, EnvZ signal output is modulated by simply adding sucrose to the bacterial growth medium. Subsequently, phospho-OmpR regulates the transcription of a number of genes, including those encoding two outer membrane porins, OmpF and OmpC. At low intracellular levels of phospho-OmpR (OmpR P), transcription of *ompF* is upregulated, whereas at higher levels of OmpR-P, transcription of *ompF* is repressed and transcription of *ompC* is activated. This results in a predominance of OmpF at low osmolarity and OmpC at higher osmolarities (Fig. 5b) [52–54]. The easily controllable nature of the input stimulus, i.e., addition of sucrose, and the well-characterized transcriptional output made the EnvZ/OmpR osmosensing circuit an ideal choice for examining aromatic tuning within an SHK.

To analyze steady-state signal output from EnvZ/OmpR osmosensing circuits containing aromatically tuned receptors, a two-color fluorescent reporter strain was employed. This *E. coli* strain possesses transcriptional fusions of *cfp* to *ompC* and of *yfp* to *ompF* within its chromosome (Fig. 5b) [55]. Quantifying the ratio of CFP to YFP fluorescence provides a rapid and sensitive measure of the ratio of *ompC* to *ompF* transcription, which estimates the intracellular level of phosphoylated OmpR. Therefore, the ratio of CFP to YFP can be employed to estimate signal output from the aromatically tuned EnvZ variants. To confirm this, we previously demonstrated that the transcription of CFP and YFP behaved as expected upon addition of sucrose to the growth medium (Fig. 5c) [7].

Due to the proposal of non-piston signaling mechanisms, it was difficult to predict what pattern of signal outputs would be observed upon aromatic tuning of an SHK. In the case of EnvZ, a Trp-Leu-Phe triplet was moved from five residues into the membrane (WLF-5) through two residues out of the membrane (WLF + 2) in single residue steps (Fig. 5d). In this case, the surface of TM2 that the aromatic residues reside upon was found to be the major determinant of EnvZ signal output rather

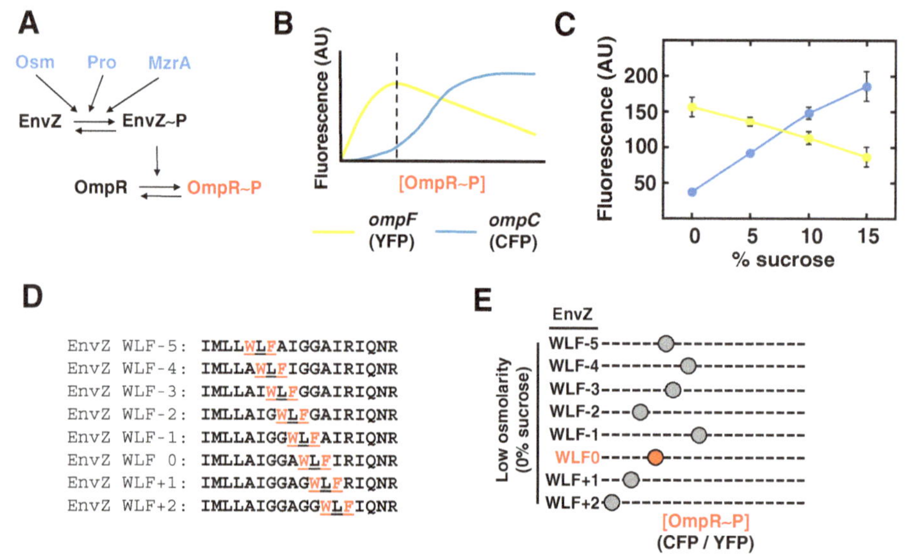

Fig. 5 Non-incremental tuning of signal output from the major osmosensor of *E. coli* (EnvZ). **a** EnvZ is bifunctional and possesses both kinase and phosphatase activity. The ratio of these activities is modulated by several factors including the presence of extracellular osmolarity [66, 67], procaine [68] or MzrA [69, 70]. OmpR serves as the cognate RR of EnvZ and the intracellular level of phosphorylated OmpR (OmpR-P) in governed by EnvZ activity. **b** OmpR-P levels control transcription of *ompF* and *ompC*, which can be monitored by employing an *E. coli* strain that contains a transcriptional fusion of *yfp* to *ompF* (*yellow*) and of *cfp* to *ompC* (*blue*) [7]. Intracellular levels of OmpR-P (*red*) can thus be estimated by monitoring the CFP/YFP ratio. **c** Experimental results demonstrating that increasing sucrose levels in the growth medium resulted in increasing levels of CFP, decreasing levels of YFP and an increase in the CFP/YFP ratio. The black dashed line in Fig. 5B represents signal output from this strain upon growth in medium with no additional sucrose. **d** When aromatic tuning was performed in EnvZ, a Trp-Leu-Phe triplet (*red*) was repositioned within the C-terminal region of TM2. **e** The gray-filled circles on the dashed lines indicate the estimated OmpR-P levels in cells expressing one of the aromatically tuned variants. Aromatic tuning in EnvZ resulted in a pattern of signal output that did not correlate with the absolute vertical position of the aromatic residues, as was the case with Tar, but rather appeared approximately helical in distribution suggesting that the surface of TM2 that the residues were located upon was of greater importance [33]. However, the key outcome is that aromatic tuning was still successful with respect to modulating EnvZ signal output in a stimulus-independent manner

than their absolute vertical position as was the case with Tar (Fig. 5E) [7]. These results clearly demonstrated that even though a different mechanism of TM communication has been proposed, aromatic tuning remained successful in modulating signal output. Similar results were obtained with a Trp-Tyr-Ala triplet that is more similar to the Trp-Tyr tandem employed during tuning of Tar suggesting that in the case of EnvZ, the composition of the residues that are moved is not a major determinant in the pattern of signal output [7]. However, additional experimentation is required to support the concept that disturbance of a particular helical surface is the cause of changes in EnvZ signal output. It should also be noted that these different patterns of signal output modulation are not surprising given that the concept of various signaling mechanisms being employed by different subclasses of bacterial receptors has been recently put forward [56–58]. Therefore, regardless of whichever signaling mechanism is employed by a receptor, aromatic tuning may be successful in modulating signal output.

Glycosylation mapping experiments with TM2 of EnvZ demonstrated that moving the Trp-Leu-Phe triplet resulted in almost no change in MGD. An exception was observed for WLF-3, which is most likely a specific case where the positive charge within the side-chain of Arg-182 would not be able to interact with the headgroups of the membrane any longer due to it being displaced very far out of the membrane. Thus, in the case of most aromatically tuned EnvZ segments, changes in MGD are very small and not steadily increasing in contrast to what was observed with the aromatically tuned Tar TM2 segments [33]. Initial CG-MD analysis with these aromatically tuned EnvZ segments also does not support incremental displacement (B. Hall, personal communication).

Employment of aromatic tuning within a wide variety of membrane-spanning receptors

Aromatic tuning was successful in modulating signal output from both Tar and EnvZ, however, a difference in the pattern of signal outputs was observed [7, 26]. This pattern of signal outputs shows that even though aromatic tuning did not displace the TM2 helix of EnvZ [33], it was still effective in modulating signal output from the full-length receptor in vivo. Aromatic tuning has also been successfully employed in two different Tar-EnvZ chimeric receptors (Lehning and Draheim, unpublished observations), when a Trp-Tyr tandem was used, and within a NarX-Tar chimera (Reinhard and Draheim, unpublished observations), when a single Trp residue was moved.

Published sequence alignments demonstrate that aromatic residues are often found at the cytoplasmic end of the final transmembrane helix within bacterial membrane-spanning receptors [27, 59] suggesting that aromatic tuning could be useful for research groups working with other two-component circuits. It is important to note that the majority of aromatically tuned Tar, EnvZ and chimeric receptor variants retain the ability to respond to stimulus suggesting that their signal output is biased but not locked in either a stimulus-deprived or a stimulus-saturated conformation. In this regard, aromatic tuning is advantageous compared to deletion of entire SHKs [60] or substitution of the conserved His residue involved in autophosphorylation and phosphotransfer because these methods may result in complete loss of kinase or phosphatase activity, which has been shown to result in non-physiological cross-talk between various two-component signaling pathways within a cell [61, 62]. Thus, by apparently biasing, rather than abolishing activity and stimulus-perception, aromatic tuning represents a significant improvement over previous attempts at stimulus-independent modulation of signal output. Even though these initial attempts have been successful, we have only observed aromatic tuning in membrane-spanning receptors that possess two TM helices. However, we feel strongly that in the case of SHKs containing only two TM helices, that aromatic tuning will facilitate stimulus-independent modulation in the majority of cases. Furthermore, we are currently expanding our range of targets to include SHKs that possess more than two TMs such as AgrC from *Staphylococcus aureus* that has recently been shown to possess seven TMs [63].

In addition to rectifying issues with signal output or stimulus-perception, we believe that aromatic tuning has other potential uses. For example, the input stimulus has been identified for only a small fraction of SHKs, thereby making it challenging to investigate downstream signaling pathways without extensive mutagenesis and subsequent screening for changes in phenotypic output. Aromatic tuning could be employed to circumvent these limitations and allow manipulation of signal output from a specifically targeted SHK. One potential use would be the rapid assignment of downstream physiological and developmental processes to particular SHKs. Each SHK within an organism could be independently subjected to aromatic tuning and subsequent monitoring for the phenotype of interest. If the appearance of the phenotype correlated with aromatic tuning of a particular SHK, this would suggest that the desired phenotype is governed by the aromatically tuned receptor. In addition to assignment of phenotypes to specific SHKs, aromatic tuning could be employed to facilitate induction of medically relevant bacterial phenotypes in the absence of complex host-pathogen interactions, thus reducing the burden of complex interkingdom laboratory model systems. Finally, aromatic tuning of individual SHKs could be coupled with pre-existing transcriptional reporter libraries [64] rather than observable phenotypes, to rapidly unravel the signaling pathways governed by each SHK. This would facilitate rapid and cost-effective organism-level signal pathway mapping.

Conclusion

The aromatic residues described here are found at the cytoplasmic boundary of the final TM helix within many bacterial membrane-spanning receptors, suggesting that aromatic tuning may useful in a wide-range of applications involving synthetic microbiology. Here, we provide a brief synopsis of the data supporting aromatic tuning, its initial successes and the advantages of this method over those previously employed for modulating TCS signal output. We have shown that aromatic tuning works within two well-characterized systems and hope that it is presented in sufficient detail to provide other research groups with the opportunity to employ the methodology within their TCSs of interest.

Abbreviations

TCS: Two-component signaling circuit; ORF: Open reading frame; SHK: Sensor histidine kinase; RR: Response regulator; WALP: Trp-flanked peptides with a poly-Ala-Leu core; YALP: Tyr-flanked peptides with a poly-Ala-Leu core; CD: Circular dichroism spectroscopy; NMR: Nuclear magnetic resonance spectroscopy; OST: Oligosaccharyltransferase; MGD: Minimum glycosylation distance; TM: Transmembrane; TM2: Second transmembrane helix; CCW: Counterclockwise; CW: Clockwise; CG-MD: Course-grained molecular dynamics simulations; CA: Domain responsible for catalytic ATPase activity; DHp: Domain responsible for dimerization and histidylphosphotransfer.

Competing interests

The authors declare that they have no competing interests.

Authors' contributions

RY and RRD conceived and planned the structure of the manuscript, wrote the manuscript and approved the final version of manuscript.

Acknowledgements

RY was generously supported by the Indonesian Endowment Fund for Education (S-1999/LPDP; 3/2014).

Author details

[1]Division of Pharmacy, Durham University, Queen's Campus, Stockton-on-Tees TS17 6BH, England, UK. [2]School of Pharmacy and Biomedical Sciences, University of Portsmouth, St. Michael's Building, White Swan Road, Portsmouth PO1 2DT, England, UK.

References

1. Stock AM, Robinson VL, Goudreau PN. Two-component signal transduction. Annu Rev Biochem. 2000;69:183–215.
2. Ji G, Beavis R, Novick RP. Bacterial interference caused by autoinducing peptide variants. Science. 1997;276:2027–30.
3. Shin D, Lee EJ, Huang H, Groisman EA. A positive feedback loop promotes transcription surge that jump-starts Salmonella virulence circuit. Science. 2006;314:1607–9.

4. Puthiyaveetil S, Kavanagh TA, Cain P, Sullivan JA, Newell CA, Gray JC, et al. The ancestral symbiont sensor kinase CSK links photosynthesis with gene expression in chloroplasts. Proc Natl Acad Sci U S A. 2008;105:10061–6.
5. David M, Daveran ML, Batut J, Dedieu A, Domergue O, Ghai J, et al. Cascade regulation of nif gene expression in Rhizobium meliloti. Cell. 1988;54:671–83.
6. Ulrich LE, Zhulin IB. The MiST2 database: a comprehensive genomics resource on microbial signal transduction. Nucleic Acids Res. 2010;38:D401–7.
7. Norholm MH, von Heijne G, Draheim RR. Forcing the Issue: Aromatic Tuning Facilitates Stimulus-Independent Modulation of a Two-Component Signaling Circuit. ACS Synth Biol. 2015;4:474–81.
8. Wang B, Barahona M, Buck M, Schumacher J. Rewiring cell signalling through chimaeric regulatory protein engineering. Biochem Soc Trans. 2013;41:1195–200.
9. Stanton BC, Siciliano V, Ghodasara A, Wroblewska L, Clancy K, Trefzer AC, et al. Systematic transfer of prokaryotic sensors and circuits to Mammalian cells. ACS Synth Biol. 2014;3:880–91.
10. Zhu Y, Inouye M. Analysis of the role of the EnvZ linker region in signal transduction using a chimeric Tar/EnvZ receptor protein, Tez1. J Biol Chem. 2003;287:22812–9.
11. Jung K, Fried L, Behr S, Heermann R. Histidine kinases and response regulators in networks. Curr Opin Microbiol. 2012;15:118–24.
12. Mascher T, Helmann JD, Unden G. Stimulus perception in bacterial signal-transducing histidine kinases. Microbiol Mol Biol R. 2006;70:910–38.
13. Russo FD, Silhavy TJ. The essential tension: opposed reactions in bacterial two-component regulatory systems. Trends Microbiol. 1993;1:306–10.
14. Killian JA, Salemink I, de Planque MR, Lindblom G, Koeppe 2nd RE, Greathouse DV. Induction of nonbilayer structures in diacylphosphatidylcholine model membranes by transmembrane alpha-helical peptides: importance of hydrophobic mismatch and proposed role of tryptophans. Biochemistry. 1996;35:1037–45.
15. de Planque MR, Boots JW, Rijkers DT, Liskamp RM, Greathouse DV, Killian JA. The effects of hydrophobic mismatch between phosphatidylcholine bilayers and transmembrane alpha-helical peptides depend on the nature of interfacially exposed aromatic and charged residues. Biochemistry. 2002;41:8396–404.
16. de Planque MR, Bonev BB, Demmers JA, Greathouse DV, Koeppe 2nd RE, Separovic F, et al. Interfacial anchor properties of tryptophan residues in transmembrane peptides can dominate over hydrophobic matching effects in peptide-lipid interactions. Biochemistry. 2003;42:5341–8.
17. de Planque MR, Greathouse DV, Koeppe 2nd RE, Schafer H, Marsh D, Killian JA. Influence of lipid/peptide hydrophobic mismatch on the thickness of diacylphosphatidylcholine bilayers. A 2H NMR and ESR study using designed transmembrane alpha-helical peptides and gramicidin A. Biochemistry. 1998;37:93–45.
18. Nilsson I, Saaf A, Whitley P, Gafvelin G, Waller C, von Heijne G. Proline-induced disruption of a transmembrane alpha-helix in its natural environment. J Mol Biol. 1998;284:1165–75.
19. Braun P, von Heijne G. The aromatic residues Trp and Phe have different effects on the positioning of a transmembrane helix in the microsomal membrane. Biochemistry. 1999;38:9778–82.
20. Hessa T, Meindl-Beinker NM, Bernsel A, Kim H, Sato Y, Lerch-Bader M, et al. Molecular code for transmembrane-helix recognition by the Sec61 translocon. Nature. 2007;450:1026–30.
21. Welch M, Oosawa K, Aizawa S, Eisenbach M. Phosphorylation-dependent binding of a signal molecule to the flagellar switch of bacteria. Proc Natl Acad Sci U S A. 1993;90:8787–91.
22. Lai RZ, Manson JM, Bormans AF, Draheim RR, Nguyen NT, Manson MD. Cooperative signaling among bacterial chemoreceptors. Biochemistry. 2005;44:14298–307.
23. Draheim RR. The role of protein-membrane interactions in modulation of signaling by bacterial chemoreceptors. Texas A&M University: Ph.D. dissertation; 2007.
24. Ravid S, Matsumura P, Eisenbach M. Restoration of flagellar clockwise rotation in bacterial envelopes by insertion of the chemotaxis protein CheY. Proc Natl Acad Sci U S A. 1986;83:7157–61.
25. Hess JF, Oosawa K, Kaplan N, Simon MI. Phosphorylation of three proteins in the signaling pathway of bacterial chemotaxis. Cell. 1988;53:79–87.
26. Draheim RR, Bormans AF, Lai RZ, Manson MD. Tuning a bacterial chemoreceptor with protein-membrane interactions. Biochemistry. 2006;45:14655–64.
27. Draheim RR, Bormans AF, Lai RZ, Manson MD. Tryptophan residues flanking the second transmembrane helix (TM2) set the signaling state of the Tar chemoreceptor. Biochemistry. 2005;44:1268–77.
28. Falke JJ, Hazelbauer GL. Transmembrane signaling in bacterial chemoreceptors. Trends Biochem Sci. 2001;26:257–65.
29. Falke JJ, Erbse AH. The piston rises again. Structure. 2009;17:1149–51.
30. Miller AS, Falke JJ. Side chains at the membrane-water interface modulate the signaling state of a transmembrane receptor. Biochemistry. 2004;43:1763–70.
31. Isaac B, Gallagher GJ, Balazs YS, Thompson LK. Site-directed rotational resonance solid-state NMR distance measurements probe structure and mechanism in the transmembrane domain of the serine bacterial chemoreceptor. Biochemistry. 2002;41:3025–36.
32. Hall BA, Armitage JP, Sansom MSP. Transmembrane Helix Dynamics of Bacterial Chemoreceptors Supports a Piston Model of Signalling. PLoS Comput Biol. 2011;7:e1002204.
33. Botelho SC, Enquist K, von Heijne G, Draheim RR. Differential repositioning of the second transmembrane helices from E. coli Tar and EnvZ upon moving the flanking aromatic residues. Biochim Biophys Acta. 2015;1848:615–21.
34. Ferris HU, Dunin-Horkawicz S, Mondejar LG, Hulko M, Hantke K, Martin J, et al. The mechanisms of HAMP-mediated signaling in transmembrane receptors. Structure. 2011;19:378–85.
35. Ferris HU, Dunin-Horkawicz S, Hornig N, Hulko M, Martin J, Schultz JE, et al. Mechanism of regulation of receptor histidine kinases. Structure. 2012;20:56–66.
36. Ferris HU, Zeth K, Hulko M, Dunin-Horkawicz S, Lupas AN. Axial helix rotation as a mechanism for signal regulation inferred from the crystallographic analysis of the E. coli serine chemoreceptor. J Struct Biol. 2014;186:349–56.
37. Hulko M, Berndt F, Gruber M, Linder JU, Truffault V, Schultz A, et al. The HAMP domain structure implies helix rotation in transmembrane signaling. Cell. 2006;126:929–40.
38. Inouye M. Signaling by transmembrane proteins shifts gears. Cell. 2006;126:829–31.
39. Schultz JE, Natarajan J. Regulated unfolding: a basic principle of intraprotein signaling in modular proteins. Trends Biochem Sci. 2013;38:538–45.
40. Parkinson JS. Signaling mechanisms of HAMP domains in chemoreceptors and sensor kinases. Annu Rev Microbiol. 2010;64:101–22.
41. Zhou Q, Ames P, Parkinson JS. Biphasic Control Logic of HAMP Domain Signaling in the Escherichia coli Serine Chemoreceptor. Mol Microbiol. 2011;80:596–611.
42. Kitanovic S, Ames P, Parkinson JS. Mutational analysis of the control cable that mediates transmembrane signaling in the Escherichia coli serine chemoreceptor. J Bacteriol. 2011;193:5062–72.
43. Park H, Im W, Seok C. Transmembrane signaling of chemotaxis receptor tar: insights from molecular dynamics simulation studies. Biophys J. 2011;100:2955–63.
44. Wright GA, Crowder RL, Draheim RR, Manson MD. Mutational analysis of the transmembrane helix 2-HAMP domain connection in the Escherichia coli aspartate chemoreceptor tar. J Bacteriol. 2011;193:82–90.
45. Adase CA, Draheim RR, Rueda G, Desai R, Manson MD. Residues at the cytoplasmic end of transmembrane helix 2 determine the signal output of the TarEc chemoreceptor. Biochemistry. 2013;52:2729–38.
46. Adase CA, Draheim RR, Manson MD. The residue composition of the aromatic anchor of the second transmembrane helix determines the signaling properties of the aspartate/maltose chemoreceptor Tar of Escherichia coli. Biochemistry. 2012;51:1925–32.
47. Zhou Q, Ames P, Parkinson JS. Mutational analyses of HAMP helices suggest a dynamic bundle model of input–output signalling in chemoreceptors. Mol Microbiol. 2009;73:801–14.
48. Egger LA, Park H, Inouye M. Signal transduction via the histidyl-aspartyl phosphorelay. Genes Cells. 1997;2:167–84.
49. Forst SA, Roberts DL. Signal transduction by the EnvZ-OmpR phosphotransfer system in bacteria. Res Microbiol. 1994;145:363–73.
50. Hoch JA, Silhavy TJ. Two-Component Signal Transduction. Washington, DC: Am. Soc. Microbiol. Press; 1995.
51. Mizuno T. His-Asp phosphotransfer signal transduction. J Biochem. 1998;123:555–63.
52. Forst S, Delgado J, Rampersaud A, Inouye M. In vivo phosphorylation of OmpR, the transcription activator of the ompF and ompC genes in Escherichia coli. J Bacteriol. 1990;172:3473–7.

53. Lan CY, Igo MM. Differential expression of the OmpF and OmpC porin proteins in Escherichia coli K-12 depends upon the level of active OmpR. J Bacteriol. 1998;180:171–4.
54. Russo FD, Silhavy TJ. EnvZ controls the concentration of phosphorylated OmpR to mediate osmoregulation of the porin genes. J Mol Biol. 1991;222:567–80.
55. Batchelor E, Silhavy TJ, Goulian M. Continuous control in bacterial regulatory circuits. J Bacteriol. 2004;186:7618–25.
56. Falke JJ. Piston versus Scissors: Chemotaxis Receptors versus Sensor His-Kinase Receptors in Two-Component Signaling Pathways. Structure. 2014;22:1219–20.
57. Unnerstale S, Maler L, Draheim RR. Structural characterization of AS1-membrane interactions from a subset of HAMP domains. Biochim Biophys Acta. 2011;1808:2403–12.
58. Molnar KS, Bonomi M, Pellarin R, Clinthorne GD, Gonzalez G, Goldberg SD, et al. Cys-Scanning Disulfide Crosslinking and Bayesian Modeling Probe the Transmembrane Signaling Mechanism of the Histidine Kinase. PhoQ, Structure. 2014;22:1239–51.
59. Boldog T, Hazelbauer GL. Accessibility of introduced cysteines in chemoreceptor transmembrane helices reveals boundaries interior to bracketing charged residues. Protein Sci. 2004;13:1466–75.
60. Zhou L, Lei XH, Bochner BR, Wanner BL. Phenotype microarray analysis of Escherichia coli K-12 mutants with deletions of all two-component systems. J Bacteriol. 2003;185:4956–72.
61. Groban ES, Clarke EJ, Salis HM, Miller SM, Voigt CA. Kinetic buffering of cross talk between bacterial two-component sensors. J Mol Biol. 2009;390:380–93.
62. Siryaporn A, Goulian M. Characterizing cross-talk in vivo avoiding pitfalls and overinterpretation. Methods Enzymol. 2010;471:1–16.
63. Wang L, Quan C, Xiong W, Qu X, Fan S, Hu W. New insight into transmembrane topology of Staphylococcus aureus histidine kinase AgrC. Biochim Biophys Acta. 2014;1838:988–93.
64. Zaslaver A, Bren A, Ronen M, Itzkovitz S, Kikoin I, Shavit S. A comprehensive library of fluorescent transcriptional reporters for Escherichia coli. Nat Methods. 2006;3:623–8.
65. Cluzel P, Surette M, Leibler S. An ultrasensitive bacterial motor revealed by monitoring signaling proteins in single cells. Science. 2000;287:1652–5.
66. Egger LA, Inouye M. Purification and characterization of the periplasmic domain of EnvZ osmosensor in Escherichia coli. Biochem Biophys Res Commun. 1997;231:68–72.
67. Forst S, Delgado J, Inouye M. Phosphorylation of OmpR by the osmosensor EnvZ modulates expression of the ompF and ompC genes in Escherichia coli. Proc Natl Acad Sci U S A. 1989;86:6052–6.
68. Rampersaud A, Inouye M. Procaine, a local anesthetic, signals through the EnvZ receptor to change the DNA binding affinity of the transcriptional activator protein OmpR. J Bacteriol. 1991;173:6882–8.
69. Gerken H, Charlson ES, Cicirelli EM, Kenney LJ, Misra R. MzrA: a novel modulator of the EnvZ/OmpR two-component regulon. Mol Microbiol. 2009;72:1408–22.
70. Gerken H, Misra R. MzrA-EnvZ interactions in the periplasm influence the EnvZ/OmpR two-component regulon. J Bacteriol. 2010;192:6271–8.

13

Autonomous buckling of micrometer-sized lipid-protein membrane patches constructed by *Dictyostelium discoideum*

Kei Takahashi[1] and Taro Toyota[1,2*]

Abstract

Background: The cytosol of amoeba cells controls the membrane deformation during their motion *in vivo*. To investigate such ability of the cytosol of amoeba cell, *Dictyostelium discoideum* (*Dictyostelium*), *in vitro*, we used lipids extracted from *Dictyostelium* and commercially available phospholipids, and prepared substrate-supported lipid membrane patches on the micrometer scale by spin coating.

Results: We found that the spin coater holder, which has pores (pore size = 3.1 mm) of negative pressure to hold the cover glass induced the concave surface of the cover glass. The membrane lipid patches were formed at each position in the vicinity of the holder pores and their sizes were in the range of 2.7 to 3.2×10^4 μm^2. After addition of the cytosol extracted from *Dictyostelium* to the lipid membrane patches, through time-lapse observation with a confocal laser scanning fluorescence microscope, we observed an autonomous buckling of the *Dictyostelium* lipid patches and localized behaviours of proteins found within.

Conclusion: The current method serves as the novel technique for the preparation of film patches in which the positions of patches are controlled by the holder pores without fabricating, modifying, and arranging the chemical properties of the solution components of lipids. The findings imply that lipid-binding proteins in the cytosol were adsorbed and accumulated within the *Dictyostelium* lipid patches, inducing the transformation of the cell-sized patch.

Keywords: Substrate-supported lipid membrane, Spin coating, Phosphatidylcholine, Phosphatidylinositides, *Dictyostelium discoideum*

Background

Micrometer-sized biochemical machinery has garnered much attention as an energy conversion device that converts chemical energy to kinetic energy and as a model of cell machinery that minimizes energy consumption [1-3]. The research trend on biochemical machinery has contributed to supramolecular chemistry and soft matter physics for understanding cell dynamics. Model construction for cell dynamics, such as membrane buckling [4,5], has attracted the attention of researchers because such cell dynamics are closely related to amoeba motion which highly functionalized with the coupling of cellular mechanics and sensors [6]. Conventionally, the buckling of a cell membrane is examined by using a narrow glass capillary with the inner diameter of several micrometers. The aspiration through the glass capillary attached on the cell membrane induces locally the cell membrane deformation (and even the transformation of the whole cell body). In this study, we focus on the membrane deformation controlled by its cytosol in amoeba motion *in vitro*. The *in vitro* self-organization of lipid membranes and proteins in a micrometer-sized reaction surface or two-dimensional plane is important. For example, essential combinations of proteins associated with membrane deformation [7,8] or cell division cycle [9] can generate a pattern formation and time course changes of a model cell membrane supported on a substrate. Our challenge contains the problem that the parameter space composed of the lipids and proteins related to amoeba motion is too vast to investigate often arises because each component

* Correspondence: cttoyota@mail.ecc.u-tokyo.ac.jp
[1]Department of Basic Science, Graduate School of Arts and Sciences, The University of Tokyo, 3-8-1 Komaba, Meguro, Tokyo 153-8902, Japan
[2]Research Center for Complex Systems Biology, Graduate School of Arts and Sciences, The University of Tokyo, 3-8-1 Komaba, Meguro, Tokyo 153-8902, Japan

should be completely purified and mixed in a sequential manner. Here, we overcame this obstacle by using lipids and cytosol extracted from an amoeba cell and constructing micrometer-sized substrate-contacting lipid-protein membrane patches. The constructed system enables us to not only examine the essence of the membrane-assisted biochemical machinery of amoeba motion without the cell complexity, but also to avoid the practical difficulty of incomplete encapsulation of cytosol extracts in the case of vesicle-type biochemical machinery.

The current purpose is thus to investigate the ability of the cytosolic extract of amoeba cells for deformation of their lipid membrane film *in vitro* (Figure 1). First, we focus on the cytosol and lipid extracts from *Dictyostelium discoideum* (*Dictyostelium*), because *Dictyostelium* is a type of amoeba that sprightly shows amoeba motion at room temperature (20 – 23°C). The experiment can be undergone without heating or cooling from room temperature. Second, we adopt the spin coating method to prepare a lipid membrane film constructed on a substrate because lipid mixtures extracted from living cells can hardly form well-defined membrane structures, such as single bilayer membranes, supported on a substrate. The spin coating method produces micrometer-sized patches with a thickness of several micrometer when the membrane-forming components are heterogeneous, resulting from the interaction between the organic solvent and the solutes [10]. Third, in order to trace the transformation of such lipid membrane film in a three-dimensional manner after the addition of the cytosolic extract, we performed time lapse observations using a confocal laser scanning fluorescence microscope.

Results

Micrometer-sized patches of lipid membrane film prepared by spin coating

The spin coating method for the lipid membrane film supported on a glass substrate conventionally enables us to prepare heterogeneous films using a lipid mixture [10]. After extraction of the lipid mixture from the wild-type *Dictyostelium*, AX4 cells, we observed the lipid membrane film of *Dictyostelium* prepared by the spin coating method on the millimeter scale by light microscopy. We found a lipid membrane film pattern on the cover glass (Figure 2A), which was held by the aspiration pores of the spin coat holder (Figure 2B). The lipid membrane film pattern was consisted of 25 islands (approximately 2.5 – 3 mm in diameter) which contained several patches on the micrometer scale (black spots shown in Figure 2A). In order to visualize such patches precisely, we obtained both the bright field microscopy images and the autofluorescence images of the lipid membrane film by using confocal laser scanning fluorescence microscopy. Two sets of autofluorescence images were obtained by using two irradiation lasers and the corresponding band-pass filter and dichroic mirror units (see Methods). Figure 2C shows the reconstructed images of the bright field microscopy and autofluorescence images of one of the patches. The regions between the islands were also covered with very thin lipid membranes (data not shown). Judging from the image analysis and taking into account the diffraction limit of light and the special resolution, the sizes of the patches were in the range of 2.7 to 3.2×10^4 μm^2 and the size range was observed reproducibly.

As a reference experiment, when we dropped a chloroform solution of 1-palmitoyl-2-oleoyl-3-*sn*-glycero-3-phosphocholine (POPC) and 10 mol% of 1,2-hexadecanoyl 3 *sn* glycero-phosphatidylinositol-(3,4,5)-trisphosphate (PIP3), islands and lipid membrane patches were also formed at the positions of the holder pores after solvent evaporation (see Additional file 1: Figure S1). Based on the micrograph of the patches, the contrast of the patches was stronger than that of the thin film island, indicating that the lamellarity of the patches was greater than that of the thin film. It implies that the patches composed of PIP3 and POPC grew in a thin layer of the organic solution which covered the concave surface after spin coating. Therefore, we could establish the preparation method of film patches in which the positions of patches are controlled in both cases of natural crude lipids and purified lipids. We thus focused on the patches

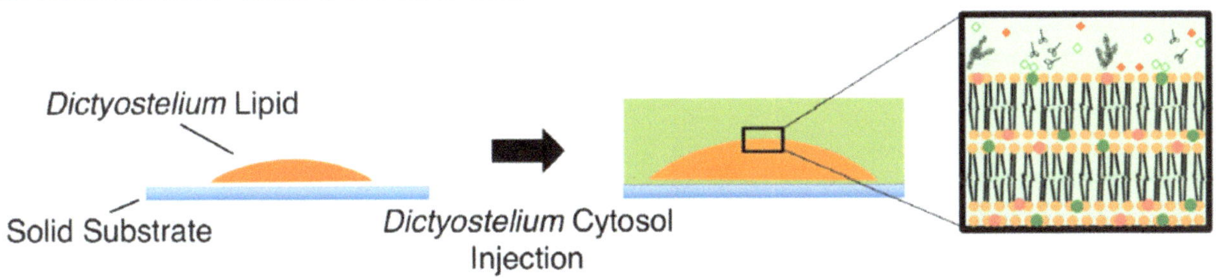

Figure 1 Schematic illustration of the lipid patch buckling induced by the cytoplasmic extract injection. Both the lipids and a cytosolic extract are extracted from *Dictyostelium*.

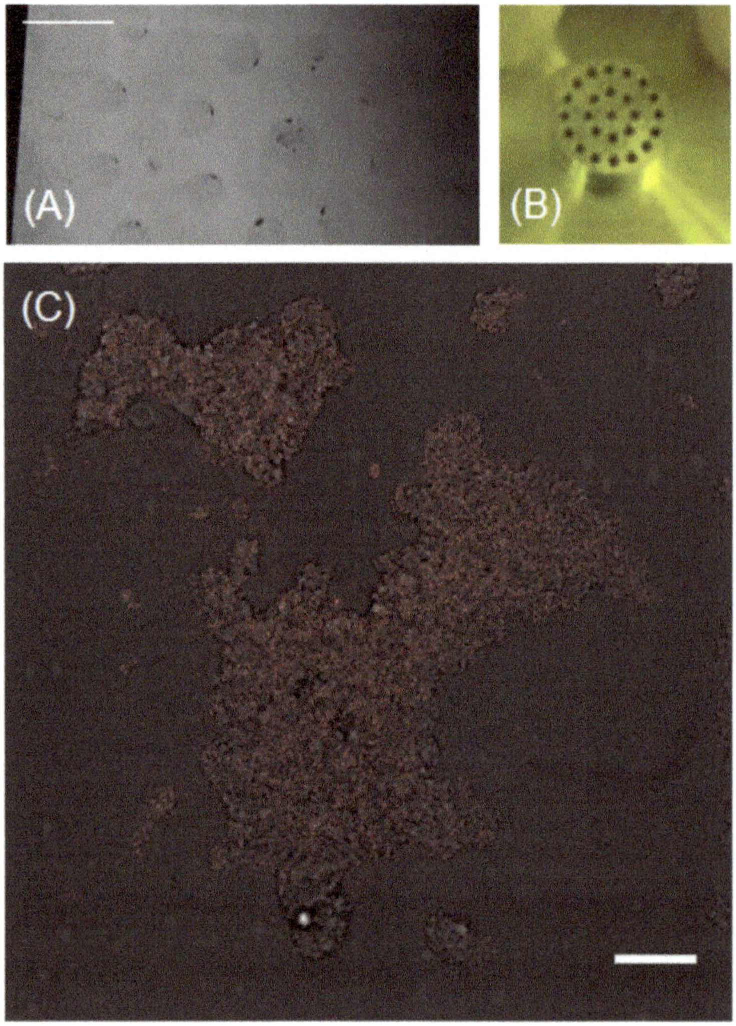

Figure 2 Microscopy images of the membrane lipid patch. (A) Bright field microscopy image of the cover glass with *Dictyostelium* lipid membrane film prepared by spin coating (bar = 3 mm). **(B)** Photograph of the cover glass holder. **(C)** Reconstructed image of the bright field microscopy image and the autofluorescence images of the typical patch of *Dictyostelium* lipid membrane film (bar = 30 μm).

by observing their deformation or the construction of micrometer-scale amoeba motion model.

Buckling of lipid membrane film after injection of cytosol extract

After setting up a chamber attaching the cover glass to a dish by a double sided tape, we injected the cytosol (25 μL) extracted from AX4 cells. As shown in Figure 3, the buckling of the lipid film patch was traced immediately after wetting the *Dictyostelium* lipid patch by PBS and following injection of the AX4 cytosol (see Additional file 2: Video 1). The *Dictyostelium* lipid patch exhibited autonomous transformation for the initial 10 min, resulting in the swelling in the x-y plane. Moreover, we observed a mesh-like structure after the transformation in the bright field microscopy images. Even though we observed the formation of vesicles, the diameter of which were several micrometer at the path of lipid membrane film immediately after the pre-wetting and the injection of cytosol, it should be noted that the cytosol induced the patch buckling and the transformation of the dome-like shape, which is several tens of micrometers in diameter. Moreover, the cytosol of genetically modified *Dictyostelium* cells, containing the Pleckstrin homology domain of cytosolic regulator of adenylyl cyclase (PH-crac) tagged with RFP (PH-crac-RFP) and the phosphatase and tensin homolog (PTEN) tagged GFP (PTEN-GFP) [11], was injected. The initial lipid membrane, with a maximum height of 10.5 μm on average (Figure 4A) under treatment with PBS for 1 min for pre-wetting started to buckle after 10 min in the chamber. In 25 min, *Dictyostelium* lipid patch stopped buckling and an equilibrium state was reached 30 min after *Dictyostelium* cytosol injection into the chamber

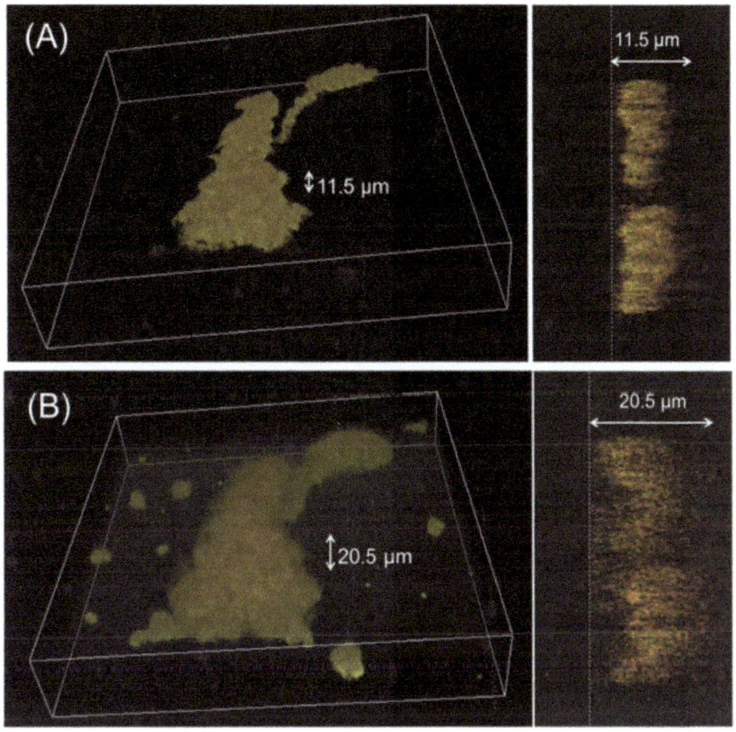

Figure 3 Microscopy image of the buckled *Dictyostelium* lipid patch after injection of the cytosol extract of *Dictyostelium* cells. 3D-merged reconstructed images of the autofluorescence images of a typical *Dictyostelium* lipid patch obtained by confocal laser scanning fluorescence microscopy with both Red and Green detection modes before **(A)** and 30 min after **(B)** injection of cytosol extract. The volume size of observation space is 210 μm x 210 μm x 37.5 μm. Each x-z plane cross section of the 3D merged reconstructed image was attached in the right column. Dashed lines correspond to the surface of the glass slide. The height indicates the distance between the surface of the glass slide and the peak top position of the autofluorescence images.

(see Additional file 3: Video 2 and Additional file 4: Video 3). The average maximum height of lipid patches was 15.5 μm as seen in 3D reconstructed images. Figure 4B shows that PH-crac-RFP and PTEN-GFP localized on micrometer-sized patch of lipid membrane, were mutually exclusive. In particular, GFP fluorescence was enriched at the edge of the patch and surrounded the RFP-fluorescing regions inside the patch (Figure 4C). Taking into account the fluorescence intensity of the confocal fluorescence images of the patch at each z-position (interval = 0.5 μm), we evaluated the change in the width and height of the patch before and 30 min after cytosol injection. By measuring the fluorescence intensity profile along the z-position in both Red and Green modes (Figures 4D,E,F, and G), we found that the height of the patch increased 30 min after cytosol injection (Figures 4E,G). The maximum thickness of the patch increased by a factor of 2.1 ± 0.3 after cytosol injection, indicating that membrane deformation and buckling dynamics took place after cytosol injection.

Then, the time course changes in the fluorescence intensity of *Dictyostelium* lipid patches in Red and Green modes were analyzed cytosols containing PH-crac-RFP and PTEN-GFP. The mean fluorescence intensity of Green and Red modes of all pixels on each image frame (Red mode, Figure 5A; Green mode, Figure 5B) increased immediately after cytosol injection ($t = 30$ s). When we compared the initial increase in fluorescence intensity in the range of 30 – 180 s, we found that PTEN-GFP was localized faster than PH-crac-RFP. As a reference experiment, we injected *Dictyostelium* cytosol containing PTEN-GFP and PH-crac-RFP to the chamber of lipid patches composed of a POPC/PIP3 membrane, and the fluorescence image obtained in the same procedure showed that PTEN-GFP localized on the patches and its localization speed was similar to that observed in the case of *Dictyostelium* lipid patch (Figure 5C). Because the lipid patch of POPC/PIP3 was stained with TexasRed-DHPE, we omitted the change in Red fluorescence in this reference experiment.

In another reference experiment, we injected *Dictyostelium* cytosol from PH-crac-RFP/PTEN-GFP co-expressing cells treated at 65°C for 1 h into the chamber of *Dictyostelium* lipid patches. Additional file 5: Figure S2 in Supporting Information shows the autofluorescence images of the *Dictyostelium* lipid patch before and 30 min after cytosol injection. Analyzed by time course change of the mean fluorescence intensity (z position; two image frames

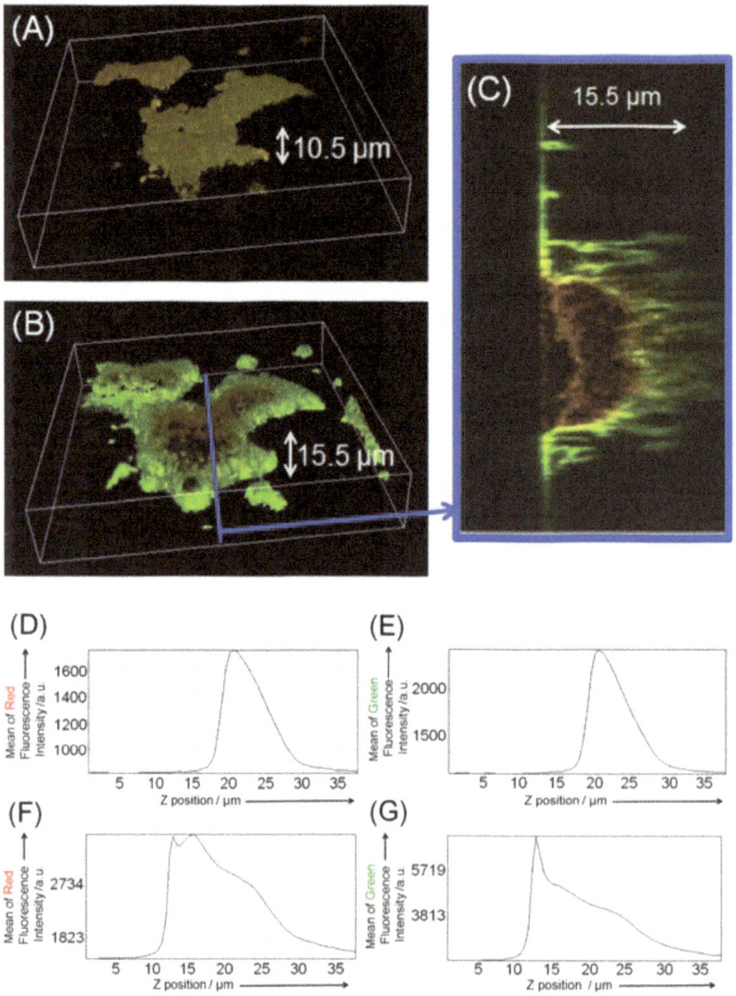

Figure 4 Microscopy image of the buckled *Dictyostelium* lipid patch after injection of the cytosol extract of PH-crac-RFP/PTEN-GFP co-expressing cells. 3D merge of reconstructed fluorescence images of a typical patch of the *Dictyostelium* lipid membrane film obtained by confocal laser scanning fluorescence microscopy, with both Red and Green fluorescence filter units, before injection of cytosol extract. **(A)** 3D merge of reconstructed fluorescence images of a typical patch of the *Dictyostelium* lipid membrane film obtained by confocal laser scanning fluorescence microscopy, with both Red and Green fluorescence filter units, before injection of cytosol extract. The volume size of observation space is 210 µm x 210 µm x 37.5 µm. The mean fluorescence intensity of each Red and Green fluorescence image was plotted along the z position in **(D)** and (E) respectively. **(B)** 3D merge of reconstructed fluorescence images and **(C)** the cross-section reconstructed fluorescence images of a typical patch of the lipid membrane film obtained by confocal laser scanning fluorescence microscopy with both Red and Green fluorescence filter units, 30 min after injection of the cytosol extract of PH-crac-RFP/PTEN-GFP co-expressing cells. The mean fluorescence intensity of each Red and Green fluorescence image was plotted along z position in **(F)** and **(G)**, respectively. The height shown in **(A-C)** indicates the distance between the surface of the glass slide and the peak top position of the autofluorescence images.

(~1 µm) above the glass slide), GFP fluorescence increased and the edge of each patch fluoresced 10 min after cytosol injection. Moreover, PTEN-GFP localization in the patch occurred but membrane thickness did not increase. When we injected the cytosol, which had been exposed to either autoclave processing or phosphate buffer treatment into the *Dictyostelium* lipid patch, no patch deformation was observed in either case. The POPC supported membrane on the cover glass remained unaltered after injection of cytosol of *Dictyostelium* cytosol from PH-crac-RFP/PTEN-GFP co-expressing cells.

It is known that F-actin polymerizes on *Dictyostelium* lipid membranes as a result of the downstream chemical reaction network of phosphatidylinositides, known as PIP signalling [12]. We used phalloidin staining to visualize the localization of F-actin polymerization and bundling in the *Dictyostelium* lipid patch in the current study. After formalin fixing, the confocal laser scanning fluorescence microscopy images of the *Dictyostelium* lipid patches showed F-actin localization and polymerization in the patches (Figure 6A). No F-actin polymerization occurred on the POPC supported membrane (Figure 6B),

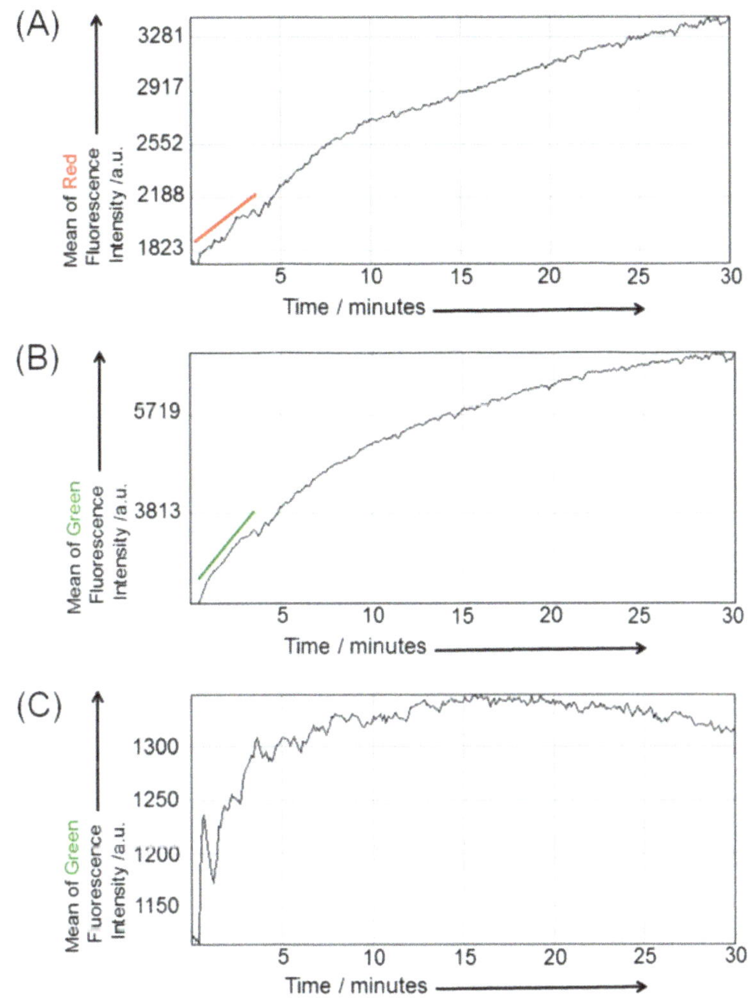

Figure 5 Time course change of fluorescence intensity of the *Dictyostelium* lipid patch. Time course change of the mean fluorescence intensity analyzed by fluorescence images (z position; two image frames (~1 μm) above the glass slide) of the *Dictyostelium* lipid patch obtained by confocal laser scanning fluorescence microscopy images with both Red **(A)** and Green **(B)** fluorescence detection filter units after injection of the cytosol extract of PH-crac-RFP/PTEN-GFP co-expressing cells. The slope in the range of 30 – 180 s is depicted in each diagram. **(C)** Time course change of the mean fluorescence intensity analyzed by fluorescence images (z position; two image frames (~1 μm) above the glass slide) of a typical patch of the POPC/PIP3 membrane film obtained by confocal laser scanning fluorescence microscopy images with Green fluorescence detection filter units after injection of the cytosol extract of PH-crac-RFP/PTEN-GFP co-expressing cells.

indicating that the *Dictyostelium* lipid patch accumulated phosphatidylinositides.

Discussion

Lipid film patches prepared by spin coating

There are several methods for lipid membrane film preparations, such as casting [13], the spin coating [10,14], Langmuir-Blodgett membrane formation [15-17], vesicle rupture [18-20], and water-in-oil droplet template methods [21]. The casting and spin coating methods are conventionally adopted in requesting membrane films composed of any kinds of lipids or mixtures. In the Langmuir-Blodgett membrane method, the formation of a lipid monolayer on a water surface is essential for the lipid molecular structure and the composition of lipid mixture. The vesicle rupture method relies on the stability of the lipid bilayer membrane for the formation of small unilamellar vesicles and giant unilamellar vesicles as the initial protocol step. The water-in-oil droplet template method is based on formation of stable water-in-oil droplets and the monolayers are stacked in layer-by-layer manner by using a microfluidic system. These last three methods have drawn much attention in terms of preparing well-defined bilayer membranes on a substrate. In the current study, the lipid mixture extracted from AX4 cells contains a various kinds of lipids as described later. In fact, previous reports on small unilamellar vesicles revealed that the components of each vesicle are dispersed even in one batch of the vesicle dispersion of lipid mixture [22,23]. And the vesicle rupture

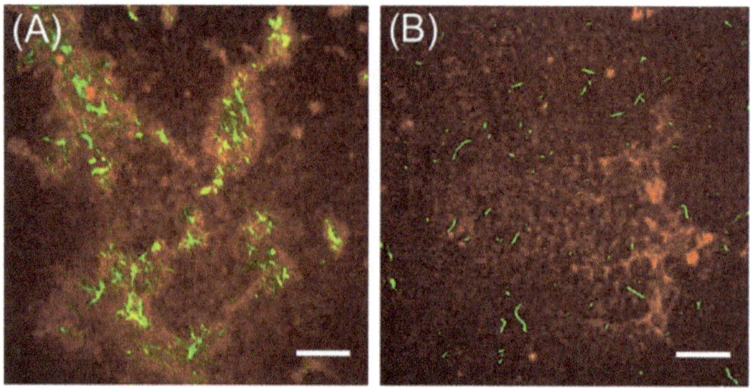

Figure 6 Microscopy images of F-actin stained by phalloidin on the membrane lipid patches. Confocal laser scanning fluorescence microscopy images of the *Dictyostelium* lipid patch **(A)** and POPC membrane film **(B)** after cytosol extract injection and following phalloidin staining (bar = 10 μm).

methods in the conventional manners did not afford such lipid membrane film using the lipid mixture extracted from AX4 cells (data not shown). The water-in-oil droplet template method is hardly applied for the bilayer membrane formation of *Dictyostelium* because the organic solvent is optimized for not dissolving the lipid component but stabilizing the emulsion droplets. Besides, the casting method, where an organic lipid solution is dropped onto a substrate and the solvent is removed resulting in the multilayer membrane structure, also has the issue of regulating membrane formation under the evaporation of the solvent. Particularly, the control of lipid membrane film formation is likely difficult due to the wetting and de-wetting process of the solvent and the solubility of lipids. The spin coating method facilitates the preparation of lipid membrane films using lipids extracted from AX4 cells and allows the control of the organic solution thickness, dissolving lipids before solvent evaporation occurs and resulting in several types of lipid membrane films on the substrate. The micrometer-sized patches of organic thin film are also formed by the spin coating method by means of micro-fabricated surfaces [24,25], chemically modified substrate surfaces [26,27], and differences in solubility between each solute in the organic solution [28]. We found that the spin coater holder, which has pores (pore size = 3.1 mm) of negative pressure to hold the cover glass induced the concave surface of the cover glass and that the membrane film was formed at each position in the vicinity of the holder pores. As far as we know, this process may serve as the fourth technique for the preparation of film patches in which the positions of patches are controlled by the holder pores without fabricating, modifying, and arranging the chemical properties of the solution components. After spin coating, the concave surface of the cover glass was covered by a thin layer of the organic solution, and then the further evaporation of the solvent resulted in the island formation with large thin film at each concave surface. The diameter of the island was approximately 3 mm in the current study. The sizes of the lipid film patches at each island were in the range of 2.7 to 3.2×10^4 μm^2. This process resembles the patch formation from heterogeneous lipid solutions in the spin coating method [29-31].

As reported in previous articles [32-34], lipids extracted from *Dictyostelium discoideum* contain phosphatidylcholines, phosphatidylethanolamines, phosphatidic acids, and phosphatidylinositides. According to the thin layer chromatograms reported in these articles, phosphatidylcholines and phosphatidylethanolamines have a higher affinity to organic solvents such as chloroform compared to phosphatidic acids and phosphatidylinositides. These results suggest that the *Dictyostelium* lipid patches formed on the cover glass contain the core structure of low-solubility lipids (phosphatidic acids, phosphatidylinositides, and so on) and the film layers of high-solubility lipids (phosphatidylcholines, phosphatidylethanolamines, and so on). Moreover, the size of the patch is influenced by the initial step of *Dictyostelium* lipid extraction, namely the volume ratio of $CHCl_3/CH_3OH$ (2:1, v/v) and the number of AX4 cells. We could systematically optimize the ratio as mentioned in the 'Methods' to obtain a patch size in the range of 2.7 to 3.2×10^4 μm^2. The thickness of the patches was 10.5 μm on average, judging by 3D constructed images of the autofluorescence images obtained by confocal laser scanning fluorescence microscopy.

Autonomous buckling of the *Dictyostelium* lipid patch

The buckling of lipid membrane films to form micrometer-sized dome-like shapes from plain lamella is conventionally observed in the so-called electroformation method of giant vesicles [35,36]. Under the application of an alternative voltage, after the injection of water or buffered solution into the lipid membrane film, the swelling process of the lipid membrane film is controlled by the electrophoresis of the membrane and the addition of water and water-soluble substances into the intermembrane space. Thus, the control of the external voltage and time interval can afford the formation of the dome-like and spherical shapes of giant unilamellar vesicles. In our study, the buckling of the lipid membrane film patch composed of *Dictyostelium* lipid autonomously took place under no external forces or stimuli after cytosol injection. We interpret the buckling mechanism of the lipid membrane film patch as follows.

The cytosol of *Dictyostelium discoideum* contains various types of membrane-binding proteins and/or lipid-conversion enzymes such as phospholipases, lipases, and phosphokinases. Since PH-crac and PTEN play important roles in membrane deformation during amoeba motion [37], we focused on their adsorbing process onto the *Dictyostelium* lipid patch by means of PTEN tagged by GFP (PTEN-GFP) and PH-crac tagged by RFP (PH-crac-RFP). As shown in Figure 5, in the initial stage immediately after cytosol injection, PTEN-GFP adsorbed more rapidly than PH-crac-RFP. Since PTEN-GFP adsorbed onto the POPC patch, which contained 10 mol% PIP3, it is implied that PTEN likely adsorbed onto the lipid membrane film. It is known that in living *Dictyostelium discoideum* cells, PTEN is one of the phosphokinases for phosphatidylinositides and tends to bind to the cell membrane during the resting mode of cells [38]. On the other hand, PH-crac tends to stay in the cytosol of resting cells and forms a lipid-protein complex with phosphatidylinositides when it adsorbs to the cell membrane [39,40]. Because the amount of phosphatidylinositides is much smaller than that of other phospholipids in *Dictyostelium* lipid [41,42], the initial adsorbing process of PH-crac on the *Dictyostelium* lipid patch is slower than that of PTEN. When the amount of adsorbed PH-crac gradually increases and the lipid-protein complex formation begins in the *Dictyostelium* lipid patch, it likely accumulates inside of the *Dictyostelium* lipid patch. This is probably because the phosphatidylinositides are organized primarily inside the patch and also, PH-crac competes with PTEN, which adsorbs and covers the *Dictyostelium* lipid patch to bind to the phosphatidylinositides. As a result, mutually exclusive distributions of PTEN-GFP and PH-crac-RFP were observed in the buckled *Dictyostelium* lipid patch. The *Dictyostelium* lipid patch is then buckled and grows to become a dome-like shape because of unbalanced osmotic pressure. As reference experiments, when we injected a cytosol, which had been treated at 65°C for 1 h just before use, we observed an increase in the fluorescence intensity of PTEN-GFP in the *Dictyostelium* lipid patch but no buckling (Additional file 5: Figure S2). This indicates that only the adsorbed PTEN on the patch had no potential to induce buckling. After treating the cell culture with LY294002 before the extraction, we injected the extracted cytosol to the *Dictyostelium* lipid patch and observed the buckling and dome-type shape of the patch. The result implied that the activity of phosphatidylinositol-3-kinase (PI3K), which is one of the phosphokinases for phosphatidylinositides and is inhibited by LY294002, exerts less influence on membrane buckling than PTEN and PH-crac. In order to clarify the phosphatidylinositides accumulation in the *Dictyostelium* lipid patch, we performed phalloidin staining for F-actin polymerization and bundle formation in the buckled *Dictyostelium* lipid patch because these chemical reactions occur at the sites of phosphatidylinositides aggregation accumulated in the *Dictyostelium discoideum* cell membrane [33]. Since bundles of polymerized F-actin were observed in the *Dictyostelium* lipid patch, it was confirmed that the *Dictyostelium* lipid patch contains phosphatidylinositides after cytosol injection. It implies that, among the phosphatidylinositides, 1,2-dialkyl-3-*sn*-glycerophosphatidylinositol-(4,5)-diphosphate, which is the product from 1,2-dialkyl-3-*sn*-glycerophosphatidylinositol-(3,4,5)-triphosphate by PI3K, is less effective than PIP3 for the PTEN-GFP adsorbing behaviour and PH-crac-RFP accumulation. To summarize the above experimental results, the buckling and dome-like shape of the *Dictyostelium* lipid patch after cytosol injection occur due to the formation of lipid-protein complexes at both the outer and inner regions of the patch.

Conclusions

The formation of the micrometer-sized lipid-protein patch from *Dictyostelium discoideum* in the current study brought us two achievements for constructing the membrane-assisted biochemical machinery. One achievement is the pre-organization of the *Dictyostelium* lipid patches. The reconstitution of the lipid-protein membrane film from *Escherichia coli* has been already established by using commercially available lipid mixtures [9]. This is because the lipid mixture extracted from *Escherichia coli* facilitates the formation of lipid membrane films supported on a substrate through the vesicle rupture method. There have been reports of lipid-protein patch formed via the break-off of living cells fixed on a substrate [42]. These lipid-protein patches have only directed the focus on the adsorbing and localizing behaviours of proteins on the lipid membrane film. Thus, such setup for research is not applicable in studying membrane film dynamics and deformation, including buckling, because the membrane film is anchored and/or fixed onto the substrate. In the current study, the spin coating method enabled us to form pre-organized lipid patches from *Dictyostelium discoideum* on the micrometer scale using a substrate holder with aspiration pores.

The other achievement is the construction of the lipid-protein membrane film, showings three-dimensional dynamic behaviours. The closed bilayer formation to prepare vesicles using natural lipids has been adopted to investigate the adsorbing and localizing behaviours of proteins. Recently, a method for preparing vesicles with diameters of larger than 1 μm, using phosphatidylcholines and phosphatidylinositides, was reported elsewhere [43]. Even though such vesicles have drawn attention in terms of artificial cell modelling, it is difficult to construct a membrane-assisted biochemical machinery to be tuned or modified inside after the encapsulation of the cytosol. The current 2D-system is one solution to this issue. Even though the thickness of the *Dictyostelium* lipid patch was evaluated to be 10.5 μm and the patch contained the stacked lipid layers, we observed the autonomous buckling of the lipid membrane film for the first time under the cytosol without any mechanical aspiration force. In the previous report [6], the membrane buckling of the *Dictyostelium* cell body itself was induced by the glass capillary aspiration and discussed in terms of the membrane bending energy and the work of the aspiration pressure. The current finding is thus crucial for the continuing development of the measurement of the cytosol ability for cell membrane deformation in an open system.

Methods

Cell culture

AX4 cells of the *Dictyostelium discoideum* axenic strain were cultured in modified HL5 medium under shaking at 155 rpm at 22°C [44]. The AX4 cells co-expressing the Pleckstrin homology domain of cytosolic regulator of adenylyl cyclase (PH-crac) tagged with RFP (PH-crac-RFP) and the phosphatase and tensin homolog (PTEN) tagged GFP (PTEN-GFP) were cultured in modified HL5 medium containing G418 (30 μg/mL, Wako, Japan) and Hygromycin B (60 μg/mL, Calbiochem, USA) [11].

Cytosol extraction

Cytosolic extracts were prepared from AX4 cells according to the protocol described elsewhere [45,46]. The cells, which were numbered at 7×10^6 cells mL^{-1} were harvested, washed two times with 80 mL of phosphate buffer (PB), and re-suspended in 500 μL of PB. Cells were disrupted by nitrogen decompression using a cell disruption vessel (model 4639l, Parr Instrument, IL, USA) on ice. Cell lysate was centrifuged at 16000 × g for 30 min at 4°C. The supernatant was isolated and centrifuged again. The supernatant were transferred to a microtube and kept on ice until just before use. When we examined LY294002 as an inhibitor assay, a dimethyl sulfoxide solution (2 μL) of LY294002 (final concentration of 40 μM) was added to the cell suspension of PB (250 μL) with shaking for 30 min just before use for the extraction.

Lipid extraction

Lipid extraction was performed according to the protocol of Bartles's group [47] with slight modifications. AX4 cells were harvested, washed two times with 80 mL of PB, and re-suspended in phosphate buffered saline (PBS) to prepare the cell dispersion of 4.0×10^8 cells mL^{-1}. The dispersion was quickly frozen once in liquid nitrogen and freeze-thawed in an ice water bath. Lysates were centrifuged at 50,000 × g for 45 min at 4°C. Pellets were homogenized by a pipette in Milli-Q water. The homogenate was centrifuged again, and the washed pellets were drained and dispersed in a mixed organic solvent ($CHCl_3/CH_3OH$ (2:1, v/v)). The volume of the organic solvent was 2250 μL per 4×10^8 cells. The organic dispersion was centrifuged at 2000 × g for 5 min at 4°C. The organic layer was transferred to a glass bottle by an acute Pasteur pipette and the solvent was removed by nitrogen gas blowing. The residue was suspended in 938 μL of $CHCl_3$/saturated NaCl aq (2:1, v/v) and the suspension was centrifuged at 2000 × g for 5 min at 4°C. The organic layer was transferred to a glass bottle by an acute Pasteur pipette.

Spin coating and microscopy observation

On a square-type cover glass (18 × 18 mm, thickness; 0.12-0.17 mm, MATSUNAMI, Japan), we dropped 60 μL of the organic lipid solution and it was spin coated at 300 rpm for 3 min using a spin coater (MIKASA, JAPAN). As a reference experiment, a chloroform solution (60 μL) of 1-palmitoyl-2-oleoyl-3-*sn*-glycero-3-phosphocholine (POPC, 0.45 mM, Wako, Tokyo, Japan), 1,2-hexadecanoyl-3-*sn*-glycerophosphatidylinositol (3,4,5)-trisphosphate (PIP3, 10 mol%, Avanti Polar Lipids, USA), and TexasRed®-1,2-dihexadecanoyl-*sn*-glycero-3-phosphoethanolamine triethylammonium salt (TexasRed-DHPE, 0.1 mol%, Invitrogen, USA) was dropped and spin coated in the same manner. The lipid membrane film formed on the cover glass was shaded from light and set in a desiccator, and the residual organic solvent was completely removed from the lipid membrane film under reduced pressure at room temperature for 17 h. To assemble a chamber, the cover glass was attached to a perforated dish with double-sided tape (Frame Seal, thickness; 280 μm, BIO-RAD, MA, USA). The chamber was placed on a microscope stage and the lipid membrane film was pre-wetted by 50 μL of PBS for 10 min. After injection of 25 μL of the extracted cytosol, the lipid membrane film was observed under a confocal laser scanning fluorescence and bright field microscope (A1R+, Nikon, Japan) equipped with a 60× objective lens. Red and green fluorescence images were obtained through the corresponding band-pass filter and dichroic mirror units (excitation; 562 nm, emission; 570 – 620 nm and excitation; 487 nm, emission; 500 – 550 nm). The time interval for time lapse image acquisition was set to 6 s. To determine the area of lipid membrane patch formed on the cover glass, we analyzed the histogram of each Green and Red fluorescence intensity of all pixels on the captured image frame by Image J (NIH, USA) and found two peaks in the histogram. Since one peak with lower fluorescence intensity was caused by the noise from the detector and the scattered light, we measured the area of pixels where the intensity was in the range of the other peak distribution with larger intensity.

Phalloidin staining and observation

For visualization of F-actin, we used Alexa488-labeled phalloidin (Invitrogen, USA). After injection of the cytosol has been injected for 30 min, the lipid membrane film was fixed with 3.7% formalin in PBS for 10 min. The formalin-fixed film was washed three times by 3 mL of PBS and stained by 50 μL of 5 units/mL Alexa488-labeled phalloidin dissolved in PBS for 20 min, and washed three times by PBS. The stained film was observed under a confocal laser scanning fluorescence microscope (CSU-X1, IX81, Olympus, Japan) equipped with a 100× objective lens. Red and green fluorescence images were obtained by using the corresponding band-pass filter and dichroic mirror units (excitation; 520 – 550 nm, emission; > 580 nm and excitation 470 – 490 nm, emission; 510 – 550 nm).

Additional files

Additional file 1: Figure S1. Bright field microscopy image of the cover glass with lipid membrane film of POPC with PIP3 (10 mol%) prepared by the spin coating. (bar = 3 mm).

Additional file 2: Video 1. Movie of the time-course change of *Dictyostelium* lipid patch. We captured the movie by typical bright field microscopy (time interval per frame = 6 s). The buckling motion of the lipid patch occurred remarkably for initial 10 min after AX4 cytosol injection (24 – 30 s). The size of the observation space is 210 μm x 210 μm.

Additional file 3: Video 2. Movie of the time-course change of *Dictyostelium* lipid patch. We captured the movie by the confocal laser scanning fluorescence microscopy with both Red and Green detection modes (time interval per frame = 6 s). The PTEN-GFP and PH-crac-RFP localization on the *Dictyostelium* lipid patch occurred after injection of cytosol extract of PH-crac-RFP/PTEN-GFP co-expressing cells (24 – 30 s). The size of the observation space is 210 μm x 210 μm.

Additional file 4: Video 3. Movie of the time-course change of *Dictyostelium* lipid patch. We captured the movie by the typical bright field microscopy (time interval per frame = 6 s). The buckling motion occurred remarkably for initial 10 min after injection of cytosol extract of PH-crac-RFP/PTEN-GFP co-expressing cells (24 – 30 s). The size of the observation space is 210 μm x 210 μm.

Additional file 5: Figure S2. 3D-merged reconstructed images of the autofluorescence images of a typical *Dictyostelium* lipid patch obtained by confocal laser scanning fluorescence microscopy with both Red and Green detection modes before (A,B) and 30 min after (C,D) injection of cytosol extract, which had been treated at 65°C for 1 h just before use, of PTEN-GFP/PH-crac-RFP co-expressing cells. The volume size of observation space is 210 μm × 210 μm × 37.5 μm. Each x-z plane cross-section of the 3D merged reconstructed image was attached in the right column (B,D). Dashed lines correspond to the surface of the glass slide. The height indicates the distance between the surface of the glass slide and the peak top position of the autofluorescence images.

Abbreviations

POPC: 1-palmitoyl-2-oleoyl-3-*sn*-glycero-3-phosphocholine; PIP3: 1,2-hexadecanoyl-3-*sn*-glycerophosphatidyl inositol-(3,4,5)-trisphosphate; PTEN: Phosphatase and tensin homolog; PTEN-GFP: PTEN tagged green fluorescent protein; PH-crac: Pleckstrin homology domain of cytosolic regulator of adenylyl cyclase; PH-crac-RFP: PH-crac tagged with RFP.

Competing interests

The authors declare that they have no competing interests.

Authors' contributions

KT made the conceptual design of the study and the experimental setup and methods, performed experiments and prepared the manuscript. TT assisted with study's conception, data interpretation and manuscript preparation. All authors read and approved the final manuscript.

Acknowledgements

The authors acknowledge Prof. Satoshi Sawai (The University of Tokyo) and Dr. Nao Shimada (The University of Tokyo) for culturing cells, assisting with experiments, and delivering fruitful discussions. This work was supported by Platform for Dynamic Approaches to Living System (KT and TT) and Grant-in-Aid for Scientific Research on Innovative Areas "Molecular Robotics" (TT, No. 24104004) from the Ministry of Education, Culture, Sports, Science and Technology, Japan.

References

1. Noireaux V, Libchaber A. A vesicle bioreactor as a step toward an artificial cell assembly. Proc Natl Acad Sci U S A. 2004;101:17669–74.
2. Szostak JW, Bartel DP, Luigi PL. Synthesizing life. Nature. 2001;409:387–90.
3. Noireaux V, Bar-Ziv R, Godefroy J, Salman H, Libchaber A. Toward an artificial cell based on gene expression in vesicles. Phys Biol. 2005;2:1–8.
4. Evans EA. Bending elastic modulus of red blood cell membrane derived from buckling instability in micropipet aspiration tests. Biophys J. 1983;43:28–30.
5. Helfer E, Harlepp S, Bourdieu L, Robert J, MacKintosh FC, Chatenay D. Buckling of actin-coated membranes under application of a local force. Phys Rev Lett. 2001;87:088103.
6. Merkel R, Simson R, Simson DA, Hohenadl M, Boulbitch A, Wallraff E, et al. A micromechanic study of cell polarity and plasma membrane cell body coupling in *Dictyostelium*. Biophys J. 2000;79:707–19.
7. Antonny B. Membrane deformation by protein coats. Current Opinion Cell Biol. 2006;18:386–94.
8. Zimmerberg J, Gawrisch K. The physical chemistry of biological membranes. Nat Chem Biol. 2006;2:564–7.
9. Loose M, Fischer-Friedrich E, Ries J, Kruse K, Schwille P. Spatial regulators for bacterial cell division self-organize into surface waves *in Vitro*. Science. 2008;320:789–92.
10. Mennicke U, Salditt T. Preparation of solid-supported lipid bilayers by spin-coating. Langmuir. 2002;18:8172–7.
11. Taniguchi D, Ishihara S, Oonuki T, Honda-Kitahara M, Kaneko K, Sawai S. Phase geometries of two-dimensional excitable waves govern self-organized morphodynamics of amoeboid cells. Proc Natl Acad Sci U S A. 2013;110:5016–21.
12. Parent CA, Devreotes PN. Molecular genetics of signal transduction in *Dictyostelium*. Annu Rev Biochem. 1996;65:411–40.
13. Banghum AD, Standish MM, Watkins JC. Diffusion of univalent ions across the lamellae of swollen phospholipids. J Mol Biol. 1965;13:238–52.
14. Billerit C, Jeffries DGM, Orwar O, Jesorka A. Formation of giant unilamellar vesicles from spin-coated lipid films by localized IR heating. Soft Matter. 2012;8:10823–6.
15. Okahata Y, Tsuruta T, Ijiro K, Ariga K. Langmuir-blodgett films of an enzyme-lipid complex for sensor membrane. Langmuir. 1988;4:1373–5.
16. Schmidt TF, Caseli L, Viitala T, Oliveira Jr ON. Enhanced activity of horseradish peroxidase in Langmuir–Blodgett films of phospholipids. Biochim Biophys Acta. 2008;1778:2291–7.
17. Mingeot-Leclercq M, Deleu M, Brasseur R, Dufrêne YF. Atomic force microscopy of supported lipid bilayers. Nat Protocols. 2008;3:1654–9.

18. Giocondi M, Yamamoto D, Lesniewska E, Milhiet P, Ando T, Grimellec CL. Surface topography of membrane domains. Biochim Biophys Acta. 2010;1798:703–18.
19. Alessandrini A, Facci P. Unraveling lipid/protein interaction in model lipid bilayers by atomic force microscopy. J Mol Recognit. 2011;24:387–96.
20. Lautscham LA, Lin CY, Auernheimer V, Naumann CA, Goldmann WH, Fabry B. Mechanosensing of cells in laminin-fuctionalized biomembrane-mimicking substrates. Biomaterials. 2014;35:3198–207.
21. Matosevic S, Paegel BM. Layer-by-layer cell membrane assembly. Nat Chem. 2013;5:958–63.
22. Larsen J, Hatzakis NS, Stamou D. Observation of inhomogeneity in the lipid composition of individual nanoscale liposomes. J Am Chem Soc. 2011;133:10685–7.
23. Elizondo E, Larsen J, Hatzakis NS, Cabrera I, Bjørnholm T, Veciana J, et al. Influence of the preparation route on the supramolecular organization of lipids in a vesicular system. J Am Chem Soc. 2012;134:1918–21.
24. Marie R, Schmid S, Johansson A, Ejsing L, Nordström M, Häfliger D, et al. Immobilisation of DNA to polymerised SU-8 photoresist. Biosens Bioelect. 2006;21:1327–32.
25. Shah K, Shin WC, Besser RS. Novel microfabrication approaches for directly patterning PEM fuel cell membranes. J Power Sources. 2003;123:172–81.
26. Ryu DY, Shin K, Drockenmuller E, Hawker CJ, Russell TP. A generalized approach to the modification of solid surfaces. Science. 2005;308:236–9.
27. Kim SO, Solak HH, Stoykovich MP, Ferrier NJ, Pablo JJ, Nealey PF. Epitaxial self-assembly of block copolymers on lithographically defined nanopatterned substrates. Nature. 2003;424:411–4.
28. Chang J, Sun B, Breiby DW, Nielsen MM, Sölling TI, Giles M, et al. Enhanced mobility of Poly(3-hexylthiophene) transistors by spin-coating from high-boiling-point solvents. Chem Mater. 2004;16:4772–6.
29. Simonsen AC, Bagatolli LA. Structure of spin-coated lipid films and domain formation in supported membranes formed by hydration. Langmuir. 2004;20:9720–8.
30. Dols-Perez A, Fumagalli L, Simonsen AC, Gomila G. Ultrathin spin-coated dioleoylphosphatidylcholine lipid layers in dry conditions: a combined atomic force microscopy and nanomechanical study. Langmuir. 2011;27:13165–72.
31. Jensen MH, Morris EJ, Simonsen AC. Domain shapes, coarsening, and random patterns in ternary membranes. Langmuir. 2007;23:8135–41.
32. Ellingson JS. Changes in the phospholipid composition in the differentiating cellular slime mold, *Dictyostelium discoideum*. Biochim Biophys Acta. 1974;337:60–7.
33. Yawn E, Zutfu A. Separation and analysis of ^{32}P-labeled phospholipids by a simple and rapid thin-layer chromatographic procedure and its application to cultured *Neuroblastoma* cells. Anal Biochem. 1977;80:430–7.
34. Weeks G, Herring FG. The lipid composition and membrane fluidity of *Dictyostelium discoideum* plasma membranes at various stages during differentiation. J Lipid Res. 1980;21:681–6.
35. Luisi PL, Walde P, editors. Giant Vesicles. Chichester: John Wiley & Sons; 2000.
36. Menger FM, Angelova MI. Giant vesicles: imitating the cytological processes of cell membranes. Acc Chem Res. 1998;31:789–97.
37. Dormann D, Weijer G, Parent CA, Devreotes PN, Weijer CJ. Visualizing PI3Kinase-mediated cell-cell signaling during *Dictyostelium* development. Current Biol. 2002;12:1178–88.
38. Iijima M, Devreotes P. Tumor suppressor PTEN mediates sensing of chemoattractant gradients. Cell. 2002;109:599–610.
39. Parent CA, Blacklock BJ, Froehlich WM, Murphy DB, Devreotes PN. G protein signaling events are activated at the leading edge of chemotactic cells. Cell. 1998;95:81–91.
40. Huang YE, Iijima M, Parent CA, Funamoto S, Firtel RA, Devreotes P. Receptor-mediated regulation of PI3Ks confines PI(3,4,5)P3 to the leading edge of chemotaxing cells. Mol Biol Cell. 2003;14:1913–22.
41. Haugwitz M, Noegel AA, Karakesisoglou J, Schleicher M. *Dictyostelium* amoebae that lack G-actin-sequestering profilins show defects in F-actin content, cytokinesis, and development. Cell. 1994;79:303–14.
42. Mager MD, Melosh NA. Single-step process to reconstitute cell membranes on solid supports. Langmuir. 2010;26:4635–8.
43. Jiangb Z, Redfernc RE, Islerd Y, Rosse AH, Gericke A. Cholesterol stabilizes fluid phosphoinositide domains. Chem Phys Lipids. 2014;182:52–61.
44. Watts DJ, Ashworth JM. Growth of myxamoebae of the cellular slime mould *Dictyostelium discoideum* in axenic culture. Biochem J. 1970;119:171–4.
45. Watarai H, Inagaki Y, Kubota N, Fuju K, Nagafune J, Yamaguchi Y, et al. Proteomic approach to the identification of cell membrane proteins. Electrophoresis. 2000;21:460–4.
46. Gregor T, Fujimoto K, Masaki N, Sawai S. The onset of collective behavior in social amoebae. Science. 2010;328:1021–5.
47. Bartles JR, Pardos BT, Frazier WA. Reconstitution of discoidin hemagglutination activity by lipid extracts of *Dictyostelium discoideum* cells. J Biol Chem. 1979;254:3156–9.

Molecular cloning using polymerase chain reaction, an educational guide for cellular engineering

Sayed Shahabuddin Hoseini[1,2] and Martin G Sauer[1,2,3*]

Abstract

Background: Over the last decades, molecular cloning has transformed biological sciences. Having profoundly impacted various areas such as basic science, clinical, pharmaceutical, and environmental fields, the use of recombinant DNA has successfully started to enter the field of cellular engineering. Here, the polymerase chain reaction (PCR) represents one of the most essential tools. Due to the emergence of novel and efficient PCR reagents, cloning kits, and software, there is a need for a concise and comprehensive protocol that explains all steps of PCR cloning starting from the primer design, performing PCR, sequencing PCR products, analysis of the sequencing data, and finally the assessment of gene expression. It is the aim of this methodology paper to provide a comprehensive protocol with a viable example for applying PCR in gene cloning.

Results: Exemplarily the sequence of the tdTomato fluorescent gene was amplified with PCR primers wherein proper restriction enzyme sites were embedded. Practical criteria for the selection of restriction enzymes and the design of PCR primers are explained. Efficient cloning of PCR products into a plasmid for sequencing and free web-based software for the consecutive analysis of sequencing data is introduced. Finally, confirmation of successful cloning is explained using a fluorescent gene of interest and murine target cells.

Conclusions: Using a practical example, comprehensive PCR-based protocol with important tips was introduced. This methodology paper can serve as a roadmap for researchers who want to quickly exploit the power of PCR-cloning but have their main focus on functional *in vitro* and *in vivo* aspects of cellular engineering.

Keywords: Cloning, Polymerase chain reaction (PCR), Recombinant DNA, Biological engineering, Educational guide, Transduction, Transfection

Background

Various techniques were introduced for assembling new DNA sequences [1-3], yet the use of restriction endonuclease enzymes is the most widely used technique in molecular cloning. Whenever compatible restriction enzyme sites are available on both, insert and vector DNA sequences, cloning is straightforward; however, if restriction sites are incompatible or if there is even no restriction site available in the vicinity of the insert cassette, cloning might become more complex. The use of PCR primers, in which compatible restriction enzyme sites are embedded, can effectively solve this problem and facilitate multistep cloning procedures.

Although PCR cloning has been vastly used in biological engineering [4-8], practical guides explaining all necessary steps and tips in a consecutive order are scarce. Furthermore, the emergence of new high-fidelity DNA polymerases, kits, and powerful software makes the process of PCR cloning extremely fast and efficient. Here we sequentially explain PCR cloning from the analysis of the respective gene sequence, the design of PCR primers, performing the PCR procedure itself, sequencing the resulting PCR products, analysis of sequencing data, and finally the cloning of the PCR product into the final vector.

* Correspondence: sauer.martin@mh-hannover.de
[1]Departments of Pediatric Hematology, Oncology and Blood Stem Cell Transplantation, Hannover, Germany
[2]Hannover Center for Transplantation Research, Hannover, Germany
Full list of author information is available at the end of the article

Results and discussion

Choosing proper restriction enzymes based on defined criteria

In order to proceed with a concise example, tdTomato fluorescent protein was cloned into an alpharetroviral vector. Consecutively, a murine leukemia cell line expressing tdTomato was generated. This cell line will be used to track tumor cells upon injection into mice in preclinical immunotherapy studies. However, this cloning method is applicable to any other gene. To begin the cloning project, the gene of interest (GOI) should be analyzed. First, we check whether our annotated sequence has a start codon (ATG, the most common start codon) and one of the three stop codons (TAA, TAG, TGA). In case the gene was previously manipulated or fused to another gene (e.g. via a 2A sequence), it happens that a gene of interest might not have a stop codon [9]. In such cases, a stop codon needs to be added to the end of your annotated sequence. It is also beneficial to investigate whether your GOI contains an open reading frame (ORF). This is important since frequent manipulation of sequences either by software or via cloning might erroneously add or delete nucleotides. We use Clone Manager software (SciEd) to find ORFs in our plasmid sequences; however, there are several free websites you can use to find ORFs including the NCBI open reading frame finder (http://www.ncbi.nlm.nih.gov/gorf/gorf.html).

The tdTomato gene contains ATG start codon and TAA stop codon (Figure 1). The size of the tdTomato gene is 716 bp.

In a next step, PCR primers that include proper restriction enzyme sites need to be designed for the amplification of the GOI. Several criteria should be considered in order to choose the optimal restriction enzymes. First, binding sites for restriction enzymes should be ideally available at a multiple cloning site within the vector. Alternatively they can be located downstream of the promoter in your vector sequence. Restriction enzymes should be single cutters (single cutters target one restriction site only within a DNA sequence) (Figure 2A). If they are double or multiple cutters, they should cut within a sequence that is not necessary for proper functioning of the vector plasmid and will finally be removed (Figure 2B). It is also possible to choose one double cutter or multiple cutter enzymes cutting the vector downstream of the promoter and also not within a vital sequence of the plasmid (Figure 2C). Double cutter or multiple cutter enzymes have two or more restriction sites on a DNA sequence, respectively. Cutting the vector with double or multiple cutters would give rise to two identical ends. In such a case, the insert cassette should also contain the same restriction enzyme sites on both of its ends. Therefore, when the insert and vector fragments are mixed in a ligation experiment, the insert can fuse to the vector in either the right orientation (from start codon to stop codon) or reversely (from stop codon to start codon). A third scenario can occur, if the vector fragment forms a self-ligating circle omitting the insert at all. Once the DNA has been incubated with restriction enzymes, dephosphorylation of the 5′ and 3′ ends of the vector plasmid using an alkaline phosphatase enzyme will greatly reduce the risk of self-ligation [10]. It is therefore important to screen a cloning product for those three products (right orientation, reverse orientation, self-ligation) after fragment ligation.

Second, due to higher cloning efficiency using sticky-end DNA fragments, it is desirable that at least one (better both) of the restriction enzymes is a so-called sticky-end cutter. Sticky end cutters cleave DNA asymmetrically generating complementary cohesive ends. In contrast, blunt end cutters cut the sequence symmetrically

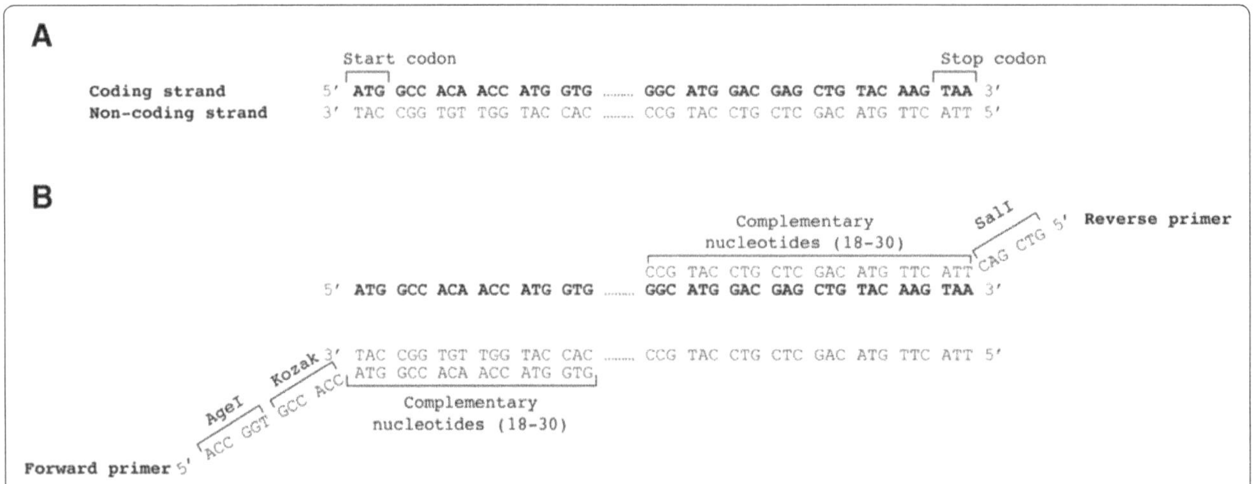

Figure 1 Overview of the start and the end of the gene of interest. (A) The nucleotide sequences at the start and the end of the tdTomato gene are shown. The coding strand nucleotides are specified in bold **(B)** The nucleotide sequences of the forward and reverse primers containing proper restriction enzyme sites and the Kozak sequence are shown.

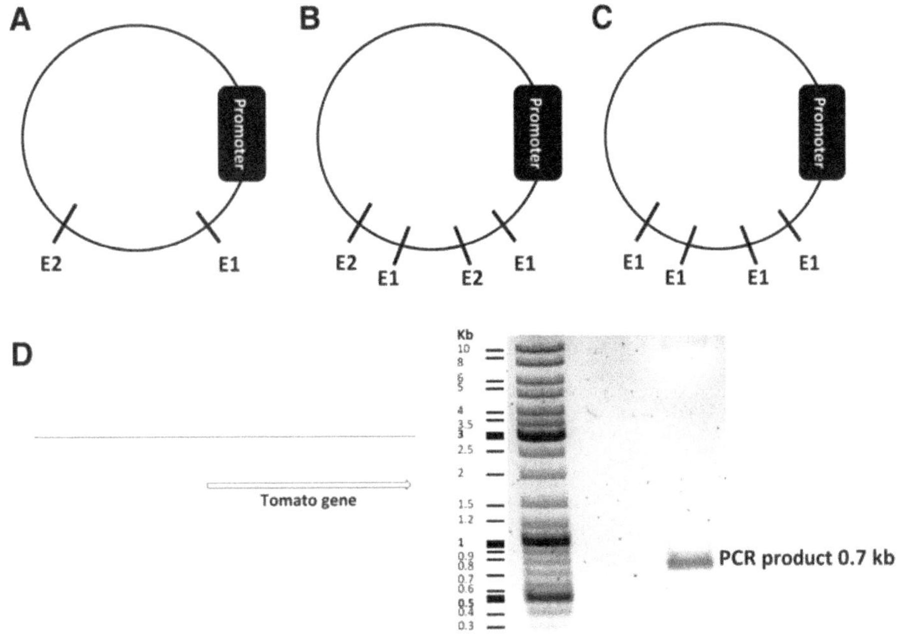

Figure 2 Choosing proper restriction enzymes based on defined criteria for PCR cloning. (A) Two single-cutter restriction enzymes (E1 and E2) are located downstream of the promoter. **(B)** E1 and E2 restriction enzymes cut the plasmid downstream of the promoter several (here two times for each enzyme) times. **(C)** The E1 restriction enzyme cuts the plasmid downstream of the promoter more than once. **(D)** The PCR product, which contains the tdTomato gene and the restriction enzyme sites, was run on a gel before being extracted for downstream applications.

leaving no overhangs. Cloning blunt-end fragments is more difficult. Nevertheless, choosing a higher insert/vector molar ratio (5 or more) and the use 10% polyethylene glycol (PEG) can improve ligation of blunt-end fragments [11].

Third, some restriction enzymes do not cut methylated DNA. Most of the strains of *E. coli* contain Dam or Dcm methylases that methylate DNA sequences. This makes them resistant to methylation-sensitive restriction enzymes [12]. Since vector DNA is mostly prepared in *E. coli*, it will be methylated. Therefore avoiding methylation-sensitive restriction enzymes is desirable; however, sometimes the isoschizomer of a methylation-sensitive restriction enzyme is resistant to methylation. For example, the *Acc*65I enzyme is sensitive while its isoschizomer *kpn*I is resistant to methylation [13]. Isoschizomers are restriction enzymes that recognize the same nucleotide sequences. If there remains no other option than using methylation-sensitive restriction enzymes, the vector DNA needs to be prepared in *dam⁻dcm⁻* *E. coli* strains. A list of these strains and also common *E. coli* host strains for molecular cloning is summarized in Table 1. Information regarding the methylation sensitivity of restriction enzymes is usually provided by the manufacturer.

Fourth, it makes cloning easier if the buffer necessary for the full functionality of restriction enzymes is the same because one can perform double restriction digest. This saves time and reduces the DNA loss during purification. It may happen that one of the restriction enzymes is active in one buffer and the second enzyme is active in twice the concentration of the same buffer. For example the *Nhe*I enzyme from Thermo Scientific is active in Tango 1X buffer (Thermo Scientific) and *Eco*R1 enzyme is active in Tango 2X buffer (Thermo Scientific). In such cases, the plasmid DNA needs to be first digested by the enzyme requiring the higher buffer concentration (here *Eco*R1). This will be followed by diluting the buffer for the next enzyme (requiring a lower concentration (here *Nhe*I)) in the same buffer. However, the emergence of universal buffers has simplified the double digest of DNA sequences [15]. In our example the vector contains the *Age*I and *Sal*I restriction sites. These enzyme sites were used for designing PCR primers (Figure 1). It is essential for proper restriction enzyme digestion that the plasmid purity is high. DNA absorbance as measured by a spectrophotometer can be used to determine the purity after purification. DNA, proteins, and solvents absorb at 260 nm, 280 nm, and 230 nm, respectively. An OD 260/280 ratio of >1.8 and an OD 260/230 ratio of 2 to 2.2 is considered to be pure for DNA samples [16]. The OD 260/280 and 260/230 ratios of our exemplary plasmid preparations were 1.89 and 2.22, respectively. We observed that the purity of the gel-extracted vector and insert DNA fragments were lower after restriction digest; ligation works even in such cases, however, better results can be expected using high-purity fragments.

Table 1 Common *E. coli* host strains in gene cloning

Application	*E. coli* strain
Transformation of large plasmids	GeneHogs, XL10 Gold, STBL4, SURE
Generation of single-stranded (ssDNA)	INV110, JM109, JS5, NM522, SCS110, SURE, STBL4, XL10 Gold, XL1 Blue, TG1
For storage of plasmids that tend to recombine	DH10B, DH5α, STBL2, STBL3, STBL4, SURE, GeneHogs, Hb101, JM109, JS5, XL10 Gold, XL1 Blue, GC5, GC10
Rapid cloning (fast cell growth)	Mach1
Cloning of unstable plasmids	SURE, STBL2, STBL3, STBL4
High efficiency cloning for library construction	XL10-Gold, MegaX, DH10B
Blue/white screening	DH10B, DH5α, MC1061, TOP10, XL1 Blue, Hb101, NM522, SCS110, STBL4, SURE, XL10 Gold, GeneHogs, INV110, JM109, JS5, GC5, GC10
For site-directed mutagenesis	XL-*mutS*, BMH 71–18 *mutS*, ES1301 *mutS*
For random mutagenesis	XL1-Red
For expression of toxic genes	ABLE C, ABLE K
General cloning and storage of plasmids	DH10B, DH5α, MC1061, TOP10, XL1-Blue, Hb101
For proliferation of plasmids encoding the ccdB toxic gene (important in Gateway cloning)	ccdB survival, DB3.1, XL1 Blue, JM109, DH5α
For generation of unmethylated DNA to be cut with methylation-sensitive restriction enzymes	JM110, ER2925, INV110, DM1, SCS110, E4109S

Data are derived from enzyme providers' data sheets, the following website, and this reference [14]. http://openwetware.org/wiki/E._coli_genotypes.

The following plasmid repository website can be useful for the selection of different vectors (viral expression and packaging, empty backbones, fluorescent proteins, inducible vectors, epitope tags, fusion proteins, reporter genes, species-specific expression systems, selection markers, promoters, shRNA expression and genome engineering): http://www.addgene.org/browse/.

A collection of cloning vectors of *E. coli* is available under the following website: http://www.shigen.nig.ac.jp/ecoli/strain/cvector/cvectorExplanation.jsp.

Designing cloning primers based on defined criteria

For PCR primer design, check the start and stop codons of your GOI. Find the sequence of the desired restriction enzymes (available on the manufacturers' websites) for the forward primer (Figure 3A). It needs to be located before the GOI (Figure 1B). The so-called Kozak sequence is found in eukaryotic mRNAs and improves the initiation of translation [17]. It is beneficial to add the Kozak sequence (GCCACC) before the ATG start codon since it increases translation and expression of the protein of interest in eukaryotes [18]. Therefore, we inserted GCCACC immediately after the restriction enzyme sequence *Age*I and before the ATG start codon. Then, the first 18 to 30 nucleotides of the GOI starting from the ATG start codon are added to the forward primer sequence. These overlapping nucleotides binding to the template DNA determine the annealing temperature (Tm). The latter is usually

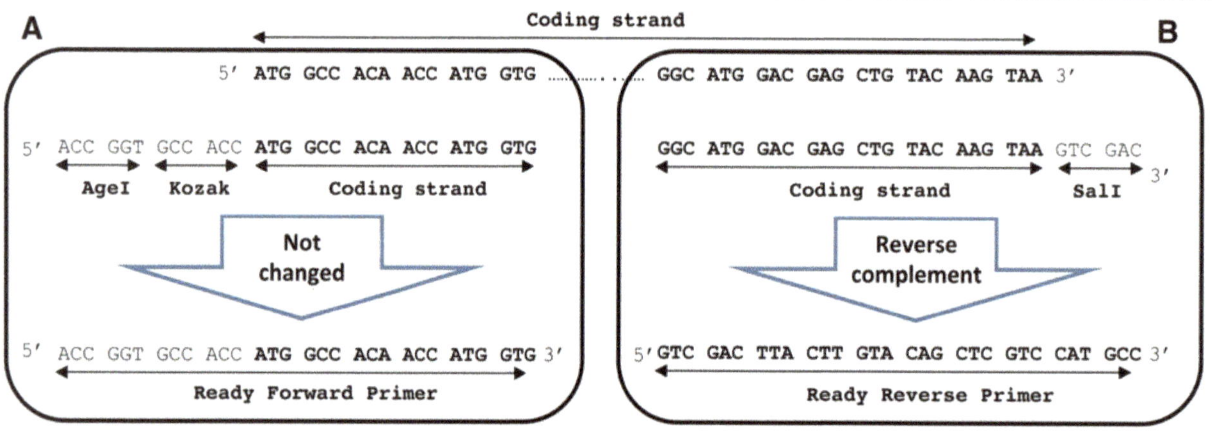

Figure 3 Designing primers based on defined criteria for PCR cloning. (A-B) Sequences of the forward and the reverse primer are depicted. The end of the coding strand is to be converted into the reverse complement format for the reverse primer design. For more information, please see the text.

higher than 60°C. Here, we use Phusion high-fidelity DNA polymerase (Thermo Scientific). You can use the following websites for determination of the optimal Tm: http://www.thermoscientificbio.com/webtools/tmc/.

https://www.neb.com/tools-and-resources/interactive-tools/tm-calculator.

The Tm of our forward primer is 66°C.

Choose the last 18 to 30 nucleotides including the stop codon of your GOI for designing the reverse primer (Figure 3B). Then calculate the Tm for this sequence which should be above 60°C and close to the Tm of the forward primer. Tm of the overlapping sequence of our reverse primer was 68°C. Then, add the target sequence of the second restriction enzyme site (in this case *Sal*I) immediately after the stop codon. Finally, convert this assembled sequence to a reverse-complement sequence. The following websites can be used to determine the sequence of the reverse primer:

http://reverse-complement.com/

http://www.bioinformatics.org/sms/rev_comp.html

This is important since the reverse primer binds the coding strand and therefore its sequence ($5' \rightarrow 3'$) must be reverse-complementary to the sequence of the coding strand (Figure 1A).

Performing PCR using proofreading polymerases

Since the PCR reaction follows logarithmic amplification of the target sequence, any replication error during this process will be amplified. The error rate of non-proofreading DNA polymerases, such as the Taq polymerase, is about 8×10^{-6} errors/bp/PCR cycle [19]; however, proofreading enzymes such as Phusion polymerase have a reported error rate of 4.4×10^{-7} errors/bp/PCR cycle. Due to its superior fidelity and processivity [20-22], the Phusion DNA polymerase was used in this example. It should be noted that Phusion has different temperature requirements than other DNA polymerases. The primer Tm for Phusion is calculated based on the Breslauer method [23] and is higher than the Tm using Taq or pfu polymerases. To have optimal results, the Tm should be calculated based on information found on the website of the enzyme providers. Furthermore, due to the higher speed of Phusion, 15 to 30 seconds are usually enough for the amplification of each kb of the sequence of interest.

After the PCR, the product needs to be loaded on a gel (Figure 2D). The corresponding band needs to be cut and the DNA extracted. It is essential to sequence the PCR product since the PCR product might include mutations. There are several PCR cloning kits available some of which are shown in Table 2. We used the pJET1.2/blunt cloning vector (Thermo Scientific, patent publication: US 2009/0042249 A1, Genbank accession number EF694056.1) and cloned the PCR product into the linearized vector. This vector contains a lethal gene (*cco47IR*) that is activated in case the vector becomes circularized. However, if the PCR product is cloned into

Table 2 Common vectors in gene cloning

Plasmid name	Advantages	Disadvantages	References
pBR322	Small size (4.4 kb), variety of cloning sites, medium copy number (15–20)		[24,25]
pUC18 and pUC19	Small size (2.7 kb), high copy number (500–700), multiple cloning site, sequencing using M13 primers	Not good if target protein is toxic or for membrane proteins, needs blue/white screen	[24,26]
pLG338	General purpose plasmid vector, size 7.3 kb, low copy number (6–8), genes coding for membrane and regulatory proteins which cannot be cloned into high-copy-number plasmids		[27]
pMiniT	Inserted PCR product disrupts a toxic minigene, no blue/white selection required, works for both blunt end and single-base overhang-containing PCR products		NEB
pCR™4Blunt-TOPO®	Inserted PCR product disrupts *ccd*B toxic gene [28], works based on topoisomerase I [29], no blue/white selection required	Needs blunt-end PCR products	Life Technologies
pDrive	Size 3.8 kb	Needs single A overhangs, not usable for proofreading DNA polymerases, needs blue/white screening	QIAGEN
StrataClone Blunt PCR Cloning Vector pSC-B-amp/kan	Uses the DNA topoisomerase I and the DNA recombination activity of Cre recombinase, up to 9 kb PCR product size	Needs blunt-end PCR products, needs special competent cells for transformation (cells expressing Cre recombinase)	Agilent Technologies
pJET1.2/blunt	Inserted PCR product disrupts the toxic gene *eco47IR*, no need for blue/white screen, works for both blunt end and single-base overhang-containing PCR products, up to 10 kb PCR product size		Thermo Scientific

Data are derived from vector providers or the cited references. pBR322, pUC18, pUC19, and pLG338 are cloning vectors and the rest are PCR sequencing vectors.

the cloning site within the lethal gene, the latter is disrupted allowing bacteria to grow colonies upon transformation. Circularized vectors not containing the PCR product express the toxic gene, which therefore kills bacteria precluding the formation of colonies. Bacterial clones are then to be cultured, plasmid DNA consecutively isolated and sequenced. The quality of isolated plasmid is essential for optimal sequencing results. We isolated the plasmid DNA from a total of 1.5 ml cultured bacteria (yield 6 μg DNA; OD 260/280 = 1.86; OD 260/230 = 2.17) using a plasmid mini-preparation kit (QIAGEN). The whole process of PCR, including cloning of the PCR product into the sequencing vector and transfection of bacteria with the sequencing vector can be done in one day. The next day, bacterial clones will be cultured overnight before being sent for sequencing.

Analysis of sequencing data

Sequencing companies normally report sequencing data as a FASTA file and also as ready nucleotide sequences via email. For sequence analysis, the following websites can be used:

http://blast.ncbi.nlm.nih.gov/Blast.cgi
http://xylian.igh.cnrs.fr/bin/align-guess.cgi

Here we will focus on the first website. On this website page, click on the "nucleotide blast" option (Figure 4A). A new window opens. By default, the "blastn" (blast nucleotide sequences) option is marked (Figure 4B). Then check the box behind "Align two or more sequences". Now two boxes will appear. In the "Enter Query Sequence" box (the upper box), insert the desired sequence of your gene of interest, which is flanked by the restriction sites you have already designed for your PCR primers. In the "Enter Subject Sequence" box (the lower box), enter the sequence or upload the FASTA file you have received from the sequencing company. Then click the "BLAST" button at the bottom of the page. After a couple of seconds, the results will be shown on another page. A part of the alignment data is shown in Figure 4C. For interpretation, the following points should be considered: 1) the number of identical nucleotides (shown under the "Identities" item) must be equal to the nucleotide number of your gene of interest. In our example, the number of nucleotides of the tdTomato gene together with those of the restriction enzyme sites and the Kozak sequence was 735. This equals the reported number (Figure 4C). 2) The sequence identity (under the "Identities" item) should be 100%. Occasionally, the sequence identity is 100% but the number of identical

Figure 4 Sequence analysis of the PCR product using the NCBI BLAST platform. (A) On the NCBI BLAST webpage, the "nucleotide blast" option is chosen (marked by the oval line). **(B)** The "blastn" option appears by default (marked by the circle). The sequence of the gene of interest (flanked by the restriction sites as previously designed for the PCR primers) and the PCR product are to be inserted to the "Enter Query Sequence" and "Enter Subject Sequence" boxes. Sequences can also be uploaded as FASTA files. **(C)** Nucleotide alignment of the first 60 nucleotides is shown. Two important items for sequence analysis are marked by oval lines.

nucleotides is lower than expected. This can happen if one or more of the initial nucleotides are absent. Remember, all sequencing technologies have an error rate. For Sanger sequencing, this error rate is reported to range from 0.001% to 1% [30-33]. Nucleotide substitution, deletion or insertion can be identified by analyzing the sequencing results [34]. Therefore, if the sequence identity does not reach 100%, the plasmid should be resequenced in order to differentiate errors of the PCR from simple sequencing errors. 3) Gaps (under the "Gaps" item) should not be present. If gaps occur, the plasmid should be resequenced.

The average length of a read, or read length, is at least 800 to 900 nucleotides for Sanger sequencing [35]. For the pJET vector one forward and one reverse primer need to be used for sequencing the complete gene. These primers can normally cover a gene size ranging up to 1800 bp. If the size of a gene is larger than 1800, an extra primer should be designed for each 800 extra nucleotides. Since reliable base calling does not start immediately after the primer, but about 45 to 55 nucleotides downstream of the primer [36], the next forward primer should be designed to start after about 700 nucleotides from the beginning of the gene. Different websites, including the following, can be used to design these primers:

http://www.ncbi.nlm.nih.gov/tools/primer-blast/
http://www.yeastgenome.org/cgi-bin/web-primer
http://www.genscript.com/cgi-bin/tools/sequencing_primer_design

Being 735 bp in length, the size of the PCR product in this example was well within the range of the pJET sequencing primers.

After choosing the sequence-verified clone, vector and insert plasmids were digested by the *Age*I and *Sal*I restriction enzymes (Figure 5). This was followed by gel purification and ligation of the fragments. Transformation of competent *E. coli* with the ligation mixture yielded several clones that were screened by restriction enzymes. We assessed eight clones, all of which contained the tdTomato insert (Figure 6). It is important to pick clones that are large. Satellite clones might not have the right construct. We used a fast plasmid mini-preparation kit (Zymo Research) to extract the plasmid from 0.6 ml bacterial suspension. The yield and purity were satisfying for restriction enzyme-based screening (2.3 μg DNA; OD 260/280 = 1.82; OD 260/230 = 1.41). For large-scale plasmid purification, a maxi-preparation kit (QIAGEN) was used to extract the plasmid from 450 ml of bacterial culture (yield 787 μg DNA; OD 260/280 = 1.89; OD 260/230 = 2.22). The expected yield of a pBR322-derived plasmid isolation from 1.5 ml and 500 ml bacterial culture is about 2-5 μg and 500-4000 μg of DNA, respectively [37].

Some plasmids tend to recombine inside the bacterial host creating insertions, deletions and recombinations [38]. In these cases, using a recA-deficient *E. coli* can be useful (Table 1). Furthermore, if the GOI is toxic, incubation of bacteria at lower temperatures (25-30°C) and using ABLE C or ABLE K strains might circumvent the problem.

Viral production and transduction of target cells

To investigate the *in vitro* expression of the cloned gene, HEK293T cells were transfected with plasmids encoding the tdTomato gene, alpharetroviral Gag/Pol, and the vesicular stomatitis virus glycoprotein (VSVG) envelope. These cells, which are derived from human embryonic kidney, are easily cultured and readily transfected [39]. Therefore they are extensively used in biotechnology and gene therapy to generate viral particles. HEK293T cells require splitting every other day using warm medium. They should not reach 100% confluency for optimal results. To have good transfection efficiency, these cells need to be cultured for at least one week to have them in log phase. Transfection efficiency was 22%, as determined based on the expression of tdTomato by fluorescence microscopy 24 hours later (Figure 7A-B). To generate a murine leukemia cell line expressing the tdTomato gene for immunotherapy studies, C1498 leukemic cells were transduced with freshly harvested virus (36 hours of transfection). Imaging studies (Figure 7C) and flow cytometric analysis (Figure 7D) four days after transduction confirmed the expression of tdTomato in the majority of the cells.

Conclusions

In this manuscript, we describe a simple and step-by-step protocol explaining how to exploit the power of PCR to clone a GOI into a vector for genetic engineering. Several PCR-based creative methods have been developed being extremely helpful for the generation of new nucleotide sequences. This includes equimolar expression of several proteins by linking their genes via a self-cleaving 2A sequence [40,41], engineering fusion proteins, as well as the use of linkers for the design of chimeric proteins [42-44]. Furthermore, protein tags [45,46] and mutagenesis (site-directed, deletions, insertions) [47] have widened the applications of biological engineering. The protocol explained in this manuscript covers for most situations of PCR-assisted cloning; however, alternative PCR-based methods are available being restriction enzyme and ligation independent [6,48-51]. They are of special interest in applications where restriction enzyme sites are lacking; nevertheless, these methods might need several rounds of PCR or occasionally a whole plasmid needs to be amplified. In such cases, the chance of PCR errors increases and necessitates sequencing of multiple clones. In conclusion, this guideline assembles a simple and straightforward protocol using resources that

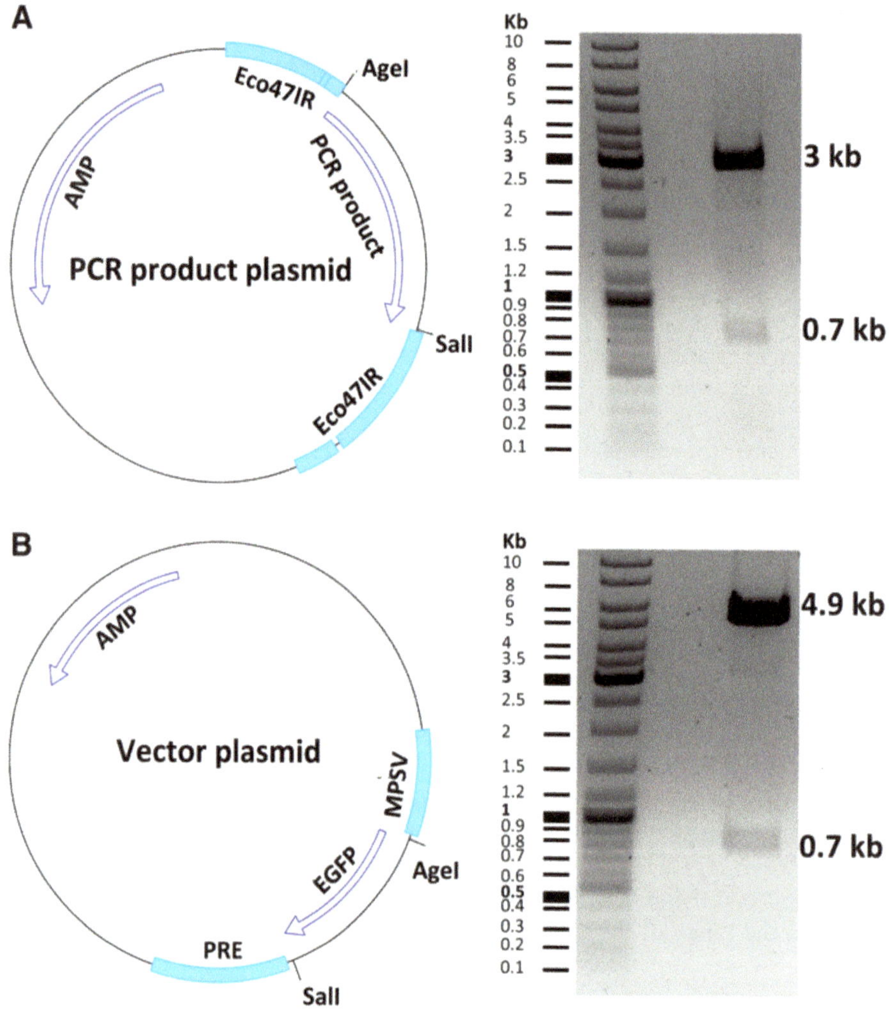

Figure 5 Vector and insert plasmid maps. (A) Illustration of the CloneJET plasmid containing the PCR product. Insertion of the PCR product in the cloning site of the plasmid disrupts the integrity of the toxic gene *eco47IR* and allows the growth of transgene positive clones. The plasmid was cut with the *AgeI* and *SalI* enzymes generating two fragments of 3 kb and 0.7 kb in size. The 0.7 kb fragment (tdTomato gene) was used as the insert for cloning. **(B)** Illustration of the vector plasmid. The plasmid was cut with the *AgeI* and *SalI* enzymes generating two fragments of 4.9 kb and 0.7 kb in size. The 4.9 kb fragment was used as the vector for cloning. AMP: Ampicillin resistance gene; PRE: posttranscriptional regulatory element; MPSV: myeloproliferative sarcoma virus promoter.

are tedious to collect on an individual basis thereby trying to minimize errors and pitfalls from the beginning.

Methods

Cell lines and media

The *E. coli* HB101 was used for the preparation of plasmid DNA. The bacteria were cultured in Luria-Bertani (LB) media. Human embryonic kidney (HEK) 293 T cells were cultured in Dulbecco's Modified Eagle medium (DMEM) supplemented with 10% heat-inactivated fetal bovine serum (FBS), 1 mM sodium pyruvate, 0.1 mM nonessential amino acids, 2 mM L-glutamine, 100 mg/ml streptomycin, and 100 units/ml penicillin. A myeloid leukemia cell line C1498 [52], was cultured in Roswell Park Memorial Institute (RPMI) 1640 medium supplemented with the same reagents used for DMEM. Cells were split every other day to keep them on log phase.

Plasmids, primers, PCR and sequencing

A plasmid containing the coding sequence of the tdTomato gene, plasmid containing an alpha-retroviral vector, and plasmids containing codon-optimized alpharetroviral gag/pol [53] were kindly provided by Axel Schambach (MHH Hannover, Germany). A forward (5′- ACCGGTG CCACCATGGCCACAACCATGGTG-3′) and a reverse (5′-GTCGACTTACTTGTACAGCTCGTCCATGCC-3′)

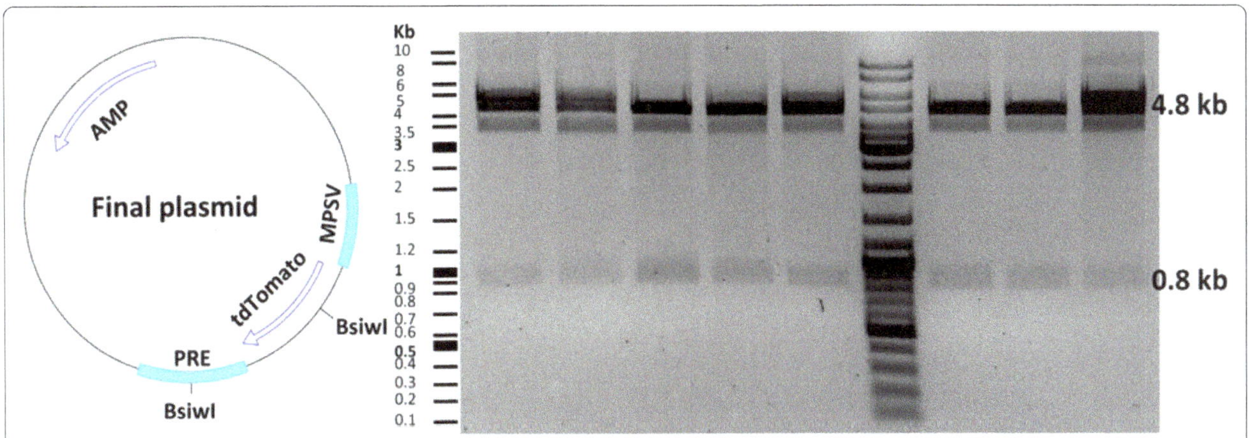

Figure 6 Screening of the final plasmid with restriction enzymes. Illustration of the final plasmid is shown. For screening, the plasmid was cut with the *BsiwI* enzyme generating two fragments of 4.8 kb and 0.8 kb in size. AMP: Ampicillin resistance gene; PRE: posttranscriptional regulatory element; MPSV: myeloproliferative sarcoma virus promoter.

primer used for the amplification of the tdTomato gene were synthesized by Eurofins Genomics (Ebersberg, Germany).

The optimal buffers for enzymes or other reagents were provided by the manufacturers along with the corresponding enzymes or inside the kits. If available by the manufacturers, the pH and ingredients of buffers are mentioned. Primers were dissolved in ultrapure water at a stock concentration of 20 pmol/µl. The template plasmid was diluted in water at a stock concentration of 50 ng/µl. For PCR, the following reagents were mixed and filled up with water to a total volume of 50 µl: 1 µl plasmid DNA (1 ng/µl final concentration), 1.25 µl of each primer (0.5 pmol/µl final concentration for each primer), 1 µL dNTP (10 mM each), 10 µl of 5X Phusion HF buffer (1X buffer provides 1.5 mM MgCl2), and 0.5 µl Phusion DNA polymerase (2U/µl, Thermo Scientific).

PCR was performed using a peqSTAR thermocycler (PEQLAB Biotechnologie) at: 98°C for 3 minutes; 25 cycles at 98°C for 10 seconds, 66°C for 30 seconds, 72°C for 30 seconds; and 72°C for 10 minutes. To prepare a 0.8% agarose gel, 0.96 g agarose (CARL ROTH) was dissolved in 120 ml 1X TAE buffer (40 mM Tris, 20 mM acetic acid, 1 mM EDTA, pH of 50X TAE: 8.4) and boiled for 4 minutes. Then 3 µl SafeView nucleic acid stain (NBS Biologicals) was added to the solution and the mixture was poured into a gel-casting tray.

DNA was mixed with 10 µl loading dye (6X) (Thermo Scientific) and loaded on the agarose gel (CARL ROTH) using 80 V for one hour in TAE buffer. The separated DNA fragments were visualized using an UV transilluminator (365 nm) and quickly cut to minimize the UV exposure. DNA was extracted from the gel slice using Zymoclean™ Gel DNA Recovery Kit (Zymo Research). The concentration of DNA was determined using a NanoDrop 2000 spectrophotometer (Thermo Scientific).

For sequence validation, the PCR product was subcloned using CloneJET PCR cloning kit (Thermo Scientific). 1 µl of blunt vector (50 ng/µl), 50 ng/µl of the PCR product, and 10 µl of 2X reaction buffer (provided in the kit) were mixed and filled with water to a total volume of 20 µl. 1 µl of T4 DNA ligase (5 U/µl) was added to the mixture, mixed and incubated at room temperature for 30 minutes. For bacterial transfection, 10 µl of the mixture was mixed with 100 µl of HB101 *E. coli* competent cells and incubated on ice for 45 minutes. Then the mixture was heat-shocked (42°C/2 minutes), put on ice again (5 minutes), filled up with 1 ml LB medium and incubated in a thermomixer (Eppendorf) for 45 minutes/37°C/450RPM. Then the bacteria were spun down for 4 minutes. The pellet was cultured overnight at 37°C on an agarose Petri dish containing 100 µg/mL of Ampicillin. The day after, colonies were picked and cultured overnight in 3 ml LB containing 100 µg/mL of ampicillin.

After 16 hours (overnight), the plasmid was isolated from the cultured bacteria using the QIAprep spin miniprep kit (QIAGEN) according to the manufacturer's instructions. 720 to 1200 ng of plasmid DNA in a total of 12 µl water were sent for sequencing (Seqlab) in Eppendorf tubes. The sequencing primers pJET1.2-forward (5′-CGACTCACTATAGGGAG-3′), and pJET1.2-reverse (5′-ATCGATTTTCCATGGCAG-3′), were generated by the Seqlab Company (Göttingen, Germany). An ABI 3730XL DNA analyzer was used by the Seqlab Company to sequence the plasmids applying the Sanger method. Sequence results were analyzed using NCBI Blast as explained in the Results and discussion section.

Manipulation of DNA fragments

For viewing plasmid maps, Clone Manager suite 6 software (SciEd) was used. Restriction endonuclease enzymes (Thermo Scientific) were used to cut plasmid

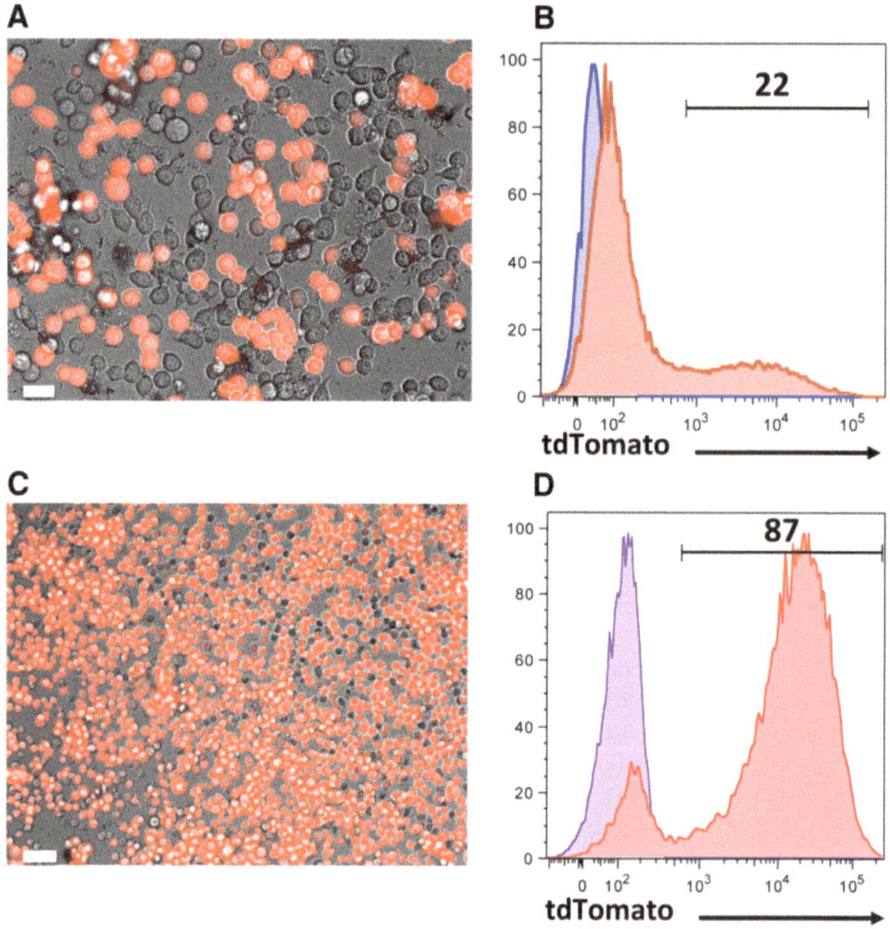

Figure 7 Assessing *in vitro* expression of the cloned gene. (A, B) HEK293T cells were transfected with Gag/Pol, VSVG, and tdTomato plasmids. The expression of the tdTomato gene was assessed using a fluorescence microscope. Fluorescent images were superimposed on a bright-field image for the differentiation of positively transduced cells. Transfection efficiency was determined based on the expression of tdTomato after 24 hours. Non-transfected HEK293T cells were used as controls (blue histogram). **(C, D)** The murine leukemia cell line C1498 was transduced with fresh virus. Four days later, transgene expression was assessed by fluorescence microscopy **(C)** and flow cytometry **(D)**. Non-transduced C1498 cells were used as controls (blue histogram). Scale bars represent 30 μm.

DNA. 5 μg plasmid DNA, 2 μl buffer O (50 mM Tris–HCl (pH 7.5 at 37°C), 10 mM MgCl2, 100 mM NaCl, 0.1 mg/mL BSA, Thermo Scientific), 1 μl *SalI* (10 U), and 1 μl *AgeI* (10 U) were mixed in a total of 20 μl water and incubated (37°C) overnight in an incubator to prevent evaporation and condensation of water under the tube lid. The next day, DNA was mixed with 4 μl loading dye (6X) (Thermo Scientific) and run on a 0.8% agarose gel at 80 V for one hour in TAE buffer. The agarose gel (120 ml) contained 3 μl SafeView nucleic acid stain (NBS Biologicals). The bands were visualized on a UV transilluminator (PEQLAB), using a wavelength of 365 nm, and quickly cut to minimize the UV damage. DNA was extracted from the gel slices using the Zymoclean™ gel DNA recovery kit (Zymo Research). The concentration of DNA was determined using a NanoDrop 2000 spectrophotometer (Thermo Scientific).

For the ligation of vector and insert fragments, a ligation calculator was designed (the Excel file available in the Additional file 1) for easy calculation of the required insert and vector volumes. The mathematical basis of the calculator is inserted into the excel spreadsheet. The size and concentration of the vector and insert fragments and the molar ratio of vector/insert (normally 1:3) must be provided for the calculation. Calculated amounts of insert (tdTomato) and vector (alpha-retroviral backbone) were mixed with 2 μl of 10X T4 ligase buffer (400 mM Tris–HCl, 100 mM MgCl2, 100 mM DTT, 5 mM ATP (pH 7.8 at 25°C), Thermo Scientific), 1 μl of T4 ligase (5 U/μl, Thermo Scientific), filled up to 20 μl using ultrapure water and incubated overnight at 16°C. The day after, HB101 *E. coli* was transfected with the ligation mixture as mentioned above. The clones were picked and consecutively cultured for one day in LB medium containing ampicillin. Plasmid DNA was isolated using Zyppy™ plasmid miniprep kit (Zymo Research) and digested with proper restriction enzymes for screening. Digested plasmids were mixed with the loading dye and run on an agarose gel as mentioned above. The separated DNA fragments were visualized using a Gel Doc™ XR+ System (BIO-RAD) and analyzed by the Image Lab™ software (BIO-RAD). The positive clone was cultured overnight in 450 ml LB medium containing ampicillin. Plasmid DNA was isolated using QIAGEN plasmid maxi kit (QIAGEN), diluted in ultrapure water and stored at −20°C for later use.

Production of viral supernatant and transduction of cells

HEK293T cells were thawed, split every other day for one week and grown in log phase. The day before transfection, 3.5×10^6 cells were seeded into tissue culture dishes (60.1 cm^2 growth surface, TPP). The day after, the cells use to reach about 80% confluence. If over confluent, transfection efficiency decreases. The following plasmids were mixed in a total volume of 450 µl ultrapure water: codon-optimized alpharetroviral gag/pol (2.5 µg), VSVG envelope (1.5 µg), and the alpharetroviral vector containing the tdTomato gene (5 µg). Transfection was performed using calcium phosphate transfection kit (Sigma-Aldrich). 50 µl of 2.5 M $CaCl_2$ was added to the plasmid DNA and the mixture was briefly vortexed. Then, 0.5 ml of 2X HEPES buffered saline (provided in the kit) was added to a 15 ml conical tube and the calcium-DNA mixture was added dropwise via air bubbling and incubated for 20 minutes at room temperature. The medium of the HEK293T cells was first replaced with 8 ml fresh medium (DMEM containing FCS and supplement as mentioned above) containing 25 µM chloroquine. Consecutively the transfection mixture was added. Plates were gently swirled and incubated at 37°C. After 12 hours, the medium was replaced with 6 ml of fresh RPMI containing 10% FCS and supplements. Virus was harvested 36 hours after transfection, passed through a Millex-GP filter with 0.22 µm pore size (Millipore), and used freshly to transduce C1498 cells. Before transduction, 24 well plates were coated with retronectin (Takara, 280 µl/well) for 2 hours at room temperature. Then, retronectin was removed and frozen for later use (it can be re-used at least five times) and 300 µl of PBS containing 2.5% bovine serum albumin (BSA) was added to the wells for 30 minutes at room temperature. To transduce C1498 cells, 5×10^4 of cells were spun down and resuspended with 1 ml of fresh virus supernatant containing 4 µg/ml protamine sulfate. The BSA solution was removed from the prepared plates and plates were washed two times with 0.5 ml PBS. Then cells were added to the wells. Plates were centrifuged at 2000RPM/32°C/90 minutes. Fresh medium was added to the cells the day after.

Flow cytometry and fluorescence microscope

For flow cytometry assessment, cells were resuspended in PBS containing 0.5% BSA and 2 mM EDTA and were acquired by a BD FACSCanto™ (BD Biosciences) flow cytometer. Flow cytometry data were analyzed using FlowJo software (Tree Star). Imaging was performed with an Olympus IX71 fluorescent microscope equipped with a DP71 camera (Olympus). Images were analyzed with AxioVision software (Zeiss). Fluorescent images were superimposed on bright-field images using adobe Photoshop CS4 software (Adobe).

Additional file

Additional file 1: Ligation calculator. To calculate the amounts of the vector and insert fragments for a ligation reaction, you need to provide the size of the vector and insert (in base pairs), the molar ration of insert/vector (normally 3 to 5), vector amount (normally 50 to 100 ng), and vector and insert fragment concentrations (ng/µl). The computational basis of this ligation calculator is mentioned in the lower box.

Abbreviations

PCR: Polymerase chain reaction; GOI: Gene of interest; ORF: Open reading frame; Tm: Melting temperature; BLAST: Basic local alignment search tool; VSVG: Vesicular stomatitis virus G glycoprotein; LB: Luria-Bertani; DMEM: Dulbecco's Modified Eagle medium; RPMI: Roswell Park Memorial Institute; BSA: Bovine serum albumin; EDTA: Ethylenediaminetetraacetic acid; FACS: Fluorescence-activated cell sorting; HEK: Human embryonic kidney; PBS: Phosphate buffered saline; FCS: Fetal calf serum; HEPES: Hydroxyethyl-piperazineethane-sulfonic acid; AMP: Ampicillin resistance gene; PRE: Posttranscriptional regulatory element; MPSV: Myeloproliferative sarcoma virus promoter.

Competing interests

The authors declare that they have no competing interests.

Authors' contributions

SSH conceived the study subject, carried out experiments and drafted the initial manuscript. MGS participated in study design and coordination and edited the manuscript. Both authors have read and approved the final manuscript.

Acknowledgments

The authors would like to thank Jessica Herbst, Abbas Behpajooh, Christian Kardinal and Juwita hübner for their fruitful discussions. We also thank Gang Xu for helping to design the cover page. This work was supported by the Deutsche Forschungsgemeinschaft, the Bundesministerium für Bildung und Forschung, the Deutsche Jose-Carreras Leukämiestiftung (grants SFB-738, IFB-TX CBT_6, DJCLS R 14/10 to M.G.S.) and the Ph.D. program Molecular Medicine of the Hannover Medical School.

Author details

[1]Departments of Pediatric Hematology, Oncology and Blood Stem Cell Transplantation, Hannover, Germany. [2]Hannover Center for Transplantation Research, Hannover, Germany. [3]Department of Pediatric Hematology and Oncology, Medizinische Hochschule Hannover, OE 6780, Carl-Neuberg-Strasse 1, 30625 Hannover, Germany.

References

1. Bryksin AV, Matsumura I. Overlap extension PCR cloning: a simple and reliable way to create recombinant plasmids. Biotechniques. 2010;48:463.
2. Hughes RA, Miklos AE, Ellington AD. Gene synthesis: methods and applications. Methods Enzymol. 2011;498:277–309.

3. Gibson DG, Young L, Chuang RY, Venter JC, Hutchison CA, Smith HO. Enzymatic assembly of DNA molecules up to several hundred kilobases. Nat Methods. 2009;6:343–U341.
4. Mitchell DB, Ruggli N, Tratschin JD. An improved method for cloning PCR fragments. PCR Methods Appl. 1992;2:81–2.
5. Radeck J, Kraft K, Bartels J, Cikovic T, Durr F, Emenegger J, et al. The Bacillus BioBrick Box: generation and evaluation of essential genetic building blocks for standardized work with Bacillus subtilis. J Biol Eng. 2013;7:29.
6. Haun RS, Serventi IM, Moss J. Rapid, reliable ligation-independent cloning of PCR products using modified plasmid vectors. Biotechniques. 1992;13:515–8.
7. Mead DA, Pey NK, Herrnstadt C, Marcil RA, Smith LM. A universal method for the direct cloning of PCR amplified nucleic acid. Biotechnology. 1991;9:657–63.
8. Holton TA, Graham MW. A simple and efficient method for direct cloning of PCR products using ddT-tailed vectors. Nucleic Acids Res. 1991;19:1156.
9. Holst J, Szymczak-Workman AL, Vignali KM, Burton AR, Workman CJ, Vignali DA. Generation of T-cell receptor retrogenic mice. Nat Protoc. 2006;1:406–17.
10. Hoffman LM, Jendrisak J. Heat-labile phosphatase simplifies the preparation of dephosphorylated vector DNA. Gene. 1990;88:97–9.
11. Pheiffer BH, Zimmerman SB. Polymer-stimulated ligation: enhanced blunt- or cohesive-end ligation of DNA or deoxyribooligonucleotides by T4 DNA ligase in polymer solutions. Nucleic Acids Res. 1983;11:7853–71.
12. Pingoud A, Wilson GG, Wende W. Type II restriction endonucleases-a historical perspective and more. Nucleic Acids Res. 2014;42:7489–527.
13. Mural RJ. Cleavage by the restriction endonuclease Asp718, an isoschizomer of KpnI, is sensitive to Escherichia coli Dcm methylation. Nucleic Acids Res. 1987;15:9085.
14. Casali N. Escherichia coli host strains. In: Casali N, Preston A, editors. E coli Plasmid Vectors. Humana Press Inc: Totowa; 2003. p. 27–48.
15. Corbett AJ, Brady JL, Lew AM. A single buffer that universally serves both restriction digestion and loading. Mol Biotechnol. 2002;21:51–5.
16. Wilfinger WW, Mackey K, Chomczynski P. Effect of pH and ionic strength on the spectrophotometric assessment of nucleic acid purity. Biotechniques. 1997;22:474–6.
17. Kozak M. An analysis of 5'-noncoding sequences from 699 vertebrate messenger RNAs. Nucleic Acids Res. 1987;15:8125–48.
18. Kozak M. Regulation of translation via mRNA structure in prokaryotes and eukaryotes. Gene. 2005;361:13–37.
19. Cline J, Braman JC, Hogrefe HH. PCR fidelity of Pfu DNA polymerase and other thermostable DNA polymerases. Nucleic Acids Res. 1996;24:3546–51.
20. Benson LM, Null AP, Muddiman DC. Advantages of Thermococcus kodakaraenis (KOD) DNA polymerase for PCR-mass spectrometry based analyses. J Am Soc Mass Spectr. 2003;14:601–4.
21. Wang Y, Prosen DE, Mei L, Sullivan JC, Finney M, Vander Horn PB. A novel strategy to engineer DNA polymerases for enhanced processivity and improved performance in vitro. Nucleic Acids Res. 2004;32:1197–207.
22. McInerney P, Adams P, Hadi MZ. Error Rate Comparison during Polymerase Chain Reaction by DNA Polymerase. Mol Biol Int. 2014. doi:10.1155/2014/287430.
23. Breslauer KJ, Frank R, Blocker H, Marky LA. Predicting DNA duplex stability from the base sequence. Proc Natl Acad Sci U S A. 1986;83:3746–50.
24. Lodge J, Lund P, Minchin S. Gene cloning principles and applications. New York: Taylor & Francis Group; 2007.
25. Bolivar F, Rodriguez RL, Greene PJ, Betlach MC, Heyneker HL, Boyer HW, et al. Construction and characterization of new cloning vehicles. II. A multipurpose cloning system. Gene. 1977;2(2):95–113.
26. Preston A. Choosing a cloning vector. In: Casali N, Preston A, editors. E coli Plasmid Vectors. Totowa: Humana Press Inc; 2003. p. 19–26.
27. Stoker NG, Fairweather NF, Spratt BG. Versatile low-copy-number plasmid vectors for cloning in Escherichia coli. Gene. 1982;18:335–41.
28. Bernard P, Gabant P, Bahassi EM, Couturier M. Positive-selection vectors using the F plasmid ccdB killer gene. Gene. 1994;148:71–4.
29. Shuman S. Novel approach to molecular cloning and polynucleotide synthesis using vaccinia DNA topoisomerase. J Biological Chem. 1994;269:32678–84.
30. Noguchi H, Park J, Takagi T. MetaGene: prokaryotic gene finding from environmental genome shotgun sequences. Nucleic Acids Res. 2006;34:5623–30.
31. Ewing B, Hillier L, Wendl MC, Green P. Base-calling of automated sequencer traces using phred. I. Accuracy assessment. Genome Res. 1998;8:175–85.
32. Keith CS, Hoang DO, Barrett BM, Feigelman B, Nelson MC, Thai H, et al. Partial sequence analysis of 130 randomly selected maize cDNA clones. Plant Physiol. 1993;101:329–32.
33. Richter DC, Ott F, Auch AF, Schmid R, Huson DH. MetaSim: a sequencing simulator for genomics and metagenomics. PLoS One. 2008;3:e3373.
34. Hoff KJ. The effect of sequencing errors on metagenomic gene prediction. BMC Genomics. 2009;10:520.
35. Treangen TJ, Salzberg SL. Repetitive DNA and next-generation sequencing: computational challenges and solutions. Nat Rev Geneti. 2012;13:36–46.
36. Prosdocimi F, Ortega JM. Accessing optimal primer distance from insert. In Silico Biol. 2005;5:469–77.
37. Engebrecht J, Heilig JS, Brent R. Isolation and purification of DNA. In: Current Protocols in Immunology. 1993.
38. Bzymek M, Lovett ST. Instability of repetitive DNA sequences: the role of replication in multiple mechanisms. Proc Natl Acad Sci U S A. 2001;98:8319–25.
39. Geisse S, Henke M. Large-scale transient transfection of mammalian cells: a newly emerging attractive option for recombinant protein production. J Struct Funct Genomics. 2005;6:165–70.
40. Szymczak-Workman AL, Vignali KM, Vignali DA. Design and construction of 2A peptide-linked multicistronic vectors. Cold Spring Harbor Protocols. 2012;2012:199–204.
41. Szymczak AL, Vignali DA. Development of 2A peptide-based strategies in the design of multicistronic vectors. Expert Opin Biol Ther. 2005;5:627–38.
42. Chen X, Zaro JL, Shen WC. Fusion protein linkers: property, design and functionality. Adv Drug Deliv Rev. 2013;65:1357–69.
43. Yang L, Ukil L, Osmani A, Nahm F, Davies J, De Souza CP, et al. Rapid production of gene replacement constructs and generation of a green fluorescent protein-tagged centromeric marker in Aspergillus nidulans. Eukaryotic Cell. 2004;3:1359–62.
44. Bhide M, Natarajan S, Hresko S, Aguilar C, Bencurova E. Rapid in vitro protein synthesis pipeline: a promising tool for cost-effective protein array design. Mol Biosyst. 2014;10:1236–45.
45. Irian S. Large-scale tag/PCR-based gene expression profiling. World J Microb Biot. 2014;30:2125–39.
46. Schneider BL, Seufert W, Steiner B, Yang QH, Futcher AB. Use of polymerase chain reaction epitope tagging for protein tagging in Saccharomyces cerevisiae. Yeast. 1995;11:1265–74.
47. Erster O, Liscovitch M. A modified inverse PCR procedure for insertion, deletion, or replacement of a DNA fragment in a target sequence and its application in the ligand interaction scan method for generation of ligand-regulated proteins. Methods Mol Biol. 2010;634:157–74.
48. Li C, Wen A, Shen B, Lu J, Huang Y, Chang Y. FastCloning: a highly simplified, purification-free, sequence- and ligation-independent PCR cloning method. BMC Biotechnol. 2011;11:92.
49. Chaudhary VK, Shrivastava N, Verma V, Das S, Kaur C, Grover P, et al. Rapid restriction enzyme-free cloning of PCR products: a high-throughput method applicable for library construction. PLoS One. 2014;9:e111538.
50. Mead DA, Pey NK, Herrnstadt C, Marcil RA, Smith LM. A universal method for the direct cloning of PCR amplified nucleic acid. Biotechnology (N Y). 1991;9:657–63.
51. Stevenson J, Krycer JR, Phan L, Brown AJ. A practical comparison of ligation-independent cloning techniques. PLoS One. 2013;8:e83888.
52. Ghosh A, Wolenski M, Klein C, Welte K, Blazar BR, Sauer MG. Cytotoxic T cells reactive to an immunodominant leukemia-associated antigen can be specifically primed and expanded by combining a specific priming step with nonspecific large-scale expansion. J Immunother. 2008;31:121–31.
53. Suerth JD, Maetzig T, Brugman MH, Heinz N, Appelt JU, Kaufmann KB, et al. Alpharetroviral self-inactivating vectors: long-term transgene expression in murine hematopoietic cells and low genotoxicity. Mol Ther. 2012;20:1022–32.

Synthetic feedback control using an RNAi-based gene-regulatory device

Ryan J Bloom, Sally M Winkler and Christina D Smolke*

Abstract

Background: Homeostasis within mammalian cells is achieved through complex molecular networks that can respond to changes within the cell or the environment and regulate the expression of the appropriate genes in response. The development of biological components that can respond to changes in the cellular environment and interface with endogenous molecules would enable more sophisticated genetic circuits and greatly advance our cellular engineering capabilities.

Results: Here we describe a platform that combines a ligand-responsive ribozyme switch and synthetic miRNA regulators to create an OFF genetic control device based on RNA interference (RNAi). We developed a mathematical model to highlight important design parameters in programming the quantitative performance of RNAi-based OFF control devices. By modifying the ribozyme switch integrated into the system, we demonstrated RNAi based OFF control devices that respond to small molecule and protein ligands, including the oncogenic protein E2F1. We utilized the OFF control device platform to build a negative feedback control system that acts as a proportional controller and maintains target intracellular protein levels in response to increases in transcription rate.

Conclusions: Our work describes a novel genetic device that increases the level of silencing from a miRNA in the presence of a ligand of interest, effectively creating an RNAi-based OFF control device. The OFF switch platform has the flexibility to be used to respond to both small molecule and protein ligands. Finally, the RNAi-based OFF switch can be used to implement a negative feedback control system, which maintains target protein levels around a set point level. The described RNAi-based OFF control device presents a powerful tool that will enable researchers to engineer homeostasis in mammalian cells.

Keywords: MicroRNA, Ribozyme, Feedback control, Synthetic biology

Background

The complexity of mammalian cells is largely due to the sophisticated control mechanisms employed to modulate the expression of endogenous genes [1]. Molecules in the cell are organized into networks, which can respond to changes in intracellular protein or metabolite levels, as well as changes in the environment, by upregulating or downregulating the expression of any number of genes [1]. These genetic circuits are essential for virtually all cellular processes including differentiation [2], replication [3], and signal transduction [4]. The ability to interface with these endogenous programs to create novel molecular networks would greatly advance our cellular engineering capabilities for a variety of applications. Central to our ability to create these new genetic circuits is the development of biological components that can respond to changes in the cellular environment and interface with endogenous molecules to enable more sophisticated synthetic circuits [5].

MicroRNAs (miRNA) are a class of small regulatory RNAs that are used to post-transcriptionally control gene expression in higher eukaryotes [6]. MiRNA biogenesis begins with longer pri-miRNAs, which are cleaved by a Microprocessor complex into shorter ~60 nucleotide pre-miRNA hairpins and subsequently exported from the nucleus. The pre-miRNA is subject to a final cleavage step in the cytoplasm by the enzyme Dicer, which creates a 22–25 nucleotide (nt) double stranded RNA [7]. One of these RNA strands, called the guide strand, is then incorporated into one of the Argonaut family of proteins to create the RNA-induced silencing complex (RISC). This complex subsequently cleaves or translationally represses the target

* Correspondence: csmolke@stanford.edu
Department of Bioengineering, Stanford University, 443 Via Ortega, MC 4245, Stanford, CA 94305, USA

transcript depending on the degree of complementarity between the miRNA and the target [6]. This powerful mechanism of gene regulation plays vital roles in diverse processes, such as development and angiogenesis, and it is estimated that RNAi regulates the majority of genes in the human genome [8].

RNA-based synthetic biology tools are powerful resources for building synthetic control systems that can detect and respond to changes in cell state or the cellular environment, and synthetic RNA-based systems have been demonstrated in mammalian cells [9-12]. Frameworks have been described for the design of miRNA- and shRNA-based molecular switches, whose biogenesis is responsive to changes in concentration of intracellular ligand molecules [13]. In general, these regulators are constructed by coupling the binding of a ligand to an RNA aptamer with the proper formation and biogenesis of the RNAi element [13]. The physical coupling of the aptamer to the RNAi element has been accomplished in several ways. Researchers have directly coupled an aptamer to a pri-miRNA, such that binding of the ligand to the aptamer sterically inhibits proper biogenesis by the Microprocessor [13]. Both direct coupling and strand displacement mechanisms have been described, where ligand binding to the aptamer changes the structure of an shRNA such that it is unable to be properly processed by Dicer [14,15].

These RNAi-based control systems are particularly useful for imparting conditional silencing over endogenous genes. Previously, we developed a computational model that predicts target gene expression levels resulting from synthetic circuits based on miRNA regulatory components [16]. This model was applied to direct the rational design of miRNA-based genetic circuits and supported an *in-vivo* protein concentration detector circuit by providing a quantitative relationship between target reporter and protein ligand levels in the cell. However, the designs in this previous work utilize a switch architecture that inhibits RNAi-based gene silencing in the presence of the ligand, thereby increasing target gene expression as a function of increasing ligand concentration and operating as ON switches. An RNAi-based control platform that can be used to decrease the expression of a target gene in response to increasing concentrations of proteins of interest has not been described. However, this type of OFF switch regulation is essential for building core control strategies such as negative feedback control. Negative feedback is a prevalent control mechanism found in diverse biological systems, such as bacterial chemotaxis [17] and vision [18], that can be utilized to achieve homeostasis.

Here we describe and characterize an OFF genetic control device based on RNAi that combines a ligand-responsive ribozyme switch and synthetic miRNA regulators. The device architecture links ribozyme cleavage to miRNA levels, such that ligand binding modulates ribozyme cleavage rates and thus miRNA-based gene silencing. The system was prototyped using a previously described theophylline-responsive ribozyme switch and a miRNA that targets a fluorescent reporter. A mathematical model of this system was developed to highlight important design parameters in programming the quantitative performance of RNAi-based OFF control devices. We demonstrated RNAi-based OFF control devices that respond to protein ligands by incorporating protein-responsive ribozyme switches to E2F1 and MS2. We utilized the OFF control device platform to build a negative feedback control system that maintains target intracellular protein levels around a set point. The negative feedback control system acts as a proportional controller, maintaining target intracellular protein levels in response to increases in transcription rate. The described RNAi-based OFF control device presents a powerful tool that will enable researchers to engineer homeostasis into mammalian cells, for example, to maintain a desired phenotype even in the presence of genetic mutations or fluctuating levels of signaling molecules or cytokines.

Results and discussion

A trans-acting genetic device that exhibits OFF control by coupling ribozyme switches and miRNAs

We developed a strategy to implement a gene-regulatory device that exhibits OFF control in the regulation of endogenous gene targets. Our platform design is inspired by a mechanism found in natural miRNA clusters, where cleavage of a transcript in a region upstream of a pri-miRNA reduces the steady-state level of the transcript encoding the pri-miRNA and thus the resulting gene silencing due to the processed miRNA [19]. We hypothesized that by imparting conditional cleavage of a transcript in a region upstream of a pri-miRNA by using a ligand-responsive ribozyme we could control the level of silencing from the encoded miRNA (Figure 1). In the proposed system design, the ribozyme cleaves in the absence of ligand, thereby reducing the steady-state level of the pri-miRNA transcript and resulting in reduced miRNA-based silencing and increased target gene expression. However, when bound to the ligand the ribozyme's cleavage rate is substantially reduced, such that higher steady-state levels of the pri-miRNA transcript will accumulate and be processed, resulting in reduced expression of the target gene.

We used a modular ligand-responsive ribozyme switch platform that supports efficient tailoring of RNA device function [20]. The ribozyme switch platform specifies physical linkages between three functional RNA components: a sensor encoded by an RNA aptamer, an actuator encoded by a hammerhead ribozyme, and a transmitter encoded by a sequence that undergoes a strand-displacement event

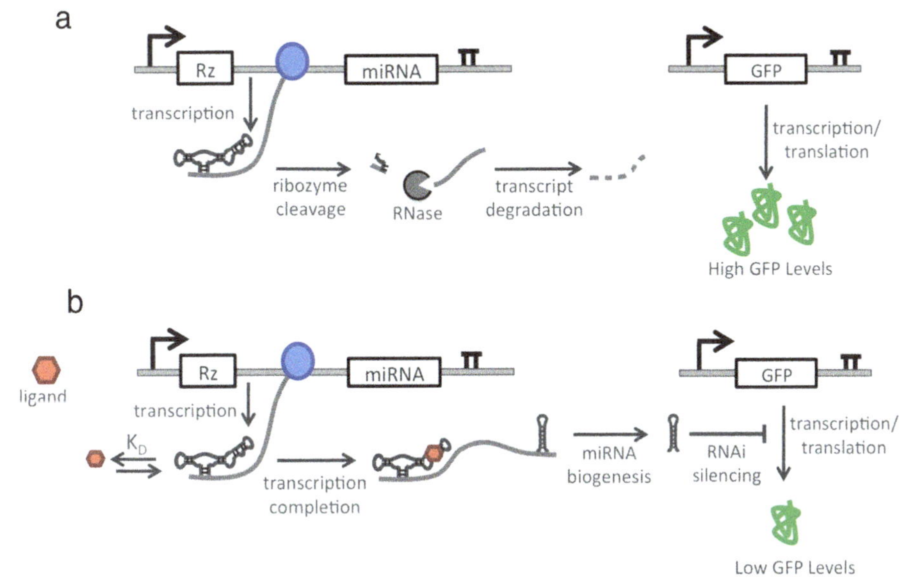

Figure 1 Schematic of the RNAi-based OFF control device architecture. A catalytic ribozyme switch is placed upstream of a pri-miRNA in the same expression cassette. **(a)** In the absence of the input ligand, the ribozyme switch cleaves once it is transcribed, aborting transcription such that the miRNA is not be transcribed, thereby resulting in high target gene expression levels. **(b)** In the presence of ligand, ribozyme cleavage is inhibited such that transcription proceeds through the pri-miRNA, which undergoes proper miRNA biogenesis. Target expression levels are lowered due to miRNA silencing.

(Additional file 1: Figure S1). The modular nature of this ribozyme design makes it relatively straightforward to modify the ligand-responsiveness of the switch by swapping in new aptamer sequences into the sensor component of the platform while maintaining the overall activity of the switch.

An expression cassette for the device was designed based on placing a ligand-responsive ribozyme upstream of a cluster of three identical GFP-targeting miRNAs (Figure 2a). The miRNA guide strands were designed to be completely complementary to the target transcript, such that their mechanism of gene silencing reflects that of an siRNA although they are processed from a primary transcript through the endogenous miRNA pathway [13]. The device was prototyped with a previously characterized theophylline-responsive ribozyme that was shown to exhibit a reduced cleavage rate as a function of increasing theophylline concentration [21]. Three versions of the OFF control device were designed in which the spacer length between the ribozyme and miRNA components was varied (500, 1500, 2500 bp) to examine the effect of spacer length on device performance. Control constructs were developed that had the theophylline-responsive ribozyme replaced with a non-cleaving ribozyme or that removed the miRNAs from the expression cassette (Additional file 1: Figure S2). A second expression cassette was constructed that encoded the expression of a GFP fluorescent reporter protein. Isogenic cell lines stably expressing both expression cassettes were generated from a Flp-In HEK293 cell line by co-transfecting a plasmid harboring both expression cassettes with a plasmid that expresses a Flp recombinase (pOG44) and selecting for stable integrants.

Device performance was quantified by measuring target protein (GFP) and mature miRNA levels in the presence and absence of ligand (Figure 2b, c). GFP levels were measured using a flow cytometry assay and mature miRNA levels were measured using a qRT-PCR assay with custom TaqMan small RNA probes. The data demonstrate that the longest spacer tested (2500 bp) resulted in the greatest change in target gene expression levels in the presence and absence of ligand (Figure 2b). qRT-PCR analysis showed that the decrease in target gene expression levels correlated with an increase in mature miRNA levels, indicating that the observed changes in target gene expression levels were due to miRNA silencing (Figure 2c). The OFF state (i.e., expression level in the presence of theophylline) measured for all constructs is similar to that from the inactive ribozyme control construct, which sets the expected lower limit for the system (Figure 2b). The ON state (i.e., expression level in the absence of theophylline) generally increased with spacer length, consistent with previous findings in the literature [19]. Overall, the device with the longest spacer length tested (2500 bp) exhibited ON and OFF states very close to the controls (Figure 2b, Additional file 1: Figure S2), indicating that this device was accessing nearly the full potential dynamic range of the system. The data indicate that longer spacer lengths may increase the time between transcription of the ribozyme

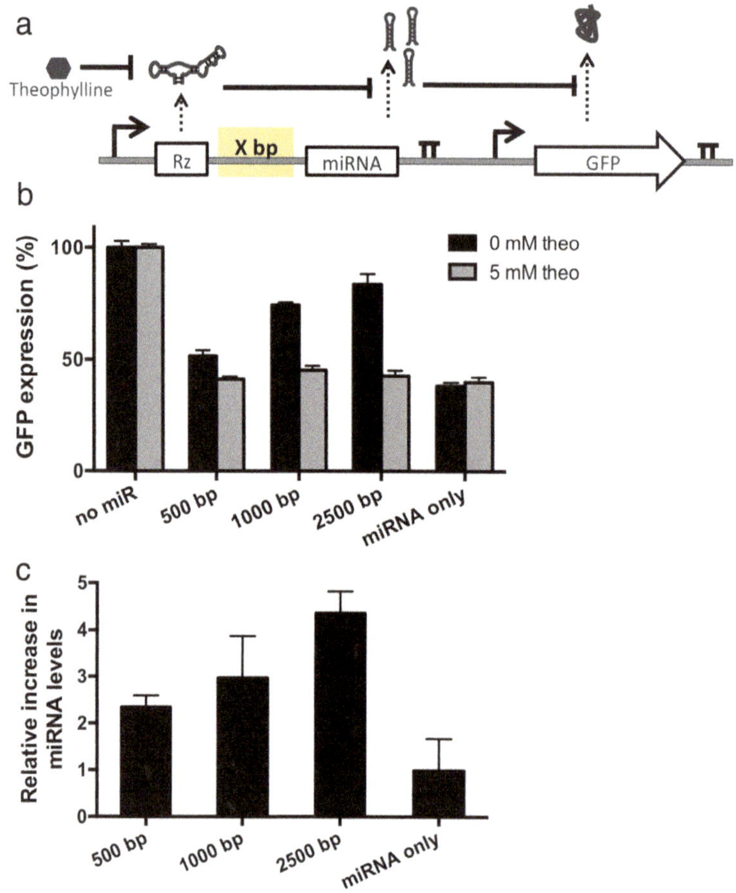

Figure 2 The effect of different spacer lengths on performance of the RNAi-based OFF control device. **(a)** Schematic of the expression cassettes used to test the OFF control device architecture. A theophylline-responsive ribozyme switch was placed at various distances upstream of a miRNA targeting GFP. GFP was expressed constitutively under a separate expression cassette. Both constructs were stably integrated into an HEK293 Flp-In cell line. **(b)** GFP expression in the absence and presence of theophylline from OFF control devices with varying spacer lengths. GFP levels were measured 48 hours after the addition of 0 or 5 mM theophylline. No miR and miRNA only represent constructs with a theophylline-responsive ribozyme switch and no downstream miRNAs (negative control) and a construct with miRNA only upstream of a GFP-targeting miRNA (positive control), respectively. Data is reported as the percent of GFP expression in a cell line with no miRNA present. Error bars represent ±1 standard deviation of at least three independent experiments. **(c)** Relative increase in miRNA levels after addition of theophylline. Relative miRNA levels are reported 48 hours after the addition of 0 or 5 mM theophylline and normalized to the miRNA levels measured from identical cell lines in the presence of theophylline. Error bars represent ±1 standard deviation of three independent experiments.

component and transcription of the downstream miRNA component, thus increasing the probability that the ribozyme will cleave and abort transcription [19] prior to the transcription and processing of the miRNA.

A computational model informs the design of the RNAi-based OFF control device

We developed a mathematical model to inform the design of the RNAi-based OFF control device to elucidate important system design parameters. The model describes the kinetics of ribozyme cleavage and miRNA formation using a series of differential equations describing the separate processes (Additional file 1: Figure S3, Additional file 1: Text S1). The transcription of the ribozyme-miRNA transcript is modeled as a two-step process in which an "immature transcript", harboring only the ribozyme component, is initially formed. This immature transcript is either cleaved and subsequently degraded or elongated into a full-length transcript, encoding both the ribozyme and miRNA components. Mature miRNAs are formed from the full-length transcript and subsequently silence target gene expression through enzymatic degradation using a term that resembles traditional Michaelis-Menten kinetics, which includes a constant (K_m) describing the affinity between the miRNA and the target transcript [22]. The presence of the input ligand will lower the ribozyme cleavage rate, resulting in an increase in the level of full-length transcript and thus increased miRNA levels. The binding kinetics between the aptamer and the ligand (k_a, k_d) as well as the cleavage rates of the ribozyme when bound and unbound to the input ligand ($k_{cleave\text{-}bound}$, k_{cleave}) were measured experimentally [23,24]. The parameters that determine the gene silencing at different miRNA levels (a, K_m) were determined previously [22]. All other model parameters were determined using parameter fitting software in MATLAB by simulating the system of equations and fitting the steady-state values to the experimental data for the theophylline-responsive OFF control device.

We used the model to predict the effects of different parameters on device function and to identify where the theophylline-responsive device fell in relation to these predictions. For example, the model predicts the quantitative increase in the change in target expression level as a function of increasing spacer length that is observed in our experimental system (Figure 3a). We used the model to interrogate the impact of miRNA-target gene affinity, K_m, on the performance of the OFF control device (Figure 3b). The model predicts an optimal value of approximately 3500 nM, corresponding to the largest dynamic range achievable by the OFF control device. At very high K_m, there is little target gene silencing from the miRNA, and an increase in miRNA levels will lead to little difference in target gene expression. At very low K_m, a large amount of miRNA silencing is observed even at low miRNA levels that saturates as miRNA levels increase [22]. Our experimental results indicate that the K_m of the theophylline-responsive control device is nearly optimal for the system (Figure 3b). Finally, we used the model to interrogate the impact of the ratio of cleavage rates between the ligand unbound and bound ribozyme switch ($k_{cleave}/k_{cleave\text{-}bound}$) on the performance of the OFF control device (Figure 3c). As this ratio increases, the model predicts an increase in the change in target gene expression levels in the presence of the ligand. In our theophylline-responsive system, the ratio of cleavage rates is approximately 5.6. The model predicts that increasing this ratio would lead to substantial increases in the change in target gene expression, indicating that this is a key parameter in the design of control devices with specified dynamic ranges.

The model can be used to identify which components of the circuit can be targeted for modification to optimize the performance of the circuit. For example, both spacer length and the ratio of ribozyme cleavage rates in the presence and absence of the ligand ($k_{cleave}/k_{cleave\text{-}bound}$) were found to be critical parameters in tuning the dynamic range of the OFF control device. High-throughput screening strategies have been described that support efficient tuning of the cleavage rates of ribozyme switches [25]. As another example, the model indicated that miRNA silencing efficiency, largely determined by K_m, determines the silencing observed at a given miRNA level. The K_m of a miRNA-target gene pair can be modified by incorporating additional target sites into the 3′ untranslated region (UTR) of the target transcript [22,26]. With the components selected for this study, the systems were limited to dynamic ranges on the order of ~2-fold. While this regulatory activity is on the order of that observed with endogenous miRNA regulatory networks [27], it is lower than many synthetic transcription-based control systems in use. Improvements in the properties of the underlying device components (i.e., ribozyme switch, miRNA) will allow for OFF control devices with improved dynamic ranges supporting broader applications.

Protein-responsive RNAi-based OFF control devices

To adapt the RNAi-based OFF control devices to respond to proteins, we replaced the small molecule-responsive ribozyme switch with ribozymes that alter their cleavage activity in the presence of specified protein ligands. The modular nature of the ribozyme switch platform allowed us to tailor the ligand responsiveness of the switch by swapping new aptamer sequences into the sensor domain of the platform [20].

We developed protein-responsive ribozyme switches to two protein ligands: the MS2 bacteriophage coat protein (MS2) and an oncogene (E2F1). Specifically, sequences for previously characterized RNA aptamers to MS2 [28] and E2F1 [29] were placed directly into the sensor domain of

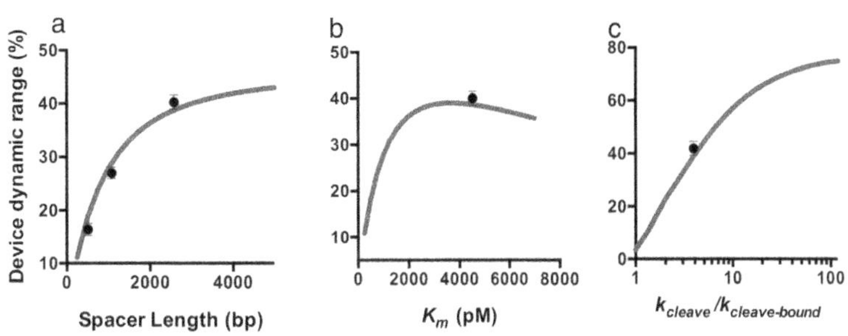

Figure 3 Parameters that affect performance of the RNAi-based OFF control device. **(a-c)** Predicted relationships between device dynamic range and the length of the spacer between the ribozyme switch and miRNA components **(a)**, miRNA-target transcript binding affinity, K_m **(b)**, and the ratio of ribozyme cleavage rates in the absence and presence of ligand, $k_{cleave}/k_{cleave\text{-}bound}$ **(c)**. Grey line: model predictions; circle: experimental results. Device dynamic range is reported as the percent change in GFP expression in the absence and presence of theophylline.

the ribozyme switch platform using a previously characterized transmitter sequence [25]. The binding and cleavage activities of the MS2- and E2F1-responsive ribozyme switches were characterized using previously described *in vitro* assays. The binding activities of the switches were measured using a surface plasmon resonance assay on a Biacore instrument [23]. The results of this binding assay indicate that the protein-responsive switches bind their cognate ligands with high affinities; the MS2-responsive switch has a K_D of 296 nM [30] and the E2F1-responsive switch has a K_D of 548 nM (Additional file 1: Figure S4a). The cleavage rate of the MS2-responsive switch was measured using a gel-based cleavage assay [23]. The results of the cleavage assay demonstrate that the MS2 ligand substantially affected the cleavage rate of the ribozyme switch, resulting in an approximate 2.5-fold decrease in cleavage rate in the presence of 2 μM MS2 [30]. End point cleavage assays were similarly performed on the E2F1-responsive switch and confirmed that the presence of ligand decreased the cleavage activity of this ribozyme switch (Additional file 1: Figure S4b, c). Taken together, the data indicate that the protein-responsive ribozyme switches bind their cognate ligands with high affinities and in a manner that disrupts cleavage activity of the ribozyme.

The validated protein-responsive ribozyme switches were placed into the OFF control device platform to build an RNAi-based control system that decreased target gene expression levels as a function of increasing MS2 and E2F1 levels. Expression cassettes were constructed with the protein-responsive ribozymes placed 2500 bp upstream of a GFP-targeting miRNA (Figure 4a). A second expression cassette was constructed that encoded the expression of a GFP fluorescent reporter protein. Isogenic cell lines stably expressing both expression cassettes were generated from a Flp-In HEK293 cell line as previously described.

Device performance was quantified by measuring target protein (GFP) and mature miRNA levels in the presence and absence of the protein ligands as previously described (Figure 4b-e). Plasmids expressing MS2 or E2F1 were transiently transfected into the stable cell lines expressing the MS2- or E2F1-responsive OFF control devices, respectively, and compared to the identical stable cell line transfected with a mock plasmid (i.e., no protein ligand expressed). The data demonstrate that GFP expression levels are reduced up to 2.1- and 1.8-fold in response to MS2 and E2F1 expression, respectively (Figure 4b,c). As with the theophylline-responsive OFF control device, the OFF states of the protein-responsive devices are similar to the positive control (i.e., miRNA only). When taking into account the differences in cleavage rates ($k_{cleave}/k_{cleave\text{-}bound}$) between the protein-responsive ribozymes and the theophylline-responsive ribozyme, with no additional refitting of other parameters, the model predicts the observed difference in gene-regulatory activities of the devices (Figure 4d). Larger changes in target gene expression correlated with larger cleavage rate ratios in the presence and absence of ligand, supporting that the cleavage rate ratio is a key design parameter for these systems. The qRT-PCR analysis indicated that the increase in GFP silencing was due to an increase in the levels of mature miRNAs within the cells (Figure 4e). Taken together, these results indicate that OFF control devices can be constructed to respond to a variety of ligands by integrating ribozyme switches that respond to the ligands of interest.

Protein-responsive RNAi-based OFF control devices support the implementation of negative feedback control systems

Feedback regulation is ubiquitous in biology and fundamental to many engineering applications. Synthetic biologists have designed a wide range of engineered biological transcriptional feedback regulation systems [31,32]. Negative and positive transcriptional feedback controllers have been used to stabilize desired output protein concentrations, accelerate system dynamics, and generate biological oscillations [33-35]. However, engineered genetic negative feedback control systems have been used to regulate the expression of transgenes; a system has not been developed that can exert feedback control over endogenous genes. This expanded capability would be a powerful tool to engineer synthetic circuits to maintain desired cell phenotypes, for example, in the presence of genetic mutations or fluctuating levels of signaling molecules or cytokines.

The RNAi-based OFF control device has the ability to regulate the levels of a target protein around a desired set point by enabling the implementation of a negative feedback control system in mammalian cells (Additional file 1: Text S1). In this system design, a ribozyme switch responsive to a protein of interest and a miRNA that targets that same protein of interest are integrated into the OFF control device architecture. Thus, an increase in levels of the input protein will result in an increase in levels of the miRNA, which will subsequently reduce levels of the protein to the desired set point level.

We first adapted our model to capture the behavior of a negative feedback system based on our protein-responsive OFF control device (Additional file 1: Text S1). In the negative feedback system model, the input is the maximum level of target gene expression, the output is the target protein levels, and our control action, or miRNA levels, regulates the ratio of output to input levels such that target protein expression remains relatively consistent independent of input level. In order to capture the behavior of a feedback control system which regulates levels of MS2-DsRed fusion protein, we included the parameters for the cleavage rate of the MS2-responsive ribozyme and a DsRed-targeting miRNA in the model (Additional file 1:

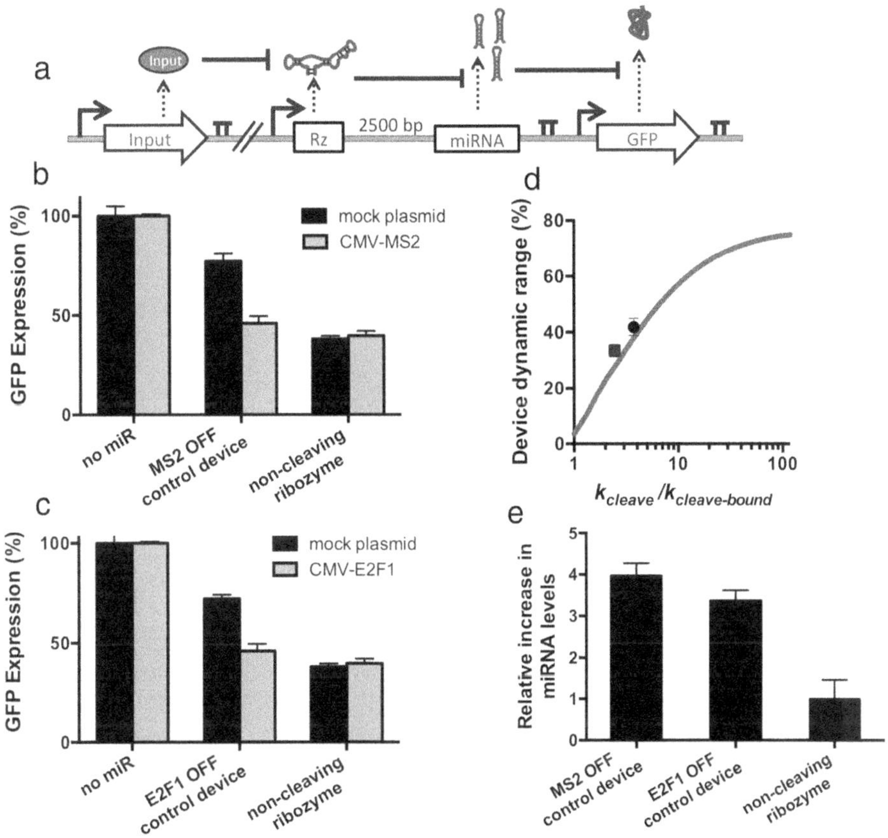

Figure 4 Protein-responsive RNAi-based OFF control devices. **(a)** Schematic overview of the genetic circuit used to characterize protein-responsive OFF control devices. The OFF control devices and GFP were stably integrated and expressed from identical expression cassettes as described in Figure 2a, except that the theophylline-responsive ribozyme was replaced with an MS2- or E2F1-responsive ribozyme switch. A plasmid encoding the expression of DsRed only (mock plasmid) or the protein ligand (MS2 or E2F1) was transfected into the cell lines to measure activity of the OFF control device. **(b-c)** GFP expression levels under the control of the MS2-responsive **(b)** or E2F1-responsive **(c)** OFF control device. No miR and miRNA only represent constructs with a protein-responsive ribozyme switch and no downstream miRNAs (negative control) and a construct with a non-cleaving ribozyme upstream of a GFP-targeting miRNA (positive control), respectively. Error bars represent ±1 standard deviation of at least three independent experiments. **(d)** Device dynamic ranges for the OFF control devices as a function of the ratio of cleavage rates ($k_{cleave}/k_{cleave\text{-}bound}$). Grey line: model; circle: theophylline-responsive device; square: MS2-responsive device. Device dynamic range is reported as the percent change in GFP expression in the absence and presence of theophylline. **(e)** Relative increase in miRNA levels after addition of protein ligand. Relative miRNA levels are reported normalized to the miRNA levels measured from identical cell lines in the absence of protein ligand. Error bars represent ±1 standard deviation of three independent experiments.

Text S1). In this way the output (MS2-DsRed protein levels) directly affects miRNA levels, and therefore at higher MS2-DsRed levels, the increase in miRNA levels will decrease levels of MS2-DsRed protein. The model predicts that in the parameter ranges observed experimentally the OFF control device exerts proportional control (P-control) over the protein of interest. Unlike in an integral controller, the output from P-type controllers is not independent of the input, but rather exhibits an offset from the desired set point. The result is that the level of our protein of interest (output) will increase linearly as the maximum level (input) increases (Figure 5a). Consistent with observations of feedback control in biological systems [18], the model predicts that the offset will decrease as the degradation of the miRNA depends less on the concentration of the miRNA. If the degradation of the miRNA is a constant rate such that it does not rely on the concentration of miRNA, then 'perfect adaptation' or integral control will be achieved. Integral control is characterized by no offset and the steady-state output protein levels are stable at a constant level, regardless of the input level (Figure 5b).

We examined the model predictions in our experimental system by modifying the OFF control device to exhibit negative feedback around the levels of the MS2 protein (Figure 5c). An expression cassette was designed in which the MS2-DsRed fusion protein was under the transcriptional control of a doxycycline-inducible CMV-TetO2 promoter and TetR transcriptional repressor pair. In a separate expression cassette, the MS2-responsive

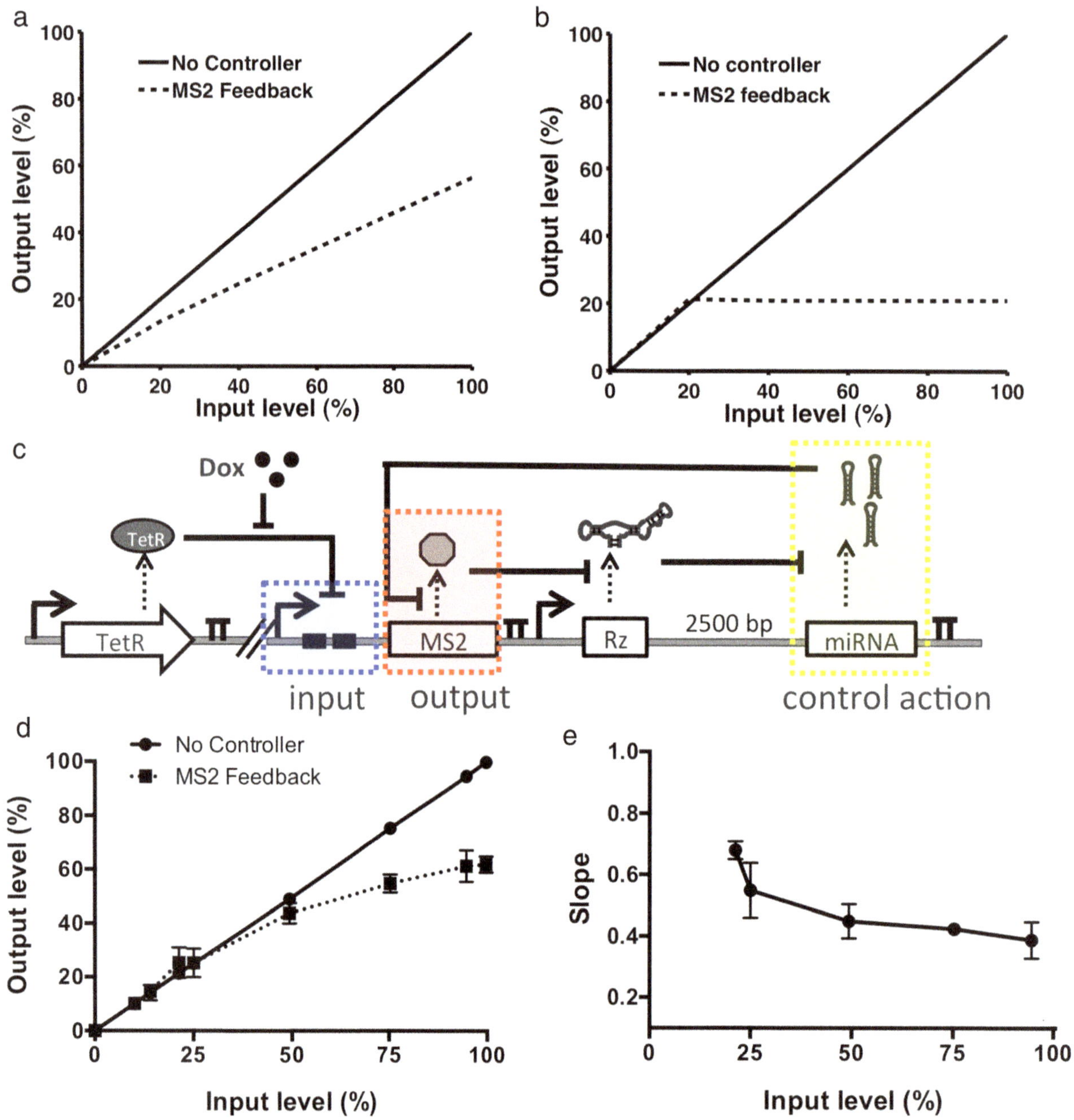

Figure 5 The protein-responsive OFF control device supports the design of a negative feedback control system over MS2 levels. **(a)** Model prediction for MS2 feedback system where miRNA level (control action) is dependent upon the level of miRNA in the cell. The output (MS2 protein) levels are shown plotted as a function of the input (maximum MS2) levels. **(b)** Model prediction for the MS2 system where the miRNA level (control action) is independent of the levels of miRNA. The output (MS2 protein) levels are shown plotted as a function of the input (maximum MS2) levels. **(c)** Schematic of the circuit used to test the MS2-responsive OFF control device exerting feedback control over MS2 levels. A doxycycline-responsive CMV-TetO2 promoter was used to control the expression of an MS2-DsRed fusion protein. An OFF control device containing an MS2-responsive ribozyme switch and a miRNA targeting DsRed were expressed in a separate expression cassette under the control of a constitutive promoter (P_{CMV}). The expression cassettes are stably integrated into a Flp-In HEK293 T-REx cell line, which expresses the TetR repressor. Input levels (maximum MS2-DsRed levels) were varied by growing the cells under different concentrations of doxycycline. The input is highlighted in red, the output (MS2-DsRed levels) is highlighted in blue, and the control action (miRNA levels) is highlighted in yellow. **(d)** Output (MS2-DsRed) levels as a function of the input (maximum MS2-DsRed transcript) levels for circuits with and without a feedback controller. The output and input are reported as the percent of the maximum DsRed levels in a cell line with no miRNA present. Error bars represent the standard deviation of three independent experiments. **(e)** The slope as a function of the input. The local slope is calculated as: $Slope_i = (Output_i - Output_{i-1}) / (Input_i - Input_{i-1})$.

OFF control device was modified such that the GFP-targeting miRNA component was replaced with a DsRed-targeting miRNA. In this negative feedback system design, the MS2-DsRed protein levels serve as the system output. As the output levels increase, the MS2 will trigger the formation of miRNAs that target the coding region of DsRed, thereby acting to reduce the levels of MS2-DsRed and serving as a form of negative feedback control. Isogenic cell lines stably expressing both expression cassettes were generated from a Flp-In HEK293 T-REx cell line as previously described. The Flp-In HEK293 T-REx cell line also stably expresses the TetR transcriptional repressor, such that the maximum level of MS2-DsRed (the input) is tightly regulated by the concentration of doxycycline in the media. In particular, the levels of MS2-DsRed were held low prior to induction and measurements were made 48 hours after induction, such that differences in the half lives of d2EGFP and MS2-DsRed (2 hours [36] versus 8 hours [37], respectively) would not affect the experimental results.

The resulting stable cell lines were used to characterize the activity of the negative feedback control device regulating MS2 levels (Figure 5c). DsRed levels were used as the output from the feedback system and were measured using a flow cytometry assay, the input (maximum MS2-DsRed level) was controlled using a doxycycline-responsive inducible promoter. Our data show that, compared to a system with no feedback control, there is a significant reduction in the levels of MS2 protein as the transcription rate is increased (Figure 5d). The local slope was also calculated as: $Slope_i = (Output_i - Output_{i-1}) / (Input_i - Input_{i-1})$ (Figure 5e). Our experimental system (Figure 5d, e) exhibits similar behavior to that predicted of a proportional controller (Figure 5b) in that the output is not independent of the input, but rather linearly increases as the input increases. Our results highlight that the RNAi-based OFF control device platform can be used to effectively exert feedback control over the levels of a protein of interest.

While there are several examples of negative feedback being implemented in synthetic circuits to alter the dynamics of the circuit [34,38], as well as in the construction of oscillators [33] and toggle switches [39], the described control systems do not have the ability to exert feedback control over endogenous genes. Our RNAi-based feedback control system behaves as a proportional controller (P-controller), where the control action (miRNA level) is proportional to the output (protein level). To implement integral feedback control the control action (miRNA level) must be solely a function of the output (protein level). Our model indicates that perfect adaptation cannot be achieved in our system as the miRNA degradation rate is a linear function of the miRNA level. Maintaining miRNA degradation at a constant level would allow us to achieve perfect adaptation [18]. However, in practice, keeping the rate of change of a molecular species independent of its own level through zeroth-order degradation kinetics is difficult to achieve in cellular systems, where molecular concentrations are governed by biochemical kinetics. Any mechanisms working to decrease the concentration of the control action species will likely need to operate at saturation with respect to the control action species in order to achieve zeroth order degradation kinetics. It is likely that many biological feedback systems operate using proportional control and may need to incorporate additional redundancies in cases where maintaining the level of a molecule within a narrow concentration range is critical.

Conclusions

Our work describes a novel genetic device that increases the level of silencing from a miRNA in the presence of a ligand of interest, effectively creating an RNAi-based OFF control device. Our OFF control device incorporates a ligand-responsive ribozyme and spacer element upstream of a miRNA regulator within the same expression cassette such that co-transcriptional cleavage of the ribozyme aborts transcription and prevents pri-miRNA formation. In the presence of ligand, the cleavage rate of the ribozyme is reduced, resulting in accumulation of pri-miRNA transcripts, which are subsequently processed into mature miRNAs that silence the target gene. Similar to previous designs [40], this coupled architecture allows the miRNA regulator to act as a genetic inverter, by inverting the signal from the ON ribozyme switch to an OFF genetic switch. We demonstrated the flexibility of the OFF switch platform by showing that it can be used to respond to both small molecule and protein ligands and developed a computational model describing the device that revealed important design parameters and parameter ranges associated with optimal performance of the control device. Finally, we used the RNAi-based OFF switch to implement a negative feedback control system, that maintained target intracellular protein levels in response to increases in transcription rate.

Our computational model revealed important design parameters and parameter ranges associated with optimal performance of these control devices. Thus, the model can be used to guide careful selection or tuning of components and/or inform how different parameters would impact the quantitative behavior of the control device. For example, the model identified spacer length, ribozyme cleavage rates in the presence and absence of ligand, and miRNA silencing efficiency as key parameters in the design of OFF RNAi-based control devices.

With the components selected for this study, the systems were limited to dynamic ranges on the order of ~2-fold. While this regulatory activity is on the order of that observed with endogenous miRNA regulatory networks [27], improvements to these activities will increase the broader utility of these devices in synthetic contexts.

Future improvements to the dynamic ranges exhibited by OFF RNAi-based control devices can be achieved through refinement of the underlying device components (i.e., ribozyme switch, miRNA).

By leveraging an RNAi-based regulatory mechanism, the described OFF control device can be adapted to exert negative feedback control over endogenous gene targets. The described control platform can be used in future cellular engineering and gene therapy applications, where the expression of key genes can be kept at desired levels even in the presence of cellular mutations or malfunctioning signaling pathways. Further development and optimization of the ribozyme switch and miRNA components will enable researchers to extend this platform for implementation into applications in health and medicine.

Methods

Plasmid construction

Modern molecular biology techniques were used to create the plasmids and cell lines used in this work. A chemically-competent *Escherichia coli* strain TOP10 was used for cloning purposes (Life Technologies, Carlsbad, CA) and was grown in LB medium (Becton, Dickinson and Company (BD), Franklin Lakes, NJ) with the indicated antibiotic concentrations. Custom oligonucleotides were synthesized by Integrated DNA Technologies (IDT, Coralville, IA) and the Stanford Protein and Nucleic Acid Facility (PAN, Stanford, CA). PCR was performed with Taq polymerase (New England Biolabs, Ipswich, MA) for products less than 1 kb and Pfu Ultra II Hotstart (Agilent Technologies, Santa Clara, CA) for products greater than 1 kb. All restriction enzymes were obtained from New England Biolabs. Plasmids were prepared from *E. coli* using Econospin columns (Epoch Life Science, Missouri City, TX) according to manufacturer's instructions. Sequencing was performed by Elim Biopharmaceuticals, Inc (Hayward, CA).

The plasmid pCS2304 was constructed by PCR amplifying a DNA fragment encoding a CMV promoter, d2EGFP coding sequence, the BGH-polyA tail, using the primers BglII CMV Fwd (5′-AATAAGATCTTAGTTATTAATA GTAATCAATTACGGGGTCATTAGT-3′) and Mfe bgh Rev (5′-TATTCAATTGTTTGCTGTTTCGTCCTCCG-3′) and cloning the fragment between the BglII and MfeI restriction sites in pcDNA5/FRT. Spacer regions of different lengths were PCR amplified from the LacZ gene using the primers BstBi-Xma-mid LacZ Fwd: (5′-ATATTCGAAAA TACCCGGGA TACTGTCGTCGTCCCCT-3′) and 500 bp LacZ rev (5′-AATAGGTACCATCGATAAT AACCG GTGTGTAGATGGGCGCATCG-3′) for the 500 bp spacer, 1000 bp LacZ rev (5′-AATAGGTACCATCGATAA TAACCGGTTTG GCCCTGTCAGAAATCCAGGGG-3′) for the 1000 bp spacer, and 2500 bp LacZ rev (5′-AA TAGGTACCATCGATAATAACCGGTAAGCGCTGTC ACCGGAATCAA-3′) for the 2500 bp spacer. These fragments were subsequently digested with BstBI and BamHI and cloned between the BstBI and BamHI restriction sites in the above intermediate plasmid to create pCS2722 (Additional file 1: Figure S5). The Rev primers used for amplifying the spacer incorporated AgeI and ClaI restriction sites on the 3′ end of the spacer, which were used for miRNA cloning in downstream steps. The resulting plasmids include two expression cassettes, one that drives the expression of d2EGFP and one that drives expression of different spacers that include restrictions sites on the 5′ end (BstBI, XmaI) and 3′ end (AgeI, ClaI) and is the base plasmid which was used in all OFF control switch studies in this work. The BstBI and XmaI restriction sites were used to clone the ribozyme switches and associated controls, and the AgeI and ClaI restriction sites were used to clone the miRNAs and associated controls. Finally, the plasmid contains FRT recombinase sites for making stable cell lines with the FLP recombinase system (Life Technologies).

We constructed pCS2934 (Additional file 1: Figure S5), which encodes MS2-DsRed expression under the control of an inducible promoter. A DNA fragment encoding CMV-TetO2-MS2-DsRed was obtained by digesting pCS2892 [22] with BglII and AvrII, and purifying the 1800 bp fragment by gel extraction. The backbone pCS2722 was digested with BglII and AvrII and the two DNA fragments were ligated together to form pCS2934, which includes an expression cassette encoding the expression of an OFF control device downstream of a cassette which encodes the expression of the MS2-DsRed fusion gene driven by the CMV-TetO2 promoter. A detailed list of all plasmids used in this work is contained in Additional file 1: Table S2.

For the construction of all OFF control devices, the miRNAs were cloned downstream of the spacer region using the AgeI and ClaI restriction sites. The ribozyme switches and controls were cloned upstream of the spacer region using the BstBI and XmaI restriction sites. All miRNA and ribozyme components were constructed by ordering two ultramer oligonucleotides (Integrated DNA Technologies), which were reverse complements of each other at their 3′ ends. The oligonucleotide pairs were subjected to five rounds of annealing and extension using Taq polymerase (New England Biolabs) in a thermocycler. The resulting DNA fragments were subsequently digested and cloned. The sequences of all miRNAs and ribozymes used in this study are presented in Additional file 1: Table S3.

Mammalian cell culture

HEK293 Flp-In T-REx cells (Life Technologies) were maintained in DMEM supplemented with 10% FBS at 37°C in a 5% CO_2-humidified incubator. Stable transfection of HEK293 Flp-In T-REx cell lines was performed using the Flp-In recombinase system (Life Technologies) according to the manufacturer's instructions to generate isogenic stable cell lines. Stable integrants were selected using 200 μg/ml hygromycin B (Life Technologies), whereas stable cell lines were maintained in 100 μg/ml hygromycin B.

Measuring fluorescent reporter expression through flow cytometry assays

miRNA induction experiments were performed by first seeding 24-well cell culture plates with stable cell lines at 2.5×10^5 cells per well. After 48 hours, cells were trypsinized and subjected to flow cytometry analysis on a Cell Lab Quanta SC MPL (Beckman Coulter, Brea, CA), and the resulting data were analyzed using the FlowJo software (Tree Star, Ashland, OR). For experiments utilizing doxycycline induction, fresh media containing 0, 0.05, 0.2, 0.8, 3.2, or 50 ng/mL doxycycline was added to the wells 24 hours after seeding and measured by flow cytometry as described above. Cells were initially gated for viability by electronic volume and side scatter. GFP and DsRed fluorescence of viable cells were measured through 525- and 670-nm bandpass filters, respectively, after excitation with a 488-nm laser. The levels of the fluorescent reporters were calculated as the median fluorescence of the stable cell population (Additional file 1: Figure S6). All fluorescent reporter levels were normalized to the corresponding fluorescent reporter level from cells harboring a control construct lacking a miRNA under the same conditions. At least two biological replicates were included in each experiment, and reported data are the averages of at least three independent experiments. Error bars represent ±1 standard deviation.

Measuring miRNA levels through Taqman assays

The levels of miRNAs in the cells were quantified using custom Taqman small RNA assays (Applied Biosystems, Carlsbad, CA) according to the manufacturer's instructions. Cells at different experimental conditions were plated in duplicate in 96-well tissue culture plates at a concentration of 10,000 cells per well. After 48 hours, the cells in one well were trypsinized and counted in a hemocytometer to obtain an accurate count of the number of cells per well for each experimental condition. The cells in the remaining well were lysed by adding 50 μl of Taqman lysis buffer and incubating for 8 minutes. The lysis reaction was stopped with the addition of 5 μl of stop buffer, resulting in a total volume of 55 μl of lysate per sample. A fixed volume of 5 μl of this lysate was used as input into the reverse transcription (RT) reaction.

Input RNA was reverse transcribed using the TaqMan miRNA Reverse Transcription Kit and custom miRNA-specific stem-loop primers (Applied BioSystems) in a small-scale RT reaction (1.387 μl of H_2O, 0.5 μl of 10X reverse transcription buffer, 0.063 μl of RNase inhibitor (20 units/μl), 0.05 μl of 100 mM dNTPs with dTTP, 0.33 μl of multiscribe reverse transcriptase, 1.67 μl of input RNA; components other than the input RNA were prepared as a larger volume master mix). The reaction was run on an iCycler iQ system (BioRAD, Hercules, CA) at 16°C for 30 min, 42°C for 30 min, and 85°C for 5 min. The RNU 6B Taqman miRNA control was used to normalize the C_t values of the experimental miRNAs and to confirm the cell quantities measured by hemocytometer counting. Real-time PCR was carried out on an iCycler iQ thermocycler (BioRAD) at 95°C for 10 minutes, followed by 40 cycles of 95°C for 15 seconds and 60°C for 1 minute. The C_t was calculated using available RT-PCR Miner software [41], which automatically detects C_t and efficiency values for each PCR reaction. The mean of the resulting C_t values for the target gene of each sample were subtracted from the mean C_t value for the control gene. The resulting values were then converted from $\log_2$ to linear scale and normalized to the value from the sample lacking any miRNA with the same concentration of ligand.

Measuring binding properties between ribozyme switches and protein ligands

The binding properties between the protein-responsive ribozyme switches or RNA aptamers and the corresponding protein ligands (i.e., MS2, E2F1) were determined according to previously described methods [24]. Briefly, ribozyme switches, aptamers, and appropriate controls were transcribed from DNA templates using the MEGA-shortscript T7 kit (Life Technologies) and purification of the transcription products was performed using the RNA Clean & Concentrator Kit (Zymo Research) according to the manufacturers' instructions. In all experiments, the HBS-N running buffer (GE Healthcare) (10 mM HEPES, 150 mM NaCl, pH 7.4) was used with no $MgCl_2$ present, to prevent cleavage of the ribozyme. A dilution series of the target was prepared in running buffer using a minimum of five concentrations for kinetic analysis. For each sample concentration, the ribozyme switch was captured onto the sample flow cell (FC2) for 30 seconds at a flow rate of 10 μL/minute, the target solution was injected over both flow cells at a flow rate of 30 μL/minute to monitor target association, and running buffer was injected over both flow cells at a flow rate of 30 μL/minute to monitor target dissociation.

Data processing and analysis were performed using Biacore X100 Evaluation Software version 2.0 (GE Healthcare, San Francisco, CA). A double-referencing method was performed to process all datasets [42,43]. Data from the sample flow cell (FC2) were referenced first by subtracting data from the reference flow cell (FC1) to correct for bulk refractive index changes, nonspecific binding, injection noise, matrix effects, and baseline drift. The reference-subtracted data (FC2-FC1) were double-referenced with a blank injection of running buffer to account for any systematic drift over the course of the injection. The double-referenced data were fit to a 1:1 binding model for kinetic analysis to calculate the k_a and k_d between the ribozyme and the protein. The overall dissociation constant, K_D, was calculated from k_a, *and* k_d.

Gel-based ribozyme cleavage assays

Gel-based ribozyme cleavage assays were performed as previously described [23]. Briefly, all gel-based ribozyme cleavage assays were performed in a physiologically relevant reaction buffer (50 µl) composed of 500 µM $MgCl_2$, 100 mM NaCl, and 50 mM Tris–HCl (pH 7.5) at 37°C. A 12-nt RNA blocking sequence that is complementary to part of the ribozyme transcript was inserted at the 5′ end of the sTRSV HHRz transcript to generate a cis-blocked ribozyme construct. In the reaction volume, 75 nM of full-length RNA was first incubated with 2.5 µM DNA activator strand (5′-AAACAACTTTGTTTGTTTCCCCC-3′), which is ordered from Integrated DNA Technologies as an oligonucleotide with standard desalting, for 2 min to activate the blocked RNA. A zero time-point aliquot was taken before initiating the self-cleavage reaction with the addition of $MgCl_2$ and the indicated concentration of input protein. Reactions were quenched after 20 min with addition of 3 volumes of RNA stop/load buffer (95% formamide, 30 mM EDTA, 0.25% bromophenol blue, 0.25% xylene cyanol) and placed on ice. Samples were heated to 95°C for 5 min, snap cooled on ice for 5 min, and size-fractionated on a denaturing (8.3 M urea) 10% polyacrylamide gel at 25 W for 45–60 min. Gels were exposed overnight on a phosphor screen and imaged on a FX Molecular Imager (BioRAD). The relative levels of the full-length transcript and cleaved products were determined by phosphorimaging analysis.

Additional file

Additional file 1: Text S1. Model description. **Figure S1.** Overview of ribozyme switch design. **Figure S2.** Non-cleaving ribozyme controls for different spacer lengths. **Figure S3**. Schematic representation of model. **Figure S4.** Characterization of E2F1-A1 riboswitch. **Figure S5.** Base plasmids used to create all plasmids stably integrated into HEK293 Flp-In cells in this work. **Figure S6**. Representative flow cytometry plots for HEK293 Flp-In cells stably expressing OFF control device constructs. **Table S1.** Description of parameters used in the model. **Table S2.** List of plasmids used in this work. **Table S3.** Sequences of ribozymes and miRNAs used in this work.

Competing interests
The authors declare that they have no competing interests.

Authors' contributions
RJB and CDS conceived of the project, designed the experiments, analyzed the results, and wrote the manuscript. RJB and SMW conducted the experiments. All authors read and approved the final manuscript.

Acknowledgements
We thank M. Mathur, M. Siddiqui for valuable comments on the transcript. This work was supported by the Defense Advanced Research Projects Agency [HR0011-11-2-0002 to CDS] the National Science Foundation [fellowship to RJB], and ARCS [fellowship to RJB]. RJB designed research, performed research, and wrote the article; SW designed the E2F1-responsive ribozyme switch; CDS designed research and wrote the article.

References
1. Lienert F, Lohmueller JJ, Garg A, Silver PA. Synthetic biology in mammalian cells: next generation research tools and therapeutics. Nat Rev Mol Cell Biol. 2014;15(2):95–107.
2. Jaenisch R, Young R. Stem cells, the molecular circuitry of pluripotency and nuclear reprogramming. Cell. 2008;132(4):567–82.
3. Hahn WC, Weinberg RA. Modelling the molecular circuitry of cancer. Nat Rev Cancer. 2002;2(5):331–41.
4. Slusarczyk AL, Lin A, Weiss R. Foundations for the design and implementation of synthetic genetic circuits. Nat Rev Genet. 2012;13(6):406–20.
5. Win MN, Liang JC, Smolke CD. Frameworks for programming biological function through RNA parts and devices. Chem Biol. 2009;16(3):298–310.
6. Bartel DP. MicroRNAs: target recognition and regulatory functions. Cell. 2009;136(2):215–33.
7. Shukla GC, Singh J, Barik S. MicroRNAs: Processing, Maturation, Target Recognition and Regulatory Functions. Mol Cell Pharmacol. 2011;3(3):83–92.
8. Lewis BP, Burge CB, Bartel DP. Conserved seed pairing, often flanked by adenosines, indicates that thousands of human genes are microRNA targets. Cell. 2005;120(1):15–20.
9. Culler SJ, Hoff KG, Smolke CD. Reprogramming cellular behavior with RNA controllers responsive to endogenous proteins. Science. 2010;330 (6008):1251–5.
10. Tigges M, Marquez-Lago TT, Stelling J, Fussenegger M. A tunable synthetic mammalian oscillator. Nature. 2009;457(7227):309–12.
11. Deans TL, Cantor CR, Collins JJ. A tunable genetic switch based on RNAi and repressor proteins for regulating gene expression in mammalian cells. Cell. 2007;130(2):363–72.
12. Leisner M, Bleris L, Lohmueller J, Xie Z, Benenson Y. Rationally designed logic integration of regulatory signals in mammalian cells. Nat Nanotechnol. 2010;5(9):666–70.
13. Beisel CL, Chen YY, Culler SJ, Hoff KG, Smolke CD. Design of small molecule-responsive microRNAs based on structural requirements for Drosha processing. Nucleic Acids Res. 2011;39(7):2981–94.
14. Beisel CL, Bayer TS, Hoff KG, Smolke CD. Model-guided design of ligand-regulated RNAi for programmable control of gene expression. Mol Syst Biol. 2008;4:224.
15. An CI, Trinh VB, Yokobayashi Y. Artificial control of gene expression in mammalian cells by modulating RNA interference through aptamer-small molecule interaction. RNA. 2006;12(5):710–6.
16. Bloom RJ, Winkler SM, Smolke CD. A quantitative framework for the forward design of synthetic miRNA circuits. Nat Methods. 2014;11(11):1147–53.
17. Yi TM, Huang Y, Simon MI, Doyle J. Robust perfect adaptation in bacterial chemotaxis through integral feedback control. Proc Natl Acad Sci U S A. 2000;97(9):4649–53.
18. Ang J, McMillen DR. Physical constraints on biological integral control design for homeostasis and sensory adaptation. Biophys J. 2013;104(2):505–15.
19. Ballarino M, Pagano F, Girardi E, Morlando M, Cacchiarelli D, Marchioni M, et al. Coupled RNA processing and transcription of intergenic primary microRNAs. Mol Cell Biol. 2009;29(20):5632–8.
20. Win MN, Smolke CD. A modular and extensible RNA-based gene-regulatory platform for engineering cellular function. Proc Natl Acad Sci U S A. 2007;104(36):14283–8.
21. Wei KY, Chen YY, Smolke CD. A yeast-based rapid prototype platform for gene control elements in mammalian cells. Biotechnol Bioeng. 2013;110(4):1201–10.
22. Bloom RJ, WInkler SM, Smolke CD. A quantitative framework for the forward design of synthetic miRNA circuits. Nat Methods. 2014; in press.
23. Kennedy AB, Liang JC, Smolke CD. A versatile cis-blocking and trans-activation strategy for ribozyme characterization. Nucleic Acids Res. 2013;41(2):e41.
24. Chang AL, McKeague M, Liang JC, Smolke CD. Kinetic and equilibrium binding characterization of aptamers to small molecules using a label-free, sensitive, and scalable platform. Anal Chem. 2014;86(7):3273–8.

25. Liang JC, Chang AL, Kennedy AB, Smolke CD. A high-throughput, quantitative cell-based screen for efficient tailoring of RNA device activity. Nucleic Acids Res. 2012;40(20):e154.
26. Mukherji S, Ebert MS, Zheng GX, Tsang JS, Sharp PA, van Oudenaarden A. MicroRNAs can generate thresholds in target gene expression. Nat Genet. 2011;43(9):854–9.
27. Kim DH, Saetrom P, Snove Jr O, Rossi JJ. MicroRNA-directed transcriptional gene silencing in mammalian cells. Proc Natl Acad Sci U S A. 2008;105(42):16230–5.
28. Hirao I, Spingola M, Peabody D, Ellington AD. The limits of specificity: an experimental analysis with RNA aptamers to MS2 coat protein variants. Mol Divers. 1998;4(2):75–89.
29. Ishizaki J, Nevins JR, Sullenger BA. Inhibition of cell proliferation by an RNA ligand that selectively blocks E2F function. Nat Med. 1996;2(12):1386–9.
30. Kennedy AB, Vowles JV, d'Espaux L, Smolke CD. Protein-responsive ribozyme switches in eukaryotic cells. Nucleic Acids Res. 2014;42(19):12306–21.
31. Purnick PE, Weiss R. The second wave of synthetic biology: from modules to systems. Nat Rev Mol Cell Biol. 2009;10(6):410–22.
32. Shimoga V, White JT, Li Y, Sontag E, Bleris L. Synthetic mammalian transgene negative autoregulation. Mol Syst Biol. 2013;9:670.
33. Elowitz MB, Leibler S. A synthetic oscillatory network of transcriptional regulators. Nature. 2000;403(6767):335–8.
34. Stapleton JA, Endo K, Fujita Y, Hayashi K, Takinoue M, Saito H, et al. Feedback control of protein expression in mammalian cells by tunable synthetic translational inhibition. ACS Synth Biol. 2012;1(3):83–8.
35. Rosenfeld N, Young JW, Alon U, Swain PS, Elowitz MB. Accurate prediction of gene feedback circuit behavior from component properties. Mol Syst Biol. 2007;3:143.
36. Corish P, Tyler-Smith C. Attenuation of green fluorescent protein half-life in mammalian cells. Protein Eng. 1999;12(12):1035–40.
37. pDsRed-Express-DR Vector Information [Brochure]. 2005. Mountain View, CA: Clontech Laboratories, Inc.
38. Becskei A, Serrano L. Engineering stability in gene networks by autoregulation. Nature. 2000;405(6786):590–3.
39. Gardner TS, Cantor CR, Collins JJ. Construction of a genetic toggle switch in Escherichia coli. Nature. 2000;403(6767):339–42.
40. Kumar D, An CI, Yokobayashi Y. Conditional RNA interference mediated by allosteric ribozyme. J Am Chem Soc. 2009;131(39):13906–7.
41. Zhao S, Fernald RD. Comprehensive algorithm for quantitative real-time polymerase chain reaction. J Comput Biol. 2005;12(8):1047–64.
42. Katsamba PS, Park S, Laird-Offringa IA. Kinetic studies of RNA-protein interactions using surface plasmon resonance. Methods. 2002;26(2):95–104.
43. Myszka DG. Improving biosensor analysis. J Mol Recognit. 1999;12(5):279–84.

Wide-dynamic-range promoters engineered for cyanobacteria

Hsin-Ho Huang and Peter Lindblad*

Abstract

Background: Cyanobacteria, prokaryotic cells with oxygenic photosynthesis, are excellent bioengineering targets to convert solar energy into solar fuels. Tremendous genetic engineering approaches and tools have been and still are being developed for prokaryotes. However, the progress for cyanobacteria is far behind with a specific lack of non-native inducible promoters.

Results: We report the development of engineered TetR-regulated promoters with a wide dynamic range of transcriptional regulation. An optimal 239 (±16) fold induction in darkness (white-light-activated heterotrophic growth, 24 h) and an optimal 290 (±93) fold induction in red light (photoautotrophic growth, 48 h) were observed with the L03 promoter in cells of the unicellular cyanobacterium *Synechocystis* sp. strain ATCC27184 (i.e. glucose-tolerant *Synechocystis* sp. strain PCC 6803). By altering only few bases of the promoter in the narrow region between the −10 element and transcription start site significant changes in the promoter strengths, and consequently in the range of regulations, were observed.

Conclusions: The non-native inducible promoters developed in the present study are ready to be used to further explore the notion of custom designed cyanobacterial cells in the complementary frameworks of metabolic engineering and synthetic biology.

Background

Promoter studies in the unicellular cyanobacterium *Synechocystis* sp. strain PCC 6803 (*Synechocystis*) mainly focus on elucidating their native transcriptional regulations to different environmental stimuli and changes. For example, some promoters are light responsive (*psa* [1], *psb* [2], and *secA* [3]), dark-inducible (*lrtA* [4]), nitrate/nitrite-inducible (*nirA* [5]), copper-ions responsive (*petE* [6]), and heavy metal-ions inducible promoters [7]. The promoters can be categorized into three different types (I, II, and III [8]) depending on how the binding motifs arranged to interact with the respective sigma factor and cognate transcription factor. Type I promoters fit the eubacterial canonical promoters [9] with the −35 and −10 elements being spaced by 17-bp. Type II promoters, with the −10 element only or including an enhancer-motif for binding an activator protein, are in majority. Type III promoters are distinct from type I and II for stringent responses. In the framework of synthetic biology, the implemented transcription regulatory system is expected to have minimal cross-talks with the host's genetic background [10]. Therefore, it is necessary to develop non-native promoters regulated by foreign transcription factor and non-metabolized ligands.

Until today, the number of non-native promoters examined in *Synechocystis* is very limited, e.g. the λP_L-derived BBa_R0040, λP_R-derived BBa_R0051, and P_{lac}-derived BBa_R0010 promoters in the BioBrick Registry (http://partsregistry.org/Promoters), and the P_{trc} promoter [11]. The three former promoters, although characterized in *Escherichia coli*, do not function in *Synechocystis*. A leaky repressed activity and a narrow range of regulation are the common problems of the strong P_{trc} promoter in the unicellular cyanobacteria *Synechocystis* [11] and *Synechococcus* [12,13]. The most commonly used systems to express and over-express genes are based on LacI-regulated *trc* and *tac* promoters [12-16]. The lack and limitations of available promoters for cyanobacterial biotechnology initiated our present work to develop non-native promoters that are fully repressed and highly induced. The regulatory system used is based on the

* Correspondence: peter.lindblad@kemi.uu.se
Microbial Chemistry, Department of Chemistry - Ångström Laboratory, Uppsala University, P.O. Box 523SE-75120, Uppsala, Sweden

foreign transcription factor TetR repressor and its ligand inducer anhydrotetracycline (aTc) [17].

Transcription initiation is a key point for regulating the gene expression [18]. Five promoter elements such as the UP element, -35 element, extended −10 element, the −10 element, and a nucleotide in two positions downstream of the −10 element are essential for the interactions with RNA polymerase (RNAP) [9]. RNAP is the key enzyme performing transcription in three stages: initiation, elongation, and termination [19]. One way to regulate transcription initiation is to repress it when a transcription factor binds to its cognate site. The transcription factor's binding creates a steric hindrance usually close to the essential promoter elements to prevent RNAP's binding to the promoter. The λP_L-derived BBa_R0040 promoter, termed the R40 promoter in the present work, serves as a template for promoter engineering. The promoter region to keep, covering the first three essential promoter elements, contains the two TetR repressor's binding sites with the consensus −35 element in between. The promoter region to engineer is the −10 element and the indicated nucleotides in the region between the −10 element and the transcription start site. We created a L promoter library (denoting its origin from the λP_L promoter [20]) to be examined in *Synechocystis*.

It is known from the studies of *Escherichia coli* [21,22] and *Thermus aquaticus* [23,24], but with no information for cyanobacteria, that the 5′-GGG-3′ located immediately downstream of the −10 element on the non-template strand of promoter DNA is crucial for transcriptional regulation and promoter strength, especially the G in the second position. In the present work we systematically change the nucleotide sequence in this region on promoter, examine its strength, and demonstrate the first functional TetR-regulated promoter library in the cyanobacterium *Synechocystis*. The potential applications of this library are discussed.

Results

Altering the −10 element and its downstream sequence makes promoters function in *Synechocystis*

To overcome the very low R40 promoter strength problem previously identified in *Synechocystis*, contrary to its high strength in *E. coli* [11], the L12 promoter was created by changing the −10 element of R40 promoter with the consensus TATAAT of *Synechocystis* [25]. In the presence of the TetR repressor and with the design of two TetR-binding sites on either R40 or L12 promoter (Figure 1), the regulation of a EYFP gene expression driven by either promoter was observed under both an induced and a repressed condition, in the presence and absence of the TetR's inducer aTc, respectively. As result (Table 1), in both induced and repressed conditions, the promoter strength of L12 is lower than the one of R40. Similarly, the L12 promoter's induction fold, which is the ratio of promoter strength in the induced to the repressed condition, is also lower than observed for R40's.

To further address the challenge of an even lower strength of the L12 promoter whose design fits a canonical promoter for cubacteria [9], we hypothesized that replacing bases in the region between the −10 element and the transcription start site (TSS) of a promoter can increase the L12 promoter strength. This region has been proved to significantly influence the promoter strength [21] but no information is available for cyanobacteria. Three parallel modifications in this region were designed (Table 1): the first line, the position(s) locating on either one or both of the second and third bases immediately downstream of −10 element of L12 promoter was/were varied with adenine or guanine or thymine or cytosine and in total 16 promoters, from L01 to L16, were constructed; the second line, a *bona fide* downstream basal promoter element GGGAgc [24] was designed immediately downstream of the −10 element of L12 and R40 promoter as L21 and L22 promoters, respectively; the third line, a −10

```
                        UP element              -35         spacer          -10
R40      BBp|X|TCCCTATCAGTGATAGAGA|TTGACA|TCCCTATCAGTGATAGA|GATACT|GAGCAC|---------------------------|downstream
L31      BBp|X|*******************|******|*****************|T***A*|*T*TGG|---------------------------|downstream
Ptrc10   BBp|X|-------------------|******|ATTAATCATCCGGCTCG|T***A*|*T*TGG|AATTGTGAGCGGATAACAATTTCACACA|downstream
J23101   BBp|-|-------------------|**T***|GCTAGC*****CC***G|T**TA*|*CTAG*|---------------------------|downstream
rnpB     BBp|- ZTGGGG***CAA*CCACAG CG*C*T ATGGCTCT*A*C*AT*G|C*C***|AGAA*A A--------------------------|downstream
               |_____________________________________________________________________________________|
                                              promoter region
```

Figure 1 Alignment of selected promoters and their flanking sequences. The reporter construct of R40 promoter (i.e. BBa_R0040) is used as a reference sequence; gap (hyphen '-') and identical base (asterisk '*') are indicated for the each promoter. 'BBp' is the BioBrick prefix (GAATTCGCGGCCGCTTCTAGAG); 'X' is the TetR repressor expressing device [BBa_J23101]-[BBa_P0440] and the 8-bp BioBrick scar (TACTAGAG); 'downstream' sequence to the initial codon ATG of the reporter EYFP gene are the 8-bp BioBrick scar, the 10-bp ribosome binding site RBS*, the 6-bp BioBrick scar, and the 3-bp initial codon (TACTAGAG|TAGTGGAGGT|TACTAG|ATG); the *tetO2* operator of the Tn*10* resistance operon was placed upstream of the −35 element and in the spacer region for the two TetR-binding sites (bases shown in magenta). Core bases of each operator for the TetR binding are underlined. The R40-derived promoters L01 to L16, L21 and L22 are compared in Table 1. The *lacO1* operator of the P_{trc1O} promoter contains the bases shown in cyan. The *rnpB* promoter is aligned with the −10 element CACACT [25] and Z is the remaining 153 bases.

Table 1 Promoter comparisons in sequence, regulation, induction and thermal opening probability pattern

Promoter	−10 element and its downstream 8 bases[a]	Regulation[b]		Induction[c]	Pattern[d]
		Induced	Repressed		
L15	TATAATG**GA**CACTA	20.1 ± 0.1	0.243 ± 0.003	79 ± 1	
L07	TATAATG**GT**CACTA	19.8 ± 0.1	0.289 ± 0.003	65 ± 1	A (17.9 ± 0.2)
L05	TATAATG**CT**CACTA	16.3 ± 0.1	0.198 ± 0.003	78 ± 1	
L13	TATAATG**CA**CACTA	15.5 ± 0.1	0.219 ± 0.003	67 ± 1	
L03	TATAATG**GC**CACTA	19.2 ± 0.1	0.220 ± 0.003	83 ± 1	
L11	TATAATG**GG**CACTA	19.1 ± 0.1	0.236 ± 0.003	77 ± 1	
L09	TATAATG**CG**CACTA	15.6 ± 0.1	2.88 ± 0.01	5.15 ± 0.01	B (16.3 ± 0.2)
noLVA_L09[e]	TATAATG**CG**CACTA	11.53 ± 0.05	0.545 ± 0.004	20.1 ± 0.1	
L01	TATAATG**CC**CACTA	11.30 ± 0.05	0.235 ± 0.003	46 ± 1	
L02	TATAATG**TC**CACTA	17.7 ± 0.1	0.214 ± 0.003	79 ± 1	
L10	TATAATG**TG**CACTA	17.6 ± 0.1	0.235 ± 0.003	71 ± 1	C (15.9 ± 0.2)
L04	TATAATG**AC**CACTA	12.3 ± 0.1	0.201 ± 0.003	58 ± 1	
L12	TATAATG**AG**CACTA	0.043 ± 0.003	0.022 ± 0.003	1.9 ± 0.3	
L16	TATAATG**AA**CACTA	17.5 ± 0.1	0.240 ± 0.003	69 ± 1	
L06	TATAATG**TT**CACTA	15.8 ± 0.1	0.219 ± 0.03	69 ± 1	D (15.1 ± 0.2)
L08	TATAATG**AT**CACTA	14.9 ± 0.1	0.255 ± 0.003	56 ± 1	
L14	TATAATG**TA**CACTA	12.1 ± 0.1	0.304 ± 0.003	38.0 ± 0.4	
L21	TATAAT***GGGA***GCTA	41.6 ± 0.2	40.1 ± 0.2	0.986 ± 0.003	E
L22	GATACT***GGGA***GCTA	0.378 ± 0.003	0.023 ± 0.004	16 ± 2	F
P_{trc1O}	TATAATGTGTGGAA	46.4 ± 0.2	43.9 ± 0.2	1.007 ± 0.003	G
L31	TATAATGTGTGGTA	3.08 ± 0.01	0.169 ± 0.003	17.3 ± 0.3	H
R40	GATACTGAGCACTA	0.272 ± 0.003	0.082 ± 0.003	3.2 ± 0.1	I
J23101	TATTATGCTAGCTA	4.57 ± 0.02	4.461 ± 0.02	0.974 ± 0.003	n.s.
rnpB	CACACTAGAAAAAT	1.00 ± 0.01 (427 ± 2)	1.00 ± 0.01 (448 ± 2)	0.95 ± 0.01	

[a], The sequences of listed promoters, excluding P_{trc1O}, J23101, and *rnpB* promoters, are identical except the region shown in this table. The consensus −10 element TATAAT is underlined. The transcription start site (TSS) is boxed. The promoter sequences are detailed (Figure 1). [b], The induced and repressed conditions are *Synechocystis* sp. strain ATCC27184 (i.e. glucose-tolerant *Synechocystis* sp. strain PCC 6803) cells in LAHG growth condition treated with and without 100 ng/mL aTc, respectively for 24 hours. The mean ± standard error of mean (s.e.m.) is relative to the strength of the *rnpB* promoter in the respective regulation condition. The value in parentheses is the experimental mean ± s.e.m. of EYFP emission per cell after subtracting the auto-fluorescence of *Synechocystis* cells containing pPMQAK1 vector only. The measurement is done by a flow cytometer to collect 50,000 events for each biological sample. [c], The induction of a promoter is the ratio of its measured strength in induced compared to in repressed condition. [d], The simulated thermal opening probability patterns (A-I) at 303 K are shown (Figure 4); n.s., not simulated. The value in a bracket is the mean ± s.e.m. under the induced condition of strengths of the promoters in a given pattern. The noLVA_L09 and L12 constructs are excluded in averaging. [e], The only difference of noLVA_L09 to L09 is the introduction of a double stop codon in the end of the *tetR* gene to cease translation of a protease LVA tag tailing in C-terminal of TetR repressor. The L09 promoter regions of both are identical.

element downstream sequence, GTGTGG, of a strong P_{trc1O} promoter [11] was replaced in the same position of the L12 promoter to generate the L31 promoter. J23101 and *rnpB* promoters were used as references when comparing promoter strength [11].

We first observed that, in the induced condition, replacing with the consensus −10 element may affect promoter strength positively or negatively depending on its adjacent downstream sequence (Table 1). It increases the L21 promoter strength by 110 (±1) -fold compared to L22 when a *bona fide* downstream basal promoter element GGGA exists; however, on the contrary, it decreases L12 promoter strength by 6 (±0.4) fold compared to R40 when its adjacent downstream sequence, GAGCACT, exists. To further gain insight of this region between −10 element and TSS of native type I promoters [8] of *Synechocystis*, the sequence conservation weighted by the number of pyrosequencing-reads from differential RNA-sequencing data [25] is GAGGG adjacent to −10 element; adenine locating 2 nt downstream from the −10 element is locally more conserved than other bases as revealed by the weighted sequence logo (Additional file 1: Figure S1). Together with the presence of the consensus −10 element TATAAT of *Synechocystis*, we argue that the even lower promoter strength of L12 compared to R40 might come from the deleterious effect of a fully consensus promoter [9].

We then confirmed our hypothesis to increase the L12 promoter strength in the induced condition (Table 1).

First, in L01 to L16 promoters, by merely changing two bases, A and G, of L12, to G and A in L15 and C and C in L01 show up to 467 (±33) fold and at least 263 (±18) fold improvement in promoter strength, respectively. For the strong L15, L07, L03, and L11 promoters, they coincide with a G locating 2 nt downstream from the −10 element. Second, in the L21 and L22 promoters, a GGGA motif largely enhances promoter strength by 967 (±68) fold from L12 to L21, but only slightly increases promoter strength by 1.4 (±0.02) fold from R40 to L22 together with a non-consensus −10 element. The L15 and L21 are 20 (±0.2) and 42 (±0.5) times stronger than the *rnpB* promoter, respectively. Third, in the L31 promoter, the −10 element downstream sequence GTGTGG of the strong P_{trc1O} promoter increases its strength by 72-fold from the L12 promoter but the increased strength is still much lower than the P_{trc1O} promoter. Taking together, the deleterious effect of a fully conserved sequence on promoter strength was removed by replacing the conserved adenine and guanine at 2 and 3 nt, respectively, downstream of the consensus −10 element with any other three nucleotides. We further concluded that guanine locating 2 nt downstream from the consensus −10 element is essential for promoter strength.

The L03 promoter is widely regulated in *Synechocystis*

The engineered L promoters are controlled by the transcriptional regulatory machinery composed of the aTc inducer, the TetR repressor, and the *tetO2* operator of the Tn*10* tetracycline resistance operon [17]. All promoters in the present study have two *tetO2* operators, except the P_{trc1O}, J23101, and *rnpB* promoters with no *tetO2* operator; the core invert repeat in the operator for TetR binding is not affected after the change of the −10 element (Figure 1). The relative amount of TetR repressors in a cell of *Synechocystis* was balanced from its constitutive expression by the J23101 promoter and its more efficient turnover by being tagged by a protease tag, LVA [11].

In the absence of inducer aTc, each of the L01 to L16 promoters excluding the L12 and L09 showed very low but detectable repressed strength which was about 10-times higher than the repressed L12 promoter strength. The L09 promoter in the L09 construct showed the repressed strength about 100-times higher than the L12 promoter's (Table 1). We hypothesize that increasing the TetR repressors by slowing down its turnover via the removal of a LVA tag from it can further repress the L09 promoter. As result, the L09 promoter can be further repressed in the noLVA_L09 construct than in the L09 construct, but the improved repressed L09 promoter strength is still 25-times higher than the L12 promoter's.

In the presence of the inducer aTc, a dynamic range of the transcription regulation on an L promoter is accessed with the induction fold: larger induction fold means wider dynamic range (Table 1). There is no dynamic range for the strong L21 promoter with two *tetO2* operators and for the promoters without *tetO2* operator such as P_{trc1O}, J23101 and *rnpB* promoters. One way to expand the dynamic range is to further lower the repressed strength of a promoter: the L09 promoter gained about 4 (±0.02) times wider range in the noLVA_L09 construct than the L09 construct. But, increasing the repression by more TetR repressors compromised the induced promoter strength. Another way to expand the dynamic range is to increase the induced strength of a promoter: the L03 promoter acquired about 44 (±7) times wider range than the L12 promoter. But, the altered intrinsic properties derived from the mutated promoter sequence compromised the repressed state of the L03 promoter. Because of the L03 promoter showing widest dynamic range among the L promoters, the dose-dependence of inducer aTc was further determined between 1 and 10 μg/mL for this promoter. Also, the induction fold of this promoter can reach about 230 (±55) fold in 72 hours when cells of *Synechocystis* are in the light-activated heterotrophic growth (LAHG) mode and activated with white light (Figure 2). Therefore, we propose that the L03 promoter is widely regulated in *Synechocystis.*

The regulation of the L03 promoter in *Synechocystis* under different physiological conditions

To examine how the environmental conditions affect the regulation of L03 promoter in cells of *Synechocystis*, the combinations of two glucose concentrations (0 and 5 mM), four aTc concentrations (0,10^2,10^3,10^4 ng/mL), three light conditions (darkness, 20 μmol m^{-2} s^{-2} red light, and 30 μmol m^{-2} s^{-2} white light) were varied and the data

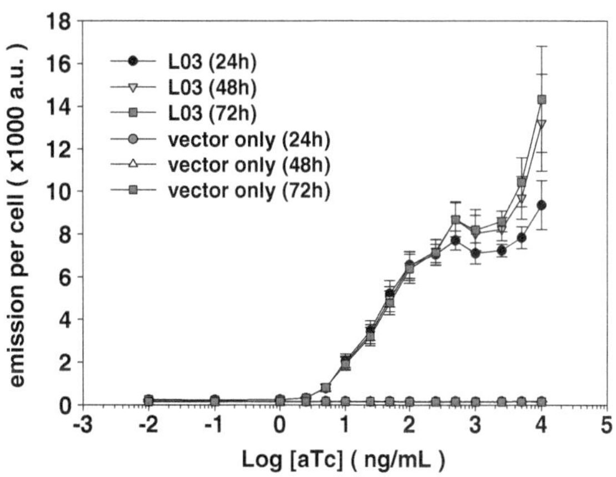

Figure 2 Dose-dependence of the inducer anhydrotetracycline (aTc) on the induction of L03 promoter in cells of *Synechocystis* sp. strain ATCC27184 (i.e. glucose-tolerant *Synechocystis* sp. strain PCC 6803) under the LAHG growth condition. The data were measured by flow cytometer to collect 50,000 events from each of three biological samples. The points in the plot represent mean ± s.e.m.a.u., arbitrary unit.

collected on three time points (24, 48, and 72 hours after treatment) (Figure 3). On a time point in a light condition, in common, EYFP expression level is higher when induced by a higher aTc concentration in the presence of glucose. However, when treated with the consistent glucose and aTc concentrations, the trend of EYFP expression with time is increasing in darkness but is decreasing under light. Furthermore, the trend decreases faster in white light than red light. This is due to the photolability of aTc and the overlapping of aTc absorption spectrum with the spectrum of photons emitted from white LEDs (Additional file 1: Figure S2). The photolability of the inducer aTc does not diminish the significance of functional L promoters for photosynthetic microorganisms. In the present study, the inducer serves as an easy-to-use ligand to develop TetR-regulated L promoters. Depending on application and purpose alternative molecules, not using aTc, to induce TetR-regulated L promoters under light are proposed in the discussion. To conclude, for the L03 promoter, the best induction in darkness is 239 (±16) fold in 24 hours under the LAHG condition with white light and the best induction in red light is 290 (±93) fold in 48 hours under photoautotrophic condition when treated with 10 μg/mL aTc.

The spontaneous thermal openings of an engineered promoter simulated by the PBD model

To understand how the thermal opening probability of a pre-melting DNA strand changes after the substitution of bases on indicated positions of a promoter, this was analyzed theoretically with the well-established Peyrard-Bishop-Dauxois (PBD) model via a Monte Carlo method [26]. Due to the sequence-independent stacking term using an averaged constant in this model, 16 L promoters can be only analyzed in four patterns categorized by hydrogen bonds of a modified base pair (Table 1).

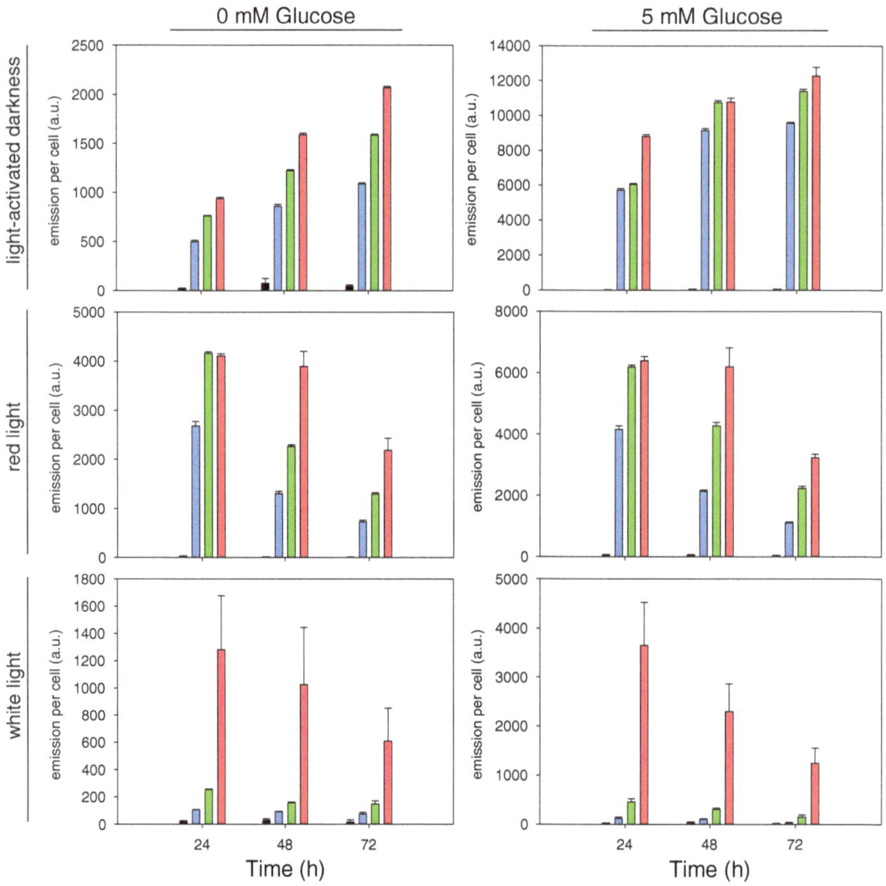

Figure 3 The L03 promoter strength measured in *Synechocystis* sp. strain ATCC27184 (i.e. glucose-tolerant *Synechocystis* sp. strain PCC 6803) cells. Cells were sampled on 24, 48, and 72 hours after induced with 0 (black bar), 10^2 (blue bar), 10^3 (green bar), and 10^4 (magenta bar) ng/mL aTc and grown under different light conditions: light-activated darkness (upper panel), red light (middle panel), and white light (lower panel). The *Synechocystis* cells were grown in kanamycin-added BG11 medium supplemented with 0 mM (left panel) or 5 mM (right panel) glucose. The EYFP emission per cell has been subtracted with auto-fluorescence of a cell containing pPMQAK1 plasmid only. a.u., arbitrary unit. The detail induction data can be found in Additional file 1: Table S2.

The simulation was successfully repeated in the present study at 303 K by showing the correspondence between the opening probability peak and the cognate site of a transcriptional regulator on corresponding positions: the two *tetO2*-operator cores (Figure 1) at the respective regions from −52 and −40 and from −27 to −15 (Figure 4A, 4B); the *lacO1* operator in the region from +2 to +22 (pattern G in Figure 4C).

For an L promoter, a locally higher opening probability peak was observed in the region between −27 and +3, where locates the core of *tetO2* operator in the spacer region, the −10 element, and a 5-nt region. But, for the P_{trc1O} promoter, the peak was observed in the region between −34 and −29, where locates the −35 element. For the R40 promoter, it showed a lack of a clear peak around the TSS. The consensus −10 element significantly increases the opening probability in the region between −27 to +3 when comparing the pattern C and I from the promoter L12 and R40 (Figure 4A) and when comparing pattern E and F from the promoter L21 and L22 (Figure 4B), respectively.

In the region between −13 and −4, the opening probability from high to low correlates with the order and number of hydrogen bonds as (2,2), (2,3), (3,2), and (3,3) of the substituted base pairs at position −6 and −5, respectively, when observing in the direction of strand separation [27] propagating from −10 element to TSS (Figure 4A). A thermal opening pattern of the promoters showing a higher averaged value in strength (Table 1) has lower opening probability in the region between +2 and +4 (Figure 4A). Similarly, the weak L31 promoter shows higher opening probability in this region when comparing to the pattern A and C of strong L promoters (Figure 4C). From the pattern G of the strong P_{trc1O} promoter, it showed higher opening probability in the flanking region of the −35 element but not of the −10 element (Figure 4C).

Evaluation of the engineered L promoters

Promoter strength is accessed by the amount of transcripts transcribed by a promoter: more transcripts mean a stronger promoter and *vice versa*. In the present work, the transcript levels are measured indirectly from the emission level of the fluorescent protein EYFP. Due to the standardized design with identical transcriptional terminators, ribosome binding sites, and stop codons arranged uniformly on the plasmid, the ratio of emission data may represent the ratio of transcript levels when comparing an engineered L promoter to the *rnpB* promoter (i.e. the relative promoter strength) under the LAHG growth condition (Table 1). The corresponding histogram of the relative promoter strengths is shown in Figure 5A. Similarly, using original data in a recent publication by Mitschke *et al.* [25] for native type I promoters and native promoters of the protein-coding genes in cells grown under photoautotrophic condition are shown in Figure 5B and Figure 5C, respectively. In the respective growth condition, the engineered L promoters, after induction, showed wider distribution in the relative promoter strength while the native promoters centered around 2-fold relative promoter strength. Interestingly, the native *rnpB* promoter, known to be unaffected by e.g. light/dark transitions, and redox perturbation [28-30], showed a consistent detection-threshold-to-promoter-strength ratio around 0.024 (±0.05) from the two totally different methods (present study and [25]). In addition, the analysis strongly indicates fully repressed states in both the L12 and L22 promoters. However, the L22 promoter has considerable range of induction whereas the L12 promoter does not (Table 1).

Discussion

In the transcriptional initiation, a pre-melting promoter DNA undergoes sequential intermediate steps with the

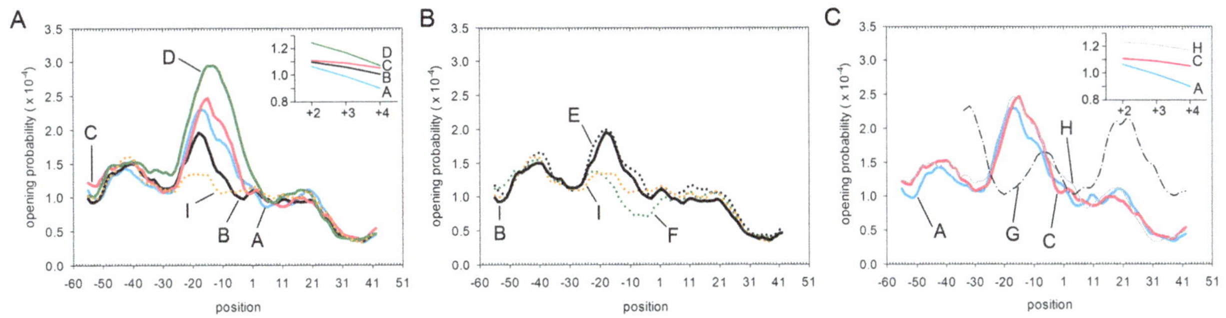

Figure 4 Spontaneous thermal opening probability simulated at 303 K by the average stacking PBD model [26] of a promoter sequence. (**A**) The pattern I (orange dot line) of the R40 promoter is compared to the four patterns, A, B, C and D (cyan line, black line, magenta line, and green line, respectively) simulated from 16 R40 -derived promoters (from L01 to L16). The grouping is shown (Table 1). (**B**) The pattern E (black dot line) and F (green dot line) of L21 and L22 promoters, respectively, are compared to the pattern B (black line) and the pattern I (orange dot line). (**C**) The patter G (black dash-dot line) and H (gray line) of P_{trc1O} and L31 promoters, respectively, are compared to pattern A (cyan line) and C (magenta line). Position +1 is the TSS of a given promoter. The opening probability at a position is read as the probability of a base pair to open at that position and its consecutive downstream 5 base pairs with the amplitude larger than 2.8 Å.

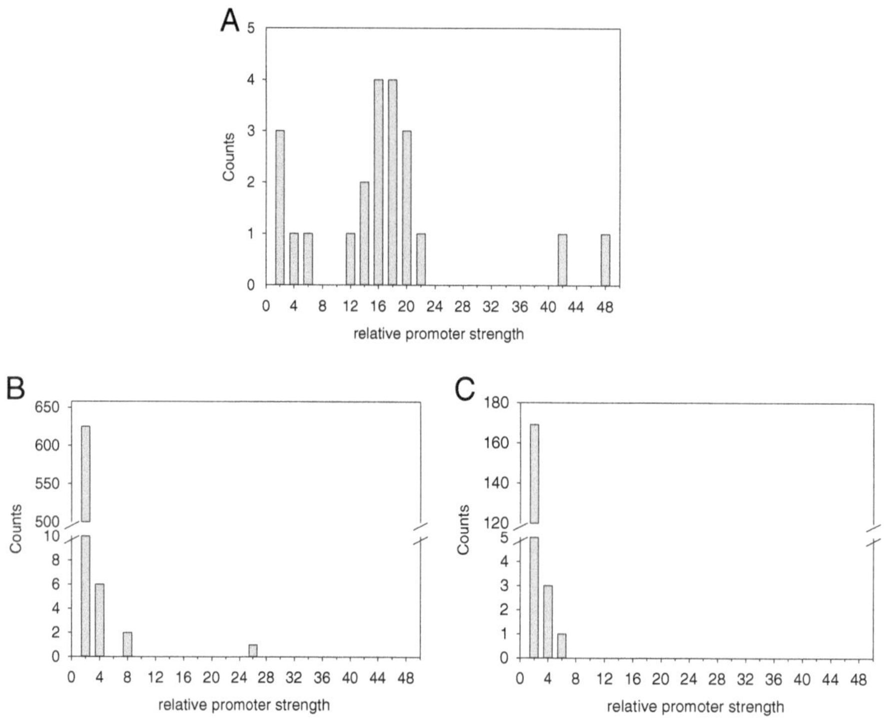

Figure 5 Evaluation of the engineered promoters (A) and the native promoters (B and C) by the relative promoter strength in *Synechocystis* PCC 6803 under the respective growth condition. (**A**) The histogram of the relative promoter strength by comparing 22 engineered promoters to the native *rnpB* promoter. The original data in EYFP emission of EYFP reporter proteins per cell are from the present work. (**B**) The histogram of the relative promoter strength by comparing 634 native type I promoters to the native non-type-I *rnpB* promoter. The original data in pyrosequencing-reads of RNA transcripts are from the supplementary Table three in the work of Mitschke *et al.* [25]. (**C**) The histogram of the relative promoter strength by comparing 173 native promoters, which transcribe protein-coding genes in high level, to the native *rnpB* promoter. The original data in pyrosequencing-reads of RNA transcripts are from the supplementary Table five in the work of Mitschke *et al.* [25].

RNA polymerase (RNAP) to form a stable open complex; RNAP binding, formation of close complex RP_c, formation of the bent and wrapped close complex I_1, formation of initial open complex I_2, and, finally formation of a stable open complex RP_o [9,31]. After the formation of RP_o, the RNAP proceeds to the transcription initiation through the DNA scrunching mechanism in which an obligatory stress intermediate forming as an extra unwinding DNA has been proposed to provide the driving force in promoter escape [32]. The abortive/productive ratio of transcription initiation may be influenced by the three competitive pathways; abortive cycle, scrunching pathway, and promoter escape–analyzed in the kinetic model of transcription initiation [33]. The RNAP escapes the promoter and enters the transcription elongation and termination [19]. In the present work, the process after the promoter escape should both be standardized in respect to the design of the construct and be well-controlled during the experiment. Therefore, the effect of the modified promoter sequence on the intrinsic thermal properties of its pre-melting state and on the productive transcriptional initiation is discussed by the simulated spontaneous thermal opening probability and the measured promoter strength, respectively.

Changing to the consensus −10 element TATAAT of *Synechocystis* clearly elevates the spontaneous thermal opening probability in the region between −27 and +3 (Figure 4B) and increases promoter strength significantly (Table 1) when comparing the L21 to the L22 promoter. A pre-melting promoter DNA with the elevated spontaneous opening probability may be more prone to expose the cognate bases for contacting with a DNA-binding protein such as σ subunit of RNAP and thus to increase the binding probability [26]. The recognition of the −10 element by σ subunit of RNAP has been proposed through the capture of two highly-conserved flipped-out bases and the extensive contact with the sugar-phosphate backbone of each nucleotide in the −10 element on the non-template strand; the recognition is coupled to strand separation and is the key step in initiating transcription [27]. Therefore, the increased L21 promoter strength may be due to a more favourable σ subunit binding compared to the L22 promoter. This is supported by similar result when comparing the mutated *lac* UV5 promoter with the wide-type *lac* promoter [26].

Changing to the consensus −10 element increases the L12 promoter's opening probability (Figure 4A) but reduces its promoter strength (Table 1) when compared to the R40 promoter. The RNAP binding on the L12 promoter might be still enhanced based on the simulated result but the transcription initiation after formation of the close complex probably was impeded [31]. Through systematic modifications in the region between the −10 element and TSS, according to our hypothesis, an essential guanine at the position 2 nt downstream from the −10 element of L15, L07, L03, and L11 promoters was observed and represents a novel discovery for cyanobacteria. Interestingly, the results are consistent with an *in vitro* study of λP_L, λP_R, *rrnB* P1 and *rrnB* P2 promoters with RNAP from *Escherichia coli* demonstrating that the base identity at this position affects the life time of the promoter-RNAP complex after formation of the close complex [21] through a non-optimal contact with the region 1.2 of the σ^{70} subunit [22]. During the formation of the open complex it has been also proposed that the base identity at this position may serve as a "check point" in the propagating process of strand separation toward the TSS [27]. During the DNA scrunching, the position at 2 nt downstream of the −10 element locates within the extra unwinding DNA turn; substituting the base identity at this position might in consequence affect the promoter escape. Taken together, we speculate that substituting the base identity in the region between the −10 element and the TSS has the effect in fine-tuning the binding probability of the pre-melting DNA to the RNAP and has effect in the propagation of strand separation, the stability of open complex, and the balance in the three competitive kinetic pathways during transcription initiation. Therefore, any removal of the deleterious effect of a fully consensus sequence causes the improved promoter strength in the engineered L promoters other than the L12 promoter in *Synechocystis.*

Our attempt to fully repress the L09 promoter by increasing TetR repressors with the noLVA_L09 construct did not succeed. Based on i) the approximate 30 bp distance between the modified bases and a *tetO2* operator upstream of the −35 element of the L09 promoter and that ii) only the repressed L09 promoter showed leakage in gene expression, we speculate that the leakage of the repressed L09 promoter was caused by a so called "long-range effect of flanking single SNPs" [34] on the *tetO2* operator.

Because the L03 promoter shows widest dynamic range, it was chosen to study the effect of the environmental conditions such as glucose and light quality/quantity on the TetR-regulation of an L promoter. The beneficial effect of glucose on gene expression controlled by the L03 promoter may result from the metabolic balance between the decreased metabolites for the Calvin cycle and the increased metabolites for the oxidative pentose phosphate pathway and glycolysis [35,36]. The light quality and light quantity affect the light-sensitive inducer aTc differently in the TetR-regulation on the L promoter. The regulationin terms of induction fold and duration time is best in darkness, followed by in red light, and in white light. Since white light exists naturally,using this inducer might limit the applications with L promoters in photosynthetic microorganisms. However, thanks to the versatile TetRsystem, the TetR repressor can be induced by e.g. an RNA aptamer [37] and the short peptide TIP [38], which should be light-insensitive molecules.

Based on our results, a simple feed-forward loop for a lasting induction of TetR repressor is proposed (Additional file 1: Figure S3). This regulation initially still needs aTc to trigger the induction of the tight TetR-regulated L22 promoter, but when aTc loses its effect to induce in white light, the induction of TetR repressor could be over-taken by e.g. the presence of the RNA aptamer or the short peptide TIP.

In order to expand the dynamic range of induction, the narrow dynamic range of the tightly regulated L22 promoter can be indirectly amplified using the T7 RNAP and its cognate T7 promoters [39,40]. To reverse the regulation on L promoter, the revTetR repressor recognizes the same cognate site *tetO2* as the TetR repressor and the role of aTc changes from inducer to co-repressor [41]. It is also possible to further incorporate other regulatory mechanisms on RNA level, such as the orthogonal riboswitch [42], genetic switchboard [43], or RNA processing [44], and on protein level such as the post-translation modifications [45] that may lead to the design and construction of a programmable genetic circuit in *Synechocystis.* The TetR-regulated promoter library developed for the cyanobacterium *Synechocystis* may be combined with a systematic promoter selection method [46] to further design and construct the genetic circuit for robust control in gene expression.

Conclusions

The novel non-native, wide-dynamic-range promoters developed in the present work demonstrates a significant step forward towards versatile purpose-driven applications with cyanobacteria in the framework of synthetic biology and metabolic engineering.

Methods

Strain and growth condition

Cells of the unicellular cyanobacterium *Synechocystis* sp. strain ATCC27184 (i.e. glucose-tolerant *Synechocystis* sp. strain PCC 6803) were grown in $BG11_0$ medium [47] supplemented with 18 mM sodium nitrate under light condition indicated and agitated with horizontal orbiting at 120 rpm. *Escherichia coli* DH5α grew in LB medium at 37°C and agitated with horizontal orbiting at 250 rpm. According to plasmid used, the final concentration of antibiotic in the medium was 50 μg/mL (kanamycin), 35 μg/mL (chloramphenicol), or 100 μg/mL (ampicillin).

Construction of promoter variants by PCR

TetR-regulated promoter variants of R40 promoter (i.e. BBa_R0040) were prepared by a PCR-based method with the universal forward primer XhoI_f_pSB2K3 and a respective reverse primer as listed in Additional file 1: Table S1. The reverse primer contains bases of the new promoter and restriction sites for further cloning. The approximate 1 kb PCR product was amplified from the BBa_pSB2K3 plasmid containing only the R40 promoter located in between the BioBrick prefix and suffix by using Phusion® Hot Start High-Fidelity DNA Polymerase (Finnzymes). After amplification, DNA polymerase was removed by GeneJET™ PCR Purification Kit and the purified PCR product was digested simultaneously with FastDigest® DpnI, XbaI and PstI. The DpnI digestion degrades the methylated template DNA, and the XbaI and PstI digestion enables the new promoter to be further cloned. The restriction enzymes were removed again with GeneJET™ PCR Purification Kit. With no need to retrieve the generated ~ 80 bp XbaI-PstI-digested DNA fragments by gel electrophoresis, the new promoter can be further cloned directly into the reporter construct with subsequent "3A assembly" method (http://partsregistry.org/Help:Assembly).

Construction of reporter and control constructs by BioBrick standard assembly and 3A assembly

The complete reporter construct [BBa_J23101]-[BBa_P0440]-[new promoter]-[RBS*]-[BBa_E0130] was prepared with high copy-number BioBrick plasmids and then cloned the whole construct into the low copy-number shuttle plasmid pPMQAK1. In step one, the TetR repressor expressing device [BBa_J23101]-[BBa_P0440] and the EYFP reporter expressing device without a promoter [RBS*]-[BBa_E0130] were prepared in parallel by the "standard assembly" method (http://partsregistry.org/Help:Assembly) and resulted in BBa_pSB1A2 plasmids. The ribosome binding site, RBS*, was previously developed in our lab [48]. In step two, to avoid potential cloning problems, the new promoter was first assembled in downstream of the TetR device with the "3A assembly" method resulting [BBa_J23101]-[BBa_P0440]-[new promoter] in the BBa_pSB1K3 plasmid. In step three, using the "3A assembly" method again, the complete reporter construct was finished in the BBa_pSB1C3 plasmid. Meanwhile, for making the control constructs, BBa_J23101, and *rnpB* promoter, respectively, were assembled in upstream of [RBS*]-[BBa_E0130] by the "3A assembly" method resulting in the BBa_pSB1C3 plasmid. In final step, the completed reporter and control constructs were cloned, respectively, into the pPMQAK1 with the restriction sites EcoRI and PstI.

LVA tag removal from the TetR repressor

The TetR repressor encoded in BBa_P0440 is tailed with a protease tag LVA for rapid degradation. This LVA tag can be removed by site-directed mutagenesis [49] introducing a double stop codon TAATAA in the 3′-end of *tetR* gene. The primers for site-directed mutagenesis are listed in Additional file 1: Table S1.

Tri-parental mating

The shuttle vector pPMQAK1 bearing different constructs was transferred into *Synechocystis* cells by tri-parental mating as described previously [11].

Anhydrotetracycline (aTc) induction in different growth modes of *Synechocystis* cells

To reduce operating error, a master *Synechocystis* cell culture was inoculated to have an absorption of 0.03 at 750 nm measured with a Plate Chameleon V Microplate Reader (Hidex) in BG11 medium with kanamycin supplemented either with or without 5 mM glucose. The master culture was distributed in aliquot to 3.5 mL each in culture wells of a 6-well tissue culture plate (SARSTEDT, 1.83.1839). The plates were agitated with horizontal orbiting at 120 rpm under 30 μmol photons m^{-2} s^{-1} of white light for 24 hours before aTc induction. For aTc induction, 3.5 μL 1000-fold stock of aTc was respectively added to the 3.5 mL cultures to reach the following final concentrations (ng/mL): 10^{-2}, 10^{-1}, 10^{0}, 2.5, 5, 10^{1}, 2.5x 10^{1}, 5x 10^{1}, 10^{2}, 2.5x 10^{2}, 5x 10^{2}, 10^{3}, 2.5x 10^{3}, 5x 10^{3}, 10^{4}, and 10^{5}. Since the aTc stock was prepared in 50% (v/v) ethanol, for 0 ng/mL aTc group, 3.5 μL 50% ethanol was added into the cell culture. After aTc induction, the cultures were exposed to either 30 μmol photons m^{-2} s^{-1} of white light, 20 μmol photons m^{-2} s^{-1} of red light, or in darkness. The reduced photon density of the red light aims for less negative effect on the aTc stability compared to in white light. The light source is a home-built LED light panel: 12 modules arranged in a 2-by-6 array and each module arranging 9 LEDs evenly distributed above each well of a 6-well plate. The spectrum of a LED is shown (Additional file 1: Figure S2) and the photon density was measured using a light sensor (Skye instruments) placed between the LEDs and the surface of the cell cultures. The cultures under different light conditions were stacked on the same shaker for identical 120 rpm agitation. The cultures were sampled every 24 hours by withdrawing 50 μL culture volume and diluted it with 950 μL fresh medium for determining specific growth rate and subsequent flow cytometry analysis. When sampling every

24 hours under the room light, it provides the white-light -activation condition for the culture growing in darkness as the LAHG growth mode [50].

Transcription start site determination

The total RNA in cells of *Synechocystis* was prepared by a PGTX-based method [51]. The transcription start site was determined by 5′ RACE System for Rapid Amplification of cDNA Ends kit (Invitrogen, 18374–058) using GSP1, GSP2 and nested GSP primers as detailed in Additional file 1: Table S1.

Flow cytometry

The emission from reporter EYFP expressed in a single *Synechocystis* cell was measured by BD™ LSRII flow cytometer with BD FACSDiva™ Software. *Synechocystis* cells harboring pPMQAK1 only were used as negative control and cells bearing reporter construct of P_{trc1O} promoter were used as positive control. The cytometer settings were: 786 volt for FSC, 412 volt for SSC, 676 volt for Alexa Fluor 488, 624 volt for PerCP-Cy5-5 and the laser is at 488 nm. In total 50,000 events were collected. The data analysis was done with FlowJo 7.6.5 software (©Tree Star, Inc.). First, a singlet cell population was gated through the plots of FSC-W against FSC-H, SSC-W against SSC-H and the histogram of PerCP-Cy5 -5 channel for auto-fluorescence from Chlorophyll a and it is usually ~90% of total population. Second, the mean and standard deviation of EYFP emission per cell from singlet cell population were obtained from a histogram of the Alexa Fluor 488 channel.

Statistical evaluation

The difference between the promoter strengths indicated, reported as fold (±s.e.m.) were tested with the unpaired t-test with one-tail ($\alpha = 0.05$) and $n > 45{,}000$. Indicated comparisons are all significantly different.

The simulation of spontaneous thermal openings of DNA strand

All parameter values used and calculation of thermal opening probability are the same as the Peyrard-Bishop-Dauxois model simulated with Monte Carlo method [26], except the use of temperature 303 K in present study. The potential energy sum of the hydrogen bonding of a base pair and the π-stacking interaction between two adjacent base pairs was simulated to identify the equilibrated configurations of a dsDNA (double-stranded DNA) fluctuating at the temperature 303 K. The spontaneous thermal opening probability was averaged over 2,000 realizations. The simulation algorithm was programed with the Matlab. The 2,000 different seeds of the random number generator, Mersenne Twister, were used for generating pseudorandom numbers in the respective 2,000 realizations of a promoter sequence; keep using this set of seeds to analyze other promoter sequences for comparisons. The simulation was accomplished by submitting the parallel distributed jobs to the cluster computing resource in Uppsala University.

Additional file

Additional file 1: Supplementary Data for Wide-Dynamic-Range Promoters Engineered for Cyanobacteria. Contains the table of primers used in the study, the table of measured values in Figure 3, the figure of the weighted sequence logo, the figure of spectra comparison, and the figure of proposed feed-forward loop.

Competing interests

The authors declare that they have no competing interests.

Authors' contributions

HHH and PL conceived of the research; HHH designed the study and performed the experiment; HHH and PL analyzed the data and wrote the manuscript. Both authors read and approved the final manuscript.

Acknowledgements

We would like to acknowledge the support from the Science for Life Laboratory BioVis Technology Platform (Uppsala University) for the flow cytometry measurements and the support from the Swedish National Infrastructure for Computing (SNIC) through Uppsala Multidisciplinary Center for Advanced Computational Science (UPPMAX) for the cluster computing resource (projects p2012093 and s00112-183). Our research received funding from the Swedish Energy Agency; the Knut and Alice Wallenberg Foundation (project MoSE); and the European Union Seventh Framework Programme (FP7/2007-2013) under grant agreement number 308518 (CyanoFactory).

References

1. Eriksson J, Salih GF, Ghebramedhin H, Jansson C: **Deletion mutagenesis of the 5′ psbA2 region in Synechocystis 6803: identification of a putative cis element involved in photoregulation.** *Mol Cell Biol Res Commun* 2000, **3**:292–298.
2. Muramatsu M, Hihara Y: **Coordinated high-light response of genes encoding subunits of photosystem I is achieved by AT-rich upstream sequences in the cyanobacterium Synechocystis sp. strain PCC 6803.** *J Bacteriol* 2007, **189**:2750–2758.
3. Mazouni K, Bulteau S, Cassier-Chauvat C, Chauvat F: **Promoter element spacing controls basal expression and light inducibility of the cyanobacterial secA gene.** *Mol Microbiol* 1998, **30**:1113–1122.
4. Imamura S, Asayama M, Shirai M: **In vitro transcription analysis by reconstituted cyanobacterial RNA polymerase: roles of group 1 and 2 sigma factors and a core subunit, RpoC2.** *Genes Cells* 2004, **9**:1175–1187.
5. Ivanikova NV, McKay RML, Bullerjahn GS: **Construction and characterization of a cyanobacterial bioreporter capable of assessing nitrate assimilatory capacity in freshwaters.** *Limnol Oceanogr: Methods* 2005, **3**:86–93.
6. Briggs LM, Pecoraro VL, McIntosh L: **Copper-induced expression, cloning, and regulatory studies of the plastocyanin gene from the cyanobacterium Synechocystis sp. PCC 6803.** *Plant Mol Biol* 1990, **15**:633–642.
7. Peca L, Kos PB, Mate Z, Farsang A, Vass I: **Construction of bioluminescent cyanobacterial reporter strains for detection of nickel, cobalt and zinc.** *FEMS Microbiol Lett* 2008, **289**:258–264.
8. Imamura S, Asayama M: **Sigma factors for cyanobacterial transcription.** *Gene Regul Syst Bio* 2009, **3**:65–87.
9. Hook-Barnard IG, Hinton DM: **Transcription initiation by mix and match elements: flexibility for polymerase binding to bacterial promoters.** *Gene Regul Syst Bio* 2007, **1**:275–293.

10. Nandagopal N, Elowitz MB: **Synthetic biology: integrated gene circuits.** *Science* 2011, **333**:1244–1248.
11. Huang HH, Camsund D, Lindblad P, Heidorn T: **Design and characterization of molecular tools for a Synthetic Biology approach towards developing cyanobacterial biotechnology.** *Nucleic Acids Res* 2010, **38**:2577–2593.
12. Nair U, Ditty JL, Min H, Golden SS: **Roles for sigma factors in global circadian regulation of the cyanobacterial genome.** *J Bacteriol* 2002, **184**:3530–3538.
13. Mutsuda M, Michel KP, Zhang X, Montgomery BL, Golden SS: **Biochemical properties of CikA, an unusual phytochrome-like histidine protein kinase that resets the circadian clock in Synechococcus elongatus PCC 7942.** *J Biol Chem* 2003, **278**:19102–19110.
14. Ng WO, Zentella R, Wang Y, Taylor JS, Pakrasi HB: **PhrA, the major photoreactivating factor in the cyanobacterium Synechocystis sp. strain PCC 6803 codes for a cyclobutane-pyrimidine-dimer-specific DNA photolyase.** *Arch Microbiol* 2000, **173**:412–417.
15. Kim EJ, Kim JS, Rhee HJ, Lee JK: **Growth arrest of Synechocystis sp. PCC6803 by superoxide generated from heterologously expressed Rhodobacter sphaeroides chlorophyllide a reductase.** *FEBS Lett* 2009, **583**:219–223.
16. Weyman PD, Vargas WA, Tong Y, Yu J, Maness PC, Smith HO, Xu Q: **Heterologous expression of Alteromonas macleodii and Thiocapsa roseopersicina [NiFe] hydrogenases in Synechococcus elongatus.** *PLoS One* 2011, **6**:e20126.
17. Bertram R, Hillen W: **The application of Tet repressor in prokaryotic gene regulation and expression.** *Microb Biotechnol* 2008, **1**:2–16.
18. Browning DF, Busby SJ: **The regulation of bacterial transcription initiation.** *Nat Rev Microbiol* 2004, **2**:57–65.
19. Herbert KM, Greenleaf WJ, Block SM: **Single-molecule studies of RNA polymerase: motoring along.** *Annu Rev Biochem* 2008, **77**:149–176.
20. Kincade JM, de Haseth PL: **Bacteriophage lambda promoters pL and pR: sequence determinants of in vivo activity and of sensitivity to the DNA gyrase inhibitor, coumermycin.** *Gene* 1991, **97**:7–12.
21. Haugen SP, Berkmen MB, Ross W, Gaal T, Ward C, Gourse RL: **rRNA promoter regulation by nonoptimal binding of sigma region 1.2: an additional recognition element for RNA polymerase.** *Cell* 2006, **125**:1069–1082.
22. Haugen SP, Ross W, Manrique M, Gourse RL: **Fine structure of the promoter-sigma region 1.2 interaction.** *Proc Natl Acad Sci USA* 2008, **105**:3292–3297.
23. Feklistov A, Barinova N, Sevostyanova A, Heyduk E, Bass I, Vvedenskaya I, Kuznedelov K, Merkiene E, Stavrovskaya E, Klimasauskas S, *et al*: **A basal promoter element recognized by free RNA polymerase sigma subunit determines promoter recognition by RNA polymerase holoenzyme.** *Mol Cell* 2006, **23**:97–107.
24. Barinova N, Kuznedelov K, Severinov K, Kulbachinskiy A: **Structural modules of RNA polymerase required for transcription from promoters containing downstream basal promoter element GGGA.** *J Biol Chem* 2008, **283**:22482–22489.
25. Mitschke J, Georg J, Scholz I, Sharma CM, Dienst D, Bantscheff J, Voss B, Steglich C, Wilde A, Vogel J, Hess WR: **An experimentally anchored map of transcriptional start sites in the model cyanobacterium Synechocystis sp. PCC6803.** *Proc Natl Acad Sci USA* 2011, **108**:2124–2129.
26. Apostolaki A, Kalosakas G: **Targets of DNA-binding proteins in bacterial promoter regions present enhanced probabilities for spontaneous thermal openings.** *Phys Biol* 2011, **8**:026006.
27. Feklistov A, Darst SA: **Structural basis for promoter-10 element recognition by the bacterial RNA polymerase sigma subunit.** *Cell* 2011, **147**:1257–1269.
28. Alfonso M, Perewoska I, Kirilovsky D: **Redox control of psbA gene expression in the cyanobacterium Synechocystis PCC 6803. Involvement of the cytochrome b(6)/f complex.** *Plant Physiol* 2000, **122**:505–516.
29. Alfonso M, Perewoska I, Kirilovsky D: **Redox control of ntcA gene expression in Synechocystis sp. PCC 6803. Nitrogen availability and electron transport regulate the levels of the NtcA protein.** *Plant Physiol* 2001, **125**:969–981.
30. Nagarajan S, Sherman DM, Shaw I, Sherman LA: **Functions of the duplicated hik31 operons in central metabolism and responses to light, dark, and carbon sources in Synechocystis sp. strain PCC 6803.** *J Bacteriol* 2012, **194**:448–459.
31. Gries TJ, Kontur WS, Capp MW, Saecker RM, Record MT Jr: **One-step DNA melting in the RNA polymerase cleft opens the initiation bubble to form an unstable open complex.** *Proc Natl Acad Sci USA* 2010, **107**:10418–10423.
32. Revyakin A, Liu C, Ebright RH, Strick TR: **Abortive initiation and productive initiation by RNA polymerase involve DNA scrunching.** *Science* 2006, **314**:1139–1143.
33. Xue XC, Liu F, Ou-Yang ZC: **A kinetic model of transcription initiation by RNA polymerase.** *J Mol Biol* 2008, **378**:520–529.
34. Alexandrov BS, Fukuyo Y, Lange M, Horikoshi N, Gelev V, Rasmussen KO, Bishop AR, Usheva A: **DNA breathing dynamics distinguish binding from nonbinding consensus sites for transcription factor YY1 in cells.** *Nucleic Acids Res* 2012, **40**:10116–10123.
35. Takahashi H, Uchimiya H, Hihara Y: **Difference in metabolite levels between photoautotrophic and photomixotrophic cultures of Synechocystis sp. PCC 6803 examined by capillary electrophoresis electrospray ionization mass spectrometry.** *J Exp Bot* 2008, **59**:3009–3018.
36. Knoop H, Zilliges Y, Lockau W, Steuer R: **The metabolic network of Synechocystis sp. PCC 6803: systemic properties of autotrophic growth.** *Plant Physiol* 2010, **154**:410–422.
37. Steber M, Arora A, Hofmann J, Brutschy B, Suess B: **Mechanistic basis for RNA aptamer-based induction of TetR.** *Chembiochem* 2011, **12**:2608–2614.
38. Goeke D, Kaspar D, Stoeckle C, Grubmuller S, Berens C, Klotzsche M, Hillen W: **Short peptides act as inducers, anti-inducers and corepressors of Tet repressor.** *J Mol Biol* 2012, **416**:33–45.
39. Wolk CP, Elhai J, Kuritz T, Holland D: **Amplified expression of a transcriptional pattern formed during development of Anabaena.** *Mol Microbiol* 1993, **7**:441–445.
40. Temme K, Hill R, Segall-Shapiro TH, Moser F, Voigt CA: **Modular control of multiple pathways using engineered orthogonal T7 polymerases.** *Nucleic Acids Res* 2012, **40**:8773–8781.
41. Kamionka A, Bogdanska-Urbaniak J, Scholz O, Hillen W: **Two mutations in the tetracycline repressor change the inducer anhydrotetracycline to a corepressor.** *Nucleic Acids Res* 2004, **32**:842–847.
42. Dixon N, Robinson CJ, Geerlings T, Duncan JN, Drummond SP, Micklefield J: **Orthogonal riboswitches for tuneable coexpression in bacteria.** *Angew Chem Int Ed Engl* 2012, **51**:3620–3624.
43. Callura JM, Cantor CR, Collins JJ: **Genetic switchboard for synthetic biology applications.** *Proc Natl Acad Sci USA* 2012, **109**:5850–5855.
44. Qi L, Haurwitz RE, Shao W, Doudna JA, Arkin AP: **RNA processing enables predictable programming of gene expression.** *Nat Biotechnol* 2012, **30**:1002–1006.
45. Khalil AS, Collins JJ: **Synthetic biology: applications come of age.** *Nat Rev Genet* 2010, **11**:367–379.
46. Wu CH, Zhang W, Chen BS: **Multiobjective H2/Hinfinity synthetic gene network design based on promoter libraries.** *Math Biosci* 2011, **233**:111–125.
47. Stanier RY, Kunisawa R, Mandel M, Cohen-Bazire G: **Purification and properties of unicellular blue-green algae (order Chroococcales).** *Bacteriol Rev* 1971, **35**:171–205.
48. Heidorn T, Camsund D, Huang HH, Lindberg P, Oliveira P, Stensjo K, Lindblad P: **Synthetic biology in cyanobacteria engineering and analyzing novel functions.** *Methods Enzymol* 2011, **497**:539–579.
49. Fisher CL, Pei GK: **Modification of a PCR-based site-directed mutagenesis method.** *Biotechniques* 1997, **23**:570–571. 574.
50. Anderson SL, McIntosh L: **Light-activated heterotrophic growth of the cyanobacterium Synechocystis sp. strain PCC 6803: a blue-light-requiring process.** *J Bacteriol* 1991, **173**:2761–2767.
51. Pinto FL, Thapper A, Sontheim W, Lindblad P: **Analysis of current and alternative phenol based RNA extraction methodologies for cyanobacteria.** *BMC Mol Biol* 2009, **10**:79.

Dual effects of single-walled carbon nanotubes coupled with near-infrared radiation on *Bacillus anthracis* spores: inactivates spores and stimulates the germination of surviving spores

Xiuli Dong[1], Yongan Tang[2], Marvin Wu[2], Branislav Vlahovic[2] and Liju Yang[1*]

Abstract

Background: *Bacillus anthracis* is a pathogen that causes life-threatening disease--anthrax. *B. anthracis* spores are highly resistant to extreme temperatures and harsh chemicals. Inactivation of *B. anthracis* spores is important to ensure the environmental safety and public health. The 2001 bioterrorism attack involving anthrax spores has brought acute public attention and triggered extensive research on inactivation of *B. anthracis* spores. Single-walled carbon nanotubes (SWCNTs) as a class of emerging nanomaterial have been reported as a strong antimicrobial agent. In addition, continuous near infrared (NIR) radiation on SWCNTs induces excessive local heating which can enhance SWCNTs' antimicrobial effect. In this study, we investigated the effects of SWCNTs coupled with NIR treatment on *Bacillus anthracis* spores.

Results and discussion: The results showed that the treatment of 10 μg/mL SWCNTs coupled with 20 min NIR significantly improved the antimicrobial effect by doubling the percentage of viable spore number reduction compared with SWCNTs alone treatment (88% vs. 42%). At the same time, SWCNTs-NIR treatment activated the germination of surviving spores and their dipicolinic acid (DPA) release during germination. The results suggested the dual effect of SWCNTs-NIR treatment on *B. anthracis* spores: enhanced the sporicidal effect and stimulated the germination of surviving spores. Molecular level examination showed that SWCNTs-NIR increased the expression levels (>2-fold) in 3 out of 6 germination related genes tested in this study, which was correlated to the activated germination and DPA release. SWCNTs-NIR treatment either induced or inhibited the expression of 3 regulatory genes detected in this study. When the NIR treatment time was 5 or 25 min, there were 3 out of 7 virulence related genes that showed significant decrease on expression levels (>2 fold decrease).

Conclusions: The results of this study demonstrated the dual effect of SWCNTs-NIR treatment on *B. anthracis* spores, which enhanced the sporicidal effect and stimulated the germination of surviving spores. SWCNTs-NIR treatment also altered the expression of germination, regulatory, and virulence-related genes in *B. anthracis*.

Keywords: Single walled carbon nanotubes, *Bacillus anthracis* spores, Near infrared radiation, Germination, Dipicolinic acid, Gene expression

* Correspondence: lyang@nccu.edu
[1]Biomanufacturing Research Institute and Technology Enterprise (BRITE) and Department of Pharmaceutical Sciences, North Carolina Central University, Durham, NC 27707, USA
Full list of author information is available at the end of the article

Background

Bacillus anthracis is a pathogen that causes life-threatening disease--anthrax. *B. anthracis* spores are highly resistant to extreme temperatures and harsh chemicals. The dormancy of spores can remain decades or even centuries until the opportunity for infection arises. Animals can be infected by spores through inhalation and grazing. Humans may develop cutaneous anthrax through the exposure of skin to infected animals, or ingestion anthrax through eating contaminated food, or inhalation anthrax by breathing in the spores or materials contaminated with the spores.

Inactivation of *B. anthracis* spores is important to ensure the environmental safety and public health. The 2001 bioterrorism attack involving anthrax spores have brought acute public attention and triggered extensive research on inactivation of *B. anthracis* spores. Common antimicrobial agents including bleach, chlorine dioxide, hydrogen peroxide, paraformaldehyde, were used in federal decontamination responses to the bioterrorism attack [1]. Whitney et al. reviewed much of the available literature on the inactivation of *Bacillus* spores that is relevant to the inactivation of *B. anthracis* [2]. However, because the efficiencies of many of these methods were determined using *Bacillus* species other than *B. anthracis,* a number of methods and reagents that reportedly killed *B. anthracis* spores suffered from low efficacy [2]. The high lethality, resistivity, aerosolization ability, and the potential to be a bioterrorism agent, have heightened the research interest in the development of new sporicidal agents/methods for inactivating *B. anthracis* spores.

Single-walled carbon nanotubes (SWCNTs) as a class of emerging nanomaterial have been reported as a strong antimicrobial agent. Kang et al. reported the first direct evidence of SWCNTs' strong antimicrobial activity to bacterial cells [3]. Since then, a number of studies have been reported on the antimicrobial effects of carbon nanotubes (CNTs) and the following related influence factors, including CNT diameter [4], length [5], aggregation [6], concentration [6,7], CNTs' surface groups, buffer solution, and bacterial species [7], contact time and intensity [6-8]. Dong et al. reported that SWCNTs might be an effective alternative to antibiotics in dealing with drug-resistant and multidrug-resistant bacterial strains because of the physical mode of bactericidal action that SWCNTs display [9]. Our most recent study showed that SWCNTs exhibited antimicrobial activity to *B. anthracis* vegetative cells at the similar level to other bacterial species [10]. The mechanisms of SWCNT's antimicrobial activity to bacterial cells depended essentially on the direct contact resulting in cell membrane damage [3,5,7], but also included oxidative stresses due to direct oxidation of cellular components, and secondary oxidation of cellular lipid bilayer by reactive oxygen species [4,11-14].

However, only a limited number of studies have been published regarding the interactions between SWCNTs and bacterial spores. Spores are structurally different from vegetative cells. *B. anthracis* spores are encased in a thick multilayered coat, which is further surrounded by the exosporium. The exosporium is composed of a paracrystalline basal layer with a hexagonal lattice structure and a hair-like outer region [15]. Monosaccharide -functionalized SWCNTs interacted with *B. anthracis* spores with the mediation of a divalent cation such as Ca^{2+} to form significant spore aggregation [16]. SWCNTs showed high adsorption affinity towards *B. subtilis* spores [17]. The use of aggregation formed between SWCNTs and spores was investigated for the removal of *Bacillus* spores from environment [17]. Notably, these studies did not mention the antimicrobial effect of SWCNTs on *Bacillus* spores. Our recent studies have initiated the study of SWCNTs' action on *B. anthracis* spores and found that SWCNTs did not exhibit antimicrobial effect to bacterial spores in deionized water suspensions at concentrations less than 150 μg/mL [10]. Considering the protective structure of bacterial spores and the mechanisms of SWCNTs inactivating bacterial cells, it is not surprising that SWCNTs showed little antimicrobial effect to *B. anthracis* spores.

Fortunately, besides its intrinsic antimicrobial activity, SWCNTs possess other properties that can be used in combination to enhance its antimicrobial activity. For example, SWCNTs show strong absorbance to 700- to 1100-nm NIR light and then convert it into heat, while biological systems are transparent at this wavelength range [18]. Continuous NIR radiation on SWCNT-internalized cells can cause cell death because of excessive local heating to SWCNTs [19]. Recent results with bacterial cells (*Salmonella*) showed that SWCNTs-NIR treatment resulted in a 3-fold decrease in viable cells compared to SWCNTs alone [20].

In this study, we investigated the antimicrobial effects of SWCNTs-NIR treatment on the inactivation of *B. anthracis* spores and the germination of surviving *B. anthracis* spores in comparison to SWCNTs alone, and examined the alterations at the molecular level upon the SWCNTs-NIR treatment to *B. anthracis* spores, including the transcriptional expression levels of germination-related genes, regulatory genes, and virulence related genes.

Results and discussion

The NIR absorbance characteristics of SWCNTs

We first examined the UV-VIS-NIR spectrum of SWCNTs, and Figure 1 shows the UV-VIS-NIR spectrum of 20 μg/mL SWCNTs in DI-H_2O. The maximal absorption peak was at 300 nm, which corresponded to the plasmon absorption of SWCNTs. The maximal absorption wavelength of different batches of SWCNTs products varied slightly, likely because of the differences in SWCNTs'

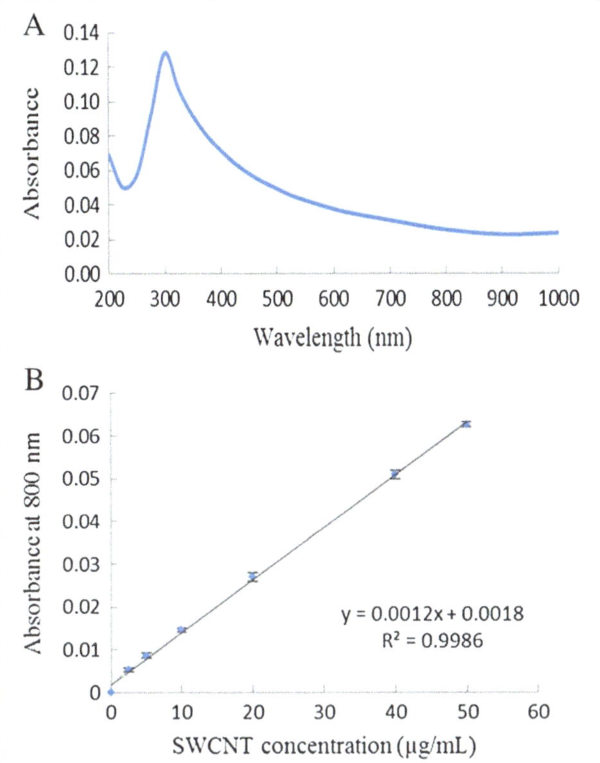

Figure 1 The optical spectrum of SWCNTs solution. (A) The UV-VIS-NIR spectrum of SWCNTs solution in DI-H_2O at a concentration of 20 μg/mL. **(B)** The linear relationship between the absorbance at 800 nm and the concentration of SWCNTs. Error bars in **(B)** are standard deviations of triplicated experimental results.

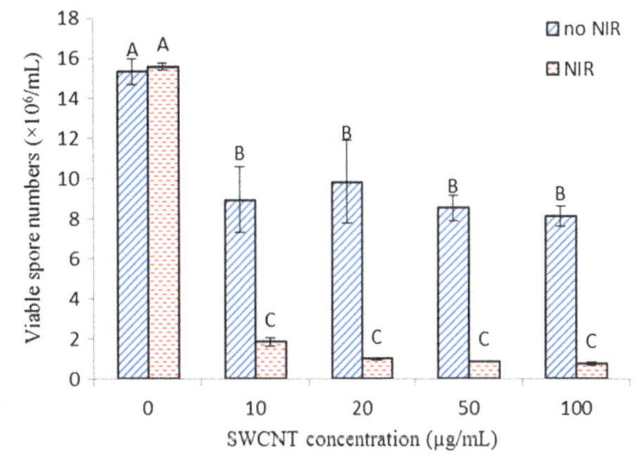

Figure 2 Viable spore numbers after being treated with SWCNTs at various concentrations and coupled with or without 20-min NIR radiation. Results are presented as the mean ± standard deviation. Different letters above the columns indicate statistically significant differences ($P < 0.05$).

composition, diameter, chiral index, chemical functionalization step, distribution of nanotube lengths, purification procedure, etc. In the NIR region (700 nm to 1000 nm), appreciable absorbance was observed. At 800 nm, a linear relationship between the optical density (OD) value and the SWCNTs concentration was represented by the regression equation of $OD = 0.0012\ C_{SWCNTs} + 0.0018$ ($R^2 = 0.9986$) (Figure 1B). This observation was consistent with previous studies [19,20].

The effect of SWCNTs-NIR treatment on inactivation of *B. anthracis* spores

The effect of SWCNTs-NIR treatment on *B. anthracis* spores was evaluated using the reduction of viable spores upon treatments determined by the traditional plating method. Figure 2 shows the reductions in viable spore number in the samples treated with SWCNTs alone and SWCNTs-NIR treatment. Compared to the SWCNTs alone treatment, the addition of SWCNTs (10-100 μg/mL) significantly reduced viable *B. anthracis* spore numbers from 1.53×10^7 spores/mL to $(8.15\text{-}9.85) \times 10^6$ spores/mL ($P < 0.05$). However, the viable spore numbers did not decrease significantly with the increasing SWCNTs concentrations from 10 to 100 μg/mL($P > 0.05$). With treatment of 100 μg/mL SWCNTs, the killing rate was about 46.9%, which was much lower than those of SWCNTs' inactivation rates to *Bacillus* vegetative cells and other bacterial cells previously reported which ranged from 80% to >99% depending on different conditions [3,4,7]. The results indicate that SWCNTs alone are not as effective in killing spores as in killing vegetative bacterial cells. Considering the protective structure of spores, the result is not unexpected.

In contrast to SWCNTs alone, SWCNTs with NIR significantly reduced the viable spore numbers ($P < 0.05$) compared to the SWCNTs-treated samples and the untreated sample, while a 20 min-NIR treatment without SWCNTs displayed no effect on viable spore numbers (Figure 2). For example, using 10 μg/mL SWCNTs, the viable spore number decreased 42% without NIR and decreased 88% with co-treatment of NIR. When SWCNTs concentration was increased to 100 μg/mL, SWCNTs-NIR co-treatment reduced the viable spore number by 95%. The results clearly demonstrated that co-treatment of SWCNTs with NIR enhanced the spore inactivation compared to SWCNTs alone treatment, although SWCNTs-NIR treatment did not exhibit a strong dependence on SWCNTs' concentration.

We also examined whether the duration of NIR treatment affected the efficiency of the co-treatment of SWCNTs-NIR. Figure 3 shows the reductions in viable spore numbers in the samples that received SWCNTs-NIR co-treatments with NIR treatment time varying from 5 to 30 min. Although the viable spore numbers in all the samples received SWCNTs-NIR co-treatment were significantly lower than that in the SWCNTs alone treated samples ($P < 0.05$), no significant difference was observed in the viable spore numbers among all the SWCNTs-NIR treated samples. The possible explanation of the observed

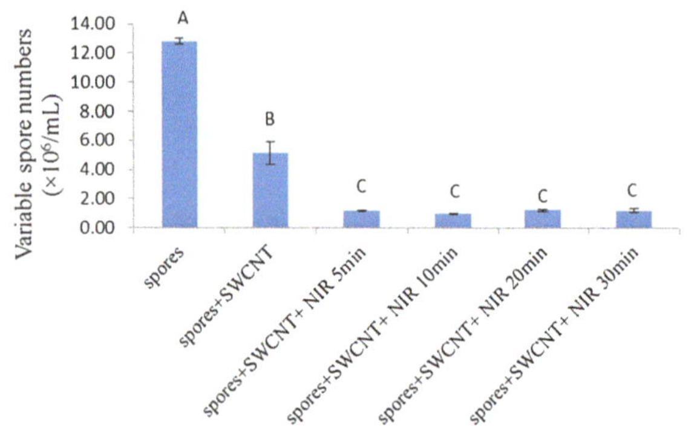

Figure 3 Effects of the co-treatment of SWCNTs-NIR on the reduction of viable *B. anthracis* spore numbers. Spores were treated with or without 20 µg/mL SWCNTs, and with or without NIR 5-30 min radiation. Results are presented as the mean ± standard deviation. Different letters above the bars indicate statistically significant differences (P < 0.05).

results is that the effect of NIR reached its maximal level rapidly (within 5 min). Most likely, the effect of NIR on the enhancement of spore inactivation is the unique optical-thermal property of SWCNTs which rapidly convert NIR radiation into heat resulting in intensive local heating in spores, similar to the mechanism when it acted on cancer cells or bacterial cells [19,20]. We thus examined the temperature increase in the bulk solution. The temperature of 10 and 100 µg/mL SWCNTs suspensions increased from 23°C to a final temperature 25.4 and 26.6°C in 5 min, respectively. Although the temperature only slightly increased in the bulk suspensions, the temperatures on the SWCNTs attached to the spores could be much higher. Panchakesan et al. reported that localized heating of SWCNTs on exposure to 200 mW/cm^2 of 800 nm NIR could lead to the explosions of SWCNT sheets in water due to the rapid temperature rise of the water molecules present in the nanopores between bundles to more than 100°C [21]. Explosions of loosely packed SWCNTs in air under laser treatment were reported by Ajayan et al. [22]. Further, a study has shown that carbon nanohorns (CNH), a family of carbon-based nanomaterials with similar atomic structure to CNTs, have the similar mechanism of NIR laser-triggered heating—optically stimulated electronic excitations being rapidly transferred to molecule vibration energies and causing local heating [23]. The heat is generated because of the high absorbance of SWCNTs in the NIR region ($\varepsilon \approx 1.03 \times 10^6$ M^{-1} cm^{-1} to 7.9×10^7 M^{-1} cm^{-1}), and such high absorbance is due to the electronic transitions between the first or second van Hove singularities of the nanotubes [19,24,25]. Heat is also generated in SWCNTs due to the electron-phonon coupling causing molecular vibrations [21]. The solution temperature reflected the extent of local heating of SWCNTs by NIR treatment. It was possible that the extreme local heat of SWCNTs in close contact with spores damaged some components of spores and thus enhanced the antimicrobial activity beyond the intrinsic antimicrobial activity of SWCNTs.

The results indicated that SWCNTs alone or SWCNTs-NIR treatment exhibited the antimicrobial activity on *B. anthracis* spores at the concentration 10 µg/ml SWCNTs, but their activity did not further increase with increasing concentration of SWCNTs from 10 µg/mL to 100 µg/mL. Previous results [26] using scanning electron microscopy (SEM) test showed that in this concentration range, spores were not saturated with SWCNTs in terms of surface coverage/interactions. In the SWCNTs-NIR treatment, the inactivation efficiency did not increase with the NIR treatment time beyond 5 min treatment time. Both observations were different from the observation on SWCNTs-NIR treatment to bacterial cells [20]. This may be due to the more protective structure of spores than bacterial cells and the limited capacity/function of SWCNTs and SWCNTs-NIR in the inactivation of spores. Further research will be necessary for more detailed mechanisms.

Effect of SWCNTs-NIR treatment on spore germination

The growth of SWCNTs-NIR pretreated B. anthracis spores

The growth of SWCNTs-NIR treated spores was examined by measuring optical density (OD) at 600 nm of the culture. Figure 4 shows the growth curves of *B. anthracis* spores in nutrient broth after the spores were pre-treated with SWCNTs (10 and 100 µg/mL) and SWCNTs with 20 min NIR treatments. The samples started exponential growth phases at similar suggesting that the treatments did not affect the lag phases, even though the viable spore number in these samples with different treatments were different (as shown in Figure 2). If the ODs at a given time during the exponential growth phase were compared, the samples treated with SWCNTs or SWCNTs-NIR showed slightly higher values than the control samples, indicating that both the SWCNTs treatment and SWCNTs-NIR

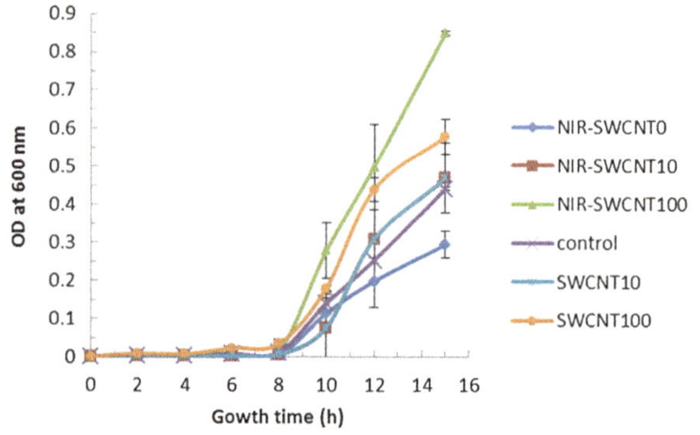

Figure 4 Growth curves of *B. anthracis* spores in nutrient broth after they were treated with SWCNTs at various concentrations, coupled with or without 20-min NIR. Error bars present the standard deviation of triplicated experimental results.

treatment promoted the more rapid growth of surviving spores in nutrient broth, and the higher concentration of SWCNTs (100 μg/mL) was more effective at activating on growth. The possible reason for this growth activation in surviving spores was that partial physical damages on the spore walls by SWCNTs may trigger spore germination and local heating on spores by SWCNTs-NIR treatment. In *B. subtilis* spores, mechanical abrasion can trigger germination by activation of either of the spore's cortex-lytic enzymes—CwlJ and SleB [27]. These enzymes are present in the spore in an active form but do not act until triggered by mechanical damage to spores or by germination stimuli, Ca^{2+}-DPA for CwlJ or some other stimulus for SleB [27-29]. Since SWCNTs can partially damage the spore coat, making it more permeable [26], damage on the spores due to SWCNTs treatment may actually triggered spore germination and rapid growth. On the other hand, previous study demonstrated that brief heat treatment at a sublethal temperature on spores significantly increased the germination percentage of spores [30,31]. Under natural conditions, the activation of dormant spores in water occurred slowly, but a heat activation step most often potentiated germination of >90% of spore populations [30]. SWCNTs-NIR treatment caused local heat to spores, during which some surviving spores experienced sublethal heat and actually were stimulated to germinate by the heat. The stimulated germination of spores by SWCNTs-NIR was actually confirmed later by the observation of increased transcriptional expression levels of germination related genes. However, spores treated with NIR showed slightly lower ODs during the exponential phase compared to the control sample, even though NIR treatment did not reduce the viable spore compared to the control (as shown in Figure 2). The reason for this observation is not clear.

DPA release from B. anthracis spores treated with SWCNTs and SWCNTs-NIR

Spore germination is the first crucial step in the return of *B. anthracis* spore to vegetative growth leading to its pathogenicity. The germination process includes several steps: release of spore H^+, release of dipicolinic acid (DPA) and its associate divalent cations, replacement of DPA by water, hydrolysis of the spore's cortex, and swelling of the spore core [29]. DPA release is one of the early events that can be used to evaluate the germination of spores. To evaluate the effect of SWCNTs concentration on spore germination, the kinetics of relative fluorescence unit (RFU) changes based on DPA releases were monitored by the Tb-DPA method suitable for rapidly detecting spore germination immediately after treatment [32].

Figure 5A shows the kinetics of DPA release from the control and the SWCNTs-treated spore samples (4.5×10^7 spores) during one hour germination. In all the samples, DPA contents quickly released in the first 5 min germination, and reached maximal DPA release after about 48 min germination. When comparing the maximal level of released DPA, the control released the highest concentration of DPA, while SWCNTs-treated samples released slightly lower concentrations of DPA. The DPA release of SWCNTs-treated samples did not display a clear SWCNTs concentration dependent trend. As it has been reported that the proportion of DPA released during germination was an indicator to test the percentage of viable spores capable of germination [33], the observed DPA release was consistent with the viable spore numbers in these treated samples which were not significantly different (as shown in Figure 2). When comparing the DPA release in the fast release period (within 5 min), the amount of DPA release at a given time showed a SWCNTs concentration-dependence. Figure 5B shows that the log value of RFU obtained from Tb-DPA method at 3 min of germination is correlated with SWCNTs concentration, following a linear equation $\text{Log RFU} = -0.0061\ C_{SWCNTs} + 4.9249$ ($R^2 = 0.935$, $P < 0.05$). This result indicated that SWCNTs treatment

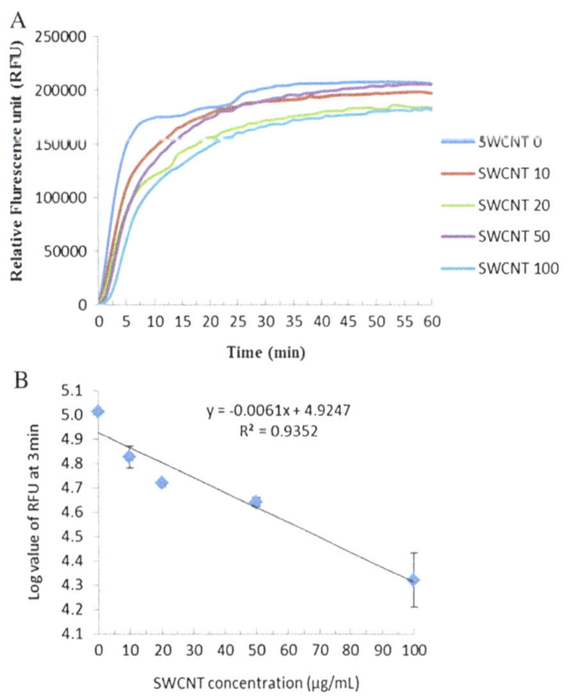

Figure 5 The kinetics of DPA release from *B. anthracis* spores treated with SWCNTs. (A) A group of curves of DPA release during the germination of *B. anthracis* spores after they were treated with different concentrations of single walled carbon nanotubes (SWCNTs). **(B)** The correlation between the log value of RFU in DPA measurement at 3 min of germination and the SWCNTs' concentration. Error bars in **(B)** present the standard deviations of triplicated experimental results.

showed a concentration-dependent effect on the rate of DPA release during the early stage germination (within 5 min) of the surviving spores.

Figure 6A shows the kinetics of DPA release from the spore samples (5.3×10^6 spores) that received the co-treatment of SWCNTs with 20 min NIR during one hour germination. Kinetically, SWCNTs-NIR treated samples took 15-20 min to reach the maximal level of DPA release, which were slightly slower than the untreated control sample (5 min shown in Figure 5A). When the maximal level of released DPA was considered, the SWCNTs-NIR treated samples only reached about half of the DPA level released by the NIR-treated sample (Figure 6A). The results indicated that difference in the kinetics and the maximal level of DPA release between the control and the SWCNTs-NIR treated samples were correlated to the viable spore numbers in the control and treated samples. However, among the SWCNTs-NIR treated samples, no correlation between RFU of DPA and SWCNTs concentration was observed (for instance, at 3 min germination, $R^2 = 0.38$, $P < 0.05$). This observation was different from samples treated with SWCNTs alone (as shown in Figure 5B), but was consistent with the viable spore numbers in SWCNTs-NIR treated samples (shown in Figure 2).

Figure 6B shows that kinetics of DPA release from *B. anthracis* spores (3.6×10^7) treated with SWCNTs (20 μg/mL) coupled with different time periods of NIR treatment, in terms of the RFU vs. germination time. All the samples reached the maximal RFU at a similar time, indicating that the NIR treatment time did not affect the kinetics of DPA release from the spore samples. The maximal DPA levels released at 60 min of germination for all the SWCNTs-NIR treated samples were very close (RFU values ranging from 136009 to 152365), and did not show a clear trend in relation to the NIR treatment time. This was correlated to the non-significantly different viable spore numbers in these samples (shown Figure 3). However, the sample treated with SWCNTs alone did not show the highest level of DPA release compared to all SWCNTs-NIR treated samples, although the viable spore number in SWCNTs-treated

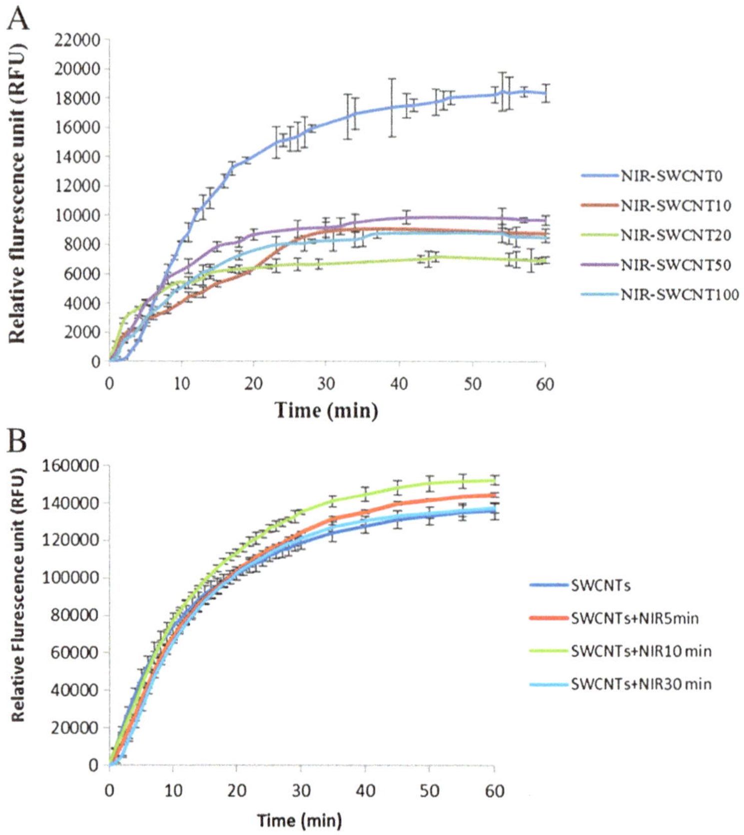

Figure 6 The kinetics of DPA release from *B. anthracis* spores treated with SWCNTs-NIR. (A) The kinetics of DPA release from *B. anthracis* spores that were treated with different concentrations of SWCNTs coupled with 20 min NIR radiation, during one hour of germination. Spore number: 5.3×10^6 spores. **(B)** The kinetics of DPA release from *B. anthracis* spores that were treated with SWCNTs (20 μg/mL) and different time of NIR treatment during one hour of germination. Spore number: 3.6×10^7 spores. Error bars present the standard deviations of triplicated experimental results.

sample was significantly higher than those in SWCNTs-NIR treated samples (shown in Figure 3). The possible reason for this result was that the SWCNTs-NIR treatment stimulated the germination and increased DPA release from the surviving spores which compensated the effect of less viable spores in the SWCNTs-NIR treated samples.

Alterations in the expression levels of germination, regulation and virulence related genes in *B. anthracis* spores treated with SWCNTs-NIR

The expressions of germination-related genes are essential for the germination step. Spore germination has been studied intensively. Recently, 31 genes required for *B. anthracis* sporulation and germination were identified [34]. A few studies have been reported on how the expression of sporulation and germination related genes in *B. anthracis* spores are affected under different physical intervention techniques, such as heat treatment, pasteurization, and microfiltration [35]. Spore germination process is initiated by germinants, which are believed to bind to germination (Ger) receptors in most cases [36,37]. The Ger receptor genes are organized mostly as tricistronic operons that encode three distinct membrane-associated proteins [37,38]. In *B. anthracis*, genome sequencing revealed six genomic (*gerA*, *gerH*, *gerK*, *gerL*, *gerS*, and *gerY*) and one plasmidic (*gerX*) *ger* receptor operons [39]. Gene *gerN* (BA1639) encodes germination protein GerN which is a homolog of a widely distributed family of cation transporters, and is required for spore germination with inosine. Among these germination related genes, we selected six genes to examine their expression levels, including *ger HC*, *gerKB*, *gerLA*, *gerLB*, *ger YA*, and *gerN*.

In order to examine the effect of SWCNTs-NIR treatment on *B. anthracis* spores at the molecular level, real-time PCR was used to test the expression levels of six germination-related genes identified by the transposon site hybridization (TraSH) assay [34], seven virulence-related genes and three regulatory genes. Table 1 shows the fold changes in the expression levels of all the detected genes in *B. anthracis* spores upon SWCNTs-NIR treatment in comparison to the untreated spores.

Upon the treatment of 20 μg/mL SWCNTs coupled with different time of NIR treatment, the expression levels of the six tested germination-related genes were changed with various folds, and the results were related to the NIR treatment time. The expression level of *gerHC* stayed the same compared to that of the untreated samples regardless of the NIR treatment time. The *gerKB* gene increased its expression level to 1.7 fold upon SWCNTs with 5 min NIR treatment, and decreased to the value of 0.26 fold of the control samples upon SWCNTs with 25 min NIR treatment. GerK acts cooperatively with GerB nutrient receptor to trigger germination with L-asparagine plus a mixture of glucose, fructose, and K + (GFK) and also with L-alanine plus GFK [29,38]. These results suggested that in the presence of SWCNTs, shorter time NIR treatment (5 min)

Table 1 Fold changes of gene expression after different treatments on *B. anthracis* Sterne spores

	Gene	Control	SWCNTs-NIR 5 min		SWCNTs-NIR 25 min	
		ΔCt*	ΔCt	Fold change**	ΔCt	Fold change
Germination related genes	*gerLA*	20.50 ± 0.84	20.31 ± 0.96	1.14	19.45 ± 0.71	2.07
	gerLB	22.61 ± 1.04	22.60 ± 0.99	1.01	21.54 ± 0.42	2.11
	gerYA	19.80 ± 0.78	20.58 ± 0.27	0.58	20.83 ± 0.30	0.49
	gerN	21.74 ± 1.32	19.51 ± 0.91	4.68	21.31 ± 0.33	1.35
	gerHC	20.84 ± 1.14	20.64 ± 0.54	1.14	20.61 ± 1.01	1.17
	gerKB	21.27 ± 1.12	20.50 ± 0.92	1.71	23.17 ± 3.60	0.26
Regulatory genes	*atxA*	22.48 ± 0.39	23.36 ± 0.47	0.54	23.64 ± 0.66	0.45
	pagR	21.85 ± 0.18	23.15 ± 0.46	0.41	22.61 ± 0.82	0.59
	GBAA1941	22.36 ± 0.30	20.75 ± 0.80	3.05	21.39 ± 0.14	1.96
Virulence related genes	*cya*	21.91 ± 0.15	23.02 ± 0.13	0.46	23.79 ± 0.41	0.27
	lef	22.74 ± 0.33	23.38 ± 0.14	0.64	24.74 ± 0.18	0.25
	pagA	21.45 ± 0.52	22.35 ± 0.24	0.54	23.07 ± 0.34	0.33
	GBAA5564	22.59 ± 0.34	22.14 ± 0.17	1.37	22.75 ± 0.22	0.89
	GBAA0797	22.62 ± 0.13	22.83 ± 0.20	0.86	22.63 ± 0.30	0.99
	purF	23.51 ± 0.74	24.92 ± 0.19	0.37	23.23 ± 0.23	1.21
	srtB	23.78 ± 0.06	26.13 ± 0.89	0.2	24.30 ± 0.10	0.7

*The ΔCt value (Ct, threshold cycle) of each tested gene was a relative expression level compared to the value of housekeeping gene 16S rRNA. The ΔCt values are expressed as the average of three replicates ± standard deviation.
**The fold change is determined as $2^{-\Delta\Delta Ct}$.

slightly activated the *gerKB* expression, but longer NIR treatment (25 min) inhibited its expression. *gerN* exhibited the similar trend to *gerKB* but with higher magnitude of changes, its expression level increased 4.7 fold upon SWCNTs with 5 min NIR treatment, and returned to 1.4 fold upon 25 min NIR treatment. *gerLA* and *gerLB* showed the same alteration trend that their expression levels almost did not change upon SWCNTs with 5 min NIR treatment but increased by 2 fold upon SWCNTs with 25 min NIR treatment. In *B. anthracis*, the GerL and GerK receptors are jointly required for the alanine germination pathway [40]. *gerYA* was the only one out of the six tested germination-related genes that decreased with NIR treatment. The *gerY* operon was found to be dispensable for recognition of all known germinant molecules in *B. anthracis* [40], but the role of GerY in *B. anthracis* spore germination is not clarified.

The detailed mechanisms of why the expression levels of some of the germination-related genes were increased, while some of them were decreases or stayed the same is not clear. -These molecular level changes might affect the germination of spores (including the release of DPA) and the lag phases in their growths for those SWCNTs-treated samples which we observed.

Besides the germination-related genes, we also examined the expression levels of three regulatory genes, *atxA*, *pag R*, GBAA1941, and seven virulence-related genes, *cya*, *lef*, *pagA*, *purF*, *srtB*, BGAA0797, and GBAA5564, upon SWCNTs-NIR treatment. These genes are important during the early infection of *B. anthracis*. Among the three regulatory genes, AtxA was isolated as a toxin gene activator 15 years ago and shown to be a *B. anthracis* master regulator [41]. AtxA controls the expression of more than a hundred genes belonging to all genetic elements, the chromosome and both virulence plasmids, including those encoding the major virulence factors [41]. AtxA is a transcriptional activator that controls the expression of three toxin genes, including the genes *pagA*, *cya*, and *lef* which encode the toxin component protective antigen (PA), edema factor (EF), and lethal factor (LF), respectively. PagR is a transcriptional repressor in *B. anthracis* that controls the chromosomal S-layer genes *eag* and *sap*, and down regulates the *pagA* gene by direct binding to their promoter regions [42]. GBAA1941 gene is a putative regulatory gene and its product shares significant homology with a family of transcriptional regulators (MarR) which includes proteins critical for control of virulence factor production, bacterial response to antibiotic and oxidative stresses and catabolism of environmental aromatic compounds [43]. GBAA1941 locus is important for *B. anthracis* pathogenesis [44]. As shown in Table 1, upon SWCNTs-NIR treatment, the expression levels of *atxA* and *pagR* decreased to about half compared to that of the untreated samples regardless of the NIR treatment time, while GBAA1941 gene increased its expression level by 3.05-and 1.96-fold at the NIR time 5 min and 25 min, respectively. The results suggested that SWCNTs-NIR treatment caused alterations in regulatory genes in *B. anthracis* spores, but the alteration levels varied in different regulatory genes depending on the functions of these genes in relation to the treatment.

In view of the expression levels of virulence genes, three toxin genes, *cya*, *lef*, and *pagA*, which are regulated by *atxA* gene decreased their expression levels accordantly to 0.46 -0.64 folds upon SWCNTs with 5 min NIR treatment, and to 0.25-0.33 folds upon 25 min NIR treatment. The alteration in the expression level of *pagA* after treatments was also correlated to the down regulation of the *pagR* gene. Gene GBAA5564 encodes a conserved hypothetical protein usually highly expressed during early infection [44]. Gene GBAA0797 encodes permease protein of the ABC transporter. The SWCNTs-NIR treatments on spores did not show significant effects on the expression levels of either GBAA5564 or GBAA0797 (varying from 0.86 to 1.37 folds). Gene *srtB* encodes sortase which is an enzyme possibly critical in the early stages of inhalation anthrax [45]. The disruption of *srtB* had no effect on the ability of the bacteria to grow in rich culture media, but mutations in this gene dramatically attenuated growth of the bacterium in J774A.1 cells [45]. Gene *purF* encodes amidophosphoribosyltransferase which is an enzyme in purine nucleotide biosynthetic process. Upon SWCNTs-NIR treatment, 5 min treatment showed significant inhibition effect on *srtB* expression in *B. anthracis* spores, decreasing the expression level to 0.2 fold (5 min) of the control samples. The SWCNTs-NIR 5 min treatment also significantly inhibited *purF* expression to 0.37 fold of the control samples, but 25 min treatment did not show obvious effects on this gene (1.21 fold).

In summary, SWCNTs-NIR treatments significantly decreased the expression levels of 3 out of 7 detected virulence-related genes and 1 out of 3 detected regulatory genes when the NIR time was 5 or 25 min. No obvious increase was observed in any of the tested virulence-related genes. However, only a very small number of genes were examined in this study.

Conclusions

We have investigated the effect of SWCNTs-NIR treatment on *B. anthracis* spores. At 10 μg/mL, SWCNTs alone only exhibited little sporicidal effect on *B. anthracis* spores, while SWCNTs coupled with 20 min NIR significantly improved the antimicrobial effect by doubling the percentage of viable spore number reduction compared with SWCNTs alone treatment (88% vs. 42%). The SWCNTs-NIR treated samples showed slightly more rapid growth, and released similar level of

DPA during germination despite the less viable spores in them. The results suggested the dual effect of SWCNTs-NIR treatment on *B. anthracis* spores, which enhanced the sporicidal effect and stimulated the germination of surviving spores. Molecular level examination showed that SWCNTs-NIR treatment altered the expression of germination, regulatory, and virulence-related genes in *B. anthracis.* The alterations in germination-related genes were correlated with the stimulated germination of surviving spores (including the release of DPA).

Materials and methods

SWCNTs

Suspensions of SWCNTs (with surface group -OH) in deionized (DI) water with varied length of 1-5 μm at concentration of 1 mg/mL were purchased from Nanolab, Inc. (Newton, MA). According to the manufacturer report, SWCNTs were produced using a chemical vapor deposition process with high yield and purity, with little or no amorphous carbon. The data of the energy-dispersive X-ray spectroscopy analysis conducted by means of SEM (SEM-EDX) provided by the manufacturer indicated that the SWCNTs contain 95.93% weight percentage of carbon and 4.07% of other elements (Na, Al, Si, S, and Fe). Based on the SEM and TEM images of SWCNTs [5], the products presented clusters in the nanotube bundles and terminals which most likely were the catalyst metals used in SWCNTs synthesis [46]. The presence of catalyst clusters in SWCNTs was observed in products used in studies by other researchers [46,47], but low concentrations of residual catalysts in CNT products were uncorrelated to CNTs' toxicity to bacterial cells [3,4,48,49].

B. anthracis spore preparations

B. anthracis spores were prepared and purified using the similar procedure as previous reported [26]. One colony of *B. anthracis* Sterne 34 F2 (Colorado Serum Company, Denver, CO) from a Luria-Bertani (LB) agar plate was used to inoculate 25 mL Disfco Sporulation Medium (DSM). The inoculum was incubated in a shaker incubator at 37°C with agitation of 225 rpm until mid-log phase, followed by adding 225 mL of pre-warmed (37°C) fresh DSM. The culture was incubated in the shaker incubator until > 95% of the culture were free spores. To purify the spores, the spore culture was first heated at 70°C for 20 min to kill the remaining vegetative cells, and then centrifuged at 8,000 × g at 4°C for 10 min. The pellet was washed with cold (4°C) DI water eight times to remove cell debris. The purified spores were suspended in 4°C DI water and stored at 4°C. Before each experiment, desired volume of the stock spores were washed with 4°C DI water twice for further use.

SWCNTs and NIR laser treatment on *B. anthracis* spores

Aliquots of 150 μL *B. anthracis* spore suspension in DI water with concentrations of ~10^5 to 10^6 cfu/mL were introduced into 1.5 mL eppendorf centrifuge tubes. The samples were added with desired volumes (2 to 20 μL) of SWCNTs, and the total volume of each sample was adjusted to 200 μL with DI H_2O with the final concentration of SWCNTs in the samples varying from 10 μg/mL to 100 μg/mL. The samples were mixed with constant rotation at 15 rpm on a rotation mixer (Model: RKDDYNAL) from Dynal Biotech, Inc. (Lake Success, NY) at room temperature for 30 min, followed by transferring the suspensions into quartz cuvettes (Fisher Scientific, Pittsburgh, PA) for NIR laser treatment. For NIR treatment, we used the same experiment setup as our previous study [20]. The Tsunami mode-locked Ti:Sapphire laser system (Newport Spectra-Physics) was used, which was pumped by the Millennia (Newport Spectra-Physics) diode pumped solid state 532 nm 15 W continuous wave laser. The resulting mode-locked Ti:sapphire laser was tunable from 700 nm to 1080 nm. An optical lens was used to focus the laser beam on the quartz cuvette in which the spore suspension with SWCNTs was placed. Continuous NIR light at 800 nm wavelength with output power of 360-410 mW was used to treat the samples for 5 to 30 min. The laser beam spot was approximately 2 mm in diameter. The laser density was in the range of 1.15-1.31 mW/cm^2. A thermocouple probe was inserted into the suspensions to monitor temperature change in the bulk solution during NIR treatment.

Determination of viable spore number and measurement of spore outgrowth

After SWCNTs or SWCNTs-NIR treatment, the viable spore number in the treated samples was determined using the traditional plating method. The treated samples were serially (1:10) diluted with DI water. 100 μL of appropriate dilutions were surface plated on LB agar plates in triplicates. Colonies were counted after 24 h of incubation of the plates at 37°C. The viable spore number was calculated as colonies forming units per mL (cfu/mL).

The outgrowths of the treated spores were examined by optical density (OD) measurement at 600 nm. 100 μL of spore suspensions were taken from the 200 μL of treated samples and added to 5 mL Nutrient broth. The samples were then incubated in a shaker incubator at 225 rpm at 37°C for growth.

Measurement of dipicolinic acid (DPA) release during spore germination

Spores (OD600 = 0.5) treated with SWCNTs or SWCNTs-NIR were germinated in a 96-well black plate with clear bottom at 37°C in 200 μL of germination solution which contained 25 mM HEPES buffer, 2 mM L-alanine, and

50 μM $TbCl_3$. DPA release was monitored by real-time measurement of fluorescence signals at 545 nm with excitation at 270 nm, which are appropriate wavelengths for the Tb^{3+} -DPA complex using the Spectra Max M5 spectrophotometer (Molecular Devices, LLC, Sunnyvale, CA) [50,51]. For each sample, the background detected at time zero was used as the blank. The percentage of DPA release at a sample time was calculated by dividing the value of relative fluorescence units (RFU) at that time by the maximum value of RFU at 1 h of germination.

RNA isolation and cDNA synthesis

RNA isolations were performed immediately after SWCNTs or SWCNTs-NIR treatment. Treated spores were collected by centrifugation at 8,000 × g at 4°C for 5 min, and washed twice with 0.5% Tween 20 solution. Spores were centrifuged and resuspended in RNAwiz buffer from RiboPure-Bacteria kit (Life Technologies, Carlsbad, CA). Zirconia beads were added and vortexed for 10 min, followed by a 5-min break. This vortexing-break step was repeated 5 times to ensure the disruption of the spores. RNA was purified using RiboPure-Bacteria kit according to the manufacturer's protocol. RNA quality and quantity were measured using NanoDrop 2000 spectrophotometer (NanoDrop Technologies Inc., Wilmington, DE). RNA samples were then treated with DNase I from the RiboPure-Bacteria kit according to the manufacturer's instruction. cDNA was synthesized using the SuperScript III First-Strand Synthesis SuperMix kit (Life Technologies, Carlsbad, CA) with random primers following the manufacturer's protocol.

Primer design and specificity test

In order to examine the effect of SWCNTs-NIR treatment on *B. anthracis* spores at the molecular level, we examined the expression levels of ten regulation and virulence-related genes and six germination related genes in *B. anthracis* Sterne. The gene sequences of the regulation and virulence- related genes were acquired from NCBI website (http://www.ncbi.nlm.nih.gov) and specific primers were designed for the analysis of transcriptional gene expression using real-time PCR (Table 2). The primer sequences for the amplification of six germination-related genes and 16S rRNA gene were acquired from a previous publication [35]. The specificities of all the primers designed in this study were tested by Blast function from Primer-Blast and further confirmed by colony PCR. Primers were synthesized and purchased from Eurofins MWG Operon (http://www.operon.com). To run the colony PCR, a single colony of *B. anthracis* Sterne on LB agar was picked and suspended in 100 μL sterilized DI-H_2O in a 1.5 μL centrifuge tube. The colony suspension was heated in a water bath with boiling water for 5 min, followed by

Table 2 Primer sequences designed in this study for the detection of specific genes through real-time PCR

Gene	Primer sequence	Product size (bp)	Description
atxA	CGGACTGTTTTATCTGCGACCTGT	127	transcriptional activator AtxA
	CACCGATATCCATCGAAAAGGAACA		
cya	TCGGATTGGGGCCCTGTAGC	80	calmodulin-sensitive adenylate cyclase
	TCGACAGCTAATTGTTGACCATGC		
lef	TCTTGCCAGCATCCGTTGATCT	129	lethal factor
	ATCGAGCCACAGCATCGTGA		
pagA	GGGTGGATACAGGCTCGAACTGG	115	protective antigen
	ACCGCCGCTATCCGCCTTTC		
pagR	ACAATGGCACATCCTATGCGTT	116	transcriptional repressor
	TGCTGGGATACAGTTGATTGTGGT		
purF	GAACTGCGCGAGCAAGGGGT	137	Amidophosphoribosyl-transferase
	GCCTCGCGAAGCATGCGAAC		
srtB	TGCTGCAGCGCTTTGAACTGT	96	sortase B
	AGTAATGGCCGAAGCACAAAACA		
GBAA1941	ACACAAGGCGGTATTTCTCGAATGT	144	MarR family transcriptional regulator
	TGCAAGTTGCTCTGGCAATGCT		
GBAA5564	TGGTTGGTTCGGCGGTTATGGT	82	hypothetical protein
	TGCCATGAATACACCAGGGCG		
GBAA0797	CCACGCGTTGCACCAAGAGC	130	ABC transporter permease
	TGGTATTCGGTGGTATTGCCGCT		

incubation at -80°C for 10 min. This step was repeated 3 times, and then centrifuged at 15,000 × g for 3 min. The supernatant was used as template DNA for PCR amplification. The PCR reaction system contained: 2 μL of 10 × PCR buffer containing Mg^{2+}, 2 μL of 2 mM dNTP mix, 0.4 μL of each 10 μM forward and reverse primer, 0.5 unit of Taq DNA polymerase (Thermo Fisher Scientific Inc, Foster City, CA), 2 μL of DNA template, and DI water to a final volume of 20 μL. The PCRs were performed in an Eppendorf Mastercycler (Eppendorf Scientific, Inc., Westbury, NY) under the following conditions: 94°C for 5 min; 32 cycles of 93°C for 30 s, 57°C for 40 s, and 72°C for 40 s; and finally 72°C for 5 min. The presence of the PCR product was confirmed by electrophoresis on 2.0% agarose gels and staining with ethidium bromide.

Real-time PCR and data analysis

Real-time PCRs were performed in 96-well plates on a 7500-Fast real-Time PCR system (Applied Biosystems, Foster City, CA, U.S.A.). Each well contained 20 μL reaction mixtures, including 10 μL of 2× Fast SYBR® Green Mster Mix (Applied Biosystems, Forster City, CA, U.S.A.), 0.4 μL of each 10 μM forward and reverse primer, 0.7 μL cDNA, and 7.8 μL DI-H_2O. Each sample was assayed in triplicates and the average value was used for data analysis. Amplification condition involved an initial denaturation at 95°C 10 min and 40 cycles of 95°C 15 s and 60°C 1 min. Fluorescence data were collected at the annealing step. A final dissociation step was performed to verity amplification specificity. Results were analyzed using 7500 Fast System SDS software v 1.4 (Applied Biosystems, Forster City, CA, U.S.A.) provided with the thermocycler. The ΔCt value (Ct, threshold cycle) of each tested gene was a relative expression level compared to the expression of housekeeping gene 16S rRNA. The fold changes of transcript abundance between the treated sample and the untreated control were calculated using the formula $2^{-\Delta\Delta CT}$ [52].

Abbreviations

CNTs: Carbon nanotubes; SWCNTs: Single walled carbon nanotubes; NIR: Near infrared; PCR: Polymerase chain reaction; DPA: Dipicolinic acid; RFU: Relative fluorescence units; UV: Ultraviolet; VIS: Visible.

Competing interests

The authors declare that they have no competing interests.

Authors' contributions

XD and YT developed the methods and performed the experiments. MW and BV provided instruments and helped to conduct the experiments. LY designed the study and supervised the work. XD and LY wrote and revised the manuscript. All authors read and approved the final manuscript.

Acknowledgements

This research was supported by the US Army Research Office (ARO) (#W911NF-10-1-0160). The project also used the facilities supported by National Science Foundation (NSF, DMR #0959679). The authors also acknowledge the support from the Golden Leaf Foundation for major research instruments and facilities. The authors thank Ms. Michelle Gignac at the Shared Materials Instrumentation Facility (SMIF) at Duke University for help on the SEM tests.

Author details

[1]Biomanufacturing Research Institute and Technology Enterprise (BRITE) and Department of Pharmaceutical Sciences, North Carolina Central University, Durham, NC 27707, USA. [2]Department of Mathematics and Physics, North Carolina Central University, Durham, NC 27707, USA.

References

1. U.S. Environmental Protection Agency (EPA): **Assessing health risks from pesticides.** 2005. www.epa.gov/pesticides/factsheets/riskassess.htm.
2. Whitney EAS, Beatty ME, Taylor TH, Weyant R, Sobel J, Arduino MJ, Ashford DA: **Inactivation of *Bacillus anthracis* spores.** *Emerg Infect Dis* 2003, **9**:623–627.
3. Kang S, Pinault M, Pfefferle LD, Elimelech M: **Single-walled carbon nanotubes exhibit strong antimicrobial activity.** *Langmuir* 2007, **23**(17):8670–8673.
4. Kang S, Herzberg M, Rodrigues DF, Elimelech M: **Antibacterial effects of carbon nanotubes: size does matter!** *Langmuir* 2008, **24**:6409–6413.
5. Yang C, Mamouni J, Tang Y, Yang L: **Antimicrobial activity of single-walled carbon nanotubes: length effect.** *Langmuir* 2010, **26**:16013–16019.
6. Liu SB, Wei L, Hao N, Fang N, Chang MW, Xu R, Yang YH, Chen Y: **Sharper and faster "Nano Darts" kill more bacteria: a study of antibacterial activity of individually dispersed pristine single-walled carbon nanotube.** *ACS Nano* 2009, **3**(12):3891–3902.
7. Arias LR, Yang LJ: **Inactivation of bacterial pathogens by carbon nanotubes in suspensions.** *Langmuir* 2009, **25**(5):3003–3012.
8. Liu S, Ng AK, Xu R, Wei J, Tan CM, Yang Y, Chen Y: **Antibacterial action of dispersed single-walled carbon nanotubes on Escherichia coli and *Bacillus subtilis* investigated by atomic force microscopy.** *Nanoscale* 2010, **2**:2744–2750.
9. Dong L, Henderson A, Field C: **Antimicrobial activity of single-walled carbon nanotubes suspended in different surfactants.** *J Nanotechnol* 2012. doi:10.1155/2012/928924.
10. Aferchich K, Lilly M, Yang L: **Effect of single-walled carbon nanotubes on *Bacillus anthracis* cell growth, sporulation, and spore germination.** *J Nanosci Nanotechnol* 2012, **12**(5):3821–3830.
11. Lyon DY, Fortner JD, Sayes CM, Colvin VL, Hughe JB: **Bacterial cell association and antimicrobial activity of a C60 water suspension.** *Environ Toxicol Chem* 2005, **24**(11):2757–2762.
12. Fortner JD, Lyon DY, Sayes CM, Boyd AM, Falkner JC, Hotze EM, Alemany LB, Tao YJ, Guo W, Ausman KD, Colvin VL, Hughes JB: **C60 in water: nanocrystal formation and microbial response.** *Environ Sci Technol* 2005, **39**:4307–4316.
13. Klaine SJ, Alvares PJ, Batley GE, Fernandes TF, Handy RD, Lyon DY, Mahendra S, McLaughlin MJ, Lead JR: **Nanomaterials in the environment: behavior, fate, bioavailability, and effects.** *Environ Toxicol Chem* 2008, **27**(9):1828–1851.
14. Li Q, Mahendra S, Lyon DY, Brunet L, Liga MV, Li D, Alvarez PJ: **Antimicrobial nanomaterials for water disinfection and microbial control: potential applications and implications.** *Water Res* 2008, **42**(18):4591–4602.
15. Gerhardt P, Ribi E: **Ultrastructure of the exosporium enveloping spores of *Bacillus cereus*.** *J. Bcteriol.* 1964, **88**:1774–1789.
16. Wang H, Gu L, Lin Y, Lu F, Meziani MJ, Luo PJ, Wang W, Cao L, Sun YP: **Unique aggregation of anthrax (*Bacillus anthracis*) spores by sugar-coated single-walled carbon nanotubes.** *J Am Chem Soc* 2006, **128**:13364–13365.
17. Upadhyayula VKK, Deng S, Smith GB, Mitchell MC: **Adsorption of *Bacillus subtilis* on single-walled carbon nanotube aggregates, activated carbon and nanoCeram™.** *Water Res* 2009, **43**:148–156.
18. König K: **Multiphoton microscopy in life sciences.** *J. Microsc. (Oxford)* 2000, **200**:83–104.
19. Kam NWS, O'Connell M, Wisdom JA, Dai H: **Carbon nanotubes as multifunctional biological transporters and near-infrared agents for selective cancer cell destruction.** *Proc Natl Acad Sci U S A* 2005, **102**:11600–11605.

20. Mamouni J, Tang Y, Wu M, Vlahovic B, Yang L: **Single-walled carbon nanotubes couples with near-infrared laser for inactivation of bacterial cells.** *J Nanosci Nanotechnol* 2011, **11**:4708–4716.
21. Panchapakesan B, Lu S, Sivakumar K, Teker K, Cesarone G, Wickstrom E: **Single wall carbon nanotube nanobomb agents for killing breast cancer cells.** *Nanobiotechnology* 2005, **1**(2):133–139.
22. Ajayan PM, Terrones M, de la Guardia A, Huc V, Grobert N, Wei BQ, Lezec H, Ramanath G, Ebbesen TW: **Nanotubes in a flash—ignition and reconstruction.** *Science* 2002, **296**(5568):705–706.
23. Miyako E, Nagata H, Hirano K, Makita Y, Nakayama K, Hirotsu T: **Near-infrared laser-triggered carbon nanohorns for selective elimination of microbes.** *Nanotechnology* 2007, **18**:475103–475109.
24. O'Connell MJ, Bachilo SM, Huffman CB, Moore V, Strano MS, Haroz E, Rialon KL, Boul PJ, Noon WH, Kittrell C, Ma J, Hauge RH: **Band gap fluorescence from individual single-walled carbon nanotubes.** *Science* 2002, **26**(297):593–596.
25. Bachilo SM, Strano MS, Kittrell C, Hauge RH, Smalley RE, Weisman RB: **Structure-assigned optical spectra of single-walled carbon nanotubes.** *Science* 2002, **298**:2361–2366.
26. Lilly M, Dong X, McCoy E, Yang L: **Inactivation of *Bacillus anthracis* spores by single-walled carbon nanotubes coupled with oxidizing antimicrobial chemicals.** *Environ Sci Technol* 2012, **46**:13417–13424.
27. Jones CA, Padula NL, Setlow P: **Effect of mechanical abrasion on the viability, disruption and germination of spores of *Bacillus subtilis*.** *J Appl Microbiol* 2005, **99**:1484–1494.
28. Bagyan I, Setlow P: **Localization of the cortex lytic enzyme CwlJ in spores of *Bacillus subtilis*.** *J Bacteriol* 2002, **184**:1289–1294.
29. Setlow P: **Spore germination.** *Curr Opin Microbiol* 2003, **6**:550–556.
30. Keynan A, Evenchik Z: **Activation**. In *The Bacterial Spore.* Edited by Gould GW, Hurst A. New York: Academic Press Inc; 1969:359–396.
31. Foster SJ, Johnstone K: **The trigger mechanism of bacterial spore germination. *in* Regulation of procaryotic development**. In *Regulation of Prokaryotic Development.* Edited by Smith I, Slepecky RA, Setlow P. New York: Plenum Publishing Corporation; 1989:89–108.
32. Yi X, Setlow P: **Studies of the commitment step in the germination of spores of Bacillus species.** *J Bacteriol* 2010, **192**:3424–3433.
33. Shafaat HS, Ponce A: **Applications of a rapid endospore viability assay for monitoring UV inactivation and characterizing arctic ice cores.** *Appl Environ Microbiol* 2006, **72**:6808–6814.
34. Day WA Jr, Rasmussen SL, Carpenter BM, Peterson SN, Friedlander AM: **Microarray analysis of transposon insertion mutations in *Bacillus anthracis*: global identification of genres required for sporulation and germination.** *J Bacteriol* 2007, **189**:3296–3301.
35. Liu Y, Ream A: **Sporulation and germination gene expression analysis of *Bacillus anthracis* Sterne spores in skim milk under heat and different intervention techniques.** *J Food Sci* 2009, **74**:M120–124.
36. Paidhungat M, Setlow P: **Role of Ger proteins in nutrient and nonnutrient triggering of spore germination in *Bacillus subtilis*.** *J Bacteriol* 2000, **182**:2513–2519.
37. Ross C, Abel-Santos E: **The Ger receptor family in sporulating bacteria.** *Curr Issues Mol Biol* 2010, **12**:147–158.
38. McCann KP, Robinson C, Sammons RL, Smith DA, Corfe BM: **Alanine germinant receptors of *Bacillus subtilis*.** *Lett. Appl. Bacteriol.* 1996, **23**:290–294.
39. Read TD, Peterson SN, Tourasse N, Baillie LW, Paulsen IT, Nelson KE, Tettelin H, Fouts DE, Eisen JA, Gill SR, Holtzapple EK, Okstad OA, Helgason E, Rilstone J, Wu M, Kolonay JF, Beanan MJ, Dodson RJ, Brinkac LM, Gwinn M, DeBoy RT, Madpu R, Daugherty SC, Durkin AS: **The genome sequence of *Bacillus anthracis* Ames and comparison to closely related bacteria.** *Nature* 2003, **423**:81–86.
40. Fisher C, Hanna PC: **Characterization of *Bacillus anthracis* germinant receptors in vitro.** *J Bacteriol* 2005, **187**:8055–8062.
41. Fouet A: **AtxA, a *Bacillus anthracis* global virulence regulator.** *Res Microbiol* 2010, **161**:735–742.
42. Zhao H, Volkov A, Veldore VH, Hoch JA, Varughese KI: **Crystal structure of the transcriptional repressor PagR of *Bacillus anthracis*.** *Microbiology* 2010, **156**:385–391.
43. Wilkinson SP, Grove A: **Ligand-responsive transcriptional regulation by members of the MarR family of winged helix protein.** *Curr Issues Mol Biol* 2006, **8**:51–62.
44. Bergman NH, Anderson EC, Swenson EE, Janes BK, Fisher N, Niemeyer MM, Miyoshi AD, Hanna PC: **Transcriptional profiling of *Bacillus anthracis* during infection of host macrophages.** *Infect Immun* 2007, **75**:3434–3444.
45. Zink SD, Burns D: **Importance of *srtA* and *srtB* for growth of *Bacillus anthracis* in Macrophages.** *Infect Immun* 2005, **73**:5222–5228.
46. Kang S, Mauter MS, Elimelech M: **Microbial cytotoxicity of carbon-based nanomaterials: implications for river water and wastewater effluent.** *Environ Sci Technol* 2009, **43**(7):2648–2653.
47. Saleh NB, Pfefferle LD, Elimelech M: **Influence of biomacromolecules and humic acid on the aggregation kinetics of single-walled carbon nanotubes.** *Environ Sci Technol* 2010, **44**(7):2412–2418.
48. Kang S, Mauter MS, Elimelech M: **Physicochemical determinants of multiwalled carbon nanotube bacterial cytotoxicity.** *Environ Sci Technol* 2008, **42**(19):7528–7534.
49. Shvedova AA, Kisin ER, Mercer R, Murria AR, Jonson VJ, Potapovich AI, Taurina Y, Gorelik O, Arepalli S, Schwegler-Berry D, Hubbs AF, Antonini J, Evans DE, Ku BK, Ramsery D, Maynard A, Kagan VE, Castranova V, Baron P: **Unusual inflammatory and fibrogenic pulmonary responses to single-walled carbon nanotubes in mice.** *Am J Physiol Lung Cell Mol Physiol* 2005, **289**:698–708.
50. Yang WW, Ponce A: **Rapid endospore viability assay of *Clostridium sporogenes* spores.** *Int J Food Microbiol* 2009, **133**:213–216.
51. Yung PP, Ponce A: **Fast sterility assessment by germinable-endospore biodosimetry.** *Appl Environ Microbiol* 2008, **74**:7669–7674.
52. Livak KJ, Schmittgen TD: **Analysis of relative gene expression data using real-time quantitative PCR and the $2^{-\Delta\Delta Ct}$ method.** *Methods* 2001, **25**:402–408.

Correlation between mass transfer coefficient k_La and relevant operating parameters in cylindrical disposable shaken bioreactors on a bench-to-pilot scale

Wolf Klöckner[1], Riad Gacem[1], Tibor Anderlei[2], Nicole Raven[3], Stefan Schillberg[3], Clemens Lattermann[1] and Jochen Büchs[1*]

Abstract

Background: Among disposable bioreactor systems, cylindrical orbitally shaken bioreactors show important advantages. They provide a well-defined hydrodynamic flow combined with excellent mixing and oxygen transfer for mammalian and plant cell cultivations. Since there is no known universal correlation between the volumetric mass transfer coefficient for oxygen k_La and relevant operating parameters in such bioreactor systems, the aim of this current study is to experimentally determine a universal k_La correlation.

Results: A Respiration Activity Monitoring System (RAMOS) was used to measure k_La values in cylindrical disposable shaken bioreactors and Buckingham's π-Theorem was applied to define a dimensionless equation for k_La. In this way, a scale- and volume-independent k_La correlation was developed and validated in bioreactors with volumes from 2 L to 200 L. The final correlation was used to calculate cultivation parameters at different scales to allow a sufficient oxygen supply of tobacco BY-2 cell suspension cultures.

Conclusion: The resulting equation can be universally applied to calculate the mass transfer coefficient for any of seven relevant cultivation parameters such as the reactor diameter, the shaking frequency, the filling volume, the viscosity, the oxygen diffusion coefficient, the gravitational acceleration or the shaking diameter within an accuracy range of +/− 30%. To our knowledge, this is the first k_La correlation that has been defined and validated for the cited bioreactor system on a bench-to-pilot scale.

Keywords: Dimensional analysis, Maximum oxygen transfer capacity, Orbitally shaken, Plant cell suspension cultures, Single-use, Volumetric mass transfer coefficient

Background

The success of new biopharmaceuticals highly depends on their potential to compete with established products. Parameters such as a fast time to market, cost effectiveness and manufacturing flexibility are key issues that need to be considered while maintaining product quality [1]. Using disposable equipment can help reduce investment costs and increase flexibility and process safety [2] ultimately leading to enhanced competitiveness of the products. Consequently, disposable bioreactors are increasingly being used in biotechnological production processes within the past ten years [3].

Today, one of the most popular disposable bioreactors is the WAVE cultivation system that was first described in 1976 [4,5]. Since then, various other types of disposable cultivation systems for different applications have been developed. A wide range of different reactor sizes, starting with disposable screening systems in microliter scale, up to bioreactors with capacities of several cubic meters are on the market [3]. Available cultivation systems and their applications are described in several review articles [2,3,6,7].

* Correspondence: Jochen.Buechs@avt.rwth-aachen.de
[1]AVT. Biochemical Engineering, RWTH Aachen University, Worringerweg 1, Aachen 52074, Germany
Full list of author information is available at the end of the article

Despite the increasing usage of disposable cultivation systems, only few scientific publications have reported about their characterization with respect to power input, oxygen transfer and mixing performance [8]. By contrast, stirred stainless steel bioreactors have been extensively characterized and optimized over the past 60 years [3]. A direct design transfer from stainless steel to stirred disposable bioreactor systems is not feasible because of different properties of the applied materials. For instance, it is not possible to keep geometric similarity between a stirrer made of stainless steel and a disposable stirrer, due to the reduced strength and rigidity of the disposable polymer material. New agitation concepts for disposable bioreactors are desired to simplify their design and make them more cost-efficient [6]. Surface aerated reactors without complex built-in components fulfil the requirement for a cost-efficient reactor design [9]. In these systems, oxygen transfer and power input are either introduced by a wave, rocking or shaking motion of the bioreactor [3]. Orbitally shaken bioreactors are advantageous due to their well-defined liquid distribution that allows a precise characterization of power input and oxygen transfer. The magnitude of volumetric power input in these systems is comparable to that of conventional stirred tank reactors [10,11], indicating excellent mixing characteristics also for liquids at elevated viscosity.

In the 50 mL scale, the orbitally shaken TubeSpin Bioreactor was developed and tested as a high throughput system for mammalian cell cultivation [12,13]. On a larger scale, different studies reported about power input [10,11,14,15], mixing properties [16] and scale-up performance [15,17-20] of disposable orbitally shaken reactors. In several studies the volumetric mass transfer coefficient k_La was identified as one of the main parameters for successful scale-up and transfer of cultivation conditions from stirred tank to disposable shaken reactors [13,18,21,22]. Zhang et al. [23] also described a helical track attached on the inner reactor wall as a potential means to increase the mass transfer coefficient. However, a scale- and volume-independent k_La correlation for cylindrical orbitally shaken bioreactors has not been described so far. Thus, the aim of this current study is to experimentally determine such a k_La correlation in cylindrical disposable shaken bioreactors with volumes from 2 L up to 200 L and apply it to the cultivation of suspended plant cells.

Results and discussion

Dimensional analysis

Dimensional analysis according to Buckingham's π - Theorem was used to define a scale- and volume-independent k_La correlation [24,25]. According to the rules of Buckingham's theory, the resulting correlation is restricted to systems with geometric similarity. Influencing variables for k_La in cylindrical orbitally shaken bioreactors and their corresponding units are the volumetric mass transfer coefficient for oxygen k_La $\left[\frac{1}{s}\right]$, the reactor diameter d [m], the shaking diameter d_0 [m], the shaking frequency n $\left[\frac{1}{s}\right]$, the liquid volume V_L [m^3], the diffusion coefficient for oxygen $\mathbb{D}_{O2}$ $\left[\frac{m^2}{s}\right]$, the kinematic viscosity ν $\left[\frac{m^2}{s}\right]$ and the gravitational acceleration g $\left[\frac{m}{s^2}\right]$. The following dimensionless numbers were formed with the influencing variables according to the rules of Buckingham's theory:

Mass transfer number $k_La \cdot \left(\frac{\nu}{g^2}\right)^{\frac{1}{3}}$

Froude number $\frac{n^2 \cdot d_0}{g}$

Volume number $\frac{V_L}{d^3}$

Geometric number $\frac{d_0}{d}$

Galilei number $\frac{d^3 \cdot g}{\nu^2}$

Schmidt number $\frac{\nu}{\mathbb{D}_{O2}}$

This set of dimensionless numbers was also proposed by Henzler and Schedel [26] to develop a scale-independent correlation for the mass transfer coefficient in shake flasks. The applied numbers for surface aerated bioreactors differ from dimensionless numbers that are commonly used to describe the mass transfer in bubble aerated bioreactors because the principle of mass transfer differs fundamentally in both systems. In contrast to surface aerated orbitally shaken bioreactors, mass transfer in bubble aerated bioreactors depends on the amount, size, break-up and coalescence of gas bubbles. The characteristics of gas bubbles are strongly influenced by the volumetric power input, which is therefore commonly used as a parameter for k_La correlations for bubble aerated bioreactors. These aspects are not relevant for the utilized surface aerated bioreactors that are operated without bubble aeration.

The numbers are independent from one another and the set of numbers can be transformed into a different set of dimensionless numbers, with no influence on the resulting exponents of the included variables, as long as the rules of Buckingham's π -Theorem are preserved [25]. In this set of dimensionless numbers the Froude number is used to express the ratio of centrifugal to gravitational forces during shaking. The relative filling volume is considered with the Volume number and the Geometric number is used to describe the ratio of shaking diameter to reactor diameter. The height and shape of the free liquid surface during shaking is influenced by the ratio of gravitational to viscous forces, expressed by the Galilei number. The ratio of momentum transfer to diffusive mass transfer in the gas/liquid surface is described by the Schmidt number. In bioreactors without bubble aeration, the influence of the surface tension on the mass transfer coefficient decreases with increasing reactor size as already described by Doig et al. [27] and Hermann et al. [28] for microtiter plates. Consequently, the influence of the surface tension was not considered in orbitally shaken vessels with volumes from 2 L to 200 L. The occurrence of gas bubbles was avoided by using surface aeration only and the smooth surface of the inner reactor wall led to a well-defined and homogeneous liquid distribution during shaking. The following power law function was used to describe the correlation among the dimensionless numbers:

$$k_L a \cdot \left(\frac{\nu}{g^2}\right)^{\frac{1}{3}} = C \cdot \left(\frac{n^2 \cdot d_0}{g}\right)^{\alpha} \cdot \left(\frac{V_L}{d^3}\right)^{\beta} \cdot \left(\frac{d_0}{d}\right)^{\gamma} \cdot \left(\frac{d^3 \cdot g}{\nu^2}\right)^{\delta} \cdot \left(\frac{\nu}{\mathbb{D}_{O2}}\right)^{\varepsilon} \quad (1)$$

Each exponent in Eq. 1 has to be determined separately by empirically varying the corresponding dimensionless number.

Influence of the critical circulation frequency

As described by several authors, a critical circulation frequency N_c has to be exceeded to induce a rotating liquid motion in cylindrical orbitally shaken vessels [10,29]. Hardly any movement was observed between liquid bulk and reactor wall below the critical circulation frequency, resulting in very low power input and oxygen transfer values. Consequently, only $k_L a$ values that were measured at shaking frequencies above the critical circulation frequency were considered for determining the $k_L a$ correlation. The following equation was used to calculate the critical circulation frequency [10]:

$$N_c = \frac{1}{d^2} \cdot \sqrt{0.28 \cdot V_L \cdot g} \quad (2)$$

According to Eq. 2, the critical circulating frequency N_c is a function of the inner reactor diameter d, the filling volume V_L and the acceleration of gravity g.

Out-of-phase operation in orbitally shaken bioreactors

The transition from in-phase to out-of-phase operation in orbitally shaken bioreactors is accompanied by a strong decrease in power input and oxygen transfer [30-32]. Liquids with high viscosities in systems with low shaking diameters are prone to out-of-phase operation as described in several studies for shake flasks [32,33]. Thus, a minimum value for the ratio of shaking diameter (d_0) to reactor diameter (d), expressed by the Geometric number, is required to avoid an undesired out-of-phase operation in cylindrical orbitally shaken bioreactors.

Defining the $k_L a$ correlation

Values for $k_L a$ in cylindrical bioreactors with nominal volumes from 2 L to 50 L were determined according to Eq. 11 by using a RAMOS device in combination with a 0.5 M or 1 M sulfite system for OTR_{max} measurements. Values for $k_L a$ in cylindrical reactors of different size are presented in Figure 1. A decrease in $k_L a$ with increasing filling volume was observed. However, similar $k_L a$ values were obtained for the same relative filling volume in the applied scales (10 L - 50 L). This result is, on the first glance, astonishing as the volumetric transfer area (a) decreases with increasing vessel size. Similar $k_L a$ values in different scales for the same relative filling volume can be explained by an increasing value for the coefficient k_L with increasing reactor size because of a higher power input [10]. A precise characterization of the mass transfer coefficient using dimensional analysis is presented in the following section.

The influence of the Froude number on the $k_L a$ number was determined by varying the shaking frequency (n) as shown in Figure 2. A variation of the shaking frequency only affects the Froude number while all remaining dimensionless numbers are kept constant, allowing the influence of the Froude number to be separated from the influence of the other dimensionless numbers. An average exponent of $\alpha = 1.06$ was determined as shown in Figure 2. The validity of the exponent α is restricted to a range for the Froude number of between 0.013 and 0.097 as depicted in Figure 2.

The influence of the Volume number in Eq. 1 was determined by varying the filling volume, which only affects the Volume number. Thus, the resulting average exponent can be directly specified from the measuring values presented in Figure 3. An average exponent of $\beta = -1.20$ was determined for the influence of the Volume number. The range of applicability for the exponent β is restricted to values of the Volume number of between 0.18 and 0.65.

The influence of the Geometric number on the $k_L a$ number was investigated by altering the shaking diameter d_0. A modification of the shaking diameter in Eq. 1 leads to a variation of the Geometric number and the Froude number. Therefore, the influence of the Froude number on the $k_L a$ number had to be considered with the exponent $\alpha = 1.06$. This calculation was conducted by multiplying the $k_L a$ number with the inverse Froude number as stated in Figure 4. An average exponent of $\gamma = -1.06$ was determined within a range of variation for the Geometric number from 0.04 to 0.42 as shown in Figure 4. A low value for the ratio of shaking diameter (d_o) to reactor diameter (d), which is expressed with the Geometric number, can lead to an undesired out-of-phase operation. Thus, the minimum value for the Geometric number was set to 0.06 to prevent an out-

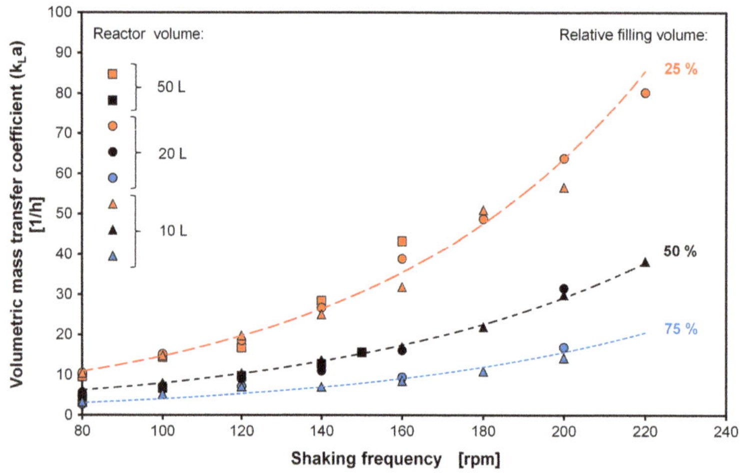

Figure 1 k_La values of 10 L, 20 L and 50 L bioreactors at 25°C determined with RAMOS using 1 M Na_2SO_3 solution with 10^{-7} M $CoSO_4$, 0.012 M phosphate buffer, initial pH = 8, data points connected with trend lines for relative filling volumes of 25%, 50% and 75%.

of-phase operation during the measurements in the present work. The final range of applicability for the exponent γ of the Geometric number was defined between 0.06 and 0.42.

Vessels of different sizes were used to vary the reactor diameter and, in this way, the Galilei number. The influence of the Volume number and the Geometric number were considered with their respective exponents and multiplied with the k_La number as stated in Figure 5. An average exponent of $\delta = -0.12$ was determined with a validity range for the Galilei number of between $6.2 \cdot 10^{10}$ and $1.88 \cdot 10^{12}$.

The variation of the Schmidt number in Eq. 1 was realized by varying the diffusion coefficient for oxygen ($\mathbb{D}_{O2}$). Sodium sulfite (Na_2SO_3) solutions with concentrations of 0.5 mol/L and 1 mol/L and measurements with deionized water, using the dynamic gassing-out method, were applied to vary $\mathbb{D}_{O2}$ as stated in Table 1. A variation of the Na_2SO_3 concentration leads to a change in the liquid viscosity which results in a change in the Galilei number and Schmidt number in Eq. 1. Thus, the influence of the Galilei number was considered during the determination of the exponent ε of the Schmidt number (Figure 6). The average value of $\varepsilon = -0.12$ can only be considered as a rough estimation because of the high experimental deviation for ε between values of -0.227 and 0.022 and the restricted range of variation for the Schmidt number between 417 and 878. However, the exponents of the Galilei number and Schmidt number are notably smaller than the exponents of the remaining numbers which implies that the influence of

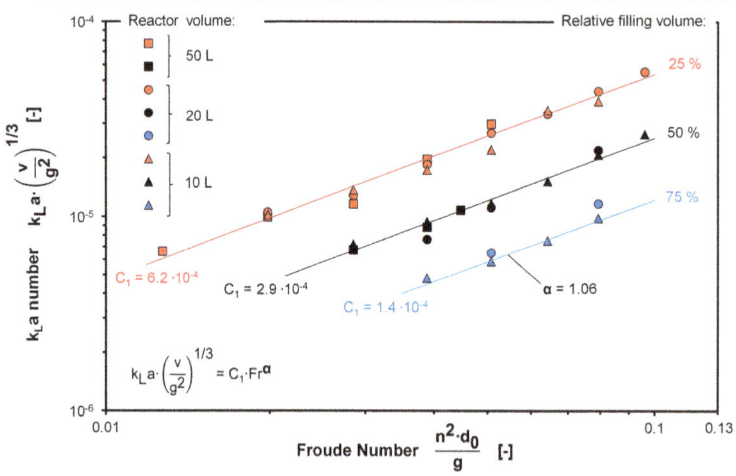

Figure 2 k_La number as a function of the Froude number for 10 L, 20 L and 50 L bioreactors at 25°C, k_La values determined with RAMOS using 1 M Na_2SO_3 solution with 10^{-7} M $CoSO_4$, 0.012 M phosphate buffer, initial pH = 8, power functions with average exponent $\alpha = 1.06$ fitted to values for relative filling volumes of 25%, 50% and 75% at shaking frequencies of 80–220 rpm with $n > N_C$ according to Eq. 2.

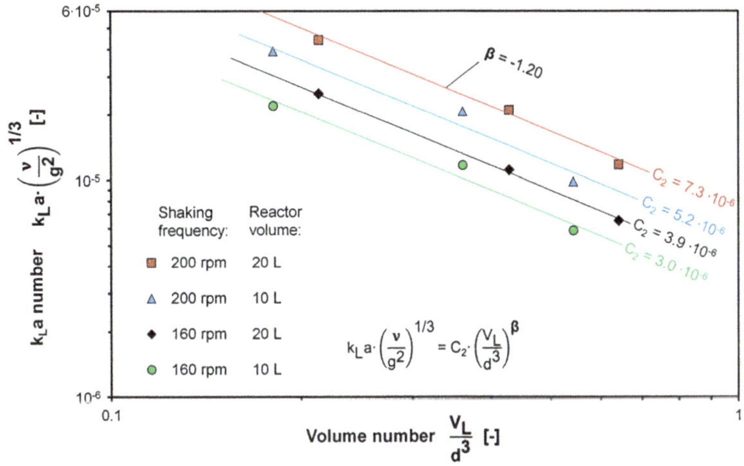

Figure 3 k_La number as a function of the Volume number for 10 and 20 L bioreactors at 25°C, relative filling volume varied between 25% and 75% at shaking frequencies of 160 rpm and 200 rpm, k_La values determined with RAMOS using 1 M Na_2SO_3 solution with 10^{-7} M $CoSO_4$, 0.012 M phosphate buffer, initial pH = 8, power functions with average exponent β = −1.20 fitted to data points.

inaccuracies in these exponents on the k_La number is also notably smaller.

With the determined exponents, the complete k_La correlation can be written as:

$$k_La \cdot \left(\frac{\nu}{g^2}\right)^{\frac{1}{3}} = 1.06 \cdot 10^{-3} \cdot \left(\frac{n^2 \cdot d_0}{g}\right)^{1.06} \cdot \left(\frac{V_L}{d^3}\right)^{-1.2} \cdot \left(\frac{d_0}{d}\right)^{-1.06} \cdot \left(\frac{d^3 \cdot g}{\nu^2}\right)^{-0.12} \cdot \left(\frac{\nu}{\mathbb{D}_{O2}}\right)^{-0.12} \tag{3}$$

The constant coefficient of $1.06 \cdot 10^{-3}$ in Eq. 3 was fitted by comparing all measured and calculated values using the method of least squares. Equation 3 can be simplified to the following final k_La correlation:

$$k_La = 1.06 \cdot 10^{-3} \cdot d^{4.3} \cdot n^{2.12} \cdot V_L^{-1.2} \cdot \nu^{-0.21} \cdot \mathbb{D}_{O2}^{0.12} \cdot g^{-0.51} \tag{4}$$

The same exponents for the Froude number and the Geometric number in Eq. 3 led to an elimination of the shaking diameter d_0 in Eq. 4. Equations 3 and 4 are only applicable for shaking frequencies that are higher than the critical frequency N_C according to Eq. 2 and are restricted to the following variation ranges for the Froude number: $0.013 < \frac{n^2 \cdot d_0}{g} < 0.097$, the Volume number: $0.18 < \frac{V_L}{d^3} < 0.65$,

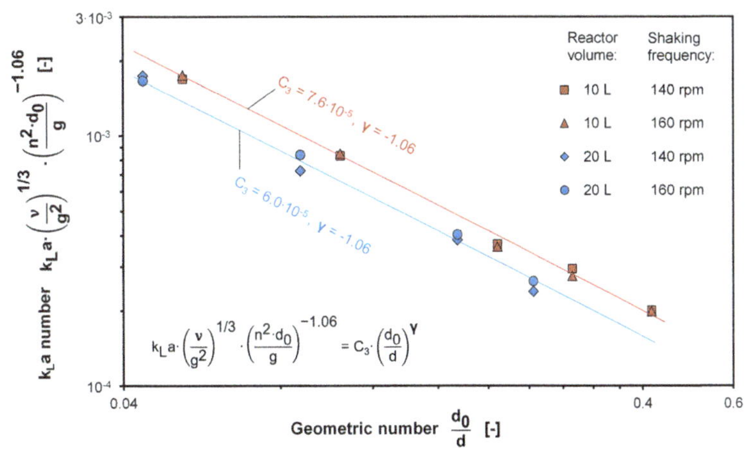

Figure 4 Product of k_La number and Froude number as a function of the Geometric number for 10 L and 20 L bioreactors at 25°C, shaking diameter d_0 varied between 1.25 cm and 10 cm at shaking frequencies of 140 rpm and 160 rpm, k_La values determined with RAMOS using 1 M Na_2SO_3 solution with 10^{-7} M $CoSO_4$, 0.012 M phosphate buffer, initial pH = 8, power functions with average exponent γ = -1.06 fitted to data points.

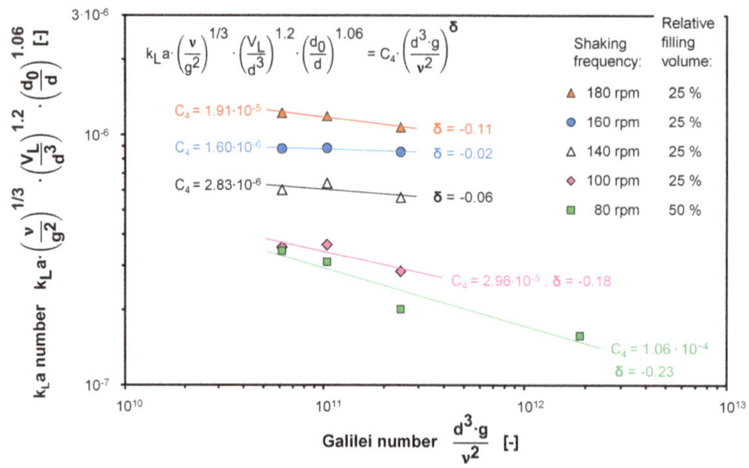

Figure 5 Product of k_La number, Volume number and Geometric number as a function of the Galilei number measured at 25°C, reactor diameter varied between 0.25 cm and 0.75 cm using 10 L, 20 L, 50 L and 200 L bioreactors, k_La values determined with RAMOS using 1 M Na_2SO_3 solution with 10^{-7} M $CoSO_4$, 0.012 M phosphate buffer, initial pH = 8, power functions fitted to data points for different shaking frequencies and relative filling volumes, average exponent δ = -0.12.

the Geometric number: $0.06 < \frac{d_0}{d} < 0.42$, the Galilei number: $6.2{\cdot}10^{10} < \frac{d^3 \cdot g}{\nu^2} < 1.88{\cdot}10^{12}$ and the Schmidt number: $417 < \frac{\nu}{\mathbb{D}_{O2}} < 878$. Exponents of the reactor diameter d and the shaking frequency n in Eq. 4 are more than double those of respective exponents of correlations for shake flasks as summarized by Klöckner and Büchs [34]. For instance, Henzler and Schedel [26] proposed the following k_La correlation for shake flasks:

$$k_La = 0.5 \cdot d^{2.03} \cdot n \cdot V_L^{-0.89} \cdot \nu^{-0.24} \cdot \mathbb{D}_{O2}^{0.5} \cdot g^{-0.13} \cdot d_0^{0.25} \quad (5)$$

An exponent of 2.03 for the influence of the reactor diameter d and 1 for the influence of the shaking frequency n were specified in Eq. 5 for shake flasks. Different characteristics with respect to the oxygen transfer are based on differences in the shape of the reactor systems. The conical shape of the shake flask wall prevents a strong expansion of the liquid surface with increasing shaking frequency and vessel size. This is not the case in cylindrical orbitally shaken bioreactors, resulting in higher exponents for the corresponding variables n and d.

Experimentally determined k_La values in this study were compared with calculated values using Eq. 4. A comparison between measured and calculated k_La values in scales from 2 L to 200 L is presented in Figure 7. Filled symbols indicate measuring values where the shaking frequency was higher than the critical frequency N_C according to Eq. 2. A range for the accuracy of +/− 30% was determined for bioreactors with volumes from 2 L to 50 L as depicted in Figure 7. Open symbols that fall within the +/− 30% accuracy range are still in-phase but close to out-of-phase operation. It was necessary to exclude these data points to define a precise validity range for the correlation. Three measurement values of the 200 L reactor system were slightly above the 30% range (Figure 7). This deviation might be caused by the oxygen-transfer enhancing influence of welded plastic seams in the 200 L bag system. This effect was not considered in Eq. 4, because most of the measurements for

Table 1 Oxygen solubility, diffusion coefficient and kinematic viscosity for the applied solutions at 25°C

Composition of the solution	Solubility (L_{O2}) [mol/(m³·bar)]	Diffusion coefficient for oxygen ($\mathbb{D}_{O2}$) [m²/s]	Kinematic viscosity (ν) [m²/s]
Deionized water	1.227	$2.14 \cdot 10^{-9}$	$0.89 \cdot 10^{-6}$
0.5 mol/L Na_2SO_3	0.844	$1.897 \cdot 10^{-9}$	$1.10 \cdot 10^{-6}$
$1 \cdot 10^{-7}$ mol/L $CoSO_4$			
0.012 mol/L phosphate buffer (pH 8)			
1 mol/L Na_2SO_3	0.561	$1.688 \cdot 10^{-9}$	$1.48 \cdot 10^{-6}$
$1 \cdot 10^{-7}$ mol/L $CoSO_4$			
0.012 mol/L phosphate buffer (pH 8)			

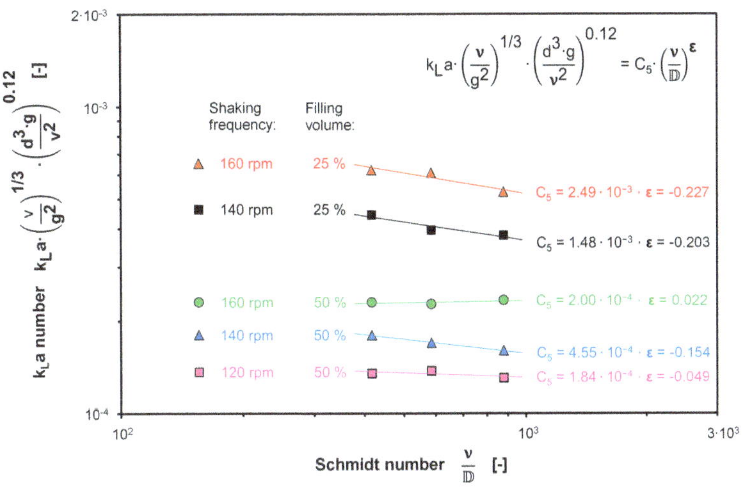

Figure 6 Product of $k_L a$ number and Galilei number as a function of the Schmidt number (Sc) measured with 20 L bioreactors at 25°C using water (Sc = 416) and sulfite solutions with 0.5 M (Sc = 580) and 1 M Na_2SO_3 (Sc = 877) with 10^{-7} M $CoSO_4$ and 0.012 M phosphate buffer, initial pH = 8, power functions fitted to data points for different shaking frequencies and filling volumes, determined average exponent ε = -0.12.

determining the exponents in Eq. 4 were conducted with disposable systems with volumes ranging from 2 L to 50 L that provide a smooth and homogeneous surface of the inner reactor wall.

The application of dimensionless numbers requires geometrical similarity. The utilized reactor systems for determining the exponents in Eq. 4 had a cylindrical reactor wall with flat bottom geometry. Thus, Eq. 4 is only applicable for such cylindrical bioreactors. The developed $k_L a$ correlation can be used to calculate suitable shaking parameters for a sufficient oxygen supply by calculating the maximum oxygen transfer capacity (OTR_{max}) according to Eq. 10. To avoid an oxygen limitation during scale-up it is essential to select shaking parameters where the OTR_{max} is higher than the oxygen uptake of the culture.

The newly developed $k_L a$ correlation was used to determine suitable cultivation conditions for a sufficient oxygen supply of *Nicotiana tabacum* cv. Bright Yellow-2 (BY-2) suspension cells. A maximum OTR value of

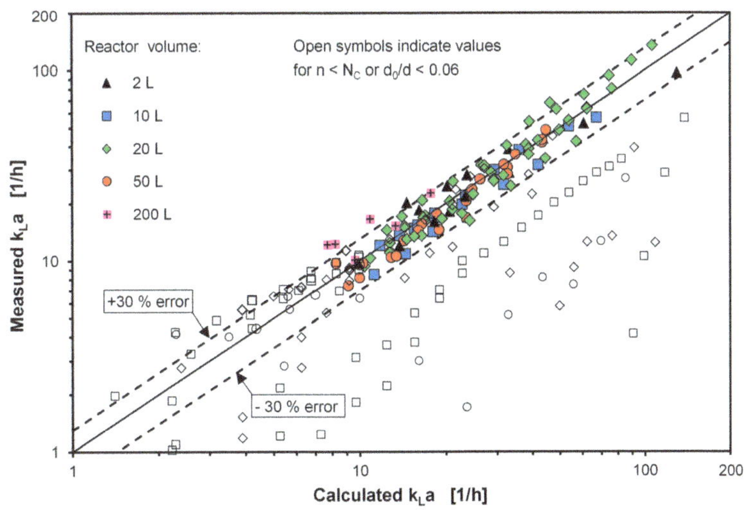

Figure 7 Comparison between measured and calculated $k_L a$ values at 25°C for Corning roller bottles (2 L), Nalgene vessels (10 L, 20 L, 50 L) and CultiBag ORB (200 L) bioreactors. Shaking frequencies of between 60 rpm and 220 rpm and relative filling volumes between 25% and 75%. Open symbols indicate values for n < N_C or d_0/d <0.06. Values measured with water and sulfite solutions with 0.5 M and 1 M Na_2SO_3, 10^{-7} M $CoSO_4$, 0.012 M phosphate buffer and initial pH = 8.

6–8 mmol/(L·h) was measured during the cultivation of tobacco BY-2 cells in 250 mL shake flasks using Murashige and Skoog medium supplemented with 30 g/L sucrose [35]. Thus, a maximum oxygen transfer capacity of OTR_{max} = 8 mmol/(L·h) is required to ensure a sufficient oxygen supply during cultivation of BY-2 cells in orbitally shaken bioreactors. As specified with Eq. 10, the OTR_{max} is a function of the k_La, the dissolved oxygen concentration at equilibrium conditions (L_{O_2}) and the oxygen partial pressure in the gas phase (p_{O_2}). The dissolved oxygen concentration at equilibrium conditions in Murashige and Skoog medium supplemented with 30 g/L sucrose at ambient air pressure and 25°C was specified by Curtis [36] as: $L_{O_2} \cdot p_{O_2}$ = 7.92 mg/L : 32 g/mol = 0.248 mmol/L. For the minimum required k_La value it follows:

$$k_La = \frac{OTR_{max}}{L_{O_2} \cdot p_{O_2}} = \frac{8\ \text{mmol/(L} \cdot \text{h)}}{0.248\ \text{mmol/L}} = 32\ \frac{1}{h} \quad (6)$$

Consequently, during scale up from shake flasks to orbitally shaken bioreactors a minimum k_La value of 32 1/h is required to avoid an oxygen limitation of tobacco BY-2 cells cultivated in Murashige and Skoog medium containing 30 g/L sucrose. The required shaking parameters were determined according to Eq. (4) with a measured kinematic viscosity of $\nu = 1.18 \cdot 10^{-6}$ m²/s, a diffusion coefficient for oxygen of $D_{O2} = 1.72 \cdot 10^{-9}$ m²/s estimated according to Jamnongwong et al. [37] for Murashige and Skoog medium and the gravitational acceleration of g = 9.81 m/s². The required shaking frequency for a 10 L reactor system (d = 0.24 m) with a filling volume of V_L = 3 L results from Eq. (4):

$$n = \left(\frac{k_La}{1.06 \cdot 10^{-3} \cdot d^{4.3} \cdot V_L^{-1.2} \cdot \nu^{-0.21} \cdot \mathbb{D}_{O2}^{0.12} \cdot g^{-0.51}}\right)^{\frac{1}{2.12}} = 156 \cdot \text{rpm} \quad (7)$$

A minimum shaking frequency of n = 156 rpm is required according to Eq. (7) to avoid an oxygen limitation at the aforementioned conditions. Figure 8 shows a comparison of the OTR signals of BY-2 cells cultivated in a 10 L orbitally shaken bioreactor and a 250 mL RAMOS shake flask.

A similar trend of the OTR signal was achieved at a shaking frequency of n = 160 rpm in the 10 L reactor system compared to the 250 mL shake flask. A plateau of the OTR signal, as it usually appears during cultivations with oxygen limitation [38], was not observed.

The minimum required shaking frequency for the cultivation of BY-2 cells in a 20 L reactor system (d = 0.286) with a filling volume of V_L = 5 L is determined analogously to Eq. (7) to: n = 146 rpm. Thus, a shaking frequency of n = 160 rpm leads to a sufficient oxygen transfer during the cultivation of BY-2 cells. Figure 9 shows a comparison of the oxygen transfer and growth parameters during the cultivation of BY-2 cells in a 20 L bioreactor system compared to a 250 mL shake flask culture. The dissolved oxygen tension (DOT) was additionally monitored in the 20 L bioreactor system (Figure 9). The DOT signal was always above 20% during the cultivation, indicating a sufficient oxygen supply. As depicted in Figure 9, a comparable trend of growth parameters between the 20 L bioreactor system and 250 mL shake flasks was observed. An increase of the filling volume in the 20 L reactor system from 5 L to 10 L at otherwise equal conditions leads to a reduced k_La value according to Eq. (7) of k_La =17 1/h. This value is significantly lower than the required k_La value of 32 1/h for the cultivation of BY-2 cells. Consequently, a filling volume of 10 L leads to an oxygen limitation at the specified conditions as shown in Figure 10. The oxygen limitation is indicated by the plateau of the OTR signal after a cultivation time of 88 h. At the same time the dissolved oxygen tension was close to zero as an additional proof for an oxygen limitation. The results demonstrated that the developed k_La correlation can be used to determine suitable cultivation conditions for a sufficient oxygen transfer at different scales.

Conclusions

A universally applicable equation was defined for calculating the mass transfer coefficient k_La in disposable cylindrical bioreactors with volumes ranging from 2 L to 200 L. Important parameters as the critical circulation frequency for the induction of liquid motion as well as in-phase operation conditions were considered during the experiments and the data evaluation. It was demonstrated that the final k_La equation can be applied to determine the OTR_{max} of cylindrical bioreactors and, in this way, enables the identification of suitable cultivation conditions such as for plant cell suspension cultures. As a result, the derived k_La equation is an essential tool for the correct application of cylindrical disposable shaken bioreactors on a bench- to pilot-scale.

Materials and methods

Cultivation and agitation systems

Table 2 presents an overview of the different geometries and material properties of the utilized disposable reactor systems and the respective shaking platform used in this study.

Model for the gas/liquid oxygen transfer

The oxygen transfer rate (OTR) from the gas phase to the liquid phase is commonly described as a function of the volumetric mass transfer coefficient (k_La) and

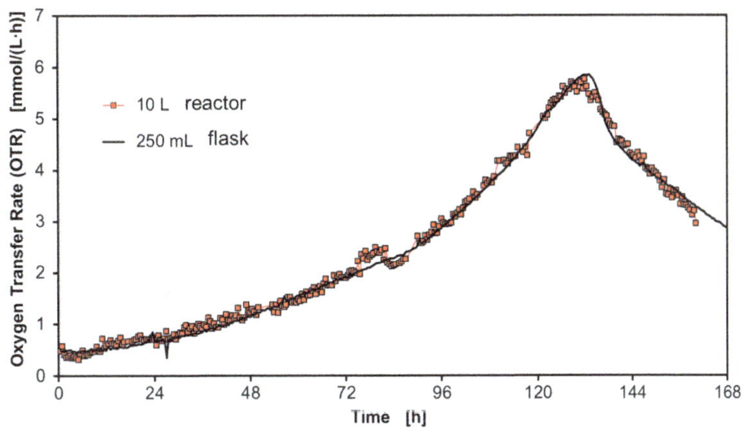

Figure 8 Oxygen Transfer Rate (OTR) during the cultivation of *N. tabacum* BY-2 MTAD cells cultivated in Murashige and Skoog medium in a 10 L bioreactor, $V_L = 3$ L, n = 160 rpm, $d_0 = 7$ cm and 250 mL shake flask, $V_L = 50$ mL, n = 180 rpm, $d_0 = 5$ cm.

the concentration difference between oxygen at the saturated gas/liquid interface (C_L^*) and oxygen in the bulk phase (C_L) as:

$$OTR = k_L a \cdot (C_L^* - C_L) \tag{8}$$

The oxygen concentration C_L^* is a function of the oxygen solubility (L_{O_2}) and the oxygen partial pressure in the gas phase (p_{O_2}) With Eq. 8, it follows:

$$OTR = k_L a \cdot (L_{O_2} \cdot p_{O_2} - C_L) \tag{9}$$

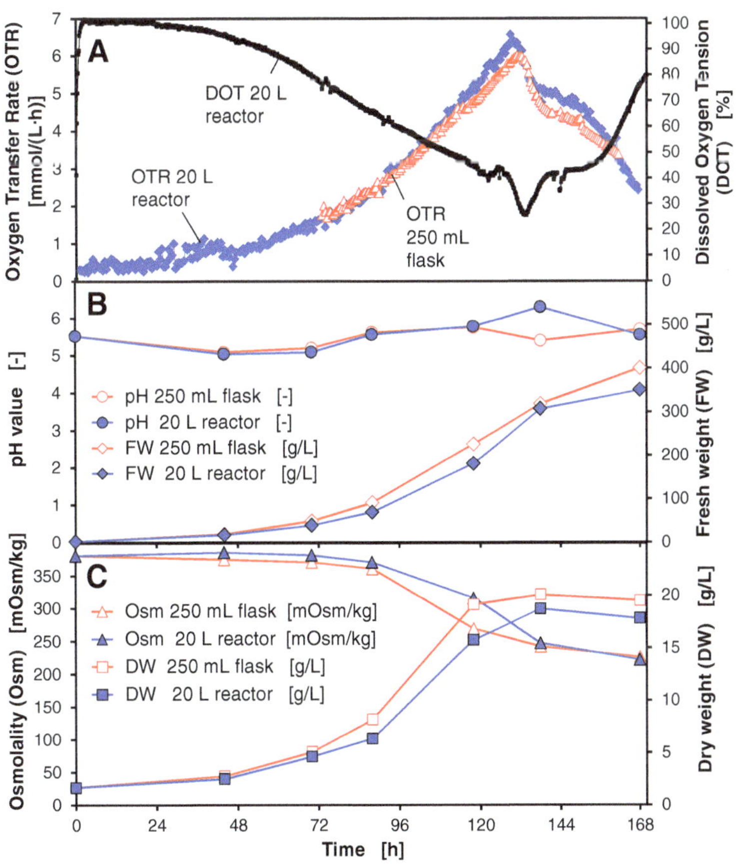

Figure 9 Comparison of oxygen transfer and growth of *N. tabacum* BY-2 MTAD cells cultivated in KNO_3 enriched Murashige and Skoog medium; 20 L bioreactor, $V_L = 5$ L, n = 160 rpm, $d_0 = 7$ cm and 250 mL shake flasks, $V_L = 50$ mL, n = 180 rpm, $d_0 = 5$ cm.

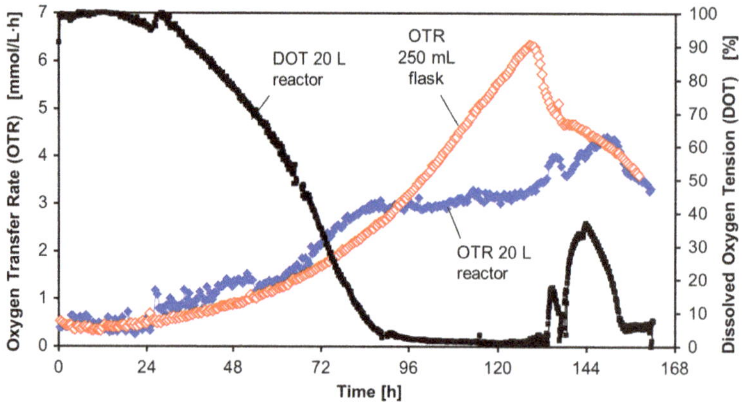

Figure 10 Oxygen Transfer Rate (OTR) and Dissolved Oxygen Tension (DOT) during the cultivation of *N. tabacum* BY-2 MTAD cells in KNO_3 enriched Murashige and Skoog medium in a 20 L reactor, $V_L = 10$ L, n = 160 rpm, $d_0 = 7$ cm and 250 mL shake flask, $V_L = 50$ mL, n = 180 rpm, $d_0 = 5$ cm.

The oxygen solubility (L_{O_2}) depends on the concentrations of salts and organic compounds in the cultivation media and can be estimated for salt solutions according to Weisenberger and Schumpe [39] or for aqueous solutions of organic compounds according to Rischbieter et al. [40]. The influence of media components on the oxygen solubility, as discussed by Curtis for plant cell cultivation media [36], depends on the specific media composition. The maximum oxygen transfer capacity (OTR_{max}) is reached at a dissolved oxygen concentration in the liquid bulk phase close to zero ($C_L \approx 0$ mol/L) [28,41]. For the OTR_{max} follows from Eq. 9:

$$OTR_{max} = k_L a \cdot L_{O_2} \cdot p_{O_2} \quad (10)$$

Equation 10 can be rewritten as:

$$k_L a = \frac{OTR_{max}}{p_{O_2} \cdot L_{O_2}} \quad (11)$$

Values for the mass transfer coefficient $k_L a$ were calculated from the measured OTR_{max} signal according to Eq. 11.

The sulfite reaction system

A sodium sulfite (Na_2SO_3) reaction can be used to reduce the dissolved oxygen in the bulk phase (C_L), and, in this way, to simulate oxygen consumption by a biological culture [28,42-44]. The main advantages of a chemical reaction compared to a biological culture are the constant and reproducible oxygen consumption without sterility problems. The cobalt catalyzed sulfite oxidation is described by the following stoichiometric equation [44]:

$$SO_3^{2-} + 1/2 \cdot O_2 \overset{Co^{2+}}{\rightarrow} SO_4^{2-} \quad (12)$$

Different reaction rates of the sodium sulfite oxidation reaction can be adjusted by varying the cobalt concentration. A non-accelerated reaction rate with a Hatta number (Ha) of Ha < 0.3 is required for $k_L a$ measurements [44]. In this range, the sulfite reaction is able to reduce the dissolved oxygen concentration (C_L) to values close to zero. An increase in the oxygen transfer rate leads to a slight increase in C_L as described by Maier et al. [42]. Thus, C_L needs to be considered during measurements with high oxygen transfer rates, as they are usually reached in bubble

Table 2 Properties of the utilized disposable reactor systems

Product name	Nominal volume	Reactor diameter (d)	Material	Shaking platform	N_c for $V_L = 20\%$ (Eq. 2)
Corning roller bottle, 850 cm^2, easy grip vent cap	2 L	116.3 mm	PS[b]	Kühner ISF1-X	147 rpm
Nalgene 10 L Clearboy	10 L	240 mm	PC[c]	Kühner SR 200-X Pilot-Shaker	77 rpm
Nalgene 20 L Carboy	20 L	286 mm	PP[a]	Kühner SR 200-X Pilot-Shaker	76 rpm
Nalgene 50 L Carboy	50 L	379 mm	PP[a]	Kühner SR 200-X Pilot-Shaker	69 rpm
Sartorius 200 L CultiBag ORB	200 L	750 mm	LDPE[d]/EVOH[e]/EVA[f]	Kühner SB 200-X OrbShake	35 rpm

[a] PP: Polypropylene, [b] PS: Polystyrene, [c] PC: Polycarbonate, [d] LDPE: Low density polyethylene, [e] EVOH: Ethyl vinyl alcohol, [f] EVA: Ethylene-vinyl acetate.

aerated stirred tank reactors [42]. As comparatively low oxygen transfer rates of less than 16 mmol/(L·h) were achieved with the disposable reactor systems in the present work, the assumption of ($C_L \approx 0$ mol/L) is, in this case, applicable for the determination of k_La values.

Sodium sulfite (Roth, Karlsruhe, Germany, purity > 98%) dissolved in deionized water and catalyzed with cobalt sulfate (Fluka, Neu-Ulm, Germany) was used for the oxidation reaction in the liquid phase. Two different sulfite concentrations of 0.5 mol/L and 1 mol/L were used to vary the diffusion coefficient for oxygen ($\mathbb{D}_{O2}$) in the liquid phase. Both solutions were buffered with 12 mmol Na_2HPO_4/NaH_2PO_4 buffer, and a pH value of 8 was adjusted with 30% (w/w) sulfuric acid. The oxygen solubility of the solutions were calculated according to Weisenberger and Schumpe [39] and oxygen diffusion coefficients according to Akita [45]. A ratio of $\mathbb{D}_{O2,Sulfite}/\mathbb{D}_{O2,Water}$ = 0.886 was determined for the 0.5 mol/L sulfite solution and a ratio of $\mathbb{D}_{O2,Sulfite}/\mathbb{D}_{O2,Water}$ = 0.788 was found for the 1 mol/L sulfite solution in agreement with values calculated by Linek and Vacek [46] using the same model proposed by Akita [45]. The viscosity was measured with an Anton Paar MCR 301 rheometer (Anton Paar GmbH, Graz, Austria). Values for the oxygen solubility, diffusion coefficients and liquid viscosity are listed in Table 1.

Adaption of the Respiration Activity Monitoring System (RAMOS)

The Respiration Activity Monitoring System (RAMOS), initially developed for shake flasks, was adapted to the different disposable bioreactor systems with volumes ranging from 2 L to 50 L to measure the OTR_{max} in combination with the sulfite reaction system. Values in the 200 L bioreactor system were determined with the dynamic gassing out method. A detailed description of the RAMOS device for shake flasks is given by Anderlei and Büchs [38] and Anderlei et al. [47]. A scheme of the modified RAMOS for cylindrical reactors is shown in Figure 11. An electrochemical oxygen partial pressure sensor (MAX-250B, Maxtec, Salt Lake City, UT, USA) and a total pressure sensor (No. 26PCAFA6D, Honeywell Sensing and Control, Golden Valley, MN, USA) were integrated in the headspace of the reactor systems. Magnetic flipper valves (Type 0332 E, Bürkert, Ingelfingen, Germany) were used to switch between measuring and rinsing phase and a mass flow controller was used to control the aeration rate during the rinsing phase (see section about aeration). Otherwise, the setup of the RAMOS device and the OTR calculation were like those for shake flasks described by Anderlei and Büchs [38]. Measurement values for OTR_{max} were used to determined k_La values according to Eq. 11.

Aeration of the 2 L Corning roller bottle reactor

The Corning roller bottle reactor (Corning Inc., Acton, MA, USA) with vent cap is equipped with a 0.2 µm membrane that is integrated in the cap to allow a sufficient aeration during cultivation. A method developed by Anderlei et al. [48] for determining the mass transfer resistance of shake flask closures was applied to measure the diffusive mass transfer through the membrane of the cultivation system. A volume flow rate of 9 mL/min was determined with open ports at 25°C. This value was used for the active aeration with a Brooks 5850 TR mass flow controller (Brooks Instrument, Ede, The Netherlands) in the RAMOS as pictured in Figure 11. An oxygen sensor was placed on top of the reactor systems to measure the oxygen partial pressure in the gas phase. No further changes of the RAMOS for shake flasks were required to measure the OTR_{max} of the Corning roller bottle cultivation system.

Aeration of cylindrical orbitally shaken bioreactors with volumes from 10 L to 200 L

The settings of the air flow rate for the active headspace aeration of the cylindrical shaken bioreactors was transferred from the RAMOS device for shake flasks and adjusted according to the nominal reactor volume. A volume flow rate of 10 mL/min is usually applied in the RAMOS device for 250 mL shake flasks to mimic the oxygen transfer through a conventional cotton plug of shake flasks with a narrow neck. Consequently, an air flow rate of 400 mL/min was used for a volume of 10 L, 800 mL/min for 20 L, 2 L/min for 50 L and 8 L/min for 200 L bioreactors, respectively. For reactor volumes from 2–50 L, the flow rate was adjusted with a Brooks 5850-E mass flow controller (Brooks Instrument) in the RAMOS as illustrated in Figure 11. A manual air flow meter was used for the dynamic gassing-out method in the 200 L reactor system. The air flow rates in different scales were high enough to keep the absolute headspace concentration of oxygen, measured with the oxygen partial pressure sensor, above 20% during all OTR measurements.

Application of the dynamic gassing-out method for k_La measurements with water

A defined and constant headspace volume is required for OTR_{max} measurements with RAMOS [38]. This could not be provided in the flexible 200 L bag reactor system. Thus, the dynamic gassing-out method, first described by Bandyopadhyay et al. [49], was used for k_La measurements in the 200 L scale. Nitrogen was used to replace the dissolved oxygen in the liquid phase, and deionized water was used as medium. Oxygen-sensitive spots (type SP-PSt3-YAU-D5-YOP) from PreSens (PreSens GmbH, Regensburg, Germany) were applied to

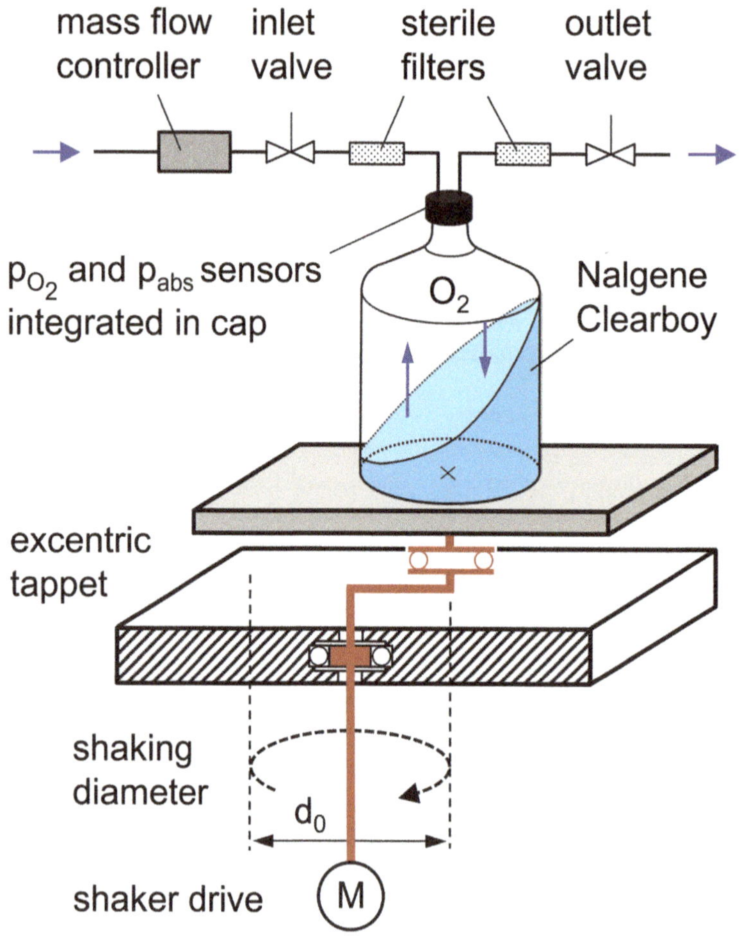

Figure 11 **Experimental setup of the RAMOS for determining k_La values in cylindrical disposable bioreactors with sensors for the oxygen partial pressure (p_{O2}) and total air pressure (p_{abs}) integrated in the reactor cap; mass flow controller, inlet and outlet valves and sterile filters used for active aeration.**

measure the dissolved oxygen tension (DOT) in the liquid phase. An electrochemical sensor (MAX-250B, Maxtec, Salt Lake City, UT, USA) was additionally used to measure the oxygen concentration in the headspace of the reactor. The reactor system was filled with deionized water according to the desired filling volume. Then, the headspace of the system was filled with nitrogen, and the reactor was shaken at 80 rpm until the DOT reached a value below 1%. Subsequently, the shaker was stopped, and the gas volume in the headspace was replaced with air until the relative oxygen concentration in the headspace reached a value above 98%. The shaker was then immediately started with the designated shaking frequency, and the DOT was recorded over time. During the required time to replace the nitrogen in the headspace with air, the diffusion of oxygen from the gas phase to the liquid phase could cause small oxygen concentration differences in the liquid phase. Therefore, measured DOT values during the first 60 s were not considered for the calculation of the mass transfer coefficient (k_La) to ensure a sufficient mixing of the liquid phase. According to Tissot et al. [16], a mixing time of 60 s can be regarded as sufficient for a 200 L reactor system using liquids with water-like viscosity and shaking frequencies above 60 rpm. During all measurements the DOT signal after the 60 s mixing step was below 4%, indicating that only small amounts of oxygen were transferred to the liquid phase during the time needed to replace the gas in the headspace. Values for k_La were calculated from the recorded DOT signal over time as described by Van Suijdam et al. [50] and recently summarized by Suresh et al. [51].

Cultivation of *Nicotiana tabacum* cv. BY-2 plant cell suspension cultures

For evaluation of the k_La correlation the transgenic *N. tabacum* cv. Bright Yellow-2 (BY-2) MTAD cell line producing the human antibody M12 was used. The generation of the transgenic BY-2 cell line is described by Raven et al. [52]. BY-2 plant cell suspension cultures were cultivated at 26°C in the dark using Murashige and Skoog media [53] with minimal organics (# M6899, Sigma Aldrich, Saint Louis, MO, USA) supplemented with 30 g/L sucrose, 0.2 g/L KH_2PO_4, 0.6 mg/L thiamine-HCl, 0.2 mg/L 2.4-Dichlorophenoxyacetic acid (2.4-D) and, where stated, additional 100 mmol/L KNO_3. The pH of the culture medium was adjusted to 5.8 with 1 mol/L KOH before autoclaving for 21 min at 121°C. Plant cell suspensions cultures were sub-cultivated weekly for cell maintenance in 250 mL Erlenmeyer flasks, filled with 50 ml cell suspension, sealed with a cotton plug and shaken at 180 rpm and 26°C in the dark. Inoculation was conducted by adding 5% (V/V) of a seven day old culture to fresh medium.

Oxygen transfer rate measurements in shake flasks

The OTR signals in 250 mL Erlenmeyer flasks were measured with a Respiration Activity Monitoring System (RAMOS). A detailed description of the device and its applications is provided by Anderlei et al. 2004 [47] and Anderlei and Büchs 2001 [38]. The non-invasive measuring system allows online monitoring of the OTR, CTR and RQ without changing the culture conditions compared to conventional 250 mL shake flasks [47]. Conventional Erlenmeyer flasks were used in addition to the RAMOS flasks to take samples during the cultivations.

Determination of fresh and dry cell weight

An electronic precision balance (SBC 31, Scaltec, Göttingen, Germany) was used to determine fresh and dry cell weight. Fresh cell weight was determined by vacuum filtration of 10 ml cell suspension for 3 min using Whatman filter paper grade 3 (# 1003–055, 55 mm diameter, Fisher Scientific GmbH, Schwerte, Germany). Prior to the filtration step the filter paper was weighted dry and after wetting with purified water. The fresh cell weight was determined from the difference in weight of the membrane with cells directly after filtration and the wet membrane without cells. The membrane with cells was dried at 105°C until the mass remained constant. The dry cell weight was determined from the difference in weight of the dried membrane with and without cells.

Determination of osmolality, dissolved oxygen tension and pH

Osmolality was determined in the supernatant with a Gonotec Osmomat 030 (Gonotec GmbH, Berlin, Germany). The device was calibrated with a two point calibration prior to each measurement. The pH value was determined with a pH510 pH meter (Eutech, Fisher Scientific GmbH, Schwerte, Germany). The dissolved oxygen tension (DOT) was measured by using oxygen sensitive sensor spots (type SP-PSt3-YAU-D5-YOP) from PreSens (PreSens GmbH, Regensburg, Germany).

Competing interests

The authors declare that they have no competing interests.

Authors' contributions

WK conducted experiments, made the data evaluation and drafted the manuscript. RG and TA provided additional data and helped to conduct the experiments. NR provided plant cell suspension cultures, helped to conduct the experiments and reviewed the manuscript. SS coordinated the experiments and reviewed the manuscript. CL calculated oxygen diffusion coefficients and solubilities of the applied solutions. JB helped to design and coordinate the study and reviewed the manuscript. All authors read and approved the final manuscript.

Acknowledgments

The authors thank the company Kühner AG for their helpful support with equipment and materials. This work has been financed within the EU Project CoMoFarm (grant agreement no.: 227420).

Author details

[1]AVT. Biochemical Engineering, RWTH Aachen University, Worringerweg 1, Aachen 52074, Germany. [2]Adolf Kühner AG, Dinkelbergstrasse 1, Birsfelden 4127, Switzerland. [3]Fraunhofer Institute for Molecular Biology and Applied Ecology (IME), Forckenbeckstrasse 6, Aachen 52074, Germany.

References

1. Oosterhuis NMG, Van den Berg HJ: **How multipurpose is a disposable bioreactor?** *Biopharm International* 2011, **24**:51–56.
2. Eibl R, Kaiser S, Lombriser R, Eibl D: **Disposable bioreactors: the current state-of-the-art and recommended applications in biotechnology.** *Appl Microbiol Biotechnol* 2010, **86**:41–49.
3. Eibl R, Löffelholz C, Eibl D: **Single-use bioreactors—an overview.** In *Single-Use Technology in Biopharmaceutical Manufacture.* Edited by Eibl R, Eibl D. Hoboken, New Jersey: John Wiley & Sons, Inc; 2010:33–51.
4. Kybal J, Vlcek V: **Simple device for stationary cultivation of microorganisms.** *Biotechnol Bioeng* 1976, **18**:1713–1718.
5. Kybal J, Sikyta B: **A device for cultivation of plant and animal cells.** *Biotechnol Lett* 1985, **7**:467–470.
6. Brod H, Vester A, Kauling J: **Opportunities and limitations of disposable technologies in biopharmaceutical processes.** *Chem Ing Tech* 2012, **84**:633–645.
7. Eibl R, Eibl D: **Disposable bioreactors in cell culture-based upstream processing.** *Bioprocess Int* 2009, **7**:18–23.
8. Rao G, Moreira A, Brorson K: **Disposable bioprocessing: the future has arrived.** *Biotechnol Bioeng* 2009, **102**:348–356.
9. Kato I, Tanaka H: **Development of a novel box-shaped shake flask with efficient gas exchange capacity.** *J Ferment Bioeng* 1998, **85**:404–409.
10. Klöckner W, Tissot S, Wurm F, Büchs J: **Power input correlation to characterize the hydrodynamics of cylindrical orbitally shaken bioreactors.** *Biochem Eng J* 2012, **65**:63–69.
11. Kato Y, Peter CP, Akgün A, Büchs J: **Power consumption and heat transfer resistance in large rotary shaking vessels.** *Biochem Eng J* 2004, **21**:83–91.
12. De Jesus MJ, Girard P, Bourgeois M, Baumgartner G, Jacko B, Amstutz H, Wurm FM: **TubeSpin satellites: a fast track approach for process development with animal cells using shaking technology.** *Biochem Eng J* 2004, **17**:217–223.
13. Zhang X, Stettler M, De Sanctis D, Perrone M, Parolini N, Discacciati M, De Jesus M, Hacker D, Quarteroni A, Wurm F: **Use of orbital shaken disposable bioreactors for mammalian cell cultures from the milliliter-scale to the 1,000-liter scale.** *Adv Biochem Eng/Biotechnol* 2010, **115**:33–53.
14. Raval K, Kato Y, Büchs J: **Comparison of torque method and temperature method for determination of power consumption in disposable shaken bioreactors.** *Biochem Eng J* 2007, **34**:224–227.
15. Raval KN, Liu C, Büchs J: **Large-scale disposable shaking bioreactors: a promising choice.** *Bioprocess Int* 2006, **4**:46. 48,50.
16. Tissot S, Farhat M, Hacker DL, Anderlei T, Kuhner M, Comninellis C, Wurm F: **Determination of a scale-up factor from mixing time studies in orbitally shaken bioreactors.** *Biochem Eng J* 2010, **52**:181–186.

17. Liu CM, Hong LN: **Development of a shaking bioreactor system for animal cell cultures.** *Biochem Eng J* 2001, **7**:121–125.
18. Zhang X, Bürki C-A, Stettler M, De Sanctis D, Perrone M, Discacciati M, Parolini N, DeJesus M, Hacker DL, Quarteroni A, Wurm FM: **Efficient oxygen transfer by surface aeration in shaken cylindrical containers for mammalian cell cultivation at volumetric scales up to 1000 L.** *Biochem Eng J* 2009, **45**:41–47.
19. Werner S, Eibl R, Lettenbauer C, Roll M, Eibl D, De Jesus M, Zhang X, Stettler M, Tissot S, Bürki C, *et al*: **Innovative, non-stirred bioreactors in scales from milliliters up to 1000 Liters for suspension cultures of cells using disposable bags and containers - a swiss contribution.** *Chimia* 2010, **64**:819–823.
20. Stettler M, Zhang X, Hacker DL, De Jesus M, Wurm FM: **Novel orbital shake bioreactors for transient production of CHO derived IgGs.** *Biotechnol Progr* 2007, **23**:1340–1346.
21. Tissot S, Michel PO, Hacker DL, Baldi L, De Jesus M, Wurm FM: **k(L)a as a predictor for successful probe-independent mammalian cell bioprocesses in orbitally shaken bioreactors.** *New Biotechnol* 2012, **29**:387–394.
22. Tissot S, Oberbek A, Reclari M, Dreyer M, Hacker DL, Baldi L, Farhat M, Wurm FM: **Efficient and reproducible mammalian cell bioprocesses without probes and controllers?** *New Biotechnol* 2011, **28**:382–390.
23. Zhang X, Stettler M, Reif O, Kocourek A, DeJesus M, Hacker DL, Wurm FM: **Shaken helical track bioreactors: Providing oxygen to high-density cultures of mammalian cells at volumes up to 1000 L by surface aeration with air.** *New Biotechnol* 2008, **25**:68–75.
24. Grimvall G: **Characteristic quantities and dimensional analysis.** *Sci Model Simul* 2008, **15**:21–39.
25. Buckingham E: **On physically similar systems; illustrations of the use of dimensional equations.** *Phys Rev* 1914, **4**:345–376.
26. Henzler HJ, Schedel M: **Suitability of the shaking flask for oxygen supply to microbial cultures.** *Bioprocess Eng* 1991, **7**:123–131.
27. Doig SD, Pickering SCR, Lye GJ, Baganz F: **Modelling surface aeration rates in shaken microtitre plates using dimensionless groups.** *Chem Eng Sci* 2005, **60**:2741–2750.
28. Hermann R, Lehmann M, Büchs J: **Characterization of gas–liquid mass transfer phenomena in microtiter plates.** *Biotechnol Bioeng* 2003, **81**:178–186.
29. Kato Y, Hiraoka S, Tada Y, Koh ST, Lee YS: **Mixing time and power consumption for a liquid in a vertical cylindrical vessel, shaken in a horizontal circle.** *Chem Eng Res Des* 1996, **74**:451–455.
30. Lotter S, Büchs J: **Utilization of specific power input measurements for optimization of culture conditions in shaking flasks.** *Biochem Eng J* 2004, **17**:195–203.
31. Peter CP, Lotter S, Maier U, Büchs J: **Impact of out-of-phase conditions on screening results in shaking flask experiments.** *Biochem Eng J* 2004, **17**:205–215.
32. Büchs J, Lotter S, Milbradt C: **Out-of-phase operating conditions, a hitherto unknown phenomenon in shaking bioreactors.** *Biochem Eng J* 2001, **7**:135–141.
33. Büchs J, Maier U, Milbradt C, Zoels B: **Power consumption in shaking flasks on rotary shaking machines: II. Nondimensional description of specific power consumption and flow regimes in unbaffled flasks at elevated liquid viscosity.** *Biotechnol Bioeng* 2000, **68**:594–601.
34. Klöckner W, Büchs J: **Advances in shaking technologies.** *Trends Biotechnol* 2012, **30**:307–314.
35. Ullisch DA, Müller CA, Maibaum S, Kirchhoff J, Schiermeyer A, Schillberg S, Roberts JL, Treffenfeldt W, Büchs J: **Comprehensive characterization of two different Nicotiana tabacum cell lines leads to doubled GFP and HA protein production by media optimization.** *J Biosci Bioeng* 2012, **113**:242–248.
36. Curtis WR: **Application of bioreactor design principles to plant micropropagation.** *Plant Cell Tiss Org* 2005, **81**:255–264.
37. Jamnongwong M, Loubiere K, Dietrich N, Hébrard G: **Experimental study of oxygen diffusion coefficients in clean water containing salt, glucose or surfactant: Consequences on the liquid-side mass transfer coefficients.** *Chem Eng J* 2010, **165**:758–768.
38. Anderlei T, Büchs J: **Device for sterile online measurement of the oxygen transfer rate in shaking flasks.** *Biochem Eng J* 2001, **7**:157–162.
39. Weisenberger S, Schumpe A: **Estimation of gas solubilities in salt solutions at temperatures from 273 K to 363 K.** *AlChE J* 1996, **42**:298–300.
40. Rischbieter E, Schumpe A, Wunder V: **Gas solubilities in aqueous solutions of organic substances.** *J Chem Eng Data* 1996, **41**:809–812.
41. Maier U, Losen M, Büchs J: **Advances in understanding and modeling the gas–liquid mass transfer in shake flasks.** *Biochem Eng J* 2004, **17**:155–167.
42. Maier B, Dietrich C, Büchs J: **Correct application of the sulphite oxidation methodology of measuring the volumetric mass transfer coefficient k(L) a under non-pressurized and pressurized conditions.** *Food Bioprod Process* 2001, **79**:107–113.
43. Funke M, Diederichs S, Kensy F, Mueller C, Büchs J: **The baffled microtiter plate: increased oxygen transfer and improved online monitoring in small scale fermentations.** *Biotechnol Bioeng* 2009, **103**:1118–1128.
44. Linek V, Vacek V: **Chemical engineering use of catalyzed sulfite oxidation kinetics for the determination of mass transfer characteristics of gas–liquid contactors.** *Chem Eng Sci* 1981, **36**:1747–1768.
45. Akita K: **Diffusivities of gases in aqueous electrolyte solutions.** *Ind Eng Chem Fund* 1981, **20**:89–94.
46. Linek V, Vacek V: **Comments on validity of measuring oxygen diffusion coefficients with polarographic oxygen electrodes.** *Biotechnol Bioeng* 1988, **31**:1010–1011.
47. Anderlei T, Zang W, Papaspyrou M, Büchs J: **Online respiration activity measurement (OTR, CTR, RQ) in shake flasks.** *Biochem Eng J* 2004, **17**:187–194.
48. Anderlei T, Mrotzek C, Bartsch S, Amoabediny G, Peter CP, Büchs J: **New method to determine the mass transfer resistance of sterile closures for shaken Bioreactors.** *Biotechnol Bioeng* 2007, **98**:999–1007.
49. Bandyopadhyay B, Humphrey AE, Taguchi H: **Dynamic measurement of volumetric oxygen transfer coefficient in fermentation systems.** *Biotechnol Bioeng* 1967, **9**:533–544.
50. Van Suijdam JC, Kossen NWF, Joha AC: **Model for oxygen transfer in a shake flask.** *Biotechnol Bioeng* 1978, **20**:1695–1709.
51. Suresh S, Srivastava VC, Mishra IM: **Techniques for oxygen transfer measurement in bioreactors: a review.** *J Chem Technol Biotechnol* 2009, **84**:1091–1103.
52. Raven N, Schillberg S, Kirchhoff J, Brändli J, Imseng N, Eibl R: **Growth of BY-2 suspension cells and plantibody production in single-use bioreactors.** In *Single-Use Technology in Biopharmaceutical Manufacture.* Edited by Eibl R, Eibl D. Hoboken, NJ, USA: John Wiley & Sons; 2010:251–261.
53. Murashige T, Skoog F: **A revised medium for rapid growth and bio assays with tobacco tissue cultures.** *Physiol Plantarum* 1962, **15**:473–497.

Optimization of 3-D organotypic primary colonic cultures for organ-on-chip applications

Asad A Ahmad[1], Yuli Wang[2], Adam D Gracz[3], Christopher E Sims[2], Scott T Magness[1,3,4] and Nancy L Allbritton[1,2*]

Abstract

Background: New advances enable long-term organotypic culture of colonic epithelial stem cells that develop into structures known as colonoids. Colonoids represent a primary tissue source acting as a potential starting material for development of an *in vitro* model of the colon. Key features of colonic crypt isolation and subsequent colonoid culture have not been systematically optimized compromising efficiency and reproducibility. Here murine crypt isolation yield and quality are optimized, and colonoid culture efficiency measured in microfabricated culture devices.

Results: An optimal incubation time of 60 min in a chelating buffer released 280,000 ± 28,000 crypts from the stroma of a single colon with 79.3% remaining intact. Mechanical agitation using an average acceleration of 1.5 × g liberated the highest quality crypts with 86% possessing well-defined lumens. Culture in 50% Matrigel resulted in the highest colonoid formation efficiency of 33 ± 5%. Immunostaining demonstrated that colonoids isolated under these conditions possessed stem/progenitor cells and differentiated cell lineages. Microfabrication substrates (glass, polystyrene, PDMS, and epoxy photoresists: SU-8 and 1002-F) were tested for compatibility with colonoid culture. PDMS promoted formation of 3-D colonoids containing stem/progenitor cells, while other substrates promoted outgrowth of a 2-D epithelial monolayer composed of differentiated cells.

Conclusion: Improved crypt isolation and 3-D colonoid culture, along with an understanding of colonic epithelial cell behavior in the presence of microfabrication substrates will support development of 'organ-on-a-chip' approaches for studies using primary colonic epithelium.

Background

Self-renewal of the colonic epithelium is driven by the proliferation of epithelial stem cells located at the base of the functional tissue subunit called the colon crypt. The rapid regeneration to renew the epithelium is driven by colonic epithelial stem cells (CESCs). Understanding the CESC biology and conditions impacting their growth and differentiation is an active area of research [1-3]. The colonic epithelium is negatively impacted by a number of inflammatory diseases, cancer and acute injuries. The high incidence of colorectal cancer (CRC) in the Western World is believed to be in part due to the high proliferation rate of the epithelial lining, and increasing evidence strongly suggests CRC may arise at the level of the stem cell [4,5]. Inflammatory bowel diseases including ulcerative colitis and Crohn's disease result from attack on the crypt cells by inflammatory infiltrates [6,7]. Due to technical challenges for the *in vitro* assessment of colonic mucosa and crypts, studies of colonic physiology and pathophysiology have been restricted primarily to *in vivo* inspection. *In vivo* studies by endoscopy or noninvasive imaging have enabled examination of living colonic tissue at a macroscopic, but not cellular scale. Histological evaluation of fixed tissue has permitted study at the cellular level, but with the loss of the rich and dynamic qualities of the living tissue.

Recent breakthroughs in the understanding of fundamental morphogenetic pathways and their contributions to intestinal homeostasis have enabled culture methods to be devised that successfully generate 3-D crypt-like cellular spheroids, or 'colonoids', from isolated crypts or purified stem cells [3,8-11]. Colonoids are sustainable *in vitro* for long periods up to a year. Rapid *ex vivo*

* Correspondence: nlallbri@unc.edu
[1]Department of Biomedical Engineering, University of North Carolina, Chapel Hill, NC, 27599 and North Carolina State University, Raleigh, NC 27695, USA
[2]Department of Chemistry, University of North Carolina, Chapel Hill, NC 27599, USA
Full list of author information is available at the end of the article

establishment of colonoids in culture is potentiated by three factors: Wnt-3A, the Wnt-agonist R-spondin1 and the BMP-antagonist Noggin. *In vivo*, Wnt-3A is essential for stem cell maintenance [8]; excess R-spondin1 induces hyperplasia; Noggin at supraphysiologic concentrations produces an expansion in crypt numbers [11]. In the colonoid-culture system, CESCs, progenitors and differentiated cell lineages are present, and their composition can be adjusted by the concentration of these and other growth factors [3]. The availability of the colonoid culture system is expected to open the door to future investigations into the CESC niche and the contribution of morphogenetic cues in crypt homeostasis and organization [12]. These organotypic culture methods will have widespread impact on studies of intestinal biology, host-pathogen interactions, neoplasia and regenerative medicine. Furthermore, a better understanding of the optimal crypt isolation and culture conditions may enable the creation of novel microscale devices to recapitulate gut function *in vitro* using primary cells.

While three-dimensional (3-D) cell culture systems better mimic the microstructure of intact organs relative to 2-D cultures, the 3-D systems still fail to fully recapitulate organ-level physiologic functions presumably due to an inability to fully control the microenvironment of the organoid. Consequently, a growing trend is to build 'organ-on-chip' devices which integrate the 3-D tissue culture systems with microdevice technologies to offer enhanced control of both surface and fluidic conditions [13-16]. However due to the difficulty in obtaining and isolating primary tissue, these 'organ-on-chip' devices often utilize tumor cell lines which are incapable of demonstrating organ-level physiologic function. For example, 'gut-on-chips' devices are frequently assembled by placing Caco-2 tumor cells within microdevices [17-19]. The Caco-2 tumor cell line has been adapted for tissue culture and poorly mimics the intestinal epithelium in terms of architecture, growth factor response, differentiation, gene expression and susceptibility to apoptosis [20,21]. A significant challenge to the 'organ-on-chip' community is the development of optimized strategies to isolate high-quality primary cells for culture within a microdevice.

Although recent work has enhanced *in vitro* intestinal culture, isolation of the crypts and propagation of the colonoids has not been systematically optimized. Existing protocols fail to quantify overall yield and viability of the isolated crypts over time. The current work focuses on maximizing the yield of viable, high-quality crypts obtained from resected colon and enhancing the overall culture efficiency to produce large numbers of living colonoids from the isolated crypts. Since the cell microenvironment impacts colon cell fate and function, further characterization of the matrix concentration and identification of biocompatible substrates for colonoid culture were also performed. Microengineered environments are increasingly used to direct tissue and stem cell organization so that commonly used materials for microfabrication (including glass, polydimethoxysilane, polystyrene and epoxy photoresists), were assessed for their ability to support colonoid formation. This paper focuses on three major points of emphasis: 1) standardization of crypt isolation protocol, 2) optimization of Matrigel concentration for colonoid formation, and 3) crypt cell interaction with various substrates. We believe this research will support future development of intestinal studies and 'organ-on-chip' endeavors.

Results and discussion

Optimization of incubation time with chelating agents to remove epithelium from basement membrane

In the initial step of crypt isolation, the colon is incubated in a buffer to chelate divalent cations and reduce disulfide bonds. Chelation of divalent cations reduces crypt-stromal adhesion by binding the calcium and magnesium ions required for receptor interactions between the basement membrane and stromal cells [3,20]. The chelation-buffer, incubation time was optimized by varying the time (30, 60, 90 min) in which the colon was placed in a standard buffer with EDTA and DTT, initially described by Booth *et al.* (Figure 1A) [20]. This buffer was chosen due to its past usage and reported high cell viability. All other isolation steps were held constant. Isolated crypts were assayed for the total yield of intact and broken crypts. The presence of intact crypts was used as an indicator of the extent of tissue trauma since these structures are easily fragmented when subjected to significant stress or harsh chemical conditions. Liberated crypts were considered to be intact if they were at least 150 μm in length. Utilization of a distal colon from the *Sox9*eGFP-CAGDsRed mouse model permitted facile evaluation of the viability of stem cell/progenitor (green plus red fluorescence) and differentiated lineages (red fluorescence) of the liberated crypts immediately after retrieval from the tissue [3]. The CAGDsRed mouse line, which ubiquitously expresses the red fluorescent protein DsRed, was bred with *Sox9*eGFP mice, which expresses eGFP under control of the *Sox9* promoter [3]. Previous work has demonstrated that the presence of the *Sox9* transcription factor is a distinguishing characteristic of colonic stem and progenitor cells.

Incubation of a single distal colon for 30 min in the EDTA-containing buffer resulted in a total yield of 139,000 ± 22,000 crypts of which 69.7% were intact. The 60-min incubation yielded 280,000 ± 28,000 crypts with 79.3% intact and 90-min incubation produced 360,000 ± 41,000 crypts with 65.9% intact. The 30-min incubation period provided the lowest yield of intact and total crypts,

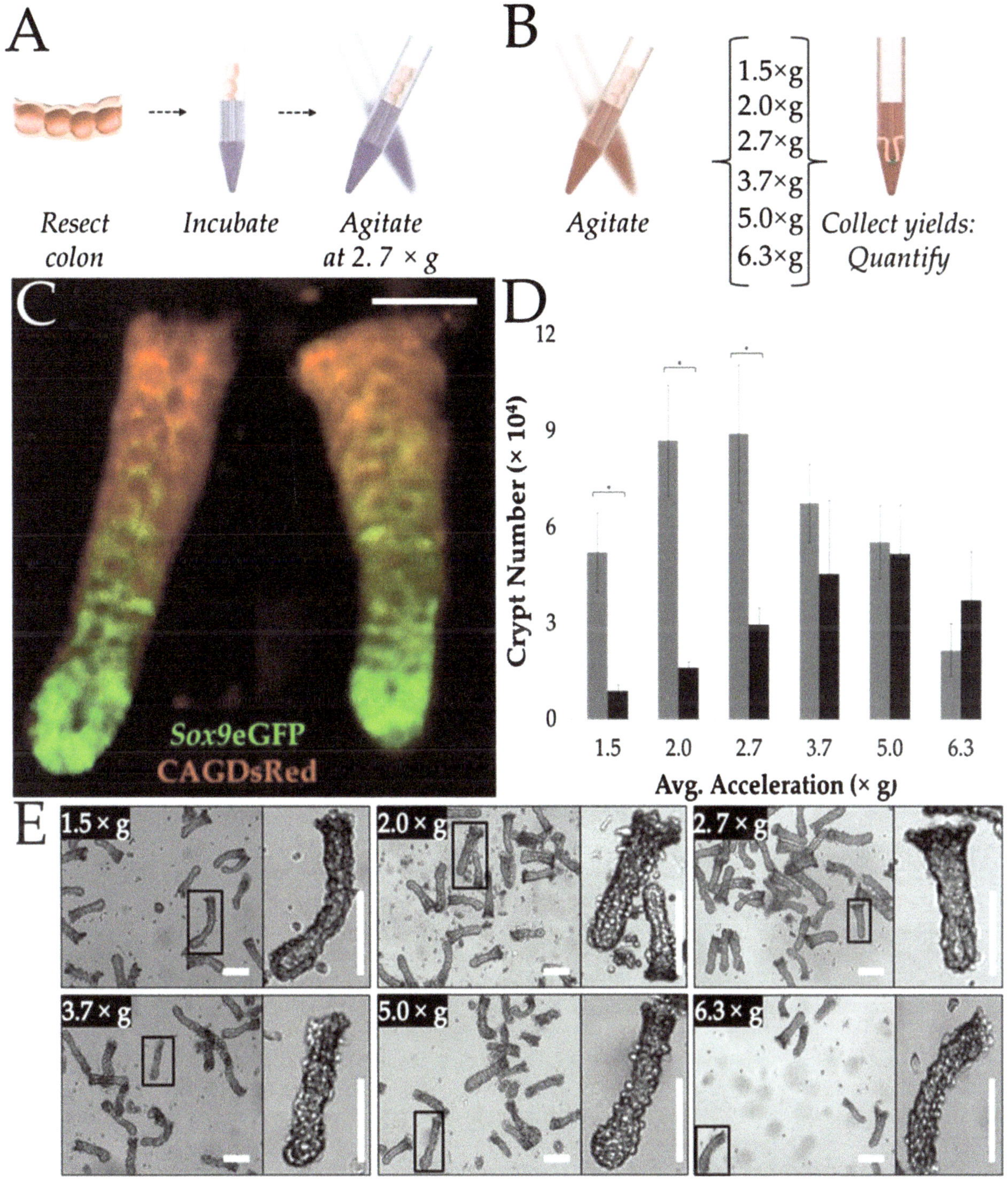

Figure 1 Isolation of crypts from a mouse colon. (A) Schematic of crypt isolation. The resected colon was incubated in chelating buffer, washed and then mechanically agitated. **(B)** Schematic of the strategy to identify the optimal acceleration intensity needed to retrieve crypts: the tissue was incubated in chelating buffer, rinsed and then sequentially agitated at different acceleration intensities. After each agitation, the crypt-rich supernatant was collected and assayed. **(C)** Fresh crypts isolated using an acceleration intensity of 1.5 × g displaying eGFP fluorescence, indicating *Sox9* expression (green). Scale bar = 40 μm. **(D)** Quantification of the number of intact and broken crypts at each of the six acceleration intensities tested. Grey bars indicate intact-crypt yield and black bars indicate broken-crypt yield. **(E)** Brightfield images of isolated crypts isolated at different accelerations intensities. Crypts isolated using an acceleration intensity of 1.5 × g display a visible lumen, indicating unperturbed crypt morphology. Scale bar = 150 μm.

probably as a result of inadequate time for the chelating agents to be effective in disrupting submucosal adhesion. Although the 90-min incubation produced a higher overall yield, the 60-min incubation retrieved a higher percentage of intact crypts and resulted in more than double the number of intact crypts relative to that after a 30-min incubation. Therefore, a 60-min incubation was chosen for all subsequent experiments.

Optimization of acceleration intensity required to release crypts

Following chelation of divalent ions and disruption of the adhesion between the epithelium and the basement membrane, mechanical agitation is use to remove crypts from the underlying tissue. Most protocols instruct "vigorous agitation of the tissue" to retrieve crypts, without quantification of the force or accelerative intensities involved [3,8,9,11,20]. To develop a reproducible protocol, varying average acceleration intensities were quantified for mechanically agitating the tissue in releasing crypts from the underlying stroma. Initially the average acceleration intensity achieved during agitation for 5 s was varied (1.5, 2.0, 2.7, 3.7, 5.0 and 6.3 × g) to optimize the agitation step (Additional file 1: Figure S1). To minimize the number of animals used, the colon was agitated for 5 s at the lowest acceleration intensity followed by settling of the tissue remnant and collection of the crypt-containing supernatant. Fresh isolation buffer was then added to the colon and the tissue was agitated at the next highest acceleration intensity after which the supernatant was again collected. This procedure was repeated until 6 crypt-containing supernatants were collected (Figure 1B). Each fraction was assayed for the total number of crypts, the number of intact crypts and crypt quality. Crypt quality was quantified by measuring the percentage of crypts that were both intact and retained an identifiable lumen. Identification of the lumen insured that the crypts possessed the basic morphology present *in vivo*. Utilization of the *Sox9*eGFP-CAGDsRed mouse model permitted verification of the quality, as only crypts possessing intact stem-cells possessed green fluorescence at the crypt base (Figure 1C). The total number of undamaged crypts with an identifiable lumen was greatest for acceleration intensities of 1.5, 2.0 and 2.7 × g (52,000 ± 15,000, 87,000 ± 17,000 and 90,000 ± 21,000 crypts, respectively) (Figure 1D). An ANOVA comparison revealed that the 1.5, 2.0 and 2.7 × g agitation acceleration intensities produced statistically different yields ($p < 0.01$) of intact versus broke crypts. The percentage of crypts with the appropriate morphology was optimal when the acceleration intensities were 1.5 × g (85.7%) and 2.0 × g (84.3%). At increased acceleration intensities (>3.7 × g), the lumens collapsed (Figure 1E). While progressively higher acceleration intensities liberated more crypts, the apparent quality of the crypts was also diminished as the acceleration intensities increased. For these measurements, the optimal compromise between crypt yield and quality was thus determined to be 1.5 and 2.0 × g. Since an untested combination of chelation-buffer incubation times and agitation conditions might have proved superior, a broad range of combinations of incubation times and acceleration intensities were assessed (Additional file 2: Table S1). Of the conditions tested, 60-min incubation in chelation buffer and an acceleration intensity of 1.5 × g yielded the greatest percentage of crypts with high-quality morphology. eGFP was expressed in 36 ± 4% of the crypt area, demonstrating that crypts isolated under these conditions possessed intact stem/proliferative cells (Additional file 3: Figure S2). Since the intended application of this work was the culture of viable crypts with formation of colonoids, these conditions were used for all subsequent experiments. When a higher yield of crypts is required without regard to quality, for example in gene expression studies, longer incubation times and greater acceleration intensities would generate significantly larger sample sizes and might be preferable.

Optimization of matrigel concentration for colonoid culture

Laminin-rich Matrigel is believed to provide the required matrix contacts for crypt cells mimicking that supplied by the underlying stroma *in vivo* [3,8,10]. Additionally, it is likely that Matrigel contains critical growth factors to maintain the crypt cells. In all past reports, crypts were cultured in 100% Matrigel, although it is unknown if this is the optimal concentration for colonoid growth. At 100%, Matrigel is extremely viscous, quick to gel and difficult to load into confined spaces such as those in microfabricated devices (*e.g.* microfluidic channels). For these reasons, four concentrations of Matrigel (25%, 50%, 75% and 100 vol% in complete culture medium (CCM) plus growth factors) were assessed for the ability to support colonoid formation. Crypts were isolated using the optimized protocol described above and were then plated on a microwell formed from native 1002-F such that the crypts remained suspended in Matrigel and not in contact with the 1002-F surface. The Matrigel-encapsulated crypts were imaged daily by brightfield and fluorescence microscopy (Figure 2A). The percentage of crypts forming colonoids was quantified as the number of budding crypts divided by the number of total crypts plated ($n = 4$ experiments for each Matrigel concentration with an average of 113 ± 32 crypts/experiment). Interestingly, 100% Matrigel was the least effective in yielding colonoid growth (18 ± 1%) after 7 days of culture in microwells (Figure 2B). 50% Matrigel supported the highest percentage of colonoid formation (33 ± 5%) followed by 75% and 25% Matrigel (23 ± 3% and 20 ± 7%, respectively) at day 7, as determined by colonoid morphology. To verify that crypts isolated under

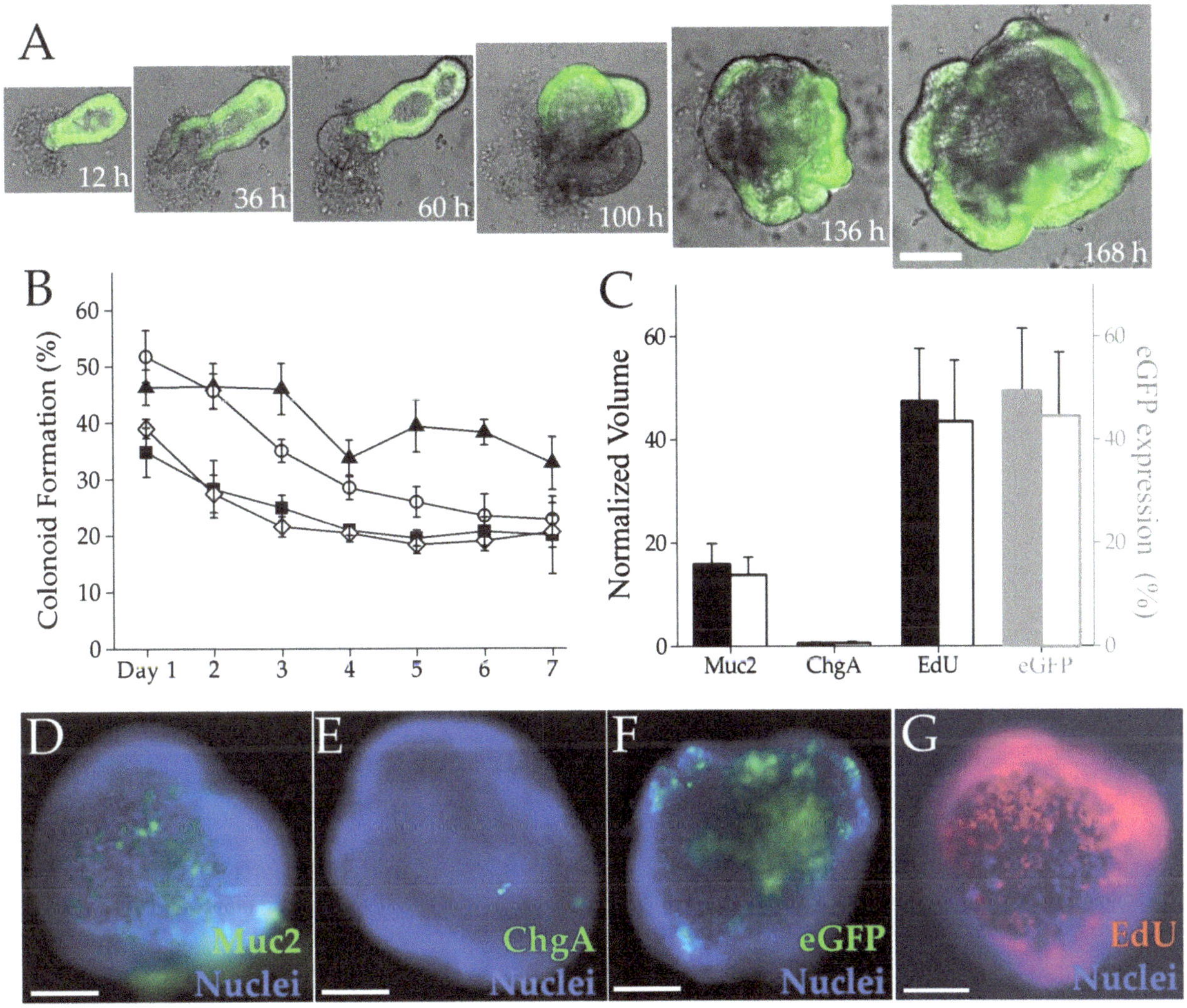

Figure 2 Effect of Matrigel concentration on *in vitro* expansion of colonic crypts into 3-D colonoids. (A) Serial overlaid brightfield and eGFP fluorescence images of the same colonoid over 1 week in culture. Scale bar = 50 µm. **(B)** Effect of Matrigel concentration on the percentage of crypts growing into colonoids over 1 week of culture. Squares, triangles, circles and diamonds represent 25%, 50%, 75% and 100% Matrigel, respectively. 50% Matrigel provides the optimum 3-D growth environment for the colonoids. **(C)** Quantification of the cell properties in the colonoids. Shown is the colonoid volume (left y axis, black) staining positive for Muc-2, ChgA or EdU divided by that positive for Hoechst 33342 when colonoids were cultured in 100% (filled bars) or 50% (open bars) Matrigel. The volume of the colonoid expressing eGFP relative to that expressing dsRed is shown on the right y-axis (grey) for colonoids cultured in 100% (filled bars) or 50% (open bars) Matrigel. **(D-E)** Colonoids were cultured for 1 week and then stained by immunohistochemistry for: **(D)** mucin-2 (goblet cell marker: green) and **(E)** chromogranin-A (enteroendocrine marker: green). **(F)** A crypt obtained from a *Sox9*eGFP-CAGDsRed mouse was cultured for 1 week and then imaged for eGFP fluorescence. **(G)** Fluorescence image of a colonoid (1 week culture) after an 8-hour EdU pulse (red). Hoescht 33442 was used as a nuclear stain (blue) in panels **C-G**. Scale bar = 75 µm.

the optimal conditions and cultured in 50% Matrigel formed colonoids which possessed all of the differentiated cell lineages, immunostaining for the post-mitotic lineage markers Muc2 (mucus-producing goblet cells) and ChgA (hormone-secreting enteroendocrine cells) was performed (Figure 2D-E). For colonoids cultured in 50% Matrigel, eGFP was expressed in 49 ± 14% of the colonoid volume compared to 49 ± 12% in 100% Matrigel, suggesting similar numbers of stem/progenitor cells at one week under both conditions (Figure 2F). The colonoid volume positive for Muc2 or ChgA was 14 ± 3 and 0.6 ± 0.2 times the volume staining positive for Hoechst 33342 when cultured in 50% Matrigel for 7 days. In the presence of 100% Matrigel for 7 days, $Muc2^+$ or $ChgA^+$ regions occupied 16 ± 4 and 0.5 ± 0.2 times more volume than that of Hoechst 33342 suggesting that the density of these differentiated cell types was similar for the two conditions. Colonoids in 50% and 100% Matrigel possessed EdU^+ positive regions (43 ± 12 and 47 ± 10 times greater in volume than that positive for Hoechst 33342) suggesting that comparable numbers of cells were actively synthesizing DNA when cultured in the two different Matrigel concentrations (Figure 2G) [22]. Thus colonoids cultured in 50% Matrigel were nearly identical to that in 100% Matrigel with respect to these measured cell properties.

Since the use of 50% Matrigel was superior to the other concentrations at forming colonoids and was also able to support both stem/proliferative and differentiated cells, 50% Matrigel was employed in all subsequent experiments. Given the high cost of Matrigel, reduction in the concentration to 50% will substantially reduce future experimental costs. The mechanistic impact of Matrigel concentration on the cells is unknown; however, the optimal concentration identified in this study suggest that 50% Matrigel may provide the optimal stiffness, the proper concentrations of growth and differentiating factors, and/or the appropriate density of extracellular matrix contacts to maximize colonoid cell growth.

Assessment of crypt interaction with microfabricated substrates

Surface biochemical properties are known to modulate the growth and differentiation of stem cells [23-26]. Thus, the property of solid surfaces in contact with the crypts is likely to impact the efficiency of colonoid formation and potentially the fate of the crypt cells. Along with the 3-D colonoids, it was noticed that the intestinal crypt cells also formed 2-D monolayers when in contact with the well substrate. Conditions promoting monolayer formation from primary cells for this monolayer have not been well described [9,11,27,28]. For this reason, five commonly used transparent, microfabrication substrates were assessed for their impact on cell growth and phenotype: glass, polystyrene, PDMS, and the epoxy photoresists SU-8 and 1002-F. Glass and polystyrene have long been the gold standard for cell culture, and devices can be microfabricated by a variety of methods from all of these materials. PDMS is the most popular material for prototyping microdevices and it can be readily microfabricated by soft lithography [13,29,30]. The epoxy-based, transparent, negative SU-8 photoresist is used in building high-aspect ratio microstructures by standard photolithography [31]. 1002-F photoresist is closely related to SU-8 in molecular structure, and prior work has demonstrated that 1002-F is biocompatible, supporting cell attachment and growth, and exhibits significantly lower autofluorescence than SU-8 [31-33]. Non-transparent or opaque substrates (*e.g.* silicon) were not assessed here due to their incompatibility with many light microscopy methods.

Crypts were cultured in contact with the microfabrication substrates and monolayer expansion efficiency was calculated by dividing the number of crypts that successfully expanded into monolayers by the total number of crypts plated. 95.5 ± 2.5% of crypts plated on glass developed into monolayers, the highest average percentage of any of the experimental materials. After one week, 46.3 ± 3.4% of crypts cultured on native PDMS substrates developed into monolayers, the lowest percentage of the experimental materials (p-value of 5.67×10^{-6}) (Figure 3A). However, monolayer-formation percentage for crypts on glass was not statistically different than that on polystyrene, 1002-F and SU-8 (p-values of 0.33, 0.10 and 0.052, respectively) (Figure 3B). Immunohistochemical staining for the goblet-cell and enteroendocrine lineages demonstrated the presence of differentiated cells throughout the monolayers on the PDMS surfaces (Figure 3C-D).

The monolayers forming on the various surveyed substrates possessed very little eGFP fluorescence, suggesting little to no *Sox9* expression. Monolayers on 1002-F and PDMS substrates possessed the most eGFP expression after a week of culture (covering 1.3 ± 0.8% and 3.7 ± 3.0% of the monolayer surface area, respectively). To determine whether cells in the monolayers were proliferating but without eGFP expression, an EdU-based cellular proliferation assay was performed on the cells grown for 7 days on PDMS substrates. EdU^+ cells were infrequent in the monolayers suggesting that most of the cells within the monolayer were not actively proliferating (compare Figure 3E to 2G). These data suggested that crypt-cells rapidly differentiated upon adherence to glass, oxidized polystyrene and epoxy photoresists. All of these materials are hydrophilic displaying charged oxygen groups on their surface. It may be necessary to avoid cell contact with these surfaces to maintain stem/progenitor cells. PDMS was an exception likely because the intrinsic hydrophobic properties of PDMS [30] discouraged surface attachment by cells. PDMS may be attractive for microfabricated devices constructed to house stem/progenitor cells. Further studies will be required to understand how the surface property of a substrate modulates the fate of crypt cells. While monolayer formation was critically dependent on the surface properties of the culture vessel, monolayer dependence on the overlaid Matrigel concentration was less pronounced as long as a concentration threshold of 50% Matrigel was utilized (Additional file 4: Figure S3).

Conclusions

The current work established a reproducible, standardized isolation protocol for isolating intact murine colonic crypts with high proliferative capacity. In a step-wise fashion, the incubation duration of the tissue in chelating buffer and mechanical acceleration intensities required for crypt release were optimized to retrieve the maximal number of high quality crypts. The concentration of Matrigel, a costly reagent used for *in vitro* expansion of intestinal stem cells, was optimized to maximize

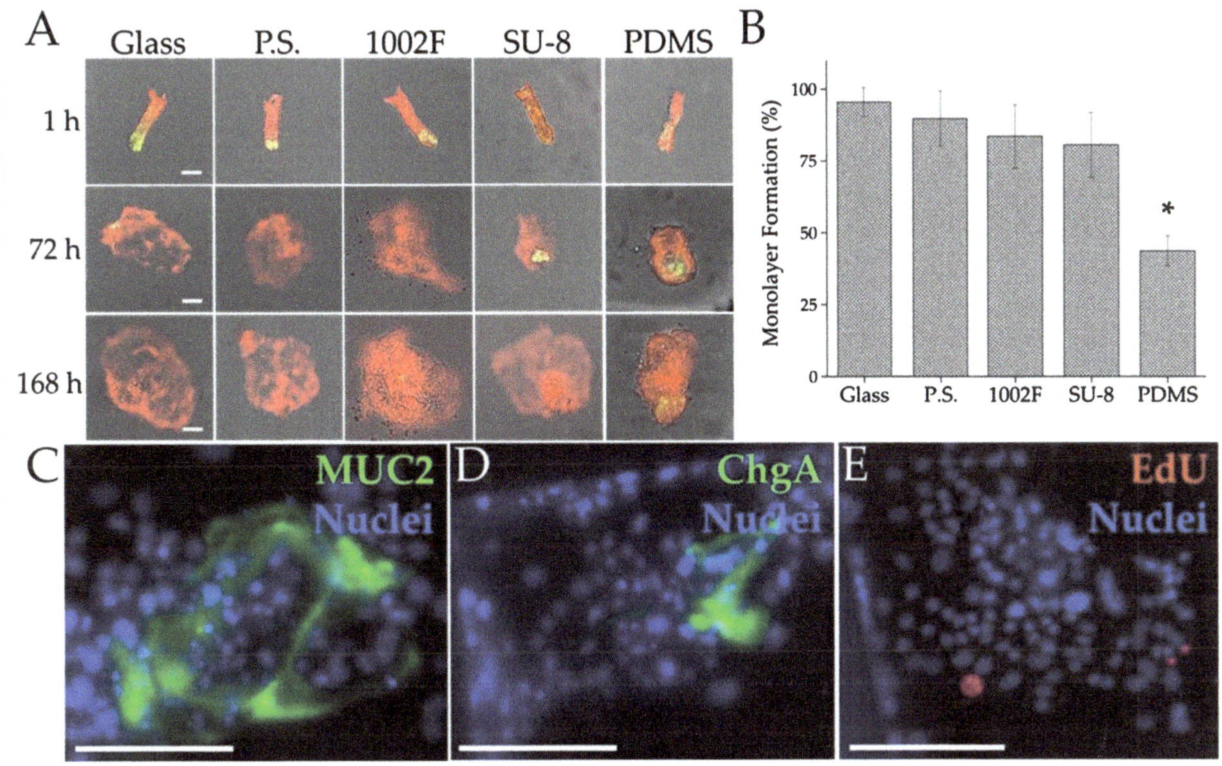

Figure 3 Crypt-substrate interaction. (A) Representative time-lapse images of monolayer formation on the experimental substrates from crypts isolated from a *Sox9*eGFP-CAGDsRed mouse. eGFP and DsRed fluorescence was overlaid on brightfield microscopy images. Upon adherence to glass, oxidized polystyrene and epoxy photoresist, crypt-cells rapidly differentiate. **(B)** Quantification of the percentage of crypts forming a monolayer when crypts were cultured on the microfabrication substrates over 1 week. **(C, D)** Whole-mount immunohistochemical staining of a monolayer after 1 week in culture. Fluorescence images are shown for: mucin-2 (green, **C**) and chromogranin-A (green, **D**). **(E)** Fluorescence image after an 8-hour EdU pulse (red). Hoescht 33442 was used as a nuclear stain (blue) in panels **C-E**. Scale bars = 50 μm.

the development of colonoids from the isolated crypts while minimizing reagent use. Crypts were isolated from a genetically engineered *Sox9*eGFP-CAGDsRed reporter mouse, enabling fluorescent measurements to be used as functional readouts of stem-cell proliferation and differentiation. The microwell provided an efficient platform for facile screening and quantification of colonoid formation while further reducing the amounts of expensive reagents including such as Wnt-3a, EGF, Noggin, and R-spondin1. Immunohistochemical staining demonstrated the presence of the differentiated intestinal cellular lineages (goblet and enteroendocrine) in these colonoids. The type of growth (2-D monolayer *vs.* 3-D colonoid) was dependent on the culture substrate properties. Crypts plated on PDMS substrates demonstrated the highest percentage of 3-D colonoid formation and most stem cells, while crypts plated on glass, polystyrene, 1002-F and SU-8 surfaces produced the highest percentages of 2-D monolayer formation with few identifiable stem cells. By standardizing the isolation process and optimizing the matrix concentrations on different surfaces, reproducible crypt isolation and robust culture protocols were established to facilitate the use of colonoid-based assays by the intestinal stem-cell community. Common microfabricated substrates were surveyed to identify substrates that are compatible with maintenance of stem and differentiated cells. This research provides a clear isolation and culture protocol for colonic crypts supporting future development of intestinal studies and 'organ-on-chip' endeavors.

Methods

Materials

N2 and B27 supplements, GlutaMAX, Advanced DMEM/F12 base media, 5-ethynyl-2′-deoxyuridine (EdU) kit and α-goat-Alexa Fluor 488 were purchased from Invitrogen (Carlsbad, CA). Y27632 Rock inhibitor, HEPES buffer, N-acetylcysteine (NAC), ethylenediaminetetraacetic acid (EDTA, 0.5 M, pH 8.0), bis-Benzimide (Hoescht 33342), and α–rabbit-Cy3 were obtained from Sigma-Aldrich (St. Louis, MO). Cell-culture-grade bovine serum albumin (BSA), dithiothreitol (DTT) and a pulse vortex-mixer were purchased from Thermo-Fisher (Fairlawn, NJ). Recombinant mouse Wnt-3a, recombinant human R-Spondin1 and recombinant mouse EGF were acquired from R&D Systems (Minneapolis, MN). Growth-factor reduced

Matrigel was obtained from BD Biosciences (Bedford, MA). Recombinant mouse Noggin was purchased from Peprotech (Rocky Hill, NJ). Sylgard 184 silicone elastomer kit was procured from Dow Corning (Midland, MI). EPON epoxy resin 1002-F (fusion solids) was purchased from Miller Stephenson Chemical Co. (Sylmar, CA). Primary antibodies α-mucin2 and α-chromogranin A were purchased from Santa Cruz Biotechnology (Santa Cruz, CA).

Transgenic mouse model and isolation of colonic crypts

The *Sox9*eGFP-CAGDsRed mouse model on a CD-1 background was used for experimental analysis. The CAGDsRed mouse line ubiquitously expresses the red fluorescent protein DsRed under the control of a chicken beta-actin promoter (CAG). The DsRed-expressing mice were bred with *Sox9*eGFP mice, which possessed the *Sox9* promoter controlling eGFP (enhanced green fluorescent protein) expression on a modified bacterial artificial chromosome (BAC) [3]. Previous work demonstrated that *Sox9* is expressed in the stem and progenitor cells of the colon so that the *Sox9*eGFP mouse possesses eGFP expression in the stem/proliferative cell compartment at the crypt base. For stem/progenitor cell quantification within fresh crypts, monolayers and colonoids, eGFP fluorescence was used as a measure of *Sox9* expression After resecting the distal colon from a 6–9 week-old mouse, the colon was cut longitudinally, flushed of its contents and washed with chilled rinse buffer (5.6 mM Na_2HPO_4, 8.0 mM KH_2PO_4, 96.2 mM NaCl, 1.6 mM KCl, 43.4 mM sucrose, 54.9 mM D-sorbitol, pH 7) [20]. The distal colon was then incubated in isolation buffer (rinse buffer + 2.0 mM EDTA + 0.5 mM DTT), for 30, 60 or 90 min at 22°C as indicated in the text. The tissue was washed by transferring to 3 separate vials containing chilled rinse buffer. The sample was then agitated at 2.7 × g for 5 seconds using a pulsing vortex mixer, unless stated otherwise in the text. The crypts were inspected by brightfield microscopy for the presence of a defined lumen. The overall yield was determined by adding a 250 μL crypt-suspension to a 12-well plate and using a 4× objective to count the number of crypts per field of view. This number was then used to calculate the number of crypts in the total volume of crypt-suspension.

Calculation of acceleration intensity

The pulsing vortex mixer (Fisher Scientific, Catalog # 02-215-375) produced acceleration profiles in the x-, y-, and z-directions. The acceleration of 6 different settings on the vortex mixer (800–1800 rpm) was measured over time using a 3-D accelerometer (Gulf Coast Data Concepts, Catalog# X16-1C). The magnitude of the x-, y-, and z-direction acceleration vectors was calculated from these measurements. The average magnitude over time was then used as the average accelerated intensity.

Culture of colonic crypts for matrigel optimization

A microwell fabricated from thick 1002-F photoresist was used for facile tracking of the colonoids to optimize the Matrigel concentration. To assess the effect of Matrigel on colonoid formation, Matrigel was diluted in complete culture medium (CCM: advanced DMEM-F12 with N2 supplement, B27 supplement, 1× GlutaMAX, 10 μM HEPES buffer, 1 μg/mL penicillin, 1 μg/mL streptomycin, 3.2 mg/mL Y27632 and 163.2 mg/mL NAC) at 4°C to yield 25, 50, 75 and 100 vol% concentrations. The microwells were sterilized with ethanol, washed × 3 in rinse buffer, and placed at 4°C before plating the crypts. A 400-μL suspension of crypts was added to each microwell (5000 crypts/mL) and the crypts were allowed to settle into the wells for 2 min. The supernatant was then carefully removed and ice-cold Matrigel (400 μL) was overlaid. The Matrigel was supplemented with the following growth factor concentrations: 5 ng/mL Wnt-3a, 50 ng/mL EGF, 100 ng/mL Noggin and 1 μg/mL R-spondin1 [3]. Matrigel was polymerized for 15 min at 37°C. After polymerization, 1.6 mL of complete culture medium was overlaid onto the Matrigel. Growth factors were replenished by direct addition to the medium every 2 days and the medium was changed every 4 days. R-spondin1 was used at 1 μg/mL for the initial plating and 500 ng/mL for the duration of the culture. Y27632 and NAC were only included in the CCM at the time of initial plating and were removed from subsequent culture media.

Culture of crypts on microfabrication substrates

Five different common microfabrication substrates were tested for the culture of crypts on their surface: PDMS, polystyrene (tissue-culture treated), glass, and the photoresists SU-8 and 1002-F. Round glass coverslips (#1, diameter = 25 mm) were spin-coated with PDMS, SU-8 or 1002-F, baked and cured, sterilized with ethanol, and placed in a 6-well plate. Before culturing, the substrates were coated with 50% Matrigel at 4°C for 8 h. A 200 μL crypt suspension (5000 crypts/mL) in Matrigel (50% in CCM unless otherwise stated) was added to each of the 6 wells. The plate was then placed at 4°C for 10 min to ensure that the crypts traveled through the liquid gel and settled onto the experimental substrate. Subsequently, the gel was polymerized at 37°C for 15 min. After polymerization, the crypts were overlaid with CCM. Growth factor and media exchange was performed as described above.

EdU analysis and immunostaining

Crypts and colonoids from wild-type mice were used for EdU analysis [22] and immunostaining. 5-ethynyl-2′-

deoxyuridine (EdU) nucleoside was used to assess proliferation as per manufacturer's instructions: EdU (10 μM) was added to the culture medium and allowed to become incorporated into the cells for 4 h. The culture medium was then removed, and the entire culture was fixed using 3.7% paraformaldehyde in phosphate-buffered saline (PBS) for 20 min at room temperature. Cells in the fixed colonoids were permeabilized using 0.5% Triton X-100 in PBS for 20 min, followed by washing × 3 with PBS containing 3% BSA. Each quadrant of the microwell was then incubated with 250 μL of click-it reaction cocktail (containing the Alexa Fluor-555 azide) for 30 min at room temperature, followed by rinsing × 3 with PBS. The samples were stored in PBS at 4°C until visualization by fluorescence microscopy. For immunostaining, colonoids were fixed, rinsed with PBS and permeabilized using 0.3% Triton X-100 in PBS for 20 min. Following rinsing × 3 with PBS containing 100 mM glycine, the colonoids were incubated in immunofluorescence (IF) wash (0.2% Triton X-100, 0.1% BSA, 0.05% Tween-20, 7.7 mM NaN_3 in PBS and 5% normal goat serum) for 90 min to block nonspecific binding. Primary-antibodies (α-chromogranin A and α-mucin2) were applied in IF wash (1:100) for 12 h at 4°C. Secondary antibodies (α-rabbit-Cy3 and α-goat-Alexa Fluor 488) were applied in IF wash (1:500) for 45 min. All nuclei were stained with bis-benzimide (10 μg/mL in PBS) using a 30 min incubation [3].

Image analysis of monolayers and freshly isolated crypts

Epifluorescence images were captured on a Nikon Eclipse TE2000-U microscope fitted with a Photometrics CoolSNAP HQ2 digital camera. Objective lenses used were 10×, 20× and 40× with numerical apertures of 0.30, 0.55 and 1.40, respectively. Prior to quantification, image acquisition and preprocessing of raw images was necessary to reduce background noise. This was done using a custom script implemented in MATLAB (MathWorks; Natick, MA) for each fluorescence image acquired. Background was first reduced using a top-hat filter followed by application of a median filter to smooth the images and further reduce noise. The images were then thresholded and 'holes' were filled to create a binary image which was used to define the image area with measurable fluorescence. The total number of pixels in this masked area was then summed for each image. Quantification for the regions of stem/proliferative cell area within monolayers and freshly isolated crypts were assessed by dividing the number of $eGFP^+$ pixels by the total number of $DsRed^+$ pixels in the image. All data points represent the average ± standard deviation of at least four separate experiments. Statistical analysis was conducted by one-way ANOVA pairwise tests. A p-value of < 0.05 was considered statistically significant.

Image analysis for the colonoids

Confocal images were captured on a Zeiss CLSM 710 Spectral Confocal Laser Scanning Microscope, using objective lenses of either 20× or 40× magnifications (numerical apertures of 0.80 and 0.95, respectively). Preprocessing of the raw images, thresholding and masking was performed for each confocal slice as described in the previous section. eGFP was quantified relative to DsRed in each slice as described in the prior section and then averaged over all slices possessing colonoids to yield the percentage colonoid volume positive for eGFP. To quantify the Muc2- and ChgA-expression or EdU-staining regions in each image slice, the number of pixels positive for these markers was divided by the number of pixels positive for Hoechst 33342. The average ratio for every slice in a sample was then calculated to yield the average volume of sample positive for Muc2, ChgA, or EdU relative to that positive for Hoechst 33342.

Additional files

Additional file 1: Figure S1. Accelerometer measurements of the acceleration vector magnitudes applied to the colonic tissue.

Additional file 2: Table S1. At different average acceleration intensities **(A)** the total number of whole crypts isolated and the **(B)** percentage of crypts isolated with intact morphology.

Additional file 3: Figure S2. Effect of increased accelerated agitation intensity on eGFP expression.

Additional file 4: Figure S3. Effect of Matrigel concentration on monolayer expansion. Data collected on native PDMS surfaces. Squares, circles, triangles, and x's represent 25%, 50%, 75% and 100% Matrigel, respectively.

Competing interests

The authors declare they have no competing interests or conflicts.

Authors' contributions

All authors were involved in the design of the experiments and writing the manuscript. AA performed the experiments. This work is in partial fulfillment of the Doctoral Degree requirements of AA. All authors read and approved final manuscript.

Acknowledgements

We would like to thank Victoria Newton for her assistance in mouse maintenance. This research was supported by NIH grants (HG004843 and EB0012549 to NLA and EB013803 to YW) and the University Cancer Research Fund (UCRF).

Author details

[1]Department of Biomedical Engineering, University of North Carolina, Chapel Hill, NC, 27599 and North Carolina State University, Raleigh, NC 27695, USA. [2]Department of Chemistry, University of North Carolina, Chapel Hill, NC 27599, USA. [3]Department of Medicine, Division of Gastroenterology and Hepatology, University of North Carolina, Chapel Hill, NC 27599, USA. [4]Department of Cell Biology and Physiology, University of North Carolina, Chapel Hill, NC 27599, USA.

References

1. Barker N, van de Wetering M, Clevers H: **The intestinal stem cell.** *Genes Dev* 2008, **22:**1856–1864.
2. Garrison AP, Helmrath MA, Dekaney CM: **Intestinal stem cells.** *J Pediatr Gastroenterol Nutr* 2009, **49:**2–7.
3. Ramalingam S, Daughtridge GW, Johnston MJ, Gracz AD, Magness ST: **Distinct levels of Sox9 expression mark colon epithelial stem cells that form colonoids in culture.** *Am J Physiol Gastrointest Liver Physiol* 2012, **302:**G10–G20.
4. Humphries A, Wright NA: **Colonic crypt organization and tumorigenesis.** *Nat Rev Cancer* 2008, **8:**415–424.
5. Medema JP, Vermeulen L: **Microenvironmental regulation of stem cells in intestinal homeostasis and cancer.** *Nature* 2011, **474:**318–326.
6. Dharmsathaphorn K, McRoberts JA, Mandel KG, Tisdale LD, Masui H: **A human colonic tumor-cell line that maintains vectoral electrolyte transport.** *Am J Physiol* 1984, **246:**G204–G208.
7. Haramis APG, Begthel H, van den Born M, Van Es J, Jonkheer S, Offerhaus GJA, Clevers H: **De novo crypt formation and juvenile polyposis on BMP inhibition in mouse intestine.** *Science* 2004, **303:**1684–1686.
8. Barker N, Van Es JH, Kuipers J, Kujala P, van den Born M, Cozijnsen M, Haegebarth A, Korving J, Begthel H, Peters PJ, Clevers H: **Identification of stem cells in small intestine and colon by marker gene Lgr5.** *Nature* 2007, **449:**1003–U1001.
9. Jung P, Sato T, Merlos-Suarez A, Barriga FM, Iglesias M, Rossell D, Auer H, Gallardo M, Blasco MA, Sancho E, Clevers H, Batlle E: **Isolation and in vitro expansion of human colonic stem cells.** *Nat Med* 2011, **17:**1225–1227.
10. Ootani A, Li XN, Sangiorgi E, Ho QT, Ueno H, Toda S, Sugihara H, Fujimoto K, Weissman IL, Capecchi MR, Kuo CJ: **Sustained in vitro intestinal epithelial culture within a Wnt-dependent stem cell niche.** *Nat Med* 2009, **15:**1–U140.
11. Sato T, Vries RG, Snippert HJ, van de Wetering M, Barker N, Stange DE, Van Es JH, Abo A, Kujala P, Peters PJ, Clevers H: **Single Lgr5 stem cells build crypt-villus structures in vitro without a mesenchymal niche.** *Nature* 2009, **459:**262–U147.
12. Clevers JC: **Lgr5 stem cells in self-renewal and cancer.** *J Pathol* 2013, **229:**S5–S5.
13. Huh D, Hamilton GA, Ingber DE: **From 3D cell culture to organs-on-chips.** *Trends Cell Biol* 2011, **21:**745–754.
14. Tandon N, Marolt D, Cimetta E, Vunjak-Novakovic G: **Bioreactor engineering of stem cell environments.** *Biotechnol Adv* 2013, **31:**1020–1031.
15. van der Meer AD, van den Berg A: **Organs-on-chips: breaking the in vitro impasse.** *Integr Biol* 2012, **4:**461–470.
16. Huh D, Torisawa Y-s, Hamilton GA, Kim HJ, Ingber DE: **Microengineered physiological biomimicry: Organs-on-Chips.** *Lab Chip* 2012, **12:**2156–2164.
17. Kim HJ, Huh D, Hamilton G, Ingber DE: **Human gut-on-a-chip inhabited by microbial flora that experiences intestinal peristalsis-like motions and flow.** *Lab Chip* 2012, **12:**2165–2174.
18. Kim HJ, Ingber DE: **Gut-on-a-Chip microenvironment induces human intestinal cells to undergo villus differentiation.** *Integr Biol* 2013, **5:**1130–1140.
19. Yu J, Peng S, Luo D, March JC: **In vitro 3D human small intestinal villous model for drug permeability determination.** *Biotechnol Bioeng* 2012, **109:**2173–2178.
20. Booth C, O'Shea JA, Potten CS: **Maintenance of functional stem cells in isolated and cultured adult intestinal epithelium.** *Exp Cell Res* 1999, **249:**359–366.
21. Basson MD, Turowski G, Emenaker NJ: **Regulation of human (Caco-2) intestinal epithelial cell differentiation by extracellular matrix proteins.** *Exp Cell Res* 1996, **225:**301–305.
22. Salic A, Mitchison TJ: **A chemical method for fast and sensitive detection of DNA synthesis in vivo.** *Proc Natl Acad Sci U S A* 2008, **105:**2415–2420.
23. Kaivosoja E, Barreto G, Levon K, Virtanen S, Ainola M, Konttinen YT: **Chemical and physical properties of regenerative medicine materials controlling stem cell fate.** *Ann Med* 2012, **44:**635–650.
24. Phillips JE, Petrie TA, Creighton FP, Garcia AJ: **Human mesenchymal stem cell differentiation on self-assembled monolayers presenting different surface chemistries.** *Acta Biomater* 2010, **6:**12–20.
25. Hakala H, Rajala K, Ojala M, Panula S, Areva S, Kellomaki M, Suuronen R, Skottman H: **Comparison of biomaterials and extracellular matrices as a culture platform for multiple, independently derived human embryonic stem cell lines.** *Tissue Eng A* 2009, **15:**1775–1785.
26. Murakami H, Masui H: **Hormonal control of human-colon carcinoma cell growth in serum-free medium.** *Proc Natl Acad Sci U S A Biol Sci* 1980, **77:**3464–3468.
27. Madara JL, Dharmsathaphorn K: **Occluding junction structure-function relationships in a cultured epithelial monolayer.** *J Cell Biol* 1985, **101:**2124–2133.
28. Madara JL, Stafford J, Dharmsathaphorn K, Carlson S: **Structural analysis of a human intestinal epithelial-cell line.** *Gastroenterology* 1987, **92:**1133–1145.
29. Duffy DC, McDonald JC, Schueller OJA, Whitesides GM: **Rapid prototyping of microfluidic systems in poly(dimethylsiloxane).** *Anal Chem* 1998, **70:**4974–4984.
30. Mata A, Fleischman AJ, Roy S: **Characterization of polydimethylsiloxane (PDMS) properties for biomedical micro/nanosystems.** *Biomed Microdevices* 2005, **7:**281–293.
31. Pai JH, Wang Y, Salazar GT, Sims CE, Bachman M, Li GP, Allbritton NL: **Photoresist with low fluorescence for bioanalytical applications.** *Anal Chem* 2007, **79:**8774–8780.
32. Wang YL, Dhopeshwarkar R, Najdi R, Waterman ML, Sims CE, Allbritton N: **Microdevice to capture colon crypts for in vitro studies.** *Lab Chip* 2010, **10:**1596–1603.
33. Wang YL, Phillips C, Xu W, Pai JH, Dhopeshwarkar R, Sims CE, Allbritton N: **Micromolded arrays for separation of adherent cells.** *Lab Chip* 2010, **10:**2917–2924.

20

Engineering biodegradable and multifunctional peptide-based polymers for gene delivery

Julie Shi, Joan G Schellinger and Suzie H Pun*

Abstract

The complex nature of *in vivo* gene transfer establishes the need for multifunctional delivery vectors capable of meeting these challenges. An additional consideration for clinical translation of synthetic delivery formulations is reproducibility and scale-up of materials. In this review, we summarize our work over the last five years in developing a modular approach for synthesizing peptide-based polymers. In these materials, bioactive peptides that address various barriers to gene delivery are copolymerized with a hydrophilic backbone of *N*-(2-hydroxypropyl) methacrylamide (HPMA) using reversible-addition fragmentation chain-transfer (RAFT) polymerization. We demonstrate that this synthetic approach results in well-defined, narrowly-disperse polymers with controllable composition and molecular weight. To date, we have investigated the effectiveness of various bioactive peptides for DNA condensation, endosomal escape, cell targeting, and degradability on gene transfer, as well as the impact of multivalency and polymer architecture on peptide bioactivity.

Keywords: Nonviral gene delivery, Peptide-polymers, Endosomal escape, Degradability, *N*-(2-hydroxypropyl) methacrylamide (HPMA)

Gene therapy has the potential to improve therapeutic outcomes for currently untreatable diseases. Viruses are naturally efficient vectors for gene therapy, but have encountered obstacles in clinical translation due to issues such as vector toxicity and immunogenicity, potential gene integration into oncogenic regions, and high production costs. Thus, nonviral materials, *e.g.* cationic lipids and polymers, have been extensively engineered as gene delivery vectors and are attractive alternatives to viral vectors because they tend to have improved safety profiles and lower costs of production [1]. However, nonviral vectors have not been as successful in attaining high transgene expression efficiencies *in vivo*. In order to enhance the gene transfer efficiency of these materials, several groups have explored the use of bioactive peptides to address various extracellular and intracellular barriers to nonviral gene delivery, such as cellular uptake, endosomal escape, cargo unpackaging, and nuclear translocation [2-5]. These multicomponent synthetic materials have been engineered to overcome these barriers for various applications [6], including delivery to neurons [7] and hepatocytes [8].

Recent advances in living polymerization techniques, such as reversible-addition fragmentation chain transfer (RAFT), have allowed the development of well-defined polymers with controlled architectures and quantitative monomer incorporation [9-11]. Peptide-polymer conjugates containing multiple peptides can be synthesized by grafting peptides to preformed polymers or by polymerization of peptide monomers. The grafting technique has been extensively reviewed elsewhere [12-14]. The Klok group demonstrated the synthesis of peptide brush copolymers by RAFT polymerization of coiled coil peptide motifs in 2010 [15].

In this review, we summarize our work in the development of copolymers consisting of *N*-(2-hydroxypropyl) methacrylamide (HPMA) and multiple pendant oligopeptides for gene delivery (Figure 1). Cationic peptide moieties were explored and optimized for DNA condensation. To improve biodegradability, environmentally-responsive linkers were incorporated into the HPMA copolymers. In addition, both buffering and membrane-disruptive peptides were evaluated for enhancing endosomal escape of the polyplexes. Throughout our work, we have also noted effects of polymer architecture and multivalency on transfection efficiency. Finally, we conclude with some therapeutic

* Correspondence: spun@uw.edu
Department of Bioengineering and Molecular Engineering & Sciences Institute, University of Washington, 3720 15th Ave NE, Seattle, WA 98195, USA

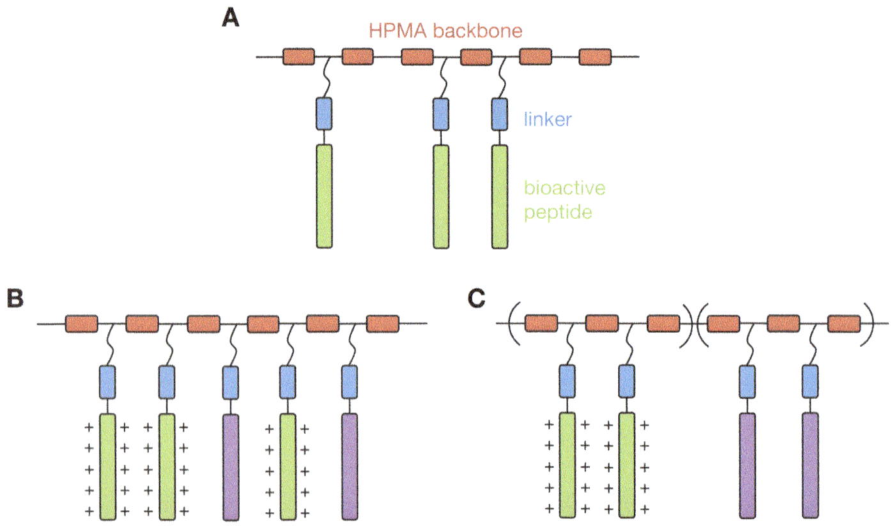

Figure 1 Schematic of peptide-functionalized HPMA copolymers. (A) Random-statistical copolymers of HPMA (red) and pendant bioactive peptides (green) with a linker (blue) between the hydrophilic backbone and the main peptide sequence. **(B)** Random-statistical copolymers of HPMA and multiple pendant peptides (cationic peptide in green; second bioactive peptide in purple). **(C)** Diblock copolymer of HPMA and multiple pendant peptides, where the cationic peptide is on one block, and a second bioactive peptide on a different block.

applications for this class of peptide-functionalized brush polymers that we have investigated.

Design of peptide-based polymers

Nucleic acid condensation using basic peptides

Cationic vectors can be used to form electrostatic complexes with anionic nucleic acid, termed "polyplexes", to protect the nucleic acid from serum and intracellular proteases. Early studies demonstrated the use of poly-L-lysine as a transfection agent to deliver plasmid DNA to achieve exogeneous protein expression [16]. Further work was conducted to optimize the oligolysine and polylysine residues for enhanced gene transfer, as well as introduce salt- and serum-stability into these carriers [17]. Since then, a number of cationic polymers have been developed for gene delivery applications [1,18]. In 1997, O'Brien-Simpson and coworkers reported a general method for the assembly of multi-peptide polymer constructs using acryloyl peptides by radical polymerization for vaccine development [19]. Thus, we hypothesized that this polymer architecture would enable a greater incorporation of peptide moieties and flexibility for multiple peptide incorporation for the development of gene delivery vectors.

As a starting point, HPMA was chosen to compose the hydrophilic backbone. HPMA has been widely used for the synthesis of polymer-drug conjugates due to its biocompatibility, as well as its synthetic versatility and flexibility since it has been used in a variety of polymerization techniques [20]. Therefore, we copolymerized short pendant oligolysine peptides with HPMA in a random-statistical brush polymer architecture using both free radical and RAFT polymerization approaches (Figure 2) [21,22]. Polymers synthesized by RAFT polymerization exhibited polydispersities closer to 1, better control over final polymer composition, and reduced toxicities [23]. We next optimized the peptide length and molecular weight of these statistical polymers generated by RAFT polymerization, and found that a peptide length of 10 lysines, a composition of 5 mmol lysine per gram polymer, and a molecular weight of ~60 kDa was optimal for high transfection efficiencies (comparable to that of branched polyethylenimine, or bPEI) and limited cytotoxicity [23]. Surprisingly, a 50% longer oligolysine peptide (K_{15}) demonstrated poor transfection efficiencies despite similar lysine composition and molecular weight. Extensive studies on the mechanism of enhanced transfection of the optimized polymer (p[HPMA-*co*-K_{10}]) over the poorly-performing polymer (p[HPMA-*co*-K_{15}]) showed that polyplexes of larger aspect ratios (more rod-like) greatly reduced cellular uptake and subsequent transfection efficiency (Shi J, Choi JL, Chou B, Johnson RN, Schellinger JG, Pun SH: Effect of polyplex morphology on cellular uptake, intracellular trafficking, and transgene expression. Submitted). Previous studies have reported that polylysine length correlates to the aspect ratio of the lysine/DNA complexes, thus highlighting the importance of understanding the biophysical interaction of plasmid DNA with various cationic moieties. In order to reduce cytotoxicity, we explored the effect of reduced charge density by alternating lysine residues with glycine spacers (p[HPMA-*co*-$(GK)_5$]) since increased charge densities were found to correlate with greater cytotoxicities [24]. Interestingly, these polymers induced similar transfection efficiencies than its pentamer

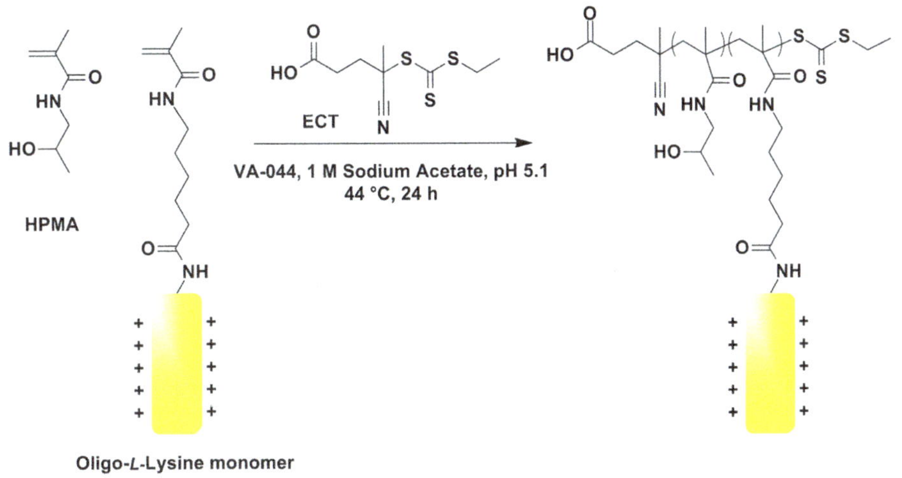

Figure 2 Synthesis of statistical HPMA-oligolysine copolymers by RAFT polymerization. HPMA and methacrylamido-functionalized oligolysine peptides are copolymerized under aqueous RAFT conditions using ethyl cyanovaleric trithiocarbonate (ECT) as the chain transfer agent (CTA) [71] and VA-044 as the initiator. The 6-carbon aminohexanoic acid (Ahx) is used as a linker between the pendant oligolysine peptide and the hydrophilic backbone.

counterpart (*p*[HPMA-*co*-K_5]) but actually were more cytotoxic [25].

Arginine is another cationic amino acid residue that has been explored as a complexation agent. Arginine has been shown to exhibit cell-penetrating capabilities, thus allowing for enhanced cellular uptake [26]. Cell-penetrating peptides (CPPs), such as the TAT protein transduction domain, contain a number of arginine residues [27,28]. Harashima and coworkers demonstrated that liposomes modified with an increasing density of octaarginine (R_8) peptides were taken up *via* macropinocytosis rather than clathrin-mediated endocytosis [29]. Therefore, arginine oligopeptides were also used as the complexation moiety in statistical HPMA brush polymers [25]. Similarly to previous reports comparing arginine motifs with lysine motifs [30-33], HPMA-oligoarginine polymers performed better in transfection efficiency when compared to a lysine analogue (~5-10 mmol lysine or arginine per gram polymer) but also caused increased cytotoxicity. Furthermore, due to challenges in synthesizing oligoarginine peptides, guanidinylation of the lysine resides with *O*-methylisourea was also explored; guanidinylation has been shown to increase transfection efficiencies of various carriers [34-37] and cellular uptake of HPMA constructs [38]. Conversion of the lysine residues in HPMA-oligolysine polymers to homoarginines resulted in higher transfection efficiencies and lower cytotoxicity than bPEI. Therefore, this method could be readily applied to other primary amine-based polymers to increase gene transfer efficiencies.

Degradability using environmentally-responsive elements

Biodegradability is a desirable attribute for the *in vivo* application of gene delivery vectors. Since higher molecular weights of polycations can lead to increased cytotoxicity [23,24,39,40], linker chemistries have been used to introduce degradability into polymeric vectors. For example, environmentally-responsive linkages such as disulfide and acid-labile bonds can enable the release of cargo in specific intracellular compartments and promote degradability [41]. Likewise, specific amino acid sequences can be enzymatically degraded by various proteases [42-44]. We have explored both of these strategies for introducing a degradable segment into these HPMA-oligolysine polymers. Due to the relatively high levels of glutathione, a reducing agent, in the cytosolic environment compared to the extracellular space [45], the incorporation of disulfide linkages into polymeric carriers has been an attractive approach to increase biodegradability. To introduce reducibility, the six-carbon linker 6-aminohexanoic acid (Ahx) was exchanged with a linker containing a disulfide bond, 3-[(2-aminoethyl)dithio] propionic acid (Aedp) [46]. These reducible polymers were less cytotoxic, but achieved less efficient transfection efficiencies compared to the non-reducible analogue. However, a mixed formulation of reducible and non-reducible polymers achieved an intermediate level of transfection efficiency and reduced cytotoxicity. The high concentration of disulfide bonds within the polymer may lead to chemical instability, which was evidenced by partial improvement in transfection efficiency in the presence of EDTA.

As an alternative approach to enhancing degradability, we explored the use of enzymatically-cleavable peptide linkers, which have been used to introduce site-specific cleavage sites for the release of drugs and peptides [47-49]. A commonly used peptide linker sequence is cathepsin B-labile [12,50]; cathepsin B is a lysosomal cysteine protease that exhibits endo- and exopeptidase activity [51].

We designed a cathepsin B-labile peptide sequence (FKFL), and introduced the linker, flanked by six-carbon spacers (Ahx), between the HPMA backbone and the pendant cationic peptide [52]. The peptides demonstrated site-specific cleavage by cathepsin B within 15 minutes, while the polymers showed complete degradation of the pendant modified oligolysine motifs within 1 hour. In contrast to the reducible polymers in which transfection efficiencies were lower with polymers containing reducible linkers, the cathepsin B-labile polymers showed similar levels of transfection and were less toxic compared to a non-degradable analogue consisting of nondegradable D-amino acids. Therefore, this work demonstrates the possibility of using enzymatically-cleavable linkers to enable site-specific release and degradability for polyplex formulations.

Endosomal escape strategies

Once internalized, polyplexes are exposed to increasingly acidic environments in endosomes and lysosomes, and eventually are degraded by lysosomal proteases. To circumvent lysosomal degradation, various strategies have been investigated to induce endosomal escape, such as the incorporation of peptide moieties that enable proton buffering [53] or interaction with lipid membranes [54]. Virally-derived peptides, such as TAT, Antennapedia (Antp), and HGP, and membrane-disruptive peptides, such as melittin, have been used to increase delivery efficiencies of cargo due to their ability to interact with lipid membranes [3]. We have used both of these approaches to enhance the endosomal escape abilities of our HPMA-oligolysine brush polymers, with varying success [55-57].

As a first approach, we used a similar endosomal escape strategy to bPEI, which is commonly used as a transfection agent due to its ability to induce endosomal escape *via* the buffering of protons at pH ~6-7 [58,59]. Several groups have mimicked this buffering strategy by incorporating histidine residues, which contain a protonable imidazole group at pH 6–7, into various gene carriers [53]. The addition of a second oligohistidine-containing peptide into statistical HPMA-oligolysine polymers demonstrated increased transfection efficiencies [22]. To further optimize the incorporation of oligohistidine peptides to increase transfection, statistical and diblock polymers were synthesized with varying amounts of oligohistidine residues incorporated into the polymer (Figure 3) [55]. Interestingly, the polymer architecture affected the buffering range of the polymer in that diblock polymers buffered in the upper endosomal pH range (pH 5.6-7.4) whereas statistical polymers buffered in the lower endosomal pH range (pH 5.1-6.6). Despite improved buffering capabilities, only the statistical polymer containing 1.4 mmol histidine per gram polymer showed slight improvements in transfection ability over its non-histidylated analogue, possibly due to preferentially trafficking of the polyplexes through the non-acidifying caveolae-mediated endocytic pathway [60]. Furthermore, endosomal buffering requires a critical concentration threshold to be a viable strategy for endosomal escape, and thus, the use of more potent strategies may be beneficial.

Hydrophobic modification of polycationic polymers is another strategy for promoting endosomal escape. Xu and Szoka proposed that upon endocytosis, lipid-modified polycation/DNA complexes form ion pairs with the negatively-charged amphiphilic endosomal membrane, resulting in membrane destabilization and release of cargo [61]. Furthermore, incorporation of hydrophobic moieties allows for physical encapsulation of the genetic payload, enhanced serum stability, increased cell viability, and targeting ability [62]. Several reviews have been published regarding the advantages of lipid or hydrophobic modification of polycationic gene carriers [63,64]. Previously, Abassi *et al.* reported various lipid-substituted polylysines as vectors for plasmid delivery and expression in skin fibroblasts. They showed that while all modified and unmodified polylysines demonstrated complete complexation with plasmid DNA, polymers with increased lipid substitution were more resistant to unpackaging with heparin treatment. Furthermore, most lipid-modified polymers exhibited increased pEGFP expression compared to native polylysine, with myristic and stearic acid lipid substituents demonstrating less cytotoxicity [65]. Inspired by this work, we prepared a panel of stearic acid-modified copolymers of HPMA and oligolysine. We grafted NHS-activated stearic acid (SA) onto the ε-amine of the lysine moiety of the previously optimized *p*(HPMA-*co*-K_{10}) copolymers at DP 190. The lipid substituent was grafted at various lipid to lysine ratios for optimization (Table 1). We then evaluated the polymers for gene transfection. There was no significant increase in gene transfection with these lipid-modified copolymers *in vitro* (Figure 4A); increasing lipid substitution also decreased transfection efficiency (Figure 4B). Interestingly, high lipid substitution showed decreased cytotoxicity. Thus, for these HPMA-oligolysine polymers, lipid substitution is not a viable method for increasing transfection efficiency.

Alternatively, membrane-active peptides have been promising for inducing endosomal escape due to their high disruptive potency [54]. We have investigated the use of melittin, a 26-mer derived from venom of the honey bee *Apis mellifera* [56], and sHGP, a shortened and optimized 15-mer peptide derived from the endodomain of the HIV gp41 region [66]. Melittin undergoes an α-helical conformational change at lipid membranes, allowing the peptide to insert into the membrane and induce pore formation [67,68]. Polymeric carriers conjugated with melittin showed significantly increased transfection *in vitro* and *in vivo* [56,69,70]. Similarly to the HPMA-oligolysine-oligohistidine diblock copolymer, melittin was grafted onto a diblock of *p*[HPMA-*co*-PDSMA]-*b*-[HPMA-*co*-K_{10}],

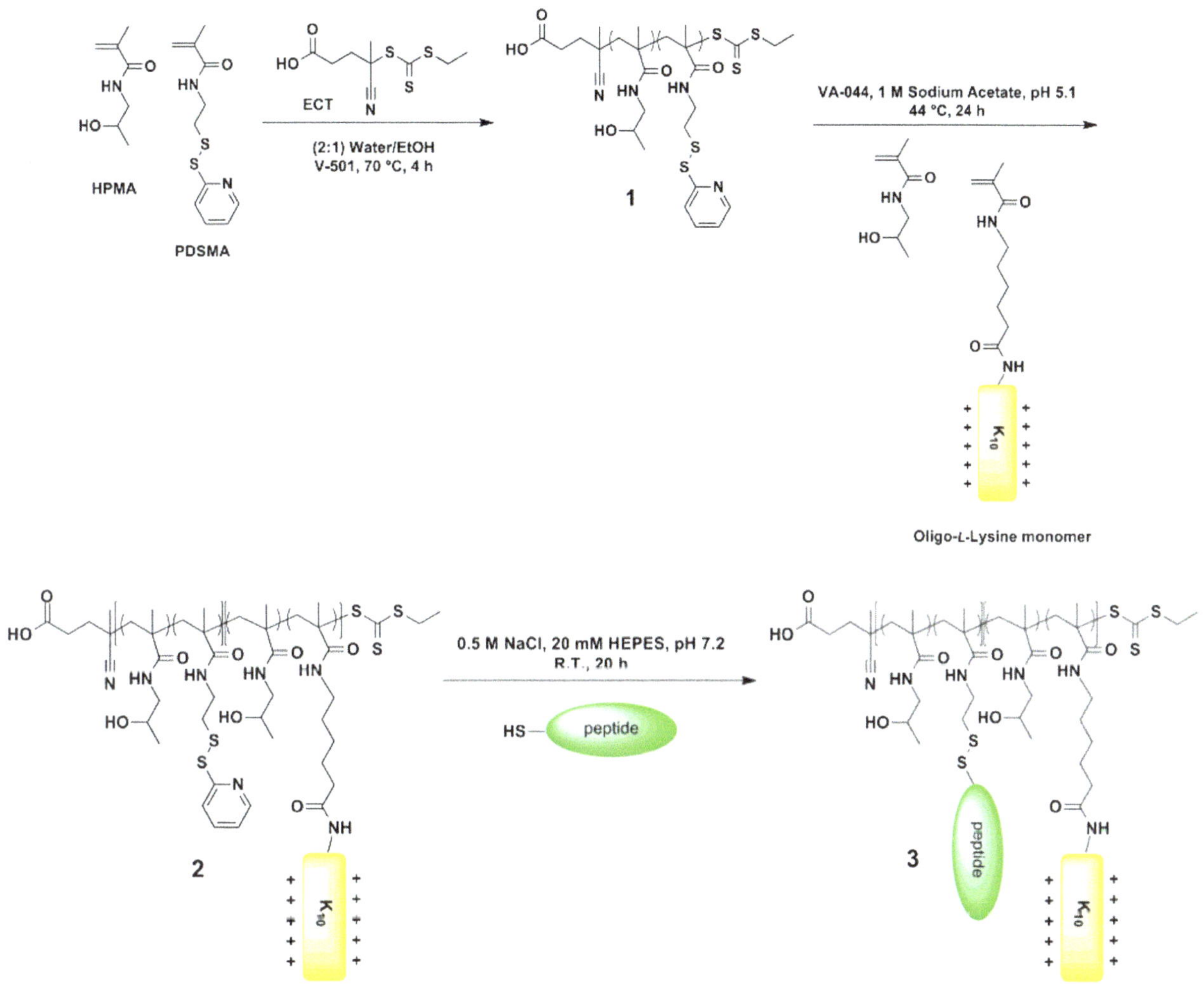

Figure 3 Synthesis of diblock HPMA copolymers containing multiple pendant peptides. RAFT polymerization of HPMA and PDSMA (molecule **1**), and then chain-extension with HPMA and oligo(L-lysine) to form the second block (molecule **2**). Disulfide exchange between the pyridyl disulfide on **2** and cysteine-functionalized oligo(L-histidine) peptides yields a diblock polymer with two bioactive peptides (molecule **3**).

using PDSMA as the point of conjugation [56]. Incorporation of melittin into these brush polymers displayed hemolytic ability and significantly improved transfection efficiency over the polymer without melittin incorporation, but also increased toxicity. To ameliorate some cytotoxicity, mixed formulations of the base polymer and the melittin-containing polymer were used for transfection and showed markedly improved toxicity profiles and even further improved transfection efficiencies to greater than that of branched PEI.

Table 1 Characterization of lipid substitution on *p*[HPMA-*co*-K_{10}] copolymers

Polymer designation	Lys:Lipid mole ratio	Lipid/HPMA*	% Substituted Lys[#]
p(HPMA-*co*-K_{10}-*g*-SA) (10:1)	10:1	0.35	14
p(HPMA-*co*-K_{10}-*g*-SA) (20:1)	20:1	0.21	8.4
p(HPMA-*co*-K_{10}-*g*-SA (40:1)	40:1	0.12	4.8

*Based on NMR integration of lipid –$C\underline{H}_2$-CO (~ 2.4 ppm) and HPMA –$C\underline{H}$(OH)-CH_3 (~ 3.7 ppm).
[#]Based on the theoretical mole ratio (~20% AhxK_{10} and 80% HPMA): 2.5 mole Lys/1 mole HPMA.

The incorporation of sHGP also helped increase the transfection efficiency of HPMA-oligolysine polymers [57]. sHGP was incorporated into HPMA-oligolysine polymers in a diblock architecture, similarly to melittin and the diblock oligohistidine-containing polymer, as well as a statistical architecture. Both polymer architectures demonstrated improved transfection efficiency compared to the base HPMA-oligolysine polymer, but the diblock polymer self-assembled into micelles, thereby sequestering the hydrophobic, membrane-lytic sHGP region. Interestingly, the diblock architecture resulted in less hemolytic activity and improved cytotoxicity profiles. Thus, these results suggest that the exposure of hemolytic regions under certain environmental conditions can lead to more efficient yet less toxic vectors. This strategy has previously been

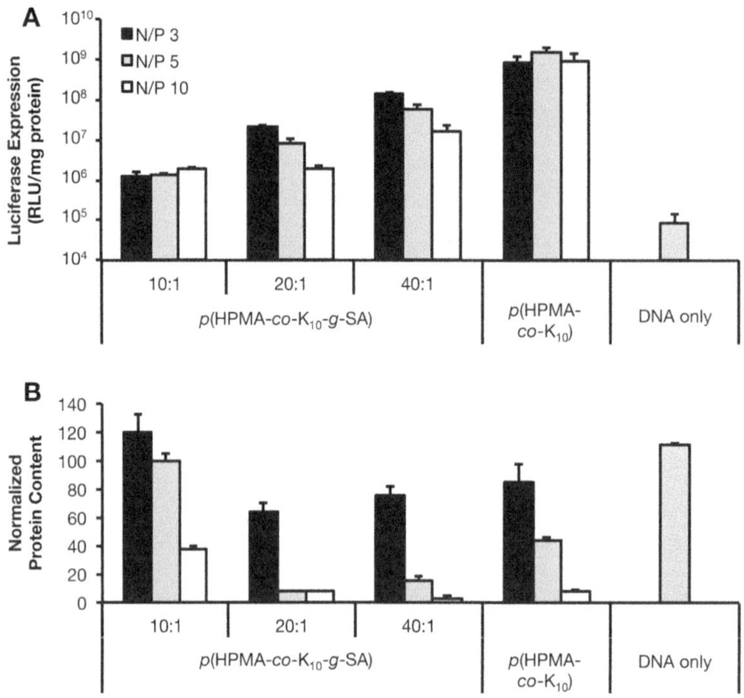

Figure 4 Transfection of stearic acid (SA)-modified *p*[HPMA-*co*-K_{10}] polyplexes in HeLa cells. HeLa cells were treated with polyplexes (containing 1 μg plasmid DNA) at charge ratios (N/P) of 3, 5, and 10 for 4 h in serum-free conditions, washed, and replenished with complete media. At 48 h post-transfection, cell lysates were assessed for **(A)** luciferase reporter gene expression and **(B)** protein content as an indicator for cytotoxicity. Data are presented as the mean ± S.D., n = 3.

used in the design of synthetic polymers for siRNA delivery, where the exposure of a hydrophobic lytic domain occurred under endosomal pH to facilitate cargo release into the cytosol [71]. To compare these different endosomal escape strategies, cells were transfected with *p*[HPMA-*co*-K_{10}] containing either oligohistidine, melittin, or sHGP peptides (Figure 5). All polymers that contained endosomal escape modalities performed better than the base polymer containing only oligolysine peptides. The statistical polymer containing oligohistidine peptides and the block polymer containing sHGP performed the best; however, the block sHGP polymer was the least toxic overall.

Design considerations of HPMA-peptide polymers for gene delivery

Salt stability is necessary for the systemic administration of polyplexes since polyplex aggregation can lead to *in vivo* toxicity [72]. The incorporation of HPMA as a hydrophilic segment has been shown to reduce the salt-induced aggregation of multiple polymeric gene delivery systems [73-75]. We have also demonstrated increased salt stability of HPMA-oligolysine polyplexes when compared to lysine peptides or polylysine polyplexes [21-23]. In our optimization studies, salt stability decreased with increasing peptide incorporation [21,22], larger polymer molecular weights, and longer oligolysine peptide lengths [23]. In particular, the latter trend was unexpected since polymers containing longer oligolysine peptides also had longer HPMA segments in the polymer backbone. The incorporation of a hydrophilic HPMA shell onto *poly*(glycidyl methacrylate-tetraethylenepentamine) (*p*[GMA-TEPA]) also reduced salt-induced aggregation of the polyplexes; in this case, longer hydrophilic segments further enhanced polyplex stability in physiological salt conditions [76]. As seen with polymers containing poly (ethylene glycol) (PEG) to reduce polyplex aggregation under saline conditions [77], polymers containing the HPMA shell also demonstrated decreased *in vitro* transfection efficiencies. These results highlight the tradeoffs of incorporating hydrophilic segments into polymers for enhanced gene delivery *in vivo*.

In order to develop improved materials for nucleic acid delivery, it is important to understand the delivery mechanisms of engineered materials. To further characterize the HPMA-oligolysine polymers, we first determined the uptake efficiency of the optimized oligolysine-containing polymer (*p*[HPMA-*co*-K_{10}]) (Shi J, Choi JL, Chou B, Johnson RN, Schellinger JG, Pun SH: Effect of polyplex morphology on cellular uptake, intracellular trafficking, and transgene expression. Submitted). Despite much lower polyplex uptake efficiencies of the optimized polymer (*p*[HPMA-*co*-K_{10}]) *vs.* bPEI [78], the HPMA-oligolysine polymer achieved similar transfection efficiencies to that of bPEI [23]. Since alternative uptake routes have been

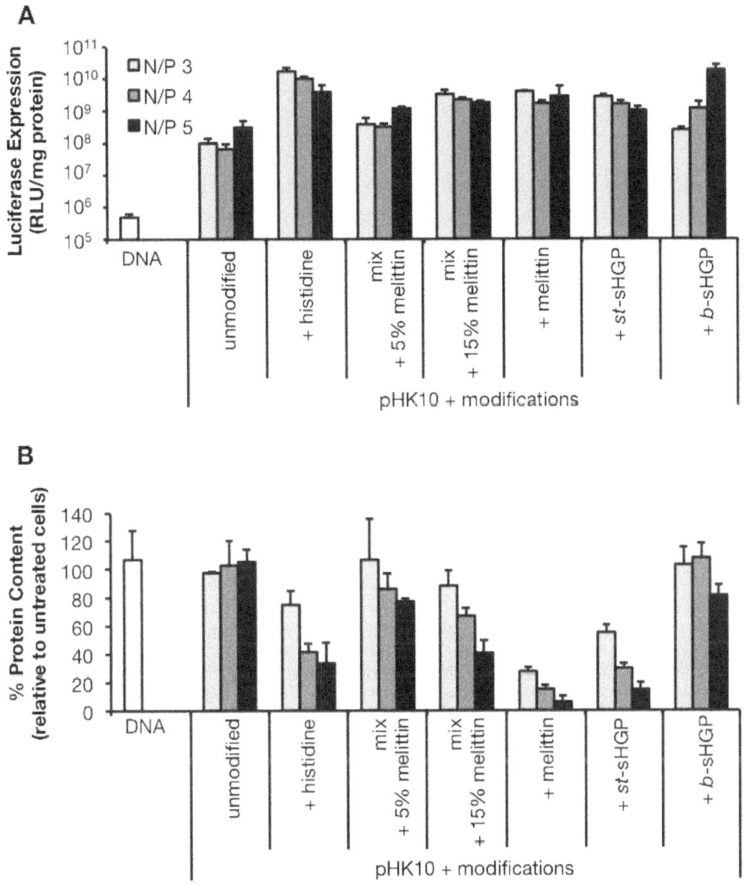

Figure 5 Comparative transfection of *p*[HPMA-*co*-K_{10}] polymers ("pHK10") modified with various endosomal escape modalities in HeLa cells. HeLa cells (3×10^4) were treated with polyplexes (containing 1 μg plasmid DNA) at charge ratios (N/P) of 3, 4, and 5 for 4 h in serum-free conditions, washed, and replenished with complete media. At 48 h post-transfection, cell lysates were assessed for **(A)** luciferase reporter gene expression and **(B)** protein content as an indicator for cytotoxicity. *pHK10 unmodified*: *p*[HPMA-*co*-K_{10}]; *pHK10 + histidine*: *p*[HPMA-*co*-K_{10}-*co*-K_5H_5] [55]; *pHK10 mix + 5% melittin*: a mixture of 95:5% (v/v) *p*[HPMA-*co*-K_{10}]: *p*[HPMA-*co*-melittin]-*b*-[HPMA-*co*-K_{10}] [56]; *pHK10 mix + 15% melittin*: a mixture of 85:15% (v/v) *p*[HPMA-*co*-K_{10}]: *p*[HPMA-*co*-melittin]-*b*-[HPMA-*co*-K_{10}]; *pHK10 + melittin*: *p*[HPMA-*co*-melittin]-*b*-[HPMA-*co*-K_{10}]; *pHK10 + st-sHGP*: *p*[HPMA-*co*-K_{10}-*co*-sHGP]; *pHK10 + b-sHGP*: *p*[HPMA-*co*-sHGP]-*b*-[HPMA-*co*-K_{10}]. Data are presented as the mean ± S.D., n = 3.

implicated in improving polyplex transfection [60], transfection of the HPMA-oligolysine copolymers was also completed with the presence of various endocytic inhibitors [55]. Transfection efficiencies were decreased in the presence of a small-molecule inhibitor for caveolae-mediated endocytosis, a pathway that circumvents the acidification process necessary for endosomal buffering. These results suggest that transfection may be more productive when HPMA-oligolysine polyplexes are routed *via* a non-acidifying endocytic route, similarly to other polycation systems [60,79,80]. Therefore, understanding the uptake pathway of various polymer formulations can aid in the rationale design of improved materials.

Therapeutic applications

Stealth coatings for adenovirus-mediated gene delivery

Adenoviruses, especially adenovirus serotype 5 (Ad5), are effective gene delivery agents since they are able to transduce both dividing and non-dividing cells, and can be produced in large titers [81,82]. However, Ad5 also induces a host immune response, limiting its use for systemic re-administration, and can only transduce cells expressing Coxsackievirus and Adenovirus Receptor (CAR) and αV integrins on their surface. To reduce the immunogenic response to the Ad5 capsid, as well as enhance the uptake of Ad5 in CAR-negative cell types, synthetic polymers using polyethylene glycol (PEG) and HPMA have been explored as stealth coatings [83]. We have also used HPMA-oligolysine brush polymers, with varying oligolysine peptide lengths, polymer molecular weights, and linker degradability properties, to electrostatically interact with the negatively-charged viral capsid to form stealth coatings [84]. A polymer with a peptide length of ten lysines was optimal for increasing transduction of Ad5 under serum and serum-free conditions; interestingly, a similar polymer was also found to be optimal for nonviral polyplex transfections

[23]. The polymer coatings also enabled transduction of CAR-negative cells, which was mediated by the presence of heparan sulfate proteoglycans (HSPGs) on the cell surface. In particular, sulfated HSPGs have been implicated in enhancing cellular uptake of cationic materials through electrostatic interactions [85]. Furthermore, the stealth coating also significantly protected the virus from neutralizing antibodies, which inhibits efficient *in vivo* viral transduction, without reducing transduction efficiency; these results demonstrated the potential applicability of HPMA-oligolysine brush polymers to enhance Ad5 transduction *in vivo*.

Delivery to the central nervous system (CNS)

The delivery of therapeutic genes to the CNS can be beneficial for the treatment of neurological disorders [86,87]. However, gene delivery to the CNS using nonviral vectors has not been successful due to the low transfection efficiencies achieved in non-dividing cells and low cellular uptake rates in terminally-differentiated neurons [88-90]. One approach to increase efficacy is to conjugate targeting ligands to polymeric vectors to enhance uptake by neuronal subtypes [7,91,92]. We previously demonstrated that conjugation of Tet1, a 12-mer peptide identified by *in vitro* phage display to bind the *Gtb1* ganglioside [93], to polyethylenimine (PEI) enhanced gene delivery to neural progenitor cells (NPCs) [94]. We found that ~0.6 Tet1 per polymer was sufficient to enhance transfection efficiency to NPCs *in vivo*. Recently, we have extended this work by exploring the effect of multivalency on targeting capability by varying ligand density of pendant Tet1 peptides in statistical HPMA-oligolysine copolymers [95]. We showed that there was an optimal ligand density for transfection of neuron-like PC-12 cells (~3 Tet1 per polymer), mainly due to increased toxicities seen with higher Tet1 ligand incorporation. Interestingly, high concentrations (> 100 μM) of the Tet1 peptide alone did not cause significant cytotoxicity, suggesting that the incorporation of hydrophobic peptides such as Tet1 can increase the toxicity of cationic polymers [63].

Another approach for increasing transfection efficiency to neurons is to enhance endosomal escape of endocytosed polyplexes [96]. We have also demonstrated that melittin-containing HPMA polymer carriers enhanced effective bulk luciferase transfection to the brain over polymers without melittin after intraventricular injection [56].

Conclusions and future directions

In summary, we have demonstrated the synthesis of cationic polymers consisting of a hydrophilic HPMA backbone with multiple pendant peptides for DNA condensation, degradability, and endosomal escape. These HPMA-peptide brush polymers have also been used to improve adenovirus transduction and targeted delivery to the central nervous system. The relatively simple polymerization strategy used to synthesize these polymers has allowed the investigation of peptide multivalency, polymer architecture, and the use of various bioactive peptides for gene delivery applications. Currently, we are applying our findings from these described studies to improve other gene delivery systems for *in vivo* delivery. More broadly, our enhanced understanding of the delivery mechanisms of these HPMA-peptide materials is applicable to other gene delivery systems. Furthermore, these versatile materials have shown potential for the general delivery of biologics, such as therapeutic peptides and proteins.

Competing interests

The authors declare that they have no competing interests.

Authors' contributions

JS compared the transfection efficiencies of polymers modified with endosomal release moieties, JGS synthesized and tested lipid-modified carriers, and JS and SHP prepared the manuscript. All authors read and approved the final manuscripts.

Acknowledgements

This work is supported by NIH/NINDS 1R01NS064404, DMR 1206426, and the Center for the Intracellular Delivery of Biologics through the Washington Life Sciences Discovery Fund Grant No. 2496490. J.S. is supported by the National Science Foundation Graduate Research Fellowship under Grant No. DGE-0718124. We thank Dr. Russell Johnson and David Chu for helpful discussion.

References

1. Pack DW, Hoffman AS, Pun S, Stayton PS: **Design and development of polymers for gene delivery.** *Nat Rev Drug Discov* 2005, **4**:581–593.
2. Bergen JM, Pun SH: **Peptide-enhanced nucleic acid delivery.** *MRS bulletin* 2005, **30**:663–667.
3. Gopal V: **Bioinspired peptides as versatile nucleic acid delivery platforms.** *J Control Release* 2013, **167**:323–332.
4. Martin ME, Rice KG: **Peptide-guided gene delivery.** *AAPS J* 2007, **9**:E18–29.
5. Levine RM, Scott CM, Kokkoli E: **Peptide functionalized nanoparticles for nonviral gene delivery.** *Soft Matter* 2013, **9**:985–1004.
6. Kogure K, Akita H, Yamada Y, Harashima H: **Multifunctional envelope-type nano device (MEND) as a non-viral gene delivery system.** *Adv Drug Deliv Rev* 2008, **60**:559–571.
7. Kwon EJ, Bergen JM, Park IK, Pun SH: **Peptide-modified vectors for nucleic acid delivery to neurons.** *J Control Release* 2008, **132**:230–235.
8. Nishikawa M, Yamauchi M, Morimoto K, Ishida E, Takakura Y, Hashida M: **Hepatocyte-targeted in vivo gene expression by intravenous injection of plasmid DNA complexed with synthetic multi-functional gene delivery system.** *Gene Ther* 2000, **7**:548–555.
9. Chu DSH, Schellinger JG, Shi J, Convertine AJ, Stayton PS, Pun SH: **Application of living free radical polymerization for nucleic acid delivery.** *Acc Chem Res* 2012, **45**:1089–1099.
10. Ahmed M, Narain R: **Progress of RAFT Based Polymers in Gene Delivery.** *Prog Polym Sci* 2013, **38**:767–790.
11. Xu F, Yang W: **Polymer vectors via controlled/living radical polymerization for gene delivery.** *Prog Polym Sci* 2011, **36**:1099–1131.
12. Kopecek J, Kopecková P, Minko T, Lu Z: **HPMA copolymer-anticancer drug conjugates: design, activity, and mechanism of action.** *Eur J Pharm Biopharm* 2000, **50**:61–81.
13. Bulmus V: **RAFT polymerization mediated bioconjugation strategies.** *Polym Chem* 2011, **2**:1463–1472.
14. Gregory A, Stenzel MH: **Progress in Polymer Science.** *Prog Polym Sci* 2012, **37**:38–105.
15. Apostolovic B, Deacon SPE, Duncan R, Klok H-A: **Hybrid polymer therapeutics incorporating bioresponsive, coiled coil peptide linkers.** *Biomacromolecules* 2010, **11**:1187–1195.

16. Wu GY, Wu CH: **Receptor-mediated in vitro gene transformation by a soluble DNA carrier system.** *J Biol Chem* 1987, **262:**4429–4432.
17. Toncheva V, Wolfert MA, Dash PR, OUPICKY D, Ulbrich K, Seymour LW, Schacht EH: **Novel vectors for gene delivery formed by self-assembly of DNA with poly(L-lysine) grafted with hydrophilic polymers.** *Biochim Biophys Acta* 1998, **1380:**354–368.
18. Mintzer MA, Simanek EE: **Nonviral vectors for gene delivery.** *Chem Rev* 2009, **109:**259–302.
19. O'Brien-Simpson NM, Ede NJ, Brown LE, Swan J, Jackson DC: **Polymerization of unprotected synthetic peptides: A view toward synthetic peptide vaccines.** *J Am Chem Soc* 1997, **119:**1183–1188.
20. Kopeček J, Kopečková P: **HPMA copolymers: origins, early developments, present, and future.** *Adv Drug Deliv Rev* 2010, **62:**122–149.
21. Burke RS, Pun SH: **Synthesis and characterization of biodegradable HPMA-oligolysine copolymers for improved gene delivery.** *Bioconjug Chem* 2010, **21:**140–150.
22. Johnson RN, Burke RS, Convertine AJ, Hoffman AS, Stayton PS, Pun SH: **Synthesis of Statistical Copolymers Containing Multiple Functional Peptides for Nucleic Acid Delivery.** *Biomacromolecules* 2010, **11:**3007–3013.
23. Johnson RN, Chu DSH, Shi J, Schellinger JG, Carlson PM, Pun SH: **HPMA-oligolysine copolymers for gene delivery: optimization of peptide length and polymer molecular weight.** *J Control Release* 2011, **155:**303–311.
24. Hwang SJ, Bellocq NC, Davis ME: **Effects of structure of beta-cyclodextrin-containing polymers on gene delivery.** *Bioconjug Chem* 2001, **12:**280–290.
25. Carlson PM, Schellinger JG, Pahang JA, Johnson RN, Pun SH: **Comparative study of guanidine-based and lysine-based brush copolymers for plasmid delivery.** *Biomat Sci* 2013, **1:**736–744.
26. El Sayed A, Futaki S, Harashima H: **Delivery of macromolecules using arginine-rich cell-penetrating peptides: ways to overcome endosomal entrapment.** *AAPS J* 2009, **11:**13–22.
27. Brooks H, Lebleu B, Vives E: **Tat peptide-mediated cellular delivery: back to basics.** *Adv Drug Deliv Rev* 2005, **57:**559–577.
28. Torchilin VP: **Tat peptide-mediated intracellular delivery of pharmaceutical nanocarriers.** *Adv Drug Deliv Rev* 2008, **60:**548–558.
29. Khalil IA, Kogure K, Futaki S, Harashima H. **High density of octaarginine stimulates macropinocytosis leading to efficient intracellular trafficking for gene expression.** *J Biol Chem* 2006, **281:**3544–3551.
30. Okuda T, Sugiyama A, Niidome T, Aoyagi H: **Characters of dendritic poly (L-lysine) analogues with the terminal lysines replaced with arginines and histidines as gene carriers in vitro.** *Biomaterials* 2004, **25:**537–544.
31. El-Sayed A, Khalil IA, Kogure K, Futaki S, Harashima H: **Octaarginine- and octalysine-modified nanoparticles have different modes of endosomal escape.** *J Biol Chem* 2008, **283:**23450–23461.
32. Naik RJ, Chandra P, Mann A, Ganguli M: **Exogenous and cell surface glycosaminoglycans alter DNA delivery efficiency of arginine and lysine homopeptides in distinctly different ways.** *J Biol Chem* 2011, **286:**18982–18993.
33. Åmand HL, Rydberg HA, Fornander LH, Lincoln P, Nordén B, Esbjörner EK: **Cell surface binding and uptake of arginine- and lysine-rich penetratin peptides in absence and presence of proteoglycans.** *Biochim Biophys Acta* 2012, **1818:**2669–2678.
34. Bromberg L, Raduyk S, Hatton TA, Concheiro A, Rodriguez-Valencia C, Silva M, Alvarez-Lorenzo C: **Guanidinylated polyethyleneimine-polyoxypropylene-polyoxyethylene conjugates as gene transfection agents.** *Bioconjug Chem* 2009, **20:**1044–1053.
35. Tziveleka L-A, Psarra A-MG, Tsiourvas D, Paleos CM: **Synthesis and characterization of guanidinylated poly(propylene imine) dendrimers as gene transfection agents.** *J Control Release* 2007, **117:**137–146.
36. Zhai X, Sun P, Luo Y, Ma C, Xu J, Liu W: **Guanidinylation: A Simple Way to Fabricate Cell Penetrating Peptide Analogue-Modified Chitosan Vector for Enhanced Gene Delivery.** *J Appl Polym Sci* 2011, **121:**3569–3578.
37. Mattheis C, Wang H, Meister C, Agarwal S: **Effect of guanidinylation on the properties of poly(2-aminoethylmethacrylate)-based antibacterial materials.** *Macromol Biosci* 2013, **13:**242–255.
38. Treat NJ, Smith D, Teng C, Flores JD, Abel BA, York AW, Huang F, McCormick CL: **Guanidine-Containing Methacrylamide (Co)polymers via aRAFT: Toward a Cell Penetrating Peptide Mimic().** *ACS Macro Lett* 2012, **1:**100–104.
39. Fischer D, Li Y, Ahlemeyer B, Krieglstein J, Kissel T: **In vitro cytotoxicity testing of polycations: influence of polymer structure on cell viability and hemolysis.** *Biomaterials* 2003, **24:**1121–1131.
40. De Wolf HK, De Raad M, Snel C, Van Steenbergen MJ, Fens MHAM, Storm G, Hennink WE: **Biodegradable poly(2-dimethylamino ethylamino) phosphazene for in vivo gene delivery to tumor cells. Effect of polymer molecular weight.** *Pharm Res* 2007, **24:**1572–1580.
41. Ganta S, Devalapally H, Shahiwala A, Amiji M: **A review of stimuli-responsive nanocarriers for drug and gene delivery.** *J Control Release* 2008, **126:**187–204.
42. Dubowchik GM, Firestone RA, Padilla L, Willner D, Hofstead SJ, Mosure K, Knipe JO, Lasch SJ, Trail PA: **Cathepsin B-labile dipeptide linkers for lysosomal release of doxorubicin from internalizing immunoconjugates: model studies of enzymatic drug release and antigen-specific in vitro anticancer activity.** *Bioconjug Chem* 2002, **13:**855–869.
43. Lévesque SG, Shoichet MS: **Synthesis of enzyme-degradable, peptide-cross-linked dextran hydrogels.** *Bioconjug Chem* 2007, **18:**874–885.
44. Grosse SM, Tagalakis AD, Mustapa MFM, Elbs M, Meng Q-H, Mohammadi A, Tabor AB, Hailes HC, Hart SL: **Tumor-specific gene transfer with receptor-mediated nanocomplexes modified by polyethylene glycol shielding and endosomally cleavable lipid and peptide linkers.** *FASEB J* 2010, **24:**2301–2313.
45. Meister A, Anderson m: **Glutathione.** *Annu Rev Biochem* 1983, **52:**711–760.
46. Shi J, Johnson RN, Schellinger JG, Carlson PM, Pun SH: **Reducible HPMA-co-oligolysine copolymers for nucleic acid delivery.** *Int J Pharm* 2012, **427:**113–122.
47. Jiang T, Olson ES, Nguyen QT, Roy M, Jennings PA, Tsien RY: **Tumor imaging by means of proteolytic activation of cell-penetrating peptides.** *Proc Natl Acad Sci U S A* 2004, **101:**17867–17872.
48. Hatakeyama H, AKITA H, Kogure K, Oishi M, Nagasaki Y, Kihira Y, Ueno M, Kobayashi H, Kikuchi H, HARASHIMA H: **Development of a novel systemic gene delivery system for cancer therapy with a tumor-specific cleavable PEG-lipid.** *Gene Ther* 2007, **14:**68 77.
49. Ulijn RV: **Enzyme-responsive materials: a new class of smart biomaterials.** *J Mater Chem* 2006, **16:**2217–2225.
50. Sutherland MSK, Sanderson RJ, Gordon KA, Andreyka J, Cerveny CG, Yu C, Lewis TS, Meyer DL, Zabinski RF, Doronina SO, Senter PD, Law C-L, Wahl AF: **Lysosomal trafficking and cysteine protease metabolism confer target-specific cytotoxicity by peptide-linked anti-CD30-auristatin conjugates.** *J Biol Chem* 2006, **201:**10540–10547.
51. Turk V, Turk B, Turk D: **Lysosomal cysteine proteases: facts and opportunities.** *EMBO J* 2001, **20:**4629–4633.
52. Chu DSH, Johnson RN, Pun SH: **Cathepsin B-sensitive polymers for compartment-specific degradation and nucleic acid release.** *J Control Release* 2012, **157:**445–454.
53. Midoux P, Pichon C, Yaouanc J-J, Jaffrès P-A: **Chemical vectors for gene delivery: a current review on polymers, peptides and lipids containing histidine or imidazole as nucleic acids carriers.** *Br J Pharmacol* 2009, **157:**166–178.
54. Wagner E: **Application of membrane-active peptides for nonviral gene delivery.** *Adv Drug Deliv Rev* 1999, **38:**279–289.
55. Shi J, Schellinger JG, Johnson RN, Choi JL, Chou B, Anghel EL, Pun SH: **Influence of Histidine Incorporation on Buffer Capacity and Gene Transfection Efficiency of HPMA-co-oligolysine Brush Polymers.** *Biomacromolecules* 2013, **14:**1961–1970.
56. Schellinger JG, Pahang JA, Johnson RN, Chu DSH, Sellers DL, Maris DO, Convertine AJ, Stayton PS, Horner PJ, Pun SH: **Melittin-grafted HPMA-oligolysine based copolymers for gene delivery.** *Biomaterials* 2013, **34:**2318–2326.
57. Schellinger JG, Pahang JA, Shi J, Pun SH: **Block Copolymers Containing a Hydrophobic Domain of Membrane-Lytic Peptides Form Micellar Structures and Are Effective Gene Delivery Agents.** *ACS Macro Lett* 2013, **2:**725–730.
58. Kichler A, Leborgne C, Coeytaux E, Danos O: **Polyethylenimine-mediated gene delivery: a mechanistic study.** *J Gene Med* 2001, **3:**135–144.
59. Akinc A, Thomas M, Klibanov AM, Langer R: **Exploring polyethylenimine-mediated DNA transfection and the proton sponge hypothesis.** *J Gene Med* 2005, **7:**657–663.
60. Reilly MJ, Larsen JD, Sullivan MO: **Polyplexes traffic through caveolae to the Golgi and endoplasmic reticulum en route to the nucleus.** *Mol Pharm* 2012, **9:**1280–1290.
61. Xu Y, Szoka FC: **Mechanism of DNA release from cationic liposome/DNA complexes used in cell transfection.** *Biochemistry* 1996, **35:**5616–5623.
62. Liu Z, Zhang Z, Zhou C, Jiao Y: **Hydrophobic modifications of cationic polymers for gene delivery.** *Prog Polym Sci* 2010, **35:**1144–1162.

63. Incani V, Lavasanifar A, Uludağ H: **Lipid and hydrophobic modification of cationic carriers on route to superior gene vectors.** *Soft Matter* 2010, **6**:2124–2138.
64. Zhu L, Mahato RI: **Lipid and polymeric carrier-mediated nucleic acid delivery.** *Expert Opin Drug Deliv* 2010, **7**:1209–1226.
65. Abbasi M, Uludağ H, Incani V, Hsu CYM, Jeffery A: **Further investigation of lipid-substituted poly(L-Lysine) polymers for transfection of human skin fibroblasts.** *Biomacromolecules* 2008, **9**:1618–1630.
66. Kwon EJ, Liong S, Pun SH: **A truncated HGP peptide sequence that retains endosomolytic activity and improves gene delivery efficiencies.** *Mol Pharm* 2010, **7**:1260–1265.
67. Juvvadi P, Vunnam S, Merrifield RB: **Synthetic melittin, its enantio, retro, and retroenantio isomers, and selected chimeric analogs: their antibacterial, hemolytic, and lipid bilayer action.** *J Am Chem Soc* 1996, **118**:8989–8997.
68. Raghuraman H, Chattopadhyay A: **Melittin: a membrane-active peptide with diverse functions.** *Biosci Rep* 2007, **27**:189–223.
69. Ogris M, Carlisle RC, Bettinger T, Seymour LW: **Melittin enables efficient vesicular escape and enhanced nuclear access of nonviral gene delivery vectors.** *J Biol Chem* 2001, **276**:47550–47555.
70. Baumhover NJ, Anderson K, Fernandez CA, Rice KG: **Synthesis and in vitro testing of new potent polyacridine-melittin gene delivery peptides.** *Bioconjug Chem* 2010, **21**:74–83.
71. Convertine AJ, Benoit DSW, Duvall CL, Hoffman AS, Stayton PS: **Development of a novel endosomolytic diblock copolymer for siRNA delivery.** *J Control Release* 2009, **133**:221–229.
72. Ogris M, Brunner S, Schüller S, Kircheis R, Wagner E: **PEGylated DNA/transferrin-PEI complexes: reduced interaction with blood components, extended circulation in blood and potential for systemic gene delivery.** *Gene Ther* 1999, **6**:595–605.
73. OUPICKY D, Howard KA, Konák C, Dash PR, Ulbrich K, Seymour LW: **Steric stabilization of poly-L-Lysine/DNA complexes by the covalent attachment of semitelechelic poly[N-(2-hydroxypropyl)methacrylamide].** *Bioconjug Chem* 2000, **11**:492–501.
74. Oupický D, Ogris M, Howard KA, Dash PR, Ulbrich K, Seymour LW: **Importance of lateral and steric stabilization of polyelectrolyte gene delivery vectors for extended systemic circulation.** *Mol Ther* 2002, **5**:463–472.
75. Šubr V, Koňák C, Laga R, Ulbrich K: **Coating of DNA/poly(L-lysine) complexes by covalent attachment of poly[N-(2-hydroxypropyl) methacrylamide].** *Biomacromolecules* 2006, **7**:122–130.
76. Wei H, Pahang JA, Pun SH: **Optimization of brush-like cationic copolymers for nonviral gene delivery.** *Biomacromolecules* 2013, **14**:275–284.
77. Hatakeyama H, Akita H, Harashima H: **A multifunctional envelope type nano device (MEND) for gene delivery to tumours based on the EPR effect: a strategy for overcoming the PEG dilemma.** *Adv Drug Deliv Rev* 2011, **63**:152–160.
78. Shi J, Chou B, Choi JL, Ta AL, Pun SH: **Investigation of Polyethylenimine/DNA Polyplex Transfection to Cultured Cells Using Radiolabeling and Subcellular Fractionation Methods.** *Mol Pharm* 2013, **10**:2145–2156.
79. Fichter KM, Ingle NP, McLendon PM, Reineke TM: **Polymeric nucleic acid vehicles exploit active interorganelle trafficking mechanisms.** *ACS Nano* 2013, **7**:347–364.
80. McLendon PM, Fichter KM, Reineke TM: **Poly(glycoamidoamine) vehicles promote pDNA uptake through multiple routes and efficient gene expression via caveolae-mediated endocytosis.** *Mol Pharm* 2010, **7**:738–750.
81. Chailertvanitkul VA, Pouton CW: **Adenovirus: a blueprint for non-viral gene delivery.** *Curr Opin Biotechnol* 2010, **21**:627–632.
82. Jang J-H, Schaffer DV, Shea LD: **Engineering biomaterial systems to enhance viral vector gene delivery.** *Mol Ther* 2011, **19**:1407–1415.
83. Kreppel F, Kochanek S: **Modification of adenovirus gene transfer vectors with synthetic polymers: a scientific review and technical guide.** *Mol Ther* 2008, **16**:16–29.
84. Wang C-HK, Chan LW, Johnson RN, Chu DSH, Shi J, Schellinger JG, Lieber A, Pun SH: **The transduction of Coxsackie and Adenovirus Receptor-negative cells and protection against neutralizing antibodies by HPMA-co-oligolysine copolymer-coated adenovirus.** *Biomaterials* 2011, **32**:9536–9545.
85. Ruponen M, Rönkkö S, Honkakoski P, Pelkonen J, Tammi M, Urtti A: **Extracellular glycosaminoglycans modify cellular trafficking of lipoplexes and polyplexes.** *J Biol Chem* 2001, **276**:33875–33880.
86. Davidson BL, Breakefield XO: **Viral vectors for gene delivery to the nervous system.** *Nat Rev Neurosci* 2003, **4**:353–364.
87. Rogers M-L, Rush RA: **Non-viral gene therapy for neurological diseases, with an emphasis on targeted gene delivery.** *J Control Release* 2012, **157**:183–189.
88. Pérez-Martínez FC, Guerra J, Posadas I, Ceña V: **Barriers to non-viral vector-mediated gene delivery in the nervous system.** *Pharm Res* 2011, **28**:1843–1858.
89. Suk JS, Suh J, Choy K, Lai SK, Fu J, Hanes J: **Gene delivery to differentiated neurotypic cells with RGD and HIV Tat peptide functionalized polymeric nanoparticles.** *Biomaterials* 2006, **27**:5143–5150.
90. Bergen JM, Pun SH: **Analysis of the intracellular barriers encountered by nonviral gene carriers in a model of spatially controlled delivery to neurons.** *J Gene Med* 2008, **10**:187–197.
91. Zeng J, Too H-P, Ma Y, Luo ESE, Wang S: **A synthetic peptide containing loop 4 of nerve growth factor for targeted gene delivery.** *J Gene Med* 2004, **6**:1247–1256.
92. Oliveira H, Fernandez R, Pires LR, Martins MCL, Simões S, Barbosa MA, Pêgo AP: **Targeted gene delivery into peripheral sensorial neurons mediated by self-assembled vectors composed of poly(ethylene imine) and tetanus toxin fragment c.** *J Control Release* 2010, **143**:350–358.
93. Liu JK, Teng Q, Garrity-Moses M, Federici T, Tanase D, Imperiale MJ, Boulis NM: **A novel peptide defined through phage display for therapeutic protein and vector neuronal targeting.** *Neurobiol Dis* 2005, **19**:407–418.
94. Kwon EJ, Lasiene J, Jacobson BE, Park I-K, Horner PJ, Pun SH: **Targeted nonviral delivery vehicles to neural progenitor cells in the mouse subventricular zone.** *Biomaterials* 2010, **31**:2417–2424.
95. Chu DSH, Schellinger JG, Bocek MJ, Johnson RN, Pun SH: **Optimization of Tet1 Ligand Density in HPMA-Co-Oligolysine Copolymers for Targeted Neuronal Gene Delivery.** *Biomaterials* 2013, **34**:9632–9637.
96. Suk JS, Suh J, Lai SK, Hanes J: **Quantifying the intracellular transport of viral and nonviral gene vectors in primary neurons.** *Exp Biol Med* 2007, **232**:461–469.

Permissions

All chapters in this book were first published in JBE, by BioMed Central; hereby published with permission under the Creative Commons Attribution License or equivalent. Every chapter published in this book has been scrutinized by our experts. Their significance has been extensively debated. The topics covered herein carry significant findings which will fuel the growth of the discipline. They may even be implemented as practical applications or may be referred to as a beginning point for another development.

The contributors of this book come from diverse backgrounds, making this book a truly international effort. This book will bring forth new frontiers with its revolutionizing research information and detailed analysis of the nascent developments around the world.

We would like to thank all the contributing authors for lending their expertise to make the book truly unique. They have played a crucial role in the development of this book. Without their invaluable contributions this book wouldn't have been possible. They have made vital efforts to compile up to date information on the varied aspects of this subject to make this book a valuable addition to the collection of many professionals and students.

This book was conceptualized with the vision of imparting up-to-date information and advanced data in this field. To ensure the same, a matchless editorial board was set up. Every individual on the board went through rigorous rounds of assessment to prove their worth. After which they invested a large part of their time researching and compiling the most relevant data for our readers.

The editorial board has been involved in producing this book since its inception. They have spent rigorous hours researching and exploring the diverse topics which have resulted in the successful publishing of this book. They have passed on their knowledge of decades through this book. To expedite this challenging task, the publisher supported the team at every step. A small team of assistant editors was also appointed to further simplify the editing procedure and attain best results for the readers.

Apart from the editorial board, the designing team has also invested a significant amount of their time in understanding the subject and creating the most relevant covers. They scrutinized every image to scout for the most suitable representation of the subject and create an appropriate cover for the book.

The publishing team has been an ardent support to the editorial, designing and production team. Their endless efforts to recruit the best for this project, has resulted in the accomplishment of this book. They are a veteran in the field of academics and their pool of knowledge is as vast as their experience in printing. Their expertise and guidance has proved useful at every step. Their uncompromising quality standards have made this book an exceptional effort. Their encouragement from time to time has been an inspiration for everyone.

The publisher and the editorial board hope that this book will prove to be a valuable piece of knowledge for researchers, students, practitioners and scholars across the globe.

List of Contributors

Mariana Leguia
Department of Bioengineering, University of California, Berkeley, CA 94720, USA
QB3: California Institute for Quantitative Biological Research, University of California, Berkeley, CA 94720, USA
Physical Biosciences Division, Lawrence Berkeley National Laboratory, Berkeley, CA 94720, USA

Jennifer AN Brophy
Department of Bioengineering, University of California, Berkeley, CA 94720, USA

Douglas Densmore
Department of Bioengineering, University of California, Berkeley, CA 94720, USA
Current address: Department of Electrical and Computer Engineering, Boston University, Boston, MA 02215, USA

Angel Asante
Department of Bioengineering, University of California, Berkeley, CA 94720, USA

J Christopher Anderson
Department of Bioengineering, University of California, Berkeley, CA 94720, USA
QB3: California Institute for Quantitative Biological Research, University of California, Berkeley, CA 94720, USA
Physical Biosciences Division, Lawrence Berkeley National Laboratory, Berkeley, CA 94720, USA

Isaac T Yonemoto
J. Craig Venter Institute, 10355 Science Center Dr, San Diego, CA 92121, USA

Christopher W Matter
J. Craig Venter Institute, 10355 Science Center Dr, San Diego, CA 92121, USA

Thao Amy Nguyen
J. Craig Venter Institute, 10355 Science Center Dr, San Diego, CA 92121, USA

Hamilton O Smith
J. Craig Venter Institute, 10355 Science Center Dr, San Diego, CA 92121, USA

Philip D Weyman
J. Craig Venter Institute, 10355 Science Center Dr, San Diego, CA 92121, USA

Jin Gu
Intercollege Graduate Degree Program in Plant Biology, The Pennsylvania State University, University Park, PA 16802, USA

Jeffrey M Catchmark
Intercollege Graduate Degree Program in Plant Biology, The Pennsylvania State University, University Park, PA 16802, USA
Department of Agricultural and Biological Engineering, The Pennsylvania State University, 109 Ag. Eng. Bldg., University Park, PA 16802, USA

Longfei Mao
Centre for Advanced Computational Solutions, Wine, Food & Molecular Bioscience Department, Lincoln University, Ellesmere Junction Road, Lincoln 7647, New Zealand

Wynand S Verwoerd
Centre for Advanced Computational Solutions, Wine, Food & Molecular Bioscience Department, Lincoln University, Ellesmere Junction Road, Lincoln 7647, New Zealand

Mahmoud E Moustafa
Department of Chemical and Life Science Engineering, Virginia Commonwealth University, Richmond, VA 23284, USA

Venkat S Gadepalli
Department of Chemical and Life Science Engineering, Virginia Commonwealth University, Richmond, VA 23284, USA

Ahmed A Elmak
Department of Chemical and Life Science Engineering, Virginia Commonwealth University, Richmond, VA 23284, USA

Woomin Lee
Department of Chemical and Life Science Engineering, Virginia Commonwealth University, Richmond, VA 23284, USA

Raj R Rao
Department of Chemical and Life Science Engineering, Virginia Commonwealth University, Richmond, VA 23284, USA

Vamsi K Yadavalli
Department of Chemical and Life Science Engineering, Virginia Commonwealth University, Richmond, VA 23284, USA

Masayuki Tsukamoto
Department of Medical Genome Sciences, Graduate School of Frontier Sciences, The University of Tokyo, Kashiwa, Japan
Manufacturing Technology Association of Biologics, Chuo-ku, Kobe, Japan

Hideki Watanabe
Biomedical Research Institute National Institute of Advanced Industrial Science and Technology (AIST), Higashi, Tsukuba, Ibaraki 305-8566, Japan

Ayako Ooishi
Biomedical Research Institute National Institute of Advanced Industrial Science and Technology (AIST), Higashi, Tsukuba, Ibaraki 305-8566, Japan

Shinya Honda
Department of Medical Genome Sciences, Graduate School of Frontier Sciences, The University of Tokyo, Kashiwa, Japan
Biomedical Research Institute National Institute of Advanced Industrial Science and Technology (AIST), Higashi, Tsukuba, Ibaraki 305-8566, Japan
Manufacturing Technology Association of Biologics, Chuo-ku, Kobe, Japan

Nicolo' Politi
Dipartimento di Ingegneria Industriale e dell'Informazione, Università degli Studi di Pavia, via Ferrata 3, Pavia, Italy
Centro di Ingegneria Tissutale, Università degli Studi di Pavia, via Ferrata 3, Pavia, Italy

Lorenzo Pasotti
Dipartimento di Ingegneria Industriale e dell'Informazione, Università degli Studi di Pavia, via Ferrata 3, Pavia, Italy
Centro di Ingegneria Tissutale, Università degli Studi di Pavia, via Ferrata 3, Pavia, Italy

Susanna Zucca
Dipartimento di Ingegneria Industriale e dell'Informazione, Università degli Studi di Pavia, via Ferrata 3, Pavia, Italy
Centro di Ingegneria Tissutale, Università degli Studi di Pavia, via Ferrata 3, Pavia, Italy

Michela Casanova
Dipartimento di Ingegneria Industriale e dell'Informazione, Università degli Studi di Pavia, via Ferrata 3, Pavia, Italy
Centro di Ingegneria Tissutale, Università degli Studi di Pavia, via Ferrata 3, Pavia, Italy

Giuseppina Micoli
Centro di Ricerche Ambientali, IRCCS Fondazione Salvatore Maugeri, via Salvatore Maugeri 10, Pavia, Italy

Maria Gabriella Cusella De Angelis
Centro di Ingegneria Tissutale, Università degli Studi di Pavia, via Ferrata 3, Pavia, Italy

Paolo Magni
Dipartimento di Ingegneria Industriale e dell'Informazione, Università degli Studi di Pavia, via Ferrata 3, Pavia, Italy
Centro di Ingegneria Tissutale, Università degli Studi di Pavia, via Ferrata 3, Pavia, Italy

Daniel Camsund
Microbial Chemistry, Department of Chemistry–Ångström Laboratory, Science for Life Laboratory, Uppsala University, P.O. Box 523, SE-75120 Uppsala, Sweden

Thorsten Heidorn
Bioforsk, Frederik A Dahls vei 20, 1432 Ås Oslo, Norway

Peter Lindblad
Microbial Chemistry, Department of Chemistry–Ångström Laboratory, Science for Life Laboratory, Uppsala University, P.O. Box 523, SE-75120 Uppsala, Sweden

Slgirim Lee
Department of Chemical and Biomolecular Engineering, Yonsei University, Seoul 120-749, Korea

Gyuhyung Jin
Department of Chemical and Biomolecular Engineering, Yonsei University, Seoul 120-749, Korea

Jae-Hyung Jang
Department of Chemical and Biomolecular Engineering, Yonsei University, Seoul 120-749, Korea

Javad Verdi
Centre for Nanotechnology and Regenerative Medicine, UCL Division of Surgery & Interventional Science, University College London (UCL), London NW3 2QG, UK
Applied Cell Sciences Department, School of Advanced Technologies in Medicine, Tehran University of Medical Sciences, Tehran, Iran

Aaron Tan
Centre for Nanotechnology and Regenerative Medicine, UCL Division of Surgery & Interventional Science, University College London (UCL), London NW3 2QG, UK
UCL Medical School, University College London (UCL), London, UK

Alireza Shoae-Hassani
Applied Cell Sciences Department, School of Advanced Technologies in Medicine, Tehran University of Medical Sciences, Tehran, Iran

Alexander M Seifalian
Centre for Nanotechnology and Regenerative Medicine, UCL Division of Surgery & Interventional Science, University College London (UCL), London NW3 2QG, UK
Royal Free London NHS Foundation Trust Hospital, London, UK

Jara Radec
Department Biology I, AG Synthetic Microbiology, Ludwig-Maximilians-Universität München, Grosshaderner Str. 2-4, D-82152 Planegg-Martinsried, Germany

Korinna Kraft
Department Biology I, AG Synthetic Microbiology, Ludwig-Maximilians-Universität München, Grosshaderner Str. 2-4, D-82152 Planegg-Martinsried, Germany

Julia Bartels
Department Biology I, AG Synthetic Microbiology, Ludwig-Maximilians-Universität München, Grosshaderner Str. 2-4, D-82152 Planegg-Martinsried, Germany

Tamara Cikovic
Department Biology I, AG Synthetic Microbiology, Ludwig-Maximilians-Universität München, Grosshaderner Str. 2-4, D-82152 Planegg-Martinsried, Germany

Franziska Dürr
Department Biology I, AG Synthetic Microbiology, Ludwig-Maximilians-Universität München, Grosshaderner Str. 2-4, D-82152 Planegg-Martinsried, Germany

Jennifer Emenegger
Department Biology I, AG Synthetic Microbiology, Ludwig-Maximilians-Universität München, Grosshaderner Str. 2-4, D-82152 Planegg-Martinsried, Germany

Simon Kelterborn
Department Biology I, AG Synthetic Microbiology, Ludwig-Maximilians-Universität München, Grosshaderner Str. 2-4, D-82152 Planegg-Martinsried, Germany

Christopher Sauer
Department Biology I, AG Synthetic Microbiology, Ludwig-Maximilians-Universität München, Grosshaderner Str. 2-4, D-82152 Planegg-Martinsried, Germany
Present affiliation: Institute of Cell and Molecular Biosciences, Newcastle University, Centre for Bacterial Cell Biology, Richardson Road, NE2 4AX Newcastle upon Tyne, UK

Georg Fritz
Department Biology I, AG Synthetic Microbiology, Ludwig-Maximilians-Universität München, Grosshaderner Str. 2-4, D-82152 Planegg-Martinsried, Germany
Ludwig-Maximilians-University Munich, Arnold Sommerfeld Center for Theoretical Physics, Theresienstr. 37, D-80333 München, Germany

Susanne Gebhard
Department Biology I, AG Synthetic Microbiology, Ludwig-Maximilians-Universität München, Grosshaderner Str. 2-4, D-82152 Planegg-Martinsried, Germany

Thorsten Mascher
Department Biology I, AG Synthetic Microbiology, Ludwig-Maximilians-Universität München, Grosshaderner Str. 2-4, D-82152 Planegg-Martinsried, Germany

Rahmi Yusuf
Division of Pharmacy, Durham University, Queen's Campus, Stockton-on-Tees TS17 6BH, England, UK

Roger R Draheim
School of Pharmacy and Biomedical Sciences,University of Portsmouth, St. Michael's Building, White Swan Road, Portsmouth PO1 2DT, England, UK

Kei Takahashi
Department of Basic Science, Graduate School of Arts and Sciences, The University of Tokyo, 3-8-1 Komaba, Meguro, Tokyo 153-8902, Japan

Taro Toyota
Department of Basic Science, Graduate School of Arts and Sciences, The University of Tokyo, 3-8-1 Komaba, Meguro, Tokyo 153-8902, Japan
Research Center for Complex Systems Biology, Graduate School of Arts and Sciences, The University of Tokyo, 3-8-1 Komaba, Meguro, Tokyo 153-8902, Japan

Sayed Shahabuddin Hoseini
Departments of Pediatric Hematology, Oncology and Blood Stem Cell Transplantation, Hannover, Germany
Hannover Center for Transplantation Research, Hannover, Germany

Martin G Sauer
Departments of Pediatric Hematology, Oncology and Blood Stem Cell Transplantation, Hannover, Germany
Hannover Center for Transplantation Research, Hannover, Germany
Department of Pediatric Hematology and Oncology, Medizinische Hochschule Hannover, OE 6780, Carl-Neuberg-Strasse 1, 30625 Hannover, Germany

Ryan J Bloom
Department of Bioengineering, Stanford University, 443 Via Ortega, MC 4245, Stanford, CA 94305, USA

Sally M Winkler
Department of Bioengineering, Stanford University, 443 Via Ortega, MC 4245, Stanford, CA 94305, USA

Christina D Smolke
Department of Bioengineering, Stanford University, 443 Via Ortega, MC 4245, Stanford, CA 94305, USA

Hsin-Ho Huang
Microbial Chemistry, Department of Chemistry - Ångström Laboratory, Uppsala University, P.O. Box 523SE-75120, Uppsala, Sweden

Peter Lindblad
Microbial Chemistry, Department of Chemistry - Ångström Laboratory, Uppsala University, P.O. Box 523SE-75120, Uppsala, Sweden

Xiuli Dong
Biomanufacturing Research Institute and Technology Enterprise (BRITE) and Department of Pharmaceutical Sciences, North Carolina Central University, Durham, NC 27707, USA

Yongan Tang
Department of Mathematics and Physics, North Carolina Central University, Durham, NC 27707, USA

Marvin Wu
Department of Mathematics and Physics, North Carolina Central University, Durham, NC 27707, USA

Branislav Vlahovic
Department of Mathematics and Physics, North Carolina Central University, Durham, NC 27707, USA

Liju Yang
Biomanufacturing Research Institute and Technology Enterprise (BRITE) and Department of Pharmaceutical Sciences, North Carolina Central University, Durham, NC 27707, USA

Wolf Klöckner
AVT. Biochemical Engineering, RWTH Aachen University, Worringerweg 1, Aachen 52074, Germany

Riad Gacem
AVT. Biochemical Engineering, RWTH Aachen University, Worringerweg 1, Aachen 52074, Germany

Tibor Anderlei
Adolf Kühner AG, Dinkelbergstrasse 1, Birsfelden 4127, Switzerland

Nicole Raven
Fraunhofer Institute for Molecular Biology and Applied Ecology (IME), Forckenbeckstrasse 6, Aachen 52074, Germany

Stefan Schillberg
Fraunhofer Institute for Molecular Biology and Applied Ecology (IME), Forckenbeckstrasse 6, Aachen 52074, Germany

Clemens Lattermann
AVT. Biochemical Engineering, RWTH Aachen University, Worringerweg 1, Aachen 52074, Germany

Jochen Büchs
AVT. Biochemical Engineering, RWTH Aachen University, Worringerweg 1, Aachen 52074, Germany

Asad A Ahmad
Department of Biomedical Engineering, University of North Carolina, Chapel Hill, NC, 27599 and North Carolina State University, Raleigh, NC 27695, USA

Yuli Wang
Department of Chemistry, University of North Carolina, Chapel Hill, NC 27599, USA

Adam D Gracz
Department of Medicine, Division of Gastroenterology and Hepatology, University of North Carolina, Chapel Hill, NC 27599, USA

Christopher E Sims
Department of Chemistry, University of North Carolina, Chapel Hill, NC 27599, USA

Scott T Magness
Department of Biomedical Engineering, University of North Carolina, Chapel Hill, NC, 27599 and North Carolina State University, Raleigh, NC 27695, USA
Department of Medicine, Division of Gastroenterology and Hepatology, University of North Carolina, Chapel Hill, NC 27599, USA
Department of Cell Biology and Physiology, University of North Carolina, Chapel Hill, NC 27599, USA

Nancy L Allbritton
Department of Biomedical Engineering, University of North Carolina, Chapel Hill, NC, 27599 and North Carolina State University, Raleigh, NC 27695, USA
Department of Chemistry, University of North Carolina, Chapel Hill, NC 27599, USA

Julie Shi
Department of Bioengineering and Molecular Engineering & Sciences Institute, University of Washington, 3720 15th Ave NE, Seattle, WA 98195, USA

Joan G Schellinger
Department of Bioengineering and Molecular Engineering & Sciences Institute, University of Washington, 3720 15th Ave NE, Seattle, WA 98195, USA

Suzie H Pun
Department of Bioengineering and Molecular Engineering & Sciences Institute, University of Washington, 3720 15th Ave NE, Seattle, WA 98195, USA